SOLVENT EFFECTS ON CHEMICAL PHENOMENA

VOLUME I

SOLVENT EFFECTS ON CHEMICAL PHENOMENA

Edward S. Amis
James F. Hinton
Department of Chemistry
University of Arkansas
Fayetteville, Arkansas

VOLUME I

ACADEMIC PRESS   New York and London   1973
A Subsidiary of Harcourt Brace Jovanovich, Publishers

ACADEMIC PRESS, INC.
111 Fifth Avenue, New York, New York 10003

United Kingdom Edition published by
ACADEMIC PRESS, INC. (LONDON) LTD.
24/28 Oval Road, London NW1

Library of Congress Cataloging in Publication Data

Amis, Edward Stephen, DATE
    Solvent effects on chemical phenomena.

    Includes bibliographical references.
    1.   Solution (Chemistry)   2.   Solvation.   3.   Sol-
vents.   I.   Hinton, James F., joint author.   II.   Title.
QD541.A54   541'.348   72-9983
ISBN 0–12–057301–6 (v. 1)

To Dr. E. A. Moelwyn-Hughes
retired general of the Solvation Army

CONTENTS

Chapter 5.   Solvent Influence on Rates and Mechanisms

PREFACE

The influences of the solvent on chemical phenomena in solution have been observed over a number of years by the authors, who have been astonished at some of the effects noted. It is intriguing that, for example, a reaction rate can be changed by several powers of ten in magnitude by merely changing the solvent medium in which the reaction occurs. The oscillatory behavior with solvent composition in mixed solvents of phenomena has been encountered numerous times, the maxima and minima in the wavelike curves for many phenomena often occurring in the same general regions of solvent composition. The marked influence of a trace of a second solvent on many chemical phenomena in solution has continued to amaze the authors.

Some of these manifestations of solvent effects have been satisfactorily explained on the basis of electrostatics, solvation of solutes, internal cohesions of the solvents, protic or dipolar aprotic nature of solvents, viscosity, and other understandable properties of solvent and solute. But many of these effects have obscure origins that cannot be elucidated by measuring microscopic properties using macroscopic probes. What is the extent of solvation of a given ion in a given solvent? The solvation of a particular ion depends not only on the extent of solvation of some given reference ion in the pertinent solvent but also on the method of measurement. What is the microscopic dielectric constant in the neighborhood of a given ion? These and other questions on the properties of solutions are in the "dark ages" of progress as to the correct answers. The tack of elucidation of solvent influence on phenomena in solution suffers not only from obscurity as to causes of such effects, but also from the enormity of the undertaking. Nevertheless, the authors, like the proverbial fools, have rushed in where angels fear to tread, and the results are recorded in this volume and are also to be published in a second volume. We hope this effort on the presentation of selected solvent effects on selected phenomena in solution will at least be a beacon on the oceans of solvent effects to serve as a guide to other sailors on the obscure and boundless mains of solution chemistry.

CONTENTS OF VOLUME II (Tentative)

Chapter 1

INTRODUCTION

Writing anything approaching a complete discussion of the theory of solvation and the influences of solvents on chemical phenomena is almost an impossible task due to its enormity, the lack of knowledge of and the agreement on the fundamental nature of the phenomena, and the variation in the magnitude and specificity of the solvation or solvent effect in any one particular phenomenon. Thus the extent of the solvation of the lithium ion may range from zero to many molecules in solvent water depending on the nature of the measurement used. The seeming conflict in this one phenomenon results from the circumstance that different methods of measurement detect different properties of the lithium ion in its relationship to water. Each of these relationships has its own significance. For example, transference measurements indicate the average number of water molecules that accompany an ion on its movement through water, even though there is constant exchange of water molecules among the various solvent layers and between the outer loosely bound layer and the free solvent. With respect to solvation, another problem is that all solvation numbers of ions are relative to the solvation of some particular reference ion which is allocated from some reasonable assumptions a given number of molecules of solvent of solvation. These and other related phenomena are discussed in Chapter 3.

1

As for solvent effects on chemical phenomena, these are so numerous and varied for any particular phenomenon that complete discussion of each effect would lead to an enormous volume of material, as shown in Chapter 5, where an attempt is made to give a rather extensive presentation of a variety of solvent effects on chemical reactions rate and mechanisms.

Notwithstanding the great magnitude of the task, there should be an organized, comprehensive, and to a reasonable extent, detailed compilation of data and discussion on solvation and the effects of solvent on chemical phenomena. This is true because solution chemistry is so important in life processes, both in the plant and animal world, and because so many laboratory and industrial chemical processes take place in solution.

For solubility and other reasons, solvents other than water prove useful in certain chemical processes. Thus for reaction rates the solvent effect on the rate is determined by the difference in the free energies, enthalpies and entropies of solvation of the reactants and of the transition states. The degree of solvation of the reactants and activated complex can influence reaction rates markedly. As an illustration, the rates of quatenary salt formations are over 10^4 times as great in nitrobenzene as in benzene since the intermediate complexes in these reactions are highly solvated by nitrobenzene but not by benzene, the activity coefficient of the complex is much less, and the entropy of its formation much greater in nitrobenzene than in benzene. The individual solvent activity coefficients of anions and cations, like individual solvation numbers, cannot be determined directly. The problem has been approached by making certain extrathermodynamic assumptions based on reasonable foundations. This problem will be discussed fully in Chapter 5. Added to these thermodynamic and solvation effects of the solvent on reaction rates are ionization, hydrogen bonding, solvalysis, electrostatics, viscosity, internal pressure, cage, and various other effects. In fact the solvation effects themselves are complicated by inner-shell solvation, outer-shell solvation, negative-solvation, dynamic solvation, and other phenomena such as the replacement of molecules in the solvation shells of solute particles of the molecules of one-solvent component by another solvent component in the case of mixed solvents, and the specific solvation of solute particles by one component in a mixed solvent.

In mixed solvents, periodicity of chemical phenomena often occur. Various explanations of such observations are extant. In particular maxima and minima are observed in many chemical phenomena in the neighborhoods of 15 and 80 weight% of the organic component in aqueous-organic solvents. Sometimes solvent mixtures containing no solute show periodicity in certain properties with change in solvent composition. Again explanations are in the literature.

The water proton chemical shift in aqueous N-methylacetamide, N-

methylformamide and *N*-ethylformamide has been observed to undergo a periodic shift depending upon the amide concentration. The initial addition of the amide to water produces a small shift to low field. This low field shift is felt to be the result of the amide molecules behaving as predominently interstitial species causing the water–water hydrogen bonds to bend less and appear to be stronger. The low field shift reaches a maximum at about 0.2 mole fraction amide. The subsequent addition of the amide to water produces a net high field shift in the water proton resonance position indicative of the breaking of solution structure. At about 0.8 mole fraction of amide the water proton resonance shifts rather drastically downfield accompanied by the broadening of the water resonance line width. These last two effects would be indicative of a restructuring process in which water appears to be tightly bound.

The oscillatory effect observed in the relative solvation of the two ions in binary electrolytes when water–alcohol is the solvent is explained by the successive replacement of the water in the different solvation layers of the two ions beginning with the outermost solvation layer of the more highly solvated ion. These and other oscillatory effects in chemical phenomena are presented extensively in Chapter 4.

Abrupt changes in phenomena are sometimes manifested when even minuscule amounts of one solvent component is added to a solution containing another solvent component. When the particular property is plotted as ordinate versus the solvent composition, there is almost a verticle drop in the property which may amount to over 30%. This is the case with the equivalent conductance of perchloric acid in anhydrous ethanol when as little as 0.3 weight% of water is added to the solution. For processes in galvanic cells similar abrupt changes occur in ion-size parameters, salting-out coefficients, and thermodynamic functions when minuscule amounts of water are added to the anhydrous methanol solvents in such galvanic cells as those represented:

$$Pt, H_2 | HI(m), X\%CH_3OH, Y\%H_2O | AgI-Ag$$

$$Pt, H_2 | HBr(m), X\%CH_3OH, Y\%H_2O | AgBr-Ag$$

Such phenomena were explained on the basis of the breaking down of solvent methanol chains and the selective solvation by solvent water of the protons. Chapter 4 gives the details of such observations.

Another intriguing result of mixing solvents is the catalytic effect exhibited by certain substances on certain reaction rates in mixed solvents which are not manifested when the reaction is observed in one of the pure solvent components. Thus Cu^{2+} and Hg^{2+} ions show no catalytic effect on the electron exchange reaction

$$U(IV) + Tl(III) \rightarrow U(IV) + Tl(I) \tag{1.1}$$

when the solvent is pure water, but these ions produce about a sevenfold increase in the rate of this reaction when the solvent is 25% water–75% by weight methanol. This enormous increase of the rate of this reaction in this solvent is not common to all cations. Even without the presence of the catalyst the rates, the mechanism, and the thermodynamic quantities for this reaction were changed in the 25% water–75% methanol as compared to the water solvent. The catalytic effect of the Cu^{2+} and Hg^{2+} ions on the rate were explained by a stepwise mechanism for the reaction in the 25% water–75% methanol solvent. Difference in the solvolytic processes resulting in the formation of different complex intermediates was proposed to explain the differences in the rates and mechanisms of the reaction in the different solvents. This reaction is discussed in Chapter 5, and the mechanisms assumed in the two solvents are presented. However, the stepwise mechanism for the catalytic effect of the Cu^{2+} and Hg^{2+} ions are not present in Chapter 5, and is, therefore, illustrated using Hg^{2+} ion as the example:

$$U(IV) + Tl(III) \rightarrow U(V) + Tl(II) \qquad (1.2)$$

$$U(V) + Hg(II) \rightarrow U(VI) + Hg(I) \qquad (1.3)$$

$$Tl(II) + Hg(I) \rightarrow Tl(I) + Hg(II) \qquad (1.4)$$

These cation effects indicate that the reaction proceeds in two one-electron steps in 25% water–75% methanol, instead of one two-electron step as was found in water. This could be due to a solvent cage effect in which the hydrogen bonding in the water–methanol is weaker than in the water solvent thus permitting the reactants to diffuse out of the cage more readily in the former solvent for reaction with other species.

Many aspects of solvent effects are yet to be satisfactorily explained, though some of the phenomena have been verified by laboratory experiments, and have been amenable to theoretical treatment. Empirical approaches have been used to supplement theoretical methods and to explain specific effects not accounted for by theory. These empirical approaches are not to be ridiculed since they in many cases correlate much data in useful manners. Differences of opinion exist as to the correct interpretation of many solvent effects. To separate, even in a qualitative manner, the various influences at work in many instances is quite frequently impossible and the methods used may be questionable. In some cases one of many concurrent effects dominates and a particular theory can be successfully applied. One dominant effect, among several concurrent ones that tend to allay or moderate each other, may be the explanation of why some theories apply as often as they do. Much is yet unrevealed about solutes and solvents in the solution state, or about solvent components in mixed solvents. "Now we see through a glass darkly." Especially is there on the edge of night, or even the dark of night, microscopic regions around the solute particles. What is really known about the dielectric constant in the immediate vicinity of an ion in a pure or a mixed solvent? How does this microscopic dielectric constant and the extent and nature of the solvation of an ion vary from ion to ion, from solvent to solvent, with composition of mixed solvent and with temperature; or how do these properties vary in the immediate vicinity of any solute particle, be it an ion, dipolar molecule, induced dipolar particle, or, if such can exist in solution, a nonpolar particle with respect to the above listed parameters? How do the electronic configuration and the reactivity of solute particles vary with change of solvent and with compositions of mixed solvents due to solvation, negative solvation, selective solvation, solvolysis, cohesion (internal pressure), viscosity, polarizability and other solvent influences? The answers to these questions are vague and often conflicting. Before we can discern solution phenomena as they actually are, better methods of probing the microscopic state of matter must be perfected. New and helpful tools such as nuclear magnetic resonance and electron spin resonance are regularly becoming available and should permit investigators in the field to find the answers to some of the questions necessary to the understanding of solution chemistry. Also, different models for solvation of ions are continually being devised and these shed light on the possibilities of structure in the microscopic solution states. Perhaps it will be necessary to treat the solvent in the immediate vicinity of the solute particles explicitly while treating the rest of the solvent as a continuum. However, in more concentrated solutions, especially of ionic substances, the whole solvent could be structured. It should be remembered, too, that polar and hydrogen-bonded solvents are themselves structured and the possible effects of added solute on this solvent structure must be recognized in any theory of solutions.

The fact that the problems of solution chemistry are so many and varied and so difficult makes this a challenging field. By applying ever increasing ingenuity in instrumentation, methods, techniques, and theoretical insight, by unstinting effort, and by unflagging ardor, the solution chemists will progressively untangle the intertwined maze of the problems of solution chemistry. It is to the present progress in this difficult field that the chapters in this book are addressed.

Chapter 2

GENERAL CONCEPTS OF SOLVATION

I. General Statements

Solvation is the attraction of solvent molecules to themselves by ions, molecules, and complex particles in solution, by micelles and other collodial particles in suspension, and by surfaces in contact with solvents. The solvent may occur in more than one sphere around the particle or on the surface, and may be firmly or loosely attached. The solvent may slowly or rapidly exchange between solvation layers and between the outer solvation sphere and the unbound solvent. An ion may break solvent structure, and this may be the predominant effect as detected, e.g., by nuclear magnetic resonance spectrometry. By other methods, e.g., transference measurements, the ion may be found to be highly solvated. These phenomena can be explained on the basis of dynamic solvation in which the ion moves through relatively unstructured solvent, accompanied by a multilayer sphere of loosely bound solvent, the molecules of which continually exchange between the outer sphere and the unbound solvent [1].

Solvation, in mixed solvents, may be selective either in limited regions of solvent composition or over the whole range of solvent composition. Thus, if the rate of a chemical reaction is independent of solvent composition and thus of the dielectric constant, and if only the interaction energy in the complex enters into the activation energy [2], there would have to be a constant dielectric constant within the region comprising the complex irrespective of the macroscopic composition of the solvent for the rate of the reaction to be the same at all dielectric constants of the solvent [3, 4]. This constant dielectric constant region in the environment of the complex can most reasonably arise from one component of the solvent comprising this environment. In other words, there is selective solvation of the complex and perhaps of its components in this solvent system. If the complex and its components are highly polar or ionic, the solvating species is perhaps the more polar component. There are instances, due to some specific interaction perhaps, that a nonpolar component of a solvent solvates a solute, even when the nonpolar component of the solvent is present at low mole fractions [5-7].

Thus, the nature, extent, and strength of bonding or interaction of solvation may vary widely depending on the solvent and solute suspensoid, or surface. In this book we shall be interested mainly in solvation phenomena exhibited by solutions. The types of solvation presented will apply to solvents and solutes.

II. Types of Solvation

The solvation types to be presented in this chapter fall into three classifications; (1) positive and negative, (2) outer sphere and inner sphere, and (3) mixed component.

A. Positive and Negative Solvation

In dealing with solvation of ions, we must not only consider the interactions between the ions and their nearest solvent neighbors, but also the interactions between the ions and the rest of the solvent molecules. Hydration is a special case of solvation in which water is the solvent. By solvation is meant the sum of all the changes which occur when ions of an electrolyte are introduced into a solution [8, 9].

Early in the development of solution chemistry, Mendelejew [10] indicated the importance of solute–solvent interaction in relation to the properties of solutions, and suggested that a theory of liquid state aggregation be developed for the study of solutions. Earlier Kablukow [11] wrote, "In my opinion, water, as it decomposes the molecules of dissolved substance, forms with the ions unstable compounds which exist in the dissociated state." Many investigations [1–40] have been concerned with the solutions of ions and have approached the subject from many standpoints including solvation numbers of individual ions, selective solvation of ions, positive and negative ion solvation, inner and outer sphere (primary and secondary) ion solvation, dynamic ion solvation, and the effect of ion solvation on various phenomena in solution.

In water we speak many times of ion hydration. This generally connotes a certain number of molecules associated with each ion, i.e., a hydration number. Darmois [13] pointed out several difficulties connected with this concept. For example, hydration numbers ranging from 71 to 1 have been assigned [14–2] to the sodium ion. The lithium ion has been found [9] to vary from 17 to 158 in solvation number. Bockris [12] has explained these large variations in solvation numbers by assuming that some methods have resulted in the determination of close solvation and other methods distant solvation. It has been stated [9] that solvation numbers are only effective quantities in the sense that the action of the ions on the solvent is considered in terms of a stable combination with an effective number of solvent molecules, however, on the contrary solvation number or degree of solvation can be considered [1, 34] as the average solvent capacity of the ion even though the solvation is dynamic.

Stokes's law has been applied to hydration of ions in solution [9, 30–32]. For motion of an ion in dilute solution Stokes's law gives the expression

$$u\eta r = \text{const} \tag{2.1}$$

where u is the velocity of the ion in infinitely dilute solution at a given temperature, η the viscosity of the medium at this temperature, and r the ionic radius. The value of the constant is independent of the temperature, but does depend on the choice of units and on the ionic charge. From experimental values of u and η, Stokes's radius, r_s, can be calculated from Eq. (2.1). These radii [31] differ in general from the corresponding crystal radius, r_c. If $r_s > r_c$, it is assumed that the ion is solvated; if $r_s \leqslant r_c$ it is assumed the ion is not solvated. The excess volume of the Stokes's value over the crystal ions has been used to calculate the solvation numbers of the ions. The decrease in Stokes's radius in a series, like that of the alkali metal ions, coincides with the increase in mobility of the ions in the same series; i.e., the more solvated the ion, the more hindered is its motion, and the lower its mobility. Extent, but not nature of solvation are indicated by such procedures.

At the same temperature, for example 18°C, it has been pointed out [22] that the temperature coefficient of the fludity of water is less than the temperature coefficient of the mobilities of hydrated ions and greater than the mobilities of nonhydrated ions. If r is assumed to be constant in eq. (2.1), we can write

$$\ln u + \ln \eta = \ln \frac{\text{const}}{r} = \ln \text{const} \tag{2.2}$$

and taking the derivative with respect T gives

$$\frac{d \ln u}{dT} + \frac{d \ln \eta}{dT} = 0 \tag{2.3}$$

or

$$\frac{1}{u}\frac{du}{dT} = -\frac{1}{\eta}\frac{d\eta}{dT} \tag{2.4}$$

However, if r_s is variable with T, as it could if the ion became more or less hydrated depending on the temperature, then from Eq. (2.1) we can write

$$\ln u + \ln \eta + \ln r_s = \ln \text{const} \tag{2.5}$$

$$\frac{d \ln u}{dT} + \frac{d \ln \eta}{dT} + \frac{d \ln r_s}{dT} = 0 \tag{2.6}$$

and

$$\frac{1}{u}\frac{du}{dT} = -\frac{1}{\eta}\frac{d\eta}{dT} - \frac{1}{r_s}\frac{dr_s}{dT} \tag{2.7}$$

If, for hydrated ions, hydration decreased in extent with rising temperature, then

$$dr_s/dT < 0 \tag{2.8}$$

and the temperature coefficient of u for the hydrated ions would be positive since also $d\eta/dT < 0$. The two terms on the right-hand side of Eq. (2.7) increase, therefore, it would appear the temperature coefficient of fluidity of water, the first term on the right-hand side of Eq. (2.7), would be less than the temperature coefficient of the mobility of the hydrated ion represented by the left-hand side of Eq. (2.7). Like reasoning would be difficult to apply to nonhydrated ions since this would require that $dr_s/dT > 0$. However, if there is a decrease in structure of water with increase of temperature, the increased fluidity of water with increasing temperature might outweigh the increase in the mobility of the nonhydrated ion. In this as well as other phenomena in solution, the properties of both solvent and solute apparently must be considered in explaining observed data. It has been noted [13], however, that it is difficult to apply Stokes's law to nonhydrated ions.

Solvation of ions is in harmony with Coulombic interaction energies between ions and dipolar molecules which are larger than those between the dipolar molecules of the solvent and also larger than hydrogen bonding energies between solvent molecules, e.g., water.

Samoilov [9], however, indicates that the exchange process of water molecules close to an ion does not involve distances required in the application of the Coulombic formulas. He believes that exchanges does not depend on total interaction energy, but rather on the change of energy with distance in the neighborhood of the ion. Between the equilibrium position of a solvent molecule closest to an ion and the next equilibrium position farther out, there is a potential energy barrier, the height of which is a function of the inter-action energy between the ion and solvent molecule, and this height determines the exchange frequency of solvent molecules. The height of the potential barriers are lowered by repulsive forces and screening charges in the interaction of closely situated particles.

Samoilov [9] bases a theory of ion hydration in aqueous solution upon the effect of the ions on the translational motion of water molecules nearest them. This theory can be extended to the solvation of ions in any solvent. If the exchange of solvent molecules nearest the ion occurs relatively infrequently, solvation of the ion is indicated. As the exchange frequency increases, solvation decreases. The special importance of neighboring particle interactions has been dwelt upon by Sementschenko [41]. Samoilov's presentation of close and distant (inner and outer sphere) solvation will be discussed later. Here it will be pointed out that his statement that kinetic properties of solutions are not related to total hydration effect, but are principally determined by close hydration, is at variance with the experience and experimental observation of the present authors. This question will be dealt with extensively in the section on inner and outer sphere solvation.

Samoilov makes the simplification that total solvation can be separated into interactions of ions with solvent molecules in their immediate vicinity and their interactions with the remaining solvent molecules.

The activated jumps of water molecules in dilute aqueous electrolyte solution are considered, but we shall generalize this dilute electrolyte solution in any suitable solvent, i.e., any solvent which due to its polar and chemical characteristics will dissolve the electrolyte in dissociated (free ion) form. Let τ be the average time which a solvent molecule close only to other solvent molecules remains in the equilibrium position, and let E be the corresponding height of the potential energy barrier. The values of τ and E are approximately equal to those for pure solvent. A solvent molecule in the vicinity of an ion will not have a residence time τ in the equilibrium position because of the energy difference between an ion and a solvent molecule. If τ_i is the residence time of a solvent molecule in the equilibrium position next to an ion, then the fact that $\tau \neq \tau_i$ is principally due to the alteration of the height of the potential barrier from E to $E + \Delta E$ necessary for the exchange of the nearest solvent molecule. The presence of the ion causes an increase of ΔE in the potential energy barrier between neighboring equilibrium positions of the molecule.

It is assumed that solvation can be quantitatively represented by the ratio τ_i/τ and ΔE. When nearest solvent molecules are firmly bound to an ion τ_i/τ

is large, and τ_i/τ approaches infinity as the limiting case of permanent combination is more and more nearly realized. For permanent combination, τ_i/τ equals infinity. For decreasing strength of the ion–solvent bond τ_i/τ becomes smaller. The following equations relate τ and E.

$$\tau = \tau_0 e^{E/RT} \tag{2.9}$$

$$\tau_i = \tau_0' e^{(E+\Delta E)/RT} \tag{2.10a}$$

where the coefficients τ_0 and τ_0' are assumed to be similar in magnitude, implying that the frequency of vibration of a solvent molecule about its equilibrium position is independent of its distance from an ion. In the region of interest, i.e., where $\Delta E = 0$, this assumption is valid since the action of ions on the thermal motions of solvent molecules is not significant. From Eqs. (2.9) and (2.10a),

$$\tau_i/\tau = \frac{\tau_0}{\tau_0'} \frac{e^{E/RT}}{e^{(E+\Delta E)/RT}} = e^{-\Delta E/RT} \tag{2.10b}$$

and τ_i/τ can be calculated if ΔE is found.

The effect of ions on the viscosity of the solvent is directly related to their effect on the translational motion of the solvent molecules, multicharged or small singly-charged ions (Mg^{2+}, Ca^{2+}, Li^+) increase viscosity; large singly charged ions (K^+, Cs^+, I^-), decrease viscosity [9]. This latter phenomenon was termed negative viscosity [9] since the large singly-charged ions increase the translational motion of the solvent molecules nearest them. For these ions, $\tau_i < \tau$ and $\Delta E < 0$. For the multicharged and small ions $\tau_i > \tau$ and $\Delta E > 0$, and the solvent molecules in their vicinity exchange less frequently than solvent molecules in the neighborhood of other solvent molecules since in this case, there is effective binding of the nearest solvent molecules in the solution to the ions. The large singly-charged ions can be said to be negatively solvated and to produce negative viscosity since they increase the mobility and rate of exchange of solvent in their vicinity over those in the neighborhood of other solvent molecules.

This effect is similar to the effect of ions on the structural temperature of water as proposed by Bernal and Fowler [42, 43]. These ions which raise the structural temperatures of water are those which show negative hydration.

Samoilov [9] considers the influence of an electrical field on the translational motion of ions in solutions. This motion consists of activated jumps from one equilibrium position to another, the average number of activated jumps in the direction of the field being greater than in the reverse direction caused by a lowering of the potential barrier to translational motion in the field direction. Thus in Fig. 2.1, the increment of the potential in the direction opposite to the field is positive, giving a total potential of $q + \psi$ where q is the

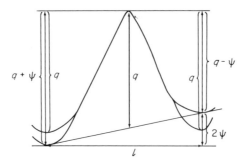

Fig. 2.1. Potential energy of an ion in the absence of an applied field, and potentials $q+\psi$ and $q-\psi$ in the directions opposite and coincident with the direction of the field.

potential energy in the absence of the field and ψ is the potential increment due to the field. In the direction of the field the potential energy is $q-\psi$. If ψ is the work required to displace an ion $\frac{1}{2}l$ cm, where l is length of an activated jump between adjacent equilibrium positions for the ion in solution, and if f is the force exerted on an ion by the electric field of l V/cm, then

$$\psi = \tfrac{1}{2}fl \tag{2.11}$$

Hence the average number of jumps in the direction of the field and in the opposite direction are

$$\vec{j} = \tfrac{1}{6}j_0 e^{-(q-\psi)/kT} \tag{2.12}$$

and

$$\overleftarrow{j} = \tfrac{1}{6}j_0 e^{-(q+\psi)/kT} \tag{2.13}$$

and the difference Δj is the number of jumps in the direction of the field:

$$\Delta j = \vec{j} - \overleftarrow{j} = \tfrac{1}{6}j_0 \exp[-q/kT]\,[\exp(\psi/kT)-\exp(-\psi/kT)] \tag{2.14}$$

Polissar [44] showed that for field strengths of 1 V/cm that $\psi \ll kT$, therefore, if $e^{\psi/kT}$ and $e^{-\psi/kT}$ are expanded by MacLauren series, and all except first-order terms in ψ be neglected, then

$$\Delta j = \tfrac{1}{6}j_0 e^{-(q/kT)(2\psi/kT)} = \tfrac{1}{6}j_0 e^{-q/kT}(fl/kT) \tag{2.15}$$

But the number of jumps j made by each ion per second is given by

$$j = j_0 e^{-q/kT} \tag{2.16}$$

Therefore,

$$\Delta j = \tfrac{1}{6}(fl/kT)j \tag{2.17}$$

Now the velocity v_i of the ion in the direction of the field is the number of jumps in the direction of the field Δj, multiplied by the length of each jump l. That is,

$$v_i = \Delta j \cdot l \tag{2.18}$$

and since ionic mobility u_i is given by

$$u_i = v_i F \tag{2.19}$$

where F is the faraday constant, we obtain

$$u_i = \tfrac{1}{6}(fF/kT) l^2 j_0 e^{-q/kT} \tag{2.20}$$

For a liquid for which the characteristic parameters for the viscuous motion of the particles are equal to those which determine the motion of ions in solution, we can proceed as follows. If a force F_η directed along a layer of a liquid causes this layer to move with a velocity c relative to another layer at a distance l from it, then

$$F_\eta = \eta S c / l \tag{2.21}$$

and dividing by S, the area of the latter gives the force per unit area f_η as

$$f_\eta = \eta c / l \tag{2.22}$$

Now l^2 is the area per particle in a layer and the force per particle is therefore

$$f_\eta l^2 = \eta c l \tag{2.23}$$

From substituting for the force f the value $f_\eta l^2$ in Eq. (2.17) we obtain

$$\Delta j = \tfrac{1}{6}(f_\eta l^2/kT) jl = \tfrac{1}{6}(f_\eta j l^3/kT) \tag{2.24}$$

But the average velocity of the particle and hence of the layer in the direction of the force is

$$c = l \, \Delta j \tag{2.25}$$

or

$$\Delta j = c/l \tag{2.26}$$

Therefore, from Eqs. (2.26) and (2.24), we have

$$c/l = \tfrac{1}{6}(f_\eta j l^3/kT) \tag{2.27}$$

Substituting the value of c/l from Eq. (2.27) into (2.22), we find

$$f_\eta = \tfrac{1}{6}\eta_i (f_\eta j l^3/kT) \tag{2.28}$$

or

$$\eta_i = 6kT/jl^3 = (6kT/j_0 l^3) e^{q/kT} \tag{2.29}$$

The product of Eq. (2.20) and (2.29) is

$$u_i \eta_i = fF/l = \text{const} \tag{2.30}$$

for water solutions since l is practically independent of temperature, and hence the right-hand side of Eq. (2.30) is approximately independent of temperature.

Equation (2.30) is the Pisarhewski–Walden rule [45] in which η_i has replaced η, the viscosity of the pure solvent. It would probably hold for solvents other than water.

For ions which are negatively hydrated, it can be assumed that activated jumps of single ions accompanied by the exchange of nearest-lying solvent molecules constitute the translational motion of ions in solution. Jumps of cells, ions plus nearest solvent molecules, need not be considered for ions exhibiting negative hydration nor for ions on the border between hydrated and nonhydrated ions. Also, the energy of activation in Eq. (2.20) and (2.29) must be the change in the potential energy barrier (exchange energy $q + \Delta q$) between equilibrium positions of the water molecules due to their proximity to ions.

From Eq. (2.30), an equation similar to Eq. (2.4) can be written in terms u_i and η_i, the mobility and viscosity which a solvent would have if the energy of activation were the same as for molecules in the neighborhood of ions. Thus

$$-\frac{1}{u_i}\frac{du_i}{dT} = \frac{1}{\eta_i}\frac{d\eta_i}{dT} \tag{2.31}$$

If in Eq. (2.29), q is taken as $q + \Delta q$, or for the mole $E + \Delta E$, we can write

$$\eta_i = B_i T e^{(E+\Delta E)/RT} \tag{2.32}$$

since $6k/(j_0 l^3)$ can be taken as a constant B_i independent of temperature. Taking logarithms of both sides of Eq. (2.32) and differentiating with respect to T yields

$$\frac{1}{\eta_i}\frac{d\eta_i}{dT} = \frac{1}{T} - \frac{E+\Delta E}{RT^2} \tag{2.33}$$

For pure solvent $\Delta E = 0$ and

$$\eta = BT e^{E/RT} \tag{2.34}$$

where $B = B_i$ according to Samiolov [9]. Thus

$$\frac{1}{\eta}\frac{d\eta}{dT} = \frac{1}{T} - \frac{E}{RT^2} \tag{2.35}$$

and from Eqs. (2.33) and (2.35)

$$\frac{1}{\eta_i}\frac{d\eta_i}{dT} = \frac{1}{\eta}\frac{d\eta}{dT} - \frac{\Delta E}{RT^2} \tag{2.36}$$

Using Eqs. (2.31) and (2.36), we find

$$\frac{1}{u_i}\frac{du_i}{dT} = -\frac{1}{\eta}\frac{d\eta}{dT} + \frac{\Delta E}{RT^2} \tag{2.37}$$

Thus for single ions

$$\Delta E > 0 \quad \text{if} \quad \frac{1}{u_i}\frac{du_i}{dT} > -\frac{1}{\eta}\frac{d\eta}{dT} \tag{2.38a}$$

$$\Delta E < 0 \quad \text{if} \quad \frac{1}{u_i}\frac{du_i}{dT} < -\frac{1}{\eta}\frac{d\eta}{dT} \tag{2.38b}$$

Samoilov [9] calculated the temperature coefficient k_i of ionic mobilities for 18 and 25°C by using the expression

$$k_i = \frac{2}{7\eta}\frac{u_i^{25}-u_i^{18}}{u_i^{25}+u_i^{18}} \tag{2.39}$$

These data are given in Table 2.1 and are compared to the crystal ionic radii in Angström units.

Table 2.1

Temperature Coefficients k_i of Ionic Mobilities of Ions and Crystallographic Radii of Ions

	Li⁺	Na⁺	K⁺	Cs⁺	Cl⁻	Br⁻	I⁻	Mg²⁺	Ca²⁺
$k_i \times 10^2$	2.65	2.44	2.17	2.12	2.16	2.15	2.13	2.54	2.54
r (Å)	0.78	0.98	1.33	1.65	1.81	1.96	2.20	0.78	1.06

For water at 21.5°C, $(1/\eta)(d\eta/dT)$ is -0.0240 [46]. Thus $\Delta E < 0$ and there is negative hydration for the K^+, Cs^+, Cl^-, Br^- and I^-, ions using Samoilov's criterion which also indicates that $\Delta E > 0$ and therefore positive hydration for the ions Mg^{2+}, Ca^{2+}, Li^+, and Na^+. However, other types of phenomena, transference for example, shows some of the negatively charged ions, K^+ for instance, in concentrations of 0.2 M KCl or greater, carrying relatively more water into the cathode region of the transference cell than Cl^- ion carries into the anode region [36], thus indicating at least loose bonding between the cation and the water solvent.

In negative solvation, the short distant distortion of the solvent structure by the ion is not compensated by the interaction of the ion with its adjacent solvent molecules. Thus there is an increase in the potential energy of the solvent molecules in the near vicinity of the ions; i.e., the potential valley in which the molecules vibrated became shallower. The large liberation of energy in solvation is, in the main, associated with distant solvation and is not in conflict with the idea of negative hydration.

It has been found [47] that the viscosity of monohydric alcohols relative to that of pure water increases with increasing concentration of KI; the relative viscosity of glycol compared to water decreases out to about 15 weight % and then increases slowly at first and then more rapidly with further addition of KI; the viscosity of glycerol relative to pure water decreases rapidly with increasing concentrations of either KI or KBr. Samoilov [9] attributes a relative viscosity $\eta/\eta_0 < 0$ to negative viscosity and points out that negative viscosity is exhibited with solvents having strong intermolecular interactions and more marked close range order as in the higher hydric alcohols where the complexity of bonding is greater. Studies of KI in butanol–glycerol mixtures confirm this view that the disturbance of the solvent by the solute gives rise to negative viscosity.

Samoilov [9] includes in his theory of negative hydration the idea that the solvent is ordered differently and consequently has a different density in the neighborhood of ions than in pure water. Thus the density of the solvent is greater in the vicinity of solvated, and less in the neighborhood of nonsolvated ions, than in pure water. Solvated cations moving toward a cathode effect a transport of solvent in this direction; while nonsolvated anions moving toward an anode effect a transport of solvent in a direction opposite to that of their motion, i.e., toward the cathode. It might be pointed out that anions are not always unsolvated [48, 34, 36, 37].

Collet's [49, 50] discussion that an ion transported influx of solvent into a cathode region causes an outflow of water carrying the reference with it, and the claim that the concentration change of the reference substance is due mainly to the transport of reference substance toward the anode rather than the water transport into the cathode space is somewhat weak. As increments of water are transported into the cathode region by cations, films of solution of the original composition cross the boundary between the cathode and central portion. The cathode and anode regions are chosen to be so large that there is no change in the concentration of the central portion during the transference experiment else that experiment is discarded. At the end of the experiment stopcocks properly placed are closed to isolate the cathode, anode and central regions. Analysis of the original solution and the solutions in the two electrode regions, allows verifications that the middle region remained unaltered, and the calculation of the quantity of each solute per given weight of solvent. Thus, the change in concentration of solute between the original and electrode portion of solution is due only to the shift of a quantity of solvent from one electrode portion to the other. Collet did not take into account the differences between average densities of water in the immediate vicinities of the ions and in pure water. According to Samoilov [9] the differences in average densities play an especially important role io the transport of water by ions.

B. NEGATIVE SOLVATION AND THE ACTIVITY COEFFICIENT OF THE SOLVENT

Activity coefficients of ions have been used [51–53] in evaluating solvation numbers. Samoilov [9] argues that a decrease in the translational motion of water molecules, i.e., a reduction in their potential energy, must lead to a decrease in the activity coefficient of water, while an increase in the translational motion of water molecules, i.e., a magnification of their potential energy, should cause an increase in the activity coefficient of water.

If μ_0, μ, and μ_i are the chemical potentials of pure water, water in an M molar solution of electrolyte, and of water in an ideal solution, respectively, a the activity of water in the solution and x its mole fraction, then

$$\mu = \mu_0 + RT \ln a \qquad (2.40)$$

$$\mu_i = \mu_0 + RT \ln x \qquad (2.41)$$

and

$$\mu - \mu_i = RT \ln a/x = RT \ln \gamma \qquad (2.42)$$

where $\gamma = a/x$ is the activity coefficient of water in the solution. Thus when $\mu < \mu_i$, i.e., when ions increase the negative potential of neighboring water, $\gamma < 1$ and the ions are positively hydrated. For negative hydration $\mu > \mu_i$ and $\gamma > 1$. The activity coefficients of water solvent in alkali metal and alkaline earth metal chlorides were calculated by Samoilov and plotted versus the molarity of the various salts. In more dilute solutions the activity coefficients of water in all the salt solvations were somewhat greater than 1, passed through a maximum and then, except for CsCl, eventually became less than one. For the metal chlorides in each series both the maxima and the points at which the activity coefficients became negative were shifted toward higher and higher salt concentrations the greater the atomic weight of the metal. The maxima and the points at which the activity coefficients became negative in general occurred over a much shorter range of salt concentrations for the alkaline earth than for the alkali metal chlorides. The shifts of the maxima and those of the cross-overs where the activity coefficients become negative toward higher concentrations with increasing atomic weights of the metals is due to the transition from hydrated cations to cations showing negative hydration.

This behavior of the activity coefficients of water with changing concentrations of salts might be contrasted to that of the behavior of the activity coefficients of electrolytes with changing electrolyte concentrations. The mean activity coefficients f of various electrolytes plotted either versus concentration or versus the square root of the ionic strength $I^{1/2}$ show minima rather than maxima values [54]. There are fewer regularities with respect to minima and points of change-over from values of less than 1 to values greater than 1 in the case of these plots than in the case of the water activity coefficient γ plots. The anions seem to influence profoundly the shapes of these curves. The curve for $CaCl_2$ shows a minimum at a $I^{1/2}$ of about 1 and crosses over to values of $f_\pm > 1$ at $I^{1/2}$ of about 2. The curve for $Ca(NO_3)_2$, on the other hand, exhibits a minimum at a $I^{1/2}$ of about 2, but shows little increase in $f_\pm$ beyond that point. The curves for LiI and HCl are quite similar giving minima at about a $I^{1/2}$ of about 0.5 and climbing steeply beyond the minima to values of $f_\pm > 1$ at values of $I^{1/2}$ between 1.0 and 1.5. The curve for KCl flattens out in the region of the minima about $I^{1/2} = 1.5$. There is no indication of a succeeding steep climb of the curve. Thus, LiI and HCl beyond $I^{1/2}$ of about 1.5 exhibit an effective concentration beyond their actual concentration, which could be explained as due to their tying up free water as water of solvation to the extent that this overcomes the decrease in effective concentration due to electrostatic attraction between the oppositely charged ions. This latter effect is dominant in lower concentration ranges of the electrolytes, which on its own or together with some other effect, such as negative solvation, dominates in the regions of KCl concentrations given by Wang [55].

Samoilov [9] pointed out that the self-diffusion coefficient of water in solutions of KCl and KI, the ions of which are only negatively hydrated, is greater than the self-diffusion coefficients of water in pure water. For these solutions the water molecules in the vicinity of the ions are more mobile than the water molecules in pure water ($\tau_i/\tau < 1$). These conclusions were supported by the data of Wang [55].

C. Negative Solvation and Isotope Equilibrium in Aqueous Electrolyte Solutions

The isotopic composition of water in electrolytic solutions studied by Feder and Taube [56] and by Taube [57] are closely related to the problem of negative hydration [9]. The exchange of ^{16}O and ^{18}O in normal isotopic composition in water $H_2^{16}O$ and $H_2^{18}O$ with CO_2 was investigated in electrolyte solutions as a function of electrolyte concentration by determining the mass spectrographic activities of the isotopic molecules $H_2^{16}O$ and $H_2^{18}O$. The equilibrium isotopic compositions of the solvent water and of the ion hydration shells, i.e., the water molecules in closest proximity of the ions, were not the same. Near some ions the $H_2^{18}O$ molecules were enriched, while in the neighborhoods of certain ions they were depleted.

Samoilov [9] wrote the equilibrium constant for the isotopic exchange of the water molecules in the vicinity of the ions, i.e., for the process

$$H_2{}^{16}O_h + H_2{}^{18}O_l \rightleftarrows H_2{}^{18}O_h + H_2{}^{16}O_l \qquad (2.43)$$

as

$$K = K_h/K_l = {}^{18}W_h{}^{16}W_l/{}^{16}W_h{}^{18}W_l \qquad (2.44)$$

where $K_h = {}^{18}W_h/{}^{16}W_h$ and $K_l = {}^{18}W_l/{}^{16}W_l$ are the isotopic composition of the hydration shells and the rest of the water in the solution respectively.

It was found [58] that the value of K depededed practically completely on the cation with little effect by the anion, and that K was determined almost exclusively by the interactions of the ions with their nearest-neighbor water molecules. The value of K was greater than 1 for Al^{3+}, Mg^{2+}, Li^+ and Ag^+ salts, K approach 1 for Na^+ salts, and K was less than 1 for K^+, $NH_4{}^+$ and Cs^+ salts. Samoilov [9] offers proof that the data on K shows negative hydration for the K^+, $NH_4{}^+$ and Cs^+ cations, that is that $\Delta E < 0$. His reasoning is as follows.

In the exchange mechanism of $H_2{}^{16}O$ and $H_2{}^{18}O$ for a water molecule to escape from an equilibrium position in the vicinity of an ion, it must attain the energy $E + \Delta E$; to leave a position not adjacent to an ion, energy E is required. Let j_0^{16} and j_0^{18} be the activated jump frequency factors for $H_2{}^{16}O$ and $H_2{}^{18}O$ and let these factors be independent of whether the molecules vibrate about positions adjacent to or far from an ion. This last assumption is equivalent to making the translational energy of water molecules depend on the presence of ions only to the extent of the alteration of the activation energies. This is justified for ions on the boundary between hydrated and non-hydrated ion, in which case the action of ions on the thermal motion of water molecules is negligible. Let $\bar{j}^{16}$ be the frequency with which $H_2{}^{16}O$ molecules per second entering the hydration shells of ions, and $\tilde{j}^{16}$ be the frequency with which $H_2{}^{16}O$ molecules leave these equilibrium positions. Let $\bar{j}^{18}$ and $\tilde{j}^{18}$ be corresponding factors for $H_2{}^{18}O$ molecules.

Then, as given by the theory for activated jumps of liquid particles [58],

$$\tilde{j}^{16} = k_n \frac{{}^{16}W_l}{{}^{16}W_l + {}^{18}W_l} j_0^{16} e^{-E/RT} \qquad (2.45)$$

$$\bar{j}^{16} = k_n \frac{{}^{16}W_h}{{}^{16}W_h + {}^{18}W_h} j_0^{16} e^{-(E+\Delta E)/RT} \qquad (2.46)$$

In the above equations k_n is a coefficient related to the coordination number of the ion in solution. In water of usual isotopic composition ${}^{16}W \gg {}^{18}W$, hence

$$\frac{{}^{16}W_l}{{}^{16}W_l + {}^{18}W_l} = \frac{{}^{16}W_h}{{}^{16}W_h + {}^{18}W_h} = 1 \qquad (2.47)$$

Therefore

$$\Delta j^{16} = \hat{j}^{16} - j^{16} = k_n j_0^{16} e^{-E/RT}(1 - e^{-\Delta E/RT}) \tag{2.48}$$

For $H_2{}^{18}O$,

$$\hat{j}^{18} = k_n \frac{{}^{18}W_1}{{}^{18}W_1 + {}^{16}W_1} j_0^{18} e^{-E/RT} \tag{2.49}$$

$$j^{18} = k_n \frac{{}^{18}W_h}{{}^{18}W_h + {}^{16}W_h} j_0^{18} e^{-(E+\Delta E)/RT} \tag{2.50}$$

But since ${}^{16}W \gg {}^{18}W$, then

$$\frac{{}^{18}W_1}{{}^{18}W_1 + {}^{16}W_1} = \frac{{}^{18}W_1}{{}^{16}W_1} = K_1$$

$$\frac{{}^{18}W_h}{{}^{18}W_h + {}^{16}W_h} = \frac{{}^{18}W_h}{{}^{16}W_h} = K_h$$

$$\hat{j}^{18} = K_1 j_0^{18} e^{-E/RT} \tag{2.51}$$

$$j^{18} = k_n K_h j_0^{18} e^{-(E+\Delta E)/RT} \tag{2.52}$$

$$\Delta j^{18} = \hat{j}^{18} - j^{18} = k_n j_0^{18} e^{-E/RT}(K_1 - K_h e^{-\Delta E/RT}) \tag{2.53}$$

The probability that a $H_2{}^{16}O$ water molecule be found in transit is the net number of $H_2{}^{16}O$ molecules entering the hydration shell, Δj^{16}, divided by the total number of $H_2{}^{16}O$ molecules, ${}^{16}W_1 + {}^{16}W_h$, which for dilute solutions is approximately equal to the number of free $H_2{}^{16}O$ molecules in the liquid ${}^{16}W_1$. The probability that an $H_2{}^{18}O$ water molecule be found in transition is the net number of $H_2{}^{18}O$ molecules entering the hydration shell, Δj^{18}, divided by the total number of $H_2{}^{18}O$ molecules, ${}^{18}W_1 + {}^{18}W_h$, which for dilute solutions is approximately equal to the number of free $H_2{}^{18}O$ molecules in the liquid ${}^{18}W_1$. At equilibrium these probabilities are equal. Thus, for dilute solutions,

$$\Delta j^{16}/{}^{16}W_1 = \Delta j^{18}/{}^{18}W_1 \tag{2.54}$$

and, therefore,

$$\Delta j^{16} = ({}^{16}W_1/{}^{18}W_1) \Delta j^{18} = K_1 \Delta j^{18} \tag{2.55}$$

Substituting Eqs. (2.48) and (2.53) into Eq. (2.55), we obtain

$$K_1 - K_h e^{-\Delta E/RT} = K_1(j_0^{16}/j_0^{18})(1 - e^{-\Delta E/RT}) \tag{2.56}$$

Dividing both sides of Eq. (2.56) by $K_1 e^{-\Delta E/RT}$, we have

$$e^{\Delta E/RT} - K_h/K_1 = (j_0^{16}/j_0^{18})(e^{(\Delta E/RT)} - 1) \tag{2.57}$$

Letting $K = K_h/K_1$, multiplying Eq. (2.57) through by -1, and transposing $e^{\Delta E/RT}$ from the left to the right-hand side of Eq. (2.57), gives us

$$K = e^{\Delta E/RT} - (j_0^{16}/j_0^{18})(e^{\Delta E/RT} - 1) = e^{\Delta E/RT}[1 - (j_0^{16}/j_0^{18})] + (j_0^{16}/j_0^{18}) \tag{2.58}$$

Nuclear magnetic resonance studies indicates that ClO_4^- ion is solvated by dioxane [5]. Polarographic [6] and thermodynamic [7] investigations also suggest solvation of ClO_4^- ion by dioxane, and that Br^-, Cl^-, and NO_3^- ions are hydrated. Of course, the nuclear magnetic resonance data could possibly be interpreted as selective solvation of ion pairs by dioxane since the studies were made in water–dioxane mixed solvents, and since ion-pairing would be enhanced in solutions containing the nonpolar, low-dielectric-constant dioxane as a solvent component. However transference [36, 37], and conductance [34] support the idea of anion solvation, including hydration.

Dynamic solvation must be considered also. Even in this situation [1, 34] due to its structure breaking effect on the solvent and at the same time its relatively loose but extensive solvation by the solvent, the ion moves through the relatively unstructured solvent accompanied by a multilayer sphere of loosely bound solvent. Between contiguous spheres and between the outer sphere and the free, unbound solvent there is continuous exchange of solvent molecules. Thus, there is an average number but not specific individual molecules of solvent associated with the particular dynamically solvated ion over a period of time for a given set of conditions such as concentration and temperature.

It could be that negative solvation is dependent on the solvent structure-breaking properties of cations resulting in less energy of activation ($\Delta E < 0$) for a jump of a solvent molecule from the hydrated shell of the ion into an unstructured mass of water than is required for the jump of a water molecule from unstructured water into the hydration shell of an ion where more structure prevails. On the other hand, for a structured solvent, ΔE could be greater than zero since for a molecule to leave the structured solvent might require a positive amount of energy compared to that required for a molecule of solvent to re-enter the mass of the solvent from the solvent shell of the ion. Thus, the concept of negative solvation might arise from a solvent structure effect on the relative translational mobility of free and ion-bound solvent molecules, rather than the action of the ion on the thermal, principally translation motion of the water molecules near the ions, and the consequent indication of the absence of the combination of a definite number of solvent molecules with an ion.

E. OUTER AND INNER SPHERE SOLVATION

Applying Helm's [60] single-coil, spin-echo method, Morgan *et al.* [61] measured for protons the nuclear magnetic resonance relaxation times in a number of aqueous solutions of symmetrical chromium III complexes. The size of the ion and its solvation were used to satisfactorily explain their data. The differences in the strengths of the hydrogen bonds formed between ligands involving these different atoms and the water solvent was caused by the

Subtracting 1 from either side of eq. (2.58), we have

$$K - 1 = e^{\Delta E/RT}[1 - (j_0^{16}/j_0^{18})] - [1 - (j_0^{16}/j_0^{18})]$$

$$= (e^{\Delta E/RT} - 1)[1 - (j_0^{16}/j_0^{18})]$$

Samoilov [9] used the data of Wang et al. [59] on self-diffusion in wa
estimate the factor j_0^{16}/j_0^{18}. From these data j_0 depends mainly on the
located at the centers of the molecules and very little on the mass of H.
values of j^0 for all of the hydrogen isotopes of water containing O^{16} are
equal, and $j_0^{16} < j_0^{18}$. Thus $j_0^{16}/j^{18} < 1$. Thus in Eq. (2.59) the last parenth
term on the right-hand side is greater than zero; and hence $K - 1$ has the
sign as the term $e^{\Delta E/RT} - 1$. When $K < 1$, $\Delta E < 0$, and negative hydrat
evidenced; Taube's work, in particular for the K^+ ion, confirms neg
hydration according to Samoilov [9], at least qualitatively. The diffe
between the corresponding equilibrium concentrations, and the sign c
difference, is evidenced by the diffusion of the isotopic forms of water,
$H_2^{18}O$, to and from the ions. Samoilov [9] also points out that ii
experiments on isotopic exchange similar effects to those of thermal diff
are operative. Thus ions which concentrate in their vicinity $H_2^{18}O$ i
cules, lowers the Bernal and Fowler [42, 43] structural temperature of v

Thus Samoilov argues that the existence of negative hydration implies
hydration of ions in aqueous solutions is an action of ions on principall
translational thermal motion as activated jumps of adjacent water molec
rather than a combination of a certain number of molecules with the ion:
also emphasizes that the concept of negative hydration is important ii
study of the structure of aqueous solutions, and particular use is made o
evidence that Na^+ shows positive while K^+ shows negative hydration.

D. DISCUSSION

While Samoilov limits his theory and applications of negative hydratio
aqueous solutions and hydration of ions, there is no obvious reason why
theory and its applications would not apply to the solvation of ions in p
solvents in general, especially if the solvents are similar to water in other t
polar properties; such as hydrogen bonding, acid–base properties, and ex
of association.

The implications that cations alone are the only ions to be considered
hydration theory and applications are certainly not generally accepta
While Samoilov's treatment of negative hydration and Taube and co-work
treatment of the equilibrium isotopic exchange of water molecules in
vicinity of ions involve only cations, yet other experimental evidence indica
that the solvation of the anion is real and important.

difference in solvation among the various Cr(III) complexes containing fluorine, oxygen, or nitrogen atoms. Second sphere coordination would result from this kind of interaction. These effects, according to the authors, would be greatest for solutions containing CrF_6^{3-}, somewhat smaller for $Cr(C_2O_4)_3^{3-}$ solutions, and least for solutions of $Cr(en)_3^{3+}$ and $Cr(CN)_6^{3-}$. The symbol (en) represents ethylenediamine.

While the oxygen atoms in the oxalate to which hydrogen bonding is attributed are relatively far from the metal atom, the orientation of the hydrogen bonded molecules with their hydrogen atoms toward the ion serves as a compensatory factor. Meaningless conclusions may be reached if efforts to correlate proton relaxation times with molecular parameters of paramagnetic ions do not take into account solvent interactions [61]. The observed relaxation times, T_1, were correlated with the formula of Bloembergen et al. [62].

$$1/T_1 = CN\mu_{eff}^2 \qquad (2.60)$$

Where C is a constant for a given temperature, N is the number of moles per liter of paramagnetic ion, and μ_{eff} is the apparent magnetic moment of the paramagnetic ion. It was found that NT_1 for a given ion at a given temperature remained constant and equal to $1/C\mu_{eff}$. The ions studied were $Cr(H_2O)_6^{3+}$, CrF_6^{3-}, $Cr(NH_3)_6^{3+}$, $Cr(C_2O_4)_3^{3-}$, $Cr(CN)_6^{3-}$, and $Cr(en)_3^{3+}$. These ions have been classified as inert complexes [63], i.e., complexes from which displacement reactions for the ligand groups are relatively slow. The oxygen atoms in the six water molecules of the aquo chromium(III) ion exchange at a measurable rate with the oxygen atoms in solvent water [64]. It is believed that the protons of the hydration sphere and the unbound water molecules exchange rapidly [61]. No case is known where hydroxylic hydrogen exchanges measurably slowly [65], and in fact this exchange may have a halflife much shorter than T_1. It is thought [61] that in the case of the hexaaquo chromium complexes, the hydrogen exchange between bound and unbound water should permit closer approach of the protons to the chromium atoms and thus magnify the relaxation of the protons. There would be no significant contribution to the signal by the small number of protons in the complex unless there was complete mixing of these protons with those of the solvent.

The effective ionic radius and the ion–solvent interaction, i.e., the proton distance of closest approach, might vary with different paramagnetic ions causing the constant C in Eq. (2.65) to vary. The radii of the complex ions $Cr(H_2O)_6^{3+} \cdot Cr(NH_3)_6^{3+}$, and CrF_6^{3-} have been estimated [61] to be of the order of 3.1–3.3 Å, and that of $Cr(en)_3^{3+}$ and $Cr(C_2O_4)_2^{3-}$ to be, respectively, 3.6 and 4.7 Å. Therefore, difference in size was not the sole factor causing variation in NT_1, though it no doubt had some influence.

The differences in ion–solvent interactions must also be considered. Adjacent water molecules would presumably hydrogen-bond with complex ions containing fluorine, oxygen, or nitrogen atoms, depending on the ionic geometry, the number of possible hydrogen bonds, and the hydrogen bond strength in a given combination of atoms. The greatest complex-solvent interaction, it was speculated, would occur for CrF_6^{3-}, with somewhat smaller effects for $Cr(C_2O_4)_3^{3-}$, and $Cr(NH_3)_6^{3+}$, and the smallest effects for $Cr(en)_3^{3+}$ and $Cr(CN)_6^{3-}$.

Experimental relaxation times were not quantitatively related to ion-size and ion-solvent interaction parameters. However, it was believed that these parameters must be considered when correlating proton relaxation times with molecular parameters of paramagnetic ions. It was suggested [61] that aging of aquated ions might profoundly influence relaxation times.

A rather highly ordered second coordination shell for Al^{3+} in anhydrous dimethyl sulfoxide (DMSO) was suggested [66] from electrostatic and hydrogen bonding considerations.

Proton NMR data indicated the $Be(DMF)_4^{2+}$ complex to be more labile than the $Al(DMF)_6^{3+}$ complex. This enhanced lability was attributed to activation enthalpy rather than to activation entropy [67]. A S_N1 (lim) solvent exchange mechanism could not be used to explain relative labilities in terms of a simple electrostatic model for the binding of dimethyl formamide (DMF) molecules to the ions. The Be(II) ion had a charge to radius ratio of 6.5 and was bound to four DMF molecules, while the Al(III) had a charge to radius ratio of 6.0 and was bound to six DMF molecules. Assuming the S_N1 mechanism, the problem was not simplified by including steric effects since scaled molecular models showed the Be(II) and Al(III) ions were about equally sterically hindered, though the former ions exhibited somewhat less entropy of activation for solvent exchange than the latter. The steric and electrostatic interactions, while no doubt influential in the primary co-ordination spheres of these ions in the solvent exchange reactions, yet other effects must be taken into account, such as the relative stabilization of these ions by ion-pair formation and coordination in the second coordination sphere of ions.

The frequency dependence, in the frequency range 2.7–28.7 MHz, of the proton spin relaxation in aqueous solutions of the $Cr(H_2O)_6^{3+}$, $Mn(H_2O)_x^{2+}$, $Co(H_2O)_x^{2+}$, and the $Nd(H_2O)_x^{2+}$ have been studied using the spin-echo method [68]. Using a theoretical model of the magnetic interaction between ion and proton in which it was assumed a random thermal process with a characteristic correlation time, T_e, of 10^{-8}–10^{-11} sec occurred, the abnormally large ratios of the spin–lattice relaxation time T_1 and spin–spin relaxation time T_2 were explained. The order of magnitude of the factor B was calculated by applying the theory to data. The equation was

$$B = N_c \mu_{\text{eff}} / r_e^6 \qquad (2.61)$$

where N_c is the number per unit volume of type i spins which can interact at one time with a spin j, μ_{eff} is the effective moment of the ion, and r_e is the distance between i and j. Among other causes, thermal motion in the third and fourth water layer about the ion was assumed to result in the large values of B obtained from these calculations as compared to other calculations.

To elucidate the relaxation mechanism in the first coordination sphere, the proton spin resonance in aqueous solutions of the paramagnetic ions Cr^{3+}, Mn^{2+}, Ni^{2+}, Cu^{2+}, and Gd^{3+} were studied [69]. The results were discussed in relation to Solomon's [70] formulation of electron–nuclear dipole–dipole interactions, and Bloembergen's [71, 72] expression for scalar coupling of electron and nuclear spins. Morgan and Nolle [69] conclude that in dilute aqueous solutions of Ni^{2+}, Cu^{2+}, and Gd^{3+} ions, the dipolar mechanism is alone operative. However, Stengle and Langford [73] believe a second sphere relaxation mechanism is involved. They point out that a dipolar mechanism, while not completely able to explain Morgan and Nolle's data on $Cr(H_2O)_6^{3+}$, does account for a major part of the relaxation. It was pointed out that dipolar interaction is probably the predominant interaction in the second sphere since this interaction decreases less rapidly with distance than any other.

Stengle and co-workers [73] have studied the effect of various paramagnetic $Cr(III)$ complexes with well defined nonlabile inner coordination spheres on the transverse relaxation times T_2 on ^{19}F nuclear magnetic resonance (NMR) signals of F^- and PF_6^- ions in aqueous solutions at constant temperature ($\pm 1°C$). Measurement of line widths at halfheights gave values of T_2. That line widths were not affected by saturation was carefully checked. According to these authors, Bloembergen and Morgan's [74] study of the mechanism of relaxation, the details of which are not understood too clearly, apparently justify the following assumptions: (a) the observed relaxation rate is given by the relaxation rate averaged over all the different nuclear environments weighted for the probability that the nucleus is in the given environment; (b) the decrease in relaxation time is predominantly influenced by those ^{19}F-containing anions in the second coordination sphere of the paramagnetic ions; (c) in the relaxation mechanism of a ^{19}F nucleus in the second coordination sphere of a $Cr(III)$ complex, the dipolar interaction is of primary and the spin echo interaction is of secondary importance under certain conditions outlined by the authors.

The solution of the Bloch equations with chemical exchange has been discussed by McConnell [75] and by Swift and Connick [76, 77]. The observed relaxation time T_2 in the case of fast exchange is given by the equation

$$\frac{1}{T_2} = \frac{P_A}{T_{2A}} + \frac{P_B}{T_{2B}} \tag{2.62}$$

where T_{2A} and T_{2B} and P_A and P_B are the relaxation times and probabilities of occurrences in the respective A and B environments. If T_{2A} refers to the relaxation time in the paramagnetic solution, then for reasonably dilute solutions P_A is nearly unity, and hence P_B will be nearly proportional to the quantity $\Delta V - \Delta V_A$, where ΔV is the observed line width at halfheight ($\Delta V = 1/\pi T_2$) and ΔV_A is the line width at halfheight in environment A ($\Delta V_A = 1/\pi T_{2A}$). T_{2A} is the relaxation time in the solution free of paramagnetic ions. Thus

$$\pi(\Delta V - \Delta V_A) = P_B/T_{2B} \qquad (2.63)$$

The law of mass action was applied by Stengle and Langford [73] to the equilibrium between the second sphere complex and the ions forming the second sphere complex. Let the Cr(III) complex ion with a fixed inner coordination sphere be represented by M, and the ^{19}F-containing anion (PF_6^- or F^- in the case of Stengle and Langford) be represented by X, then

$$M + X \rightleftarrows MX \qquad (2.64)$$

$$\alpha_1 = [MX]/C_m \qquad (2.65)$$

and

$$\beta_1 = [MX]/[M][X] \qquad (2.66)$$

where α is the degree of monocomplex formation and β is the formation constant for this complex. In the equations the concentration of a species is represented by brackets enclosing that species. The analytical concentration of the Cr(III) complex is given as C_m. By definition

$$P_B = [MX]/C_x = \alpha_1 C_m/C_x \qquad (2.67)$$

where C_x is the concentration of the ^{19}F-containing anion. From Eqs. (2.63) and (2.67), there results

$$\pi(\Delta V - \Delta V_A) = \frac{\alpha_1 C_m}{C_x T_{2B}} \qquad (2.68)$$

If for a series of Cr(III) ions T_{2B} is the same for a given anion, a plot of the experimental ΔV values for each Cr(III) ion versus C_m for that ion should give a straight line. For different Cr(III) ion species the relative slopes of such plots should be the relative α_1 values for the given anion. The data of Stengle and Langford [73] gave such straight line plots for PF_6^- and F^- anions, the order of decreasing complexation for both anions being

$$Cr(pn)_3^{3+} > Cr(en)_3^{3+} > Cr(en)_2Cl_2^+ > Cr(ox)_3^{3-} \qquad (2.69)$$

Even if T_{2B} values are not the same, provided they vary in a regular manner with the size for all Cr(III) complex ions, important quantitative data can yet be obtained. There should be a decrease in the relaxation rate $1/T_{2B}$ with increasing radius of the Cr(III) ion since the magnitude of the interaction of the paramagnetic electron and the ^{19}F nucleus depends on the inverse sixth power of the distance between the interacting centers. The relaxation time will vary with the third power of the radius with respect to tumbling correlation time for a sphere turning in a viscous medium [76]. Thus the magnitude of the dipolar interactions will be the predominant influence in the dependence of $1/T_{2B}$ on ion size, and the relaxation time will be smaller for larger ions [73]. However, an increase in the ^{19}F line width with increasing size of the Cr(III) complex was found [73], which was opposite to what would have been expected on the basis of the change of T_{2B} with ion size. Ion association was used to explain why the observed trend of line width versus $Cr(en)_3^{3+}$ complex concentration plot gave a slope which was three times that of a similar plot for the $Cr(H_2O)_6^{3+}$ complex.

For a given complex at two C_x^- values (C_m constant) the P_B ratio is given by the ratio of $\Delta V - \Delta V_A$ since the unknown quantity T_{2B} cancels out. Ratios of P_B may be calculated by assuming values of β_1 and compared with experimental values to determine approximate values of β_1. Due, however, to the larger values of C_x required to observe resonance, Stengle and Langford's data could be applied only to β_1 values between 1 and 10. Erroneous values of β_1 may be obtained when high complexing interferes with analysis in terms of monocomplexing. Activity effects may also be important. Approximate values of β_1 comparing reasonably well with recorded values [78–82] were found by Stengle and Langford. Table 2.2 contains the β_1 values for the systems referred to.

Table 2.2

β_1 Values for the Systems Referred to

System	β_1
$Cr(ox)_3^{3-}$ with PF_6^-	0
$Cr(en)_3^{3+}$ with PF_6^-	1
$Cr(en)_2Cl_2$ with F^-	3
$Cr(en)_3^{3+}$ with F^-	> 10

Stengle and Langford [73] gave an overall evaluation of NMR with respect to the study of second sphere association. In a series of complexes of very similar electronic configuration, NMR can be used to determine in a simple manner the relative degree of outer sphere complexing; but, if the association is weak, NMR gives only an approximate value of the outer sphere association constant. The NMR method, however, does detect genuine second sphere

association owing to its "short-range" character. Data suggest some broadening due to "third sphere" effects in the case of the $Cr(ox)_3^{3-}$ system even
though no ion association is likely. The effect is approximately independent of
the source of ^{19}F; it is small and yields a β_1 value of zero.

In aqueous perchloric acid solutions of Cr(III), Alei [81], using a Varian
wider line spectrometer, has observed the shifts at 8 Hz and at about 20° of
the NMR peak for ^{17}O in labile water and in ClO_4^- ion.

The chromium species in aqueous perchloric acid solutions of Cr(III) was
shown [83] to be $Cr(H_2O)_6^{3+}$. Also, it was shown that this species exchanges
slowly with unbound water in the system. This rate of exchange was found [84]
to increase with increasing concentration of anion due to the formation of an
activated complex containing both $Cr(H_2O)_6^{3+}$ and anion. It has been reasoned
[81] that such solutions should show two water peaks, one for unbound or
labile water and one for water in the $Cr(H_2O)_6^{3+}$ complex. Only the peak for
labile water was found and it was presumed that the peak due to water in
$Cr(H_2O)_6^{3+}$ complex was so strongly shifted and broadened as to escape
detection by the procedure used. It was observed that the position of the ^{17}O
resonance of the labile water compared to that in pure water shifted an increasing amount with increasing Cr(III) concentration, and it was concluded
that the labile water was subjected to a paramagnetic influence when located
in a second sphere around the $Cr(H_2O)_6^{3+}$ ion. It was thought that both labile
water and ClO_4^- ions are second sphere ligands of Cr(III), i.e., they come
into the immediate vicinity of the $Cr(H_2O)_6^{3+}$ species, because the observed
shift δ_{H_2O} in the $H_2^{17}O$ resonance of labile water is not proportional to the
ratio of $Cr(H_2O)_6^{3+}$ to labile water, and the ^{17}O resonance in ClO_4^- ion
experiences an increasing paramagnetic shift with increasing Cr(III)
concentration.

If it is assumed [81] that there are r sites in the second sphere, y of which
are occupied by water and $(r-y/n)$ of which are occupied by ClO_4^- ion
(where n is the number of water sites occupied by one ClO_4^- ion), and if K is
a constant related to the specific magnetic field experienced by an ^{17}O nucleus
in a water molecule in the second sphere of Cr(III), and $m_{Cr(III)}$ is the number
of millimoles of Cr(III) in solution, then the measured shift δ_{H_2O} in ^{17}O
resonance in labile water can be calculated using the equation

$$\delta_{H_2O} = Kym_{Cr(III)}/N \tag{2.70}$$

where N is the number of millimoles of labile water in solution
($N = m_{H_2O} - 6m_{Cr(III)}$), where m_{H_2O} is the millimoles of water in the solution).

In the second sphere, the ratio of ClO_4^- to H_2O is related to the ratio of
ClO_4^- to H_2O in the bulk solution by the expression

$$\frac{(r-y)/n}{y} = Q\frac{m_{ClO_4^-}}{N} = QM'_{ClO_4^-} \tag{2.71}$$

where Q is a constant measuring the degree to which ClO_4^- ion preferentially seeks the second sphere of $Cr(H_2O)_6^{3+}$ ion and $M'_{ClO_4^-}$ equals $m_{ClO_4^-}/N$.

The elimination of y between (2.70) and (2.71) gives

$$\frac{M'_{Cr(III)}}{\delta H_2O} = \frac{nQ}{Kr} M'_{ClO_4^-} + \frac{1}{Kr} \qquad (2.7\dot{2})$$

where $M'_{Cr(III)}$ equals $m_{Cr(III)}/N$.

A plot of the left-hand side of equation (2.72) versus $M'_{ClO_4^-}$ should result in a straight line with a ratio of the slope to the intercept of nQ. The data of Alei yielded such a plot and gave $nQ = 2$. The plot is shown in Fig. 2.2.

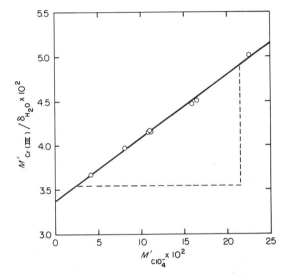

Fig. 2.2. $H_2{}^{17}O$ shifts in Cr(III)–perchlorate solutions. Intercept $= 0.0338$; slope $= (4.90 - 3.55)/(21.5 - 2.5) = 0.071$; slope/intrecept $= 2.1$.

Assuming in the same model that ClO_4^- ion exchanges rapidly between the second sphere and bulk solution, the bulk shift in ^{17}O resonance in ClO_4^-, $\delta'ClO_4^-$, can be related to $M'_{ClO_4^-}$ by the following derivable equation

$$\frac{M'_{Cr(III)}}{\delta'ClO_4^-} = \frac{nQ}{K'r} M'_{ClO_4^-} + \frac{1}{K'r} \qquad (2.73)$$

where K' is a constant appertaining to the specific magnetic field acting on the ^{17}O nucleus in the ClO_4^- ion while in the second sphere of the Cr(III). A plot of the left-hand side of Eq. (2.78) against $M'_{ClO_4^-}$ should give a straight line with a ratio of the slope to the intercept of nQ. The data of Alei met these requirements also and resulted in $nQ = 2$. See Fig. 2.3.

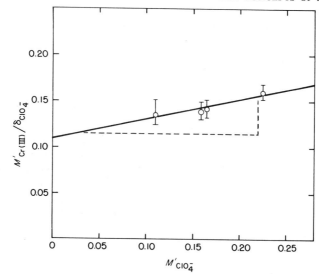

Fig. 2.3. ClO_4^- NMR shifts in Cr(III)–perchlorate solutions. Intercept = 0.110; slope = (0.155−0.115)/(0.220−0.025) = 0.205; slope/intercept = 1.9.

For Alei's data, the ratio of the intercepts of Eqs. (2.73) and (2.72), $K/K' = 3$, indicates that the specific magnetic interaction χ of $Cr(H_2O)_6^{3+}$ to produce a shift in ^{17}O resonance is three times as great for water as for ClO_4^- ion when both are in the second sphere.

Alei, by adding water enriched with ^{17}O to the solution of Cr(III) complex and ClO_4^- ion, determined in the same solution the composition of Cr(III)–H_2O complex and the rate of exchange of water between the Cr(III)–H_2O complex and labile water. The results indicated, in agreement with the data of Hunt and Taube [83], $6H_2O$ in the Cr(III) complex with a halflife $t_{1/2}$ of exchange as ~40 hours. Also in agreement with the observations of Plane and Taube [84], the halflife of exchange was found to decrease with increasing anion concentration, and lends support to the assumption of Plane and Taube that an activated complex containing both $Cr(H_2O)_6^{3+}$ and anion is formed.

Shulman [85] found that the anisotropic hyperfine and dipolar fields at the ligand vanish when averaged over the coordinates of the complex because of the symmetry of the Cr(III) octahedral complexes. Alei indicates that isotropic or contact hyperfine interaction is the source of the shifts in his work. The magnetic electrons are contained in the t_{2g} orbitals which are directed toward the centers of the 12 edges of the octahedron formed by the $Cr(H_2O)_6^{3+}$ complex. There is σ overlap with these t_{2g} orbitals by the oxygen sp orbitals in second sphere H_2O or ClO_4^-.

Proton spin relaxation data at 0°C in aqueous solutions of Cr(III) of varying acidity was given by Manley and Pollak [86].

If the reaction

$$Cr(H_2O)_6^{3+} + H_2O \rightarrow Cr(H_2O)_5OH^{2+} + H_3O^+ \qquad (2.74)$$

is assumed to be the dominant proton exchange mechanism for the hexa-aquo complex, these authors give a theoretical analysis of how the slow proton exchange between the solvent and hydration sphere [74] can be explained by identifying it with the acid dissociation of the hexaaquo complex.

At temperatures below 310°C, Würthrich and Connick [87] found additional relaxation effects other than the exchange of water molecules between the first coordination sphere of the metal ions and the bulk of the solution. These observations were made while studying the NMR relaxation of ^{17}O in aqueous solutions of vanadyl perchlorate and the rate of elimination of water molecules from the first coordination sphere. It was shown, using the model of the tumbling of loosely coordinated water molecules only through the tumbling of the whole vanadyl aquo complex, that at low temperature the additional relaxation occurred mainly via quadrupole coupling interrupted by chemical exchange of outer sphere water molecules with a lifetime of about 10^{-11} sec.

NMR relaxation data on aqueous solutions of $VOSO_4$ were explained by Hausser and Laukien [88] by assuming that at higher temperature exchange of protons from the first coordination sphere controlled the relaxation, while at lower temperature protons of water coordinated to VO^{2+} with a much shorter lifetime than those in the first coordination sphere contributed importantly to the relaxation.

Studies by Reuben and Fiat [89] of vanadyl(IV) ion in aqueous solutions using ^{17}O NMR spectroscopy yielded results similar to those of Würthrich and Connick [87]. Equatorially coordinated molecules with a residence time of $\tau_{M,eq.} = 1.35 \times 10^{-3}$ sec, determined by the method of molal shifts to be four in number were distinguished from other molecules giving rise to a shift of the solvent water peak. This shift was attributed to either the rapidly exchanging axial water molecule with an ^{17}O hyperfine coupling constant $A/h = 2.06 \times 10^6$ Hz or to interactions with molecules beyond the first coordination sphere. The rapidly exchanging axial water molecule should have a residence time of the order of 10^{-11} sec based on its contribution to the observed solvent line broadening. The enthalpy and entropy were found to be, respectively, 13.3 kcal/mole and -1.5 e.u./mole for the exchange reactions of the equatorial water molecules.

Using relaxation methods, Broersma [90] found the Al^{3+} ion to be surrounded by two or three layers of water molecules in concentrated solution, while in dilute solution, the effective solvation varied from one to two layers of water. One layer was effectively attached to Al^{3+} ion for not too concentrated solutions. In the inner layer of water surrounding trivalent ions, the lifetime for the protons was found to be 10 msec. In dilute paramagnetic solutions, the lifetime for the protons was found to range from 300 to 1 μsec, decreasing with distance to the ion increasing its magnetic moment. A value of 20 kcal/mole was found for the activation energy of the inner layer times its dependence on the distance to the center ion. Ice crystals were observed with four layers of water and a 5 msec lifetime. One layer of water for concentrated and zero layers for dilute solutions were observed for Na^+ ion solutions. Apparently the molecular motion is more strongly hampered in concentrated as compared to dilute solutions. Eyring's theory [91] of absolute reaction rates and the proton exchange times were found to agree.

The degree of dissociation of nitric acid in aluminum nitrate solutions in water as a function of concentration was calculated by Axtmann et al. [92] using an effective hydration number n of the cation to be 10. The agreement of the calculations with data showed this assumption to be correct [92, 93]. Using $n = 10$ is equivalent to assuming [93] that each ion binds 20–40 molecules of water, only four of which are strongly bound. A common ion effect due to added nitrate ion and the assumption that the water bound by the hydrated aluminum ion is unavailable for reaction with molecular nitric acid was combined in the simple theory, the data consisted of the proton magnetic resonance shifts relative to pure water of the aqueous nitric acid–aluminum nitrate solutions.

The polarization of water molecules beyond the first layer, or more obviously the formation of ion pairs, were used to explain [94] the small downward curvature of the proton magnetic resonance shift versus salt concentrations of $CaCl_2$ and $CaBr_2$. Solutions of $MgCl_2$, $ZnBr_2$, and $ZnCl_2$ show the effect to a much greater extent. Pure water was used as the reference in measuring the shifts of the salt solutions.

The expected shift of the bulk methanol OH signal and the shift of the coordinated methanol OH signal to a smaller extent toward higher field as the temperature was raised has been observed by Nakamura and Meiboom [95] for magnesium perchlorate solutions. The relative chemical shift of the two OH signals was small but significant. The extent of hydrogen bonding of the coordinated methanol molecules with the surrounding molecules in a presumed second solvation shell was one possible cause suggested for this change. Such a bond would supposedly be stronger than that between bulk methanol molecules from a consideration of the acidity of the hydrogen atoms since $Mg(CH_3OH)_6^{2+}$ has an acid dissociation constant 10^3 times greater than MeOH.

Schoolery and Alder [94] made no correction for magnetic susceptibility in calculating the molar shifts as pointed out by Shcherbakov [93], who indicated that on the whole more accurate data in measuring molar chemical shifts in electrolyte solutions was obtained by Hertz and Spalthoff [96] since they did make such corrections and also used a more accurate method in measuring δ. In addition, the latter investigators made different assumptions than did Schoolery and Alder in calculating the ion contribution to δ.

For hydrated paramagnetic salts in alcohol and acetone solutions the spin lattice (axial) relaxation time T_1 was found by Rivkind [97] to increase with the charge on the ion. This was explained on the basis of the increase in size and strength of the solvation shell.

For copper chloride and cobalt chloride in methanol solutions, the total number of molecules, N in the nth solvation sphere of either an anion or a cation of coordination number K was shown to be given by the equation [98]

$$N = K \times 2^{n-1} \qquad (2.75)$$

where 2 is the number of valance and hydrogen bonds between protons and oxygen in methanol molecules which constitute the structural network of the alcohol. The formula was applied by measuring the relaxation time T_1 of anhydrous $CoCl_2$ and $CuCl_2$ in dry methanol using the spin echo method and making optical measurements of the extinction coefficient ε at constant wavelength for each point on the T_1 in relative t units [99] versus $\log c$ curves, where c represents concentration. The meaning of relative time t should be clarified. Relative values for the spin lattice relaxation time T_1 could be found from the amplitude of the spin echo resulting from the application of a 90° pulse followed by a 180° pulse. The time interval t between pulse groups necessary to make the echo amplitude decrease to noise level was noted. The t–$\log c$ plots were made using changes in t which reflected changes in T_1. These t–$\log c$ curves show similar trends for the two salts out to 0.1 M salt solutions, t decreasing with increasing concentration of the paramagnetic ion as indicated by the following equation

$$1/T_1 = 12\pi^2\gamma^2\eta N\mu^2/5kT \qquad (2.76)$$

where $\gamma = \mu_0/Jh$ is the gyromagnetic ratio, N is the number of ions per cubic centimeter of solution, μ is the effective magnetic moment of the ion, and the other symbols have their usual significance.

A minimum in the curve at $c > 0.1$ was explained by the change in the g factor outweighing the effect on T_1 of the increased number of paramagnetic ions, thus producing a minimum in the t–$\log c$ curve by causing T_1 to begin to increase. The relationship $\mu = g\beta J$ defines the g factor where β is the Bohr magneton, and J the spin quantum number. Nearest-neighbor environment and complex formation greatly influence the g factor [100] which could change rapidly in alcoholic solutions since they may promote complex formation and solvation owing to their low dielectric constants.

The feature of most interest in the t–$\log c$ and the ε–$\log c$ curves are the distinct breaks which occur in both the $CoCl_2$ and the $CuCl_2$ systems but at different concentrations of the two salts. Figure 2.4 presents a t-$\log c$ plot for

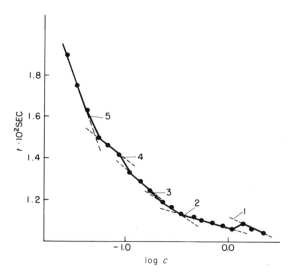

Fig. 2.4. A t–$\log c$ plot for $CoCl_2$ in methanol.

$CoCl_2$ in methanol. It was concluded by Rivkind [97, 101, 102] that short-range order in solution can influence the proton relaxation effect and determine the value of T_1; that the dominant relaxation process in solution is the change in the orientation of the electron spin of the paramagnetic ion with respect to the externally applied, static magnetic field H_0, and that the equilization of spin temperature is over the immediate surroundings of an ion and results from the continuous formation and disruption of hydrogen bonds and of the consequent proton exchange between molecules of water.

Based on these conclusions, the observed breaks in the t–$\log c$ and the ε–$\log c$ plots were ascribed [98] to abrupt variation in structure and concentration of solvated species having finite lifetimes, among other things. The effect of structure, involving a more favorable arrangement of solvent molecules in the ion field, would be evident from the short-range order of the solvation complex, that is by the magnitude of the distance between the paramagnetic ion and a proton under conditions of closest approach, and hence in the time T_1.

These ideas were tested [98] by comparing the observed concentrations at which the curves show breaks with theoretical calculations involving the total number of solvent molecules associated with each solute molecule. Coordination numbers of 8, 6, and 4 were assumed for Cl^-, Cu^{2+}, and Co^{2+} ions respectively. It was calculated, based on these assumptions, that in a $CoCl_2$ solution, every ion will have its first and most firmly held solvation sphere, or sphere of influence, filled at 1.23 M $CoCl_2$, which corresponds to the first break, point 1, in Fig. 2.4. For all points where breaks were indicated in Fig. 2.4, similar calculations were made; like calculations were made for $CuCl_2$. Table 2.3 shows the agreement of both NMR and spectral data with theory.

To the fourth sphere of influence at least, experimental and computed solvation numbers are in satisfactory agreement. The agreement becomes less satisfactory from the fourth sphere outward, because, according to the authors, the influence of thermal motion must be greater for large solvation spheres. This conclusion seems reasonable. Equation (2.75) reproduced the general formula for the number N of molecules of solvent in the nth solvation sphere of the ions. The authors included diagrams for the formation of structural lattice in methanol salt solutions.

In aqueous solutions of $CoCl_2$, the t–$\log c$ plots for HCl, NaCl, KCl, $MgCl_2$, and $CaCl_2$ showed breaks which could be used to obtain the hydration numbers of the H^+, Na^+, K^+, Mg^{2+}, Ca^{2+}, and Cl^- ions [103]. The purpose of the addition of the paramagnetic Co^{2+} ions to the electrolyte solutions listed was to reduce the relaxation time to easily measured values, about 10^{-2} sec. Aqueous solutions were amendable to the same model used in the methanol solutions. For water solutions, Eq. (2.75) had to be modified from the form used for calculating the number of solvent molecules in various solvation shells of the ions in concentrated solutions in methanol. For concentrated water solutions the equation became

$$N = K \times 3^{n-1} \tag{2.77}$$

where 3 is the number of valence and hydrogen bonds between the oxygen atoms and the protons of neighboring molecules forming the structural network of the pure solvent.

The generalized formula for any solvent showing hydrogen bonding to oxygen was given as

$$N = K \times R^{n-1} \tag{2.78}$$

Observed and calculated concentrations M of electrolytes in aqueous solutions at Mischenko points were in satisfactory agreement.

The rates of exchange of water molecules between bulk water and the first coordination sphere of Mn^{2+}, Cu^{2+}, Co^{2+}, Ni^{2+}, and Fe^{3+} ions were determined by Connick and Stover [104]. Except for the Cu^{2+} ion, their lower limits for the observed exchange were in good agreement with the first-order rate constant for the conversion for an outer sphere sulfate complex to an inner sphere complex as found by Eigen [105] using relaxation spectrum techniques.

Table 2.3

Agreement between Experimental and Calculated Values of the Number of Solvent Methanol Molecules in the Different Spheres of Influence of Co^{2+}, Cu^{2+}, and Cl^{-} Ions

Spheres of influence of the ion	$CoCl_2{}^{a}$								$CuCl_2{}^{b}$							
	Concentration (M)			Number of molecules				Concentration (M)			Number of molecules				
	NMR (spin echo)	Optical absorption spectrum	Theoretical	In all spheres experimental	Theoretical (per ion)		In all spheres theoretical	NMR (spin echo)	Optical absorption spectrum	Theoretical	In all spheres experimental	Theoretical (per ion)		In all spheres theoretical
					Co^{2+}	Cl^{-}						Cu^{2+}	Cl^{-}	
1	1.23	1.2	1.23	20	4	8	20	1.12	1.17	1.12	22	6	8	22
2	0.40	4.2	0.41	62	8	16	60	0.34	0.37	0.38	72	12	16	66
3	0.19	0.18	0.18	130	16	32	140	0.16	—	0.166	155	24	32	154
4^{c}	0.09	—	0.08	260	32	64	300	0.07	—	0.085	360	48	64	330
5	0.048	—	0.040	530	64	128	620	0.035	—	0.039	710	96	128	682

^{a} Formula for the nth solvation sphere is $4 \times 2^{n-1}$ for Co^{2+} and $8 \times 2^{n-1}$ for Cl^{-}.

^{b} Formula for the nth solvation sphere is $6 \times 2^{n-1}$ for Cu^{2+} and $8 \times 2^{n-1}$ for Cl^{-}.

^{c} Data for fourth and fifth spheres could not be confirmed by optical measurements and are only tentative.

In the case of aqueous solutions, temperature dependences of T_1 and T_2 have been reported [106, 107]. The fact that the low- and high-temperature effects represent solvation equilibria, the low-temperature effect being assigned to the secondary solvation sphere, and the high-temperature effect to the primary solvation sphere [107], has been used as one possible explanation of the temperature dependences of proton relaxation times T_1 and T_2.

From time of absorption and relaxation studies of water molecules or protons surrounding paramagnetic ions, it has been found [108] that the effective size of the atmosphere changes from one to two layers in dilute solutions to two or three layers in concentrated layers. Ice crystals with a four-layer radius and a 5-msec lifetime were found. More extensive solvation in saturated solutions was explained by assuming that molecular motion was more strongly hampered than in dilute solutions, thus permitting more extensive solvation. In general it has been assumed that solvation increases with dilution, and many types of data, e.g., transference data, support this latter assumption.

Chemical shifts of ZnO dissolved in different ratios of KOH to ZnO have been studied [109]. Due to very rapid chemical exchange among the proton containing species, only one sharp peak was observed. The change of chemical shift, $\Delta\delta$, plotted against the mole ratio of KOH to ZnO showed a break at a mole ratio of 11 and possibly another break at a mole ratio of 17 for the solution 3.89 moles in KOH. For the solution 4.84 moles in KOH, a break at a mole ratio of 11 was also evident. These breaks were probably indicative of a second coordination sphere influence according to Newman and Blomgren [109].

A study of proton relaxation in 0.1 M cobalt(III) in methanol exchange of the solvation sphere was made, and calculations carried out for the distances of closest approach $d_{OH} = 3.9$ Å for the hydroxyl proton and $d_{CH_3} = 4.6$ Å for the methyl protons in bulk methanol from the sphere formed by the paramagnetic ion and its inner coordination sphere [110]. A degree of molecular orientation in a second solvation shell might be indicated by the fact that d_{CH_3} was found to be greater than d_{OH} independent of the spin relaxation time, λ_s, of Co^{2+}. Rapid exchange at all temperatures between second solvation shell and bulk molecules had been assumed preceding these calculations, so that only average relaxation rates were observed. In methanol solutions of $Mg(ClO_4)_2$, the bulk OH signal showed the expected marked shift toward higher field as the temperature was increased; while the OH signal of the coordinated methanol, instead of remaining relatively constant, also shifted upfield but to a smaller extent [111]. It was concluded that it was not immediately clear whether this was related to the extent of hydrogen bonding of the coordinated molecules with the surrounding molecules of a presumed second solvation sphere.

In a discussion of the theoretical equations for the transverse relaxation rate, $1/T_{2M}$, where M refers to the complex, Pearson and Anderson [112] pointed out that there were two mechanisms, a dipole–dipole interaction and an isotropic contact interaction. The actual presence of an electron on the proton and the consequent coupling of the electron and nuclear spins is necessary for the latter interaction, which also necessitates contact between the molecule containing the proton being studied and the paramagnetic ion. Even in the layer next to the coordinated layer, the dipole–dipole interactions can cause some broadening of the NMR signal. This is called "outer-sphere relaxation" and must be corrected for when the broadening of the NMR signal due to dipole–dipole relaxation of coordination sphere molecules is being considered. There was no dependence of line broadening on glycine concentration in the system composed of a constant concentration of Ni^{2+} ion in aqueous solutions of glycine. The rate of exchange was so slow in Ni^{2+}-glycine complexes that only a small broadening, practically concentration independent, and due to "outer sphere relaxation" was observed.

III. Advantages of the NMR Technique for Outer Shell Coordination

It would perhaps be well to point out some advantages of the NMR technique in studying outer shell coordination. The method demonstrates the instant of formation of first, second, etc., solvation spheres or ionic spheres of influence [98]. The NMR method has been appraised by Stengle and Langford [73] as follows: (a) by it the evaluation of the relative degree of outer sphere complexation can be made for a series of complexes of very similar electronic structure; (b) because of its short range nature it detects genuine second sphere association. It yields an approximate value of the first outer sphere association constant for weak association; (c) for the particular instance of $Cr(ox)_3^{3-}$, where no ion association is likely, the method indicates some broadening due to third sphere effects.

The solvating properties of each component in mixed solvents can be observed simultaneously using NMR. Compared with other methods, for example, transference, the time required for solvation studies using NMR, is much shorter. NMR makes possible the study of the rates of exchange of free and bound solvents.

IV. Comparison of Nuclear Magnetic Resonance with Other Sources

Outer sphere and bridged activated complexes have been postulated in electron transfer reactions [113–116]. There is no interpenetration of the first coordination sphere of the exchanging species when electron exchange takes place by outer sphere mechanisms. The effects resulting from the requirement for nonequilibrium of polarization of surroundings are confined to limits beyond the first sphere of coordination. A thoughtful discussion of the contribution of solvent rearrangement to the free energy of activation of these processes has been given by Marcus [115]. Electron exchange can take place without net rearrangement of the inner coordination sphere in the cases of the following complex ions [116]. $MnO_4^- - MnO_4^{2-}$, $IrCl_6^{3-} - IrCl_6^{2-}$, $Fe(CN)_6^{4-} - Fe(CN)_6^{3-}$, $Mo(CN)_8^{3-} - Mo(CN)_8^{2-}$, $Fe(phen)_3^{2+} - Fe(phen)_3^{3+}$, $Fe(dip)_3^{2+} - Fe(dip)_3^{3+}$, $Os(dip)_3^{2+} - Os(dip)_3^{3+}$, $Co(dip)_3^{2+} - Co(dip)_3^{3+}$. Some of these ions, for example $Co(II)$, are among those found to exert outer sphere coordination by NMR techniques.

Data from Hittorf transference measurements using an iert substance to indicate solvent transfer have indicated outer solvent spheres for several ions including these for Li^+, K^+, and Na^+ [117–119]. Thus three solvation shells have been postulated for Li^+ ions in order to explain the solvation phenomena in the water–ethanol solvent system. This coincides with the three observed spheres of influence reported [98] for the Co^{2+}, Cu^{2+}, and Cl^- ions.

Aqueous solutions of $ErCl_3$ and ErI_3 and aqueous neutral and acid solutions of $FeCl_3$ have been studied [120, 121] using X-rays. It was found that there was a firmly held octahedral arrangement of water molecules around the Er^{3+} ion in $ErCl_3$ solutions. A chloride ion was bound on opposite sides of the square plane of four water molecules by one hydrogen each of two water molecules on adjacent corners of the square. There remained one hydrogen each of the four square planar water molecules and two hydrogens each of the two water molecules at the apexes of the octahedron to hydrogen bond with eight water molecules in a second solvation sphere. Actually seven water molecules were found in this sphere.

In neutral solution, the octahedral $FeCl_4(H_2O)_2^-$ ion was postulated to be surrounded by a hydration sheath about a molecule thick and interstitial Fe^{3+} ions with very low coordination numbers. As the acidity is increased, small hydrogen ions strip these outer sheath waters away from the octahedral ions, about half of which go over to the tetrahedral form. In acid solutions the octahedral and tetrahedral ions form a polymeric species, possessing alternately octahedral and tetrahedral coordination with a sharing of corners occupied by chlorides.

The upfield shift of the proton resonance in both water protons and the amide protons of lithium perchlorate solutions in mixed water–N-methylacetamide solvents indicate that the net effect of adding lithium perchlorate to aqueous–N-methylacetamide solutions is hydrogen-bonded structure breaking [122]. The fact that lithium ion is relatively highly hydrated as shown by many phenomena [117–119] and that from NMR studies many lithium salts appear to be structure breakers maintaining dynamic solvation

[122] indicates that lithium ion can effectively move as a rather highly solvated entity through unstructed water while the free solvent and the solvent layers around the lithium ion exhibit dynamic exchange of water molecules. This dynamic exchange between bound and unbound water and between different bound layers of water would be possible for highly mobile nearest neighbor molecules [123–125] and mobile, bulk molecules.

V. Calculations of Heats and Entropies of Hydration Employing Models

The thermodynamic parameters of hydration for example, heats and entropies, have been calculated [126] using three types of models [127–130]. These models are depicted in Fig. 2.5. A cycle was devised [126] to separate out various contributions of ion–solvent interactions. See Fig. 2.6. In the

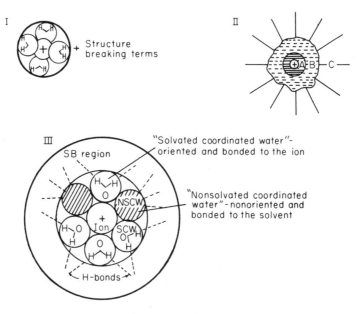

Fig. 2.5. Models for the region near an ion. (I) Bernal and Fowler [127] and Eley and Evans [128]: (i) all monovalent ions are four-coordinated; (ii) CN ≡ SN; (iii) structure-breaking terms. (II) Frank and Wen [129]: (i) two-layer model; (ii) all molecules in region "A" are immobilized, i.e., SN ≡ CN; (iii) a structure-broken region "B". (III) Bockris and Reddy [130]: (i) SN distinguished from CN; (ii) structure-broken region about a monolayer. Open circles, "solvated coordinated water"–oriented and bonded to the ion; hatched circles, "nonsolvated coordinated water"–nonoriented and bonded to the solvent.

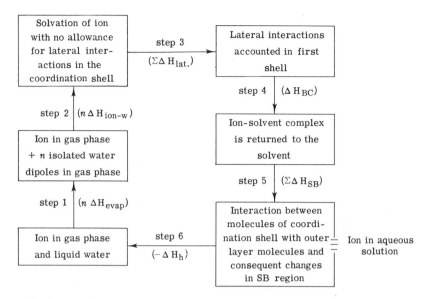

Fig. 2.6. A cycle to separate out various contributions of ion–solvent interactions.

heat and entropy calculations the models shown in Fig. 2.5 were applied using several modifications as follows: (1) No distinction is made between coordination number (CN) and solvation number (SN); (2) The CN is determined by X-rays and is not distinguished from SN; (3) The CN is taken from X-ray data and distinction is made between CN and SN. For each of the three basic models three types of structure broken (SB) regions outside the first layer are calculated. For the respective models these are designated as 1A, 1B, 1C, 2A, 2B, 2C, and 3A, 3B, 3C, respectively. The definition of CN is operational. SN is defined as the number of solvent molecules per ion which for sufficiently long time remain attached to a given ion so as to accompany it on its translational movements. When SN and CN are distinguished, the solvational coordinated water (SCW), is defined as the number of molecules of fully oriented water, i.e., $\cos\theta = 1$.

The significance of the distinction between CN and SN has been discussed [131–133]. It is pointed out that the estimation of hydration number n from apparent molal volume, using the value of electriction per mole of water, is less satisfactory, since n really measures the degree of partial loss of some property (in the present instance, volume; in other instances, compressibility or entropy) related to the solvent coordinated x-fold to the ion, rather than the true CN itself which probably does not vary as widely from (univalent) ion to ion as is indicated by the range of n values found in previous treatments of hydration. Methods of calculating component parts of the thermodynamic functions of hydration shown is Fig. 2.6 were given.

References

1. J. F. Hinton, E. S. Amis, and W. Mettetal, *Spectrochim. Acta Part A* **25**, 119 (1969).
2. J. Weiss, *Proc. Roy. Soc. Ser. A* **222**, 128 (1954).
3. D. Cohen, J. C. Sullivan, E. S. Amis, and J. C. Hindman, *J. Amer. Chem. Soc.* **78**, 1543 (1956).
4. E. S. Amis, *J. Chem. Phys.* **26**, 880 (1957).
5. J. F. Hinton, L. S. McDowell, and E. S. Amis, *Chem. Commun.* p. 76 (1966).
6. J. M. Walters and Sr. M. Rosalie, *Anal. Chem.* **37**, 45 (1965).
7. E. Grunwald, G. Baughman, and G. Kohnstam, *J. Amer. Chem. Soc.* **82**, 5801 (1960).
8. K. P. Mistchenko, *Zh. Fiz. Khim.* **26**, 1736 (1952).
9. O. Ya. Samoilov, "Structure of Aqueous Electrolyte Solutions and the Hydration of Ions," p. 74. Consultants Bureau, New York, 1965.
10. D. Mendeljew, "Works," Vol. 4, p. 443. Leningrad, 1937.
11. I. A. Kablukow, "The Modern Theories of Solution (of van't Hoff and Arrhenius) Relation to Studies of Chemical Equilibrium," p. 86. Moscow, 1891.
12. J. O'M. Bockris, *Quart. Rev. Chem. Soc.* **3**, 173 (1949).
13. G. Darmois, *C. R. Acad. Sci.* **240**, 1341 (1955).
14. E. H. Risenfeld, *Z. Phys. Chem. Stoechiom. Verwandschaftslehre* **66**, 672 (1909).
15. H. Remy, *Z. Phys. Chem. Stoechiom. Verwandschaftslehre* **89**, 529 (1915); **118**, 161 (1925); **124**, 379 (1926); **126**, 161 (1927); *Trans. Faraday Soc.* **23**, 381 (1927).
16. J. Baborovsky, *Rec. Trav. Chim. Pays-Bas* **42**, 229 (1923); *Z. Phys. Chem. Stoechiom. Verwandschaftslehre* **129**, 129 (1927); *Collect. Czech. Chem. Commun.* **11**, 542 (1938).
17. G. Jander, *Z. Phys. Chem. Abt. A* **149**, 97 (1930); *Z. Phys. Chem. Abt. A* **190**, 81 (1942); *Z. Anorg. Allg. Chem.* **222**, 113 (1935); *J. Amer. Chem. Soc.* **37**, 722 (1935).
18. H. Brintzinger, *Z. Anorg. Allg. Chem.* **223**, 101 (1935).
19. A. G. Posynski, *Zh. Fiz. Khim.* **11**, 606 (1938).
20. A. M. Azzam, *Z. Elektrochem.* **58**, 889 (1954).
21. H. Ulich, *Trans. Faraday Soc.* **23**, 392 (1927); *Z. Elektrochem.* **36**, 497 (1930); *Z. Phys. Chem.* **168**, 141 (1934).
22. E. Gleuckauf, *Trans. Faraday Soc.* **51**, 1235 (1955).
23. E. Darmois, *J. Phys. Radium* **2**, 2 (1941).
24. J. E. F. Alonso, J. Mira, and J. L. S. Lucas, *An. Real Soc. Espan. Fis. Quim.* **49 B**, 337 (1953).
25. J. I. F. Alonso and A. Hidalgo, *An. Real Soc. Espan. Fis. Quim.* **46 B**, 17 (1950).
26. E. Darmois, La Solvation des Ions. *Mem. Sci. Phys. Paris* (1946).
27. G. Journet and J. Vadon, *Bull. Soc. Chim. Fr.* **4**, 593 (1955).
28. B. E. Conway and J. O'M. Bockris, "Modern Aspects of Electrochemistry." Academic Press, New York, 1954.
29. Symposium on hydration of aqueous ions. *J. Phys. Chem.* **58**, 513–672 (1954); Symposium on solutions of electrolytes. *J. Phys. Chem.* **58**, 673–795 (1954).
30. H. Ulich and J. Eucken-Wolf, *Handb. Jahrb. Chem. Phys.* **6**, Pt. II, 17–19 (1933).
31. E. Darmois, *J. Chim. Phys.* **43**, 1 (1946).

32. H. Mukhergee, *Indian J. Phys.* **23**, 503 (1949); **24**, 137 (1950).
33. A. L. Levy, *J. Chem. Phys.* **21**, 656 (1953).
34. E. S. Amis, *Inorg. Chim. Acta* **3**, 7 (1969).
35. E. S. Amis, *J. Phys. Chem.* **60**, 428 (1956).
36. J. O. Wear, C. V. McNully, and E. S. Amis, *Inorg. Nucl. Chem.* **18**, 48 (1961).
37. J. O. Wear, C. V. McNully, and E. S. Amis, *J. Inor. Nucl. Chem.* **19**, 278 (1961).
38. N. Goldenberg and E. S. Amis, *Z. Phys. Chem. (Leipzig)* **31**, 10 (1962).
39. J. M. McIntyre and E. S. Amis, *J. Chem. Eng. Data* **13**, 371 (1968).
40. S. L. Melton and E. S. Amis, *J. Chem. Eng. Data* **13**, 439 (1968).
40a. J. F. Hinton and E. S. Amis, *Chem. Rev.* **67**, 367 (1967).
41. W. K. Sementschenko, *Zh. Neorg. Khim.* **1**, 1131 (1956).
42. J. D. Bernal and H. D. Fowler, *J. Chem. Phys.* **1**, 515 (1933).
43. J. D. Bernal and H. D. Fowler, *Usp. Fiz. Nauk*, **14**, 586 (1934).
44. M. J. Polissar, *J. Chem. Phys.* **6**, 833 (1938).
45. A. I. Brodski, "Physical Chemistry," p. 612. Moscow, Leningrad, 1948.
46. N. E. Dorsey, "Properties of Ordinary Water Substance." New York, 1940.
47. A. S. Golik, A. W. Oristschenko, and A. R. Artenitschenko, *Dok. Akad. Nauk SSSR* **6**, 543 (1954); **5**, 464 (1955).
48. J. F. Hinton, L. S. McDowell, and E. S. Amis, *Chem. Commun.* p. 776 (1966).
49. L. H. Collet, *C. R. Acad. Sci.* **237**, 252 (1953); **239**, 266 (1954).
50. L. H. Collet, *J. Phys. Radium* **16**, 159 (1955).
51. E. Glueckauf, *Trans. Faraday Soc.* **51**, 1235 (1965).
52. E. Glueckauf and G. P. Kitt, *Proc. Roy. Soc. Ser. A* **228**, 322 (1955).
53. Ya. I. Tur'yan, *Zh. Fiz. Khim.* **38** (6), 1690 (1964).
54. G. Kortum and J. O'M. Bockris, "Textbook of Electrochemistry," Vol. 1, p. 167. Amer. Elsevier, New York, 1951.
55. J. H. Wang, *J. Phys. Chem.* **58**, 686 (1954).
56. H. M. Feder and H. Taube, *J. Chem. Phys.* **20**, 1335 (1952).
57. H. Taube, *J. Phys. Chem.* **58**, 523 (1954).
58. K. Wirtz, *Z. Naturforsch. A* **3**, 672 (1948).
59. J. H. Wang, C. V. Robinson, and I. S. Edelman, *J. Amer. Chem. Soc.* **75**, 466 (1953).
60. E. L. Hahn, *Phys. Rev.* **80**, 580 (1950).
61. L. O. Morgan, A. W. Nolle, R. L. Hull, and J. Murphy, *J. Chem. Phys.* **25**, 206 (1956).
62. N. Bloembergen, E. M. Purcell, and R. V. Pound, *Phys. Rev.* **73**, 679 (1948).
63. H. Taube, *Chem. Rev.* **50**, 69 (1952).
64. J. P. Hunt and H. Taube, *J. Chem. Phys.* **18**, 757 (1950).
65. V. Gold and D. P. N. Satchell, *Quart. Rev. Chem. Soc.* **9**, 51 (1955).
66. S. Thomas and W. L. Reynolds, *J. Chem. Phys.* **44**, 3148 (1966).
67. N. A. Matwiyoff and W. G. Movius, *J. Amer. Chem. Soc.* **89**, 6077 (1967).
68. A. W. Nolle and L. O. Morgan, *J. Chem. Phys.* **26**, 642 (1957).
69. L. O. Morgan and A. W. Nolle, *J. Chem. Phys.* **31**, 365 (1959).
70. I. Solomon, *Phys. Rev.* **99**, 559 (1955).
71. N. Bloembergen, *J. Chem. Phys.* **27**, 572 (1957).
72. R. G. Codrington and N. Bloembergen, *J. Chem. Phys.* **29**, 600 (1958).
73. J. R. Stengle and C. H. Langford, *J. Phys. Chem.* **69**, 3299 (1965).
74. N. Bloembergen and L. O. Morgan, *J. Chem. Phys.* **34**, 842 (1961).
75. H. M. McConnell, *J. Chem. Phys.* **28**, 430 (1958).
76. T. J. Swift and R. E. Connick, *J. Chem. Phys.* **37**, 307 (1962).
77. T. J. Swift and R. E. Connick, *J. Chem. Phys.* **41** 2553 (1964).
78. H. Taube and F. A. Posey, *J. Amer. Chem. Soc.* **75**, 1463 (1953).

79. F. A. Posey and H. Taube, *J. Amer. Chem. Soc.* **78**, 15 (1956).
80. N. Fogel, J. Tai, and J. Yarborough, *J. Amer. Chem. Soc.* **84**, 1145 (1962).
81. M. Alei, *Inorg. Chem.* **3**, 44 (1964).
82. R. G. Pearson and F. Basolo, *J. Amer. Chem. Soc.* **78**, 4878 (1956).
83. J. P. Hunt and H. Taube, *J. Chem. Phys.* **18**, 757 (1950).
84. R. A. Plane and H. Taube, *J. Pyhs. Chem.* **56**, 33 (1952).
85. R. G. Shulman, *J. Chem. Phys.* **29**, 945 (1958).
86. C. E. Manley and V. L. Pollak, *J. Chem. Phys.* **46**, 2106 (1967).
87. K. Würthrich and R. E. Connick, *Inorg. Chem.* **6**, 583 (1967).
88. R. Hausser and G. Laukien, *Z. Phys.* **153**, 394 (1959).
89. J. Reuben and D. Fiat, *Inorg. Chem.* **6**, 579 (1967).
90. S. Broersma, *J. Chem. Phys.* **24**, 659 (1956).
91. H. Eyring, *J. Chem. Phys.* **28**, 805 (1955).
92. R. G. Axtmann, W. E. Shuler, and B. B. Murray, *J. Phys. Chem.* **64**, 57 (1960).
93. U. A. Shcherbakov, *Strukt. Khim.* **2**, (4), 484 (1961).
94. J. N. Schoolery and B. J. Alder, *J. Chem. Phys.* **23**, 805 (1955).
95. S. Nakamura and S. Meiboom, *J. Amer. Chem. Soc.* **89**, 1765 (1967).
96. H. G. Hertz and W. Spalthoff, *Z. Elektrochem.* **63**, 1096 (1959).
97. A. J. Rivkind, *Dokl. Akad. Nauk SSSR* **117**, 448 (1957).
98. P. A. Zagorets, V. I. Ermakov, and A. P. Granau, *Zh. Fiz. Khim.* **37**, 2155 (1963).
99. P. A. Zagorets, V. J. Ermokov, and A. P. Granau, *Zh. Fiz. Khim* **37**, 1413 (1963).
100. G. A. Altshuler and B. M. Kozyrev, "Elecktronnyi Paramagnitnyi Rezonanz." Fizmatgiz, Moscow, 1961.
101. A. J. Rivkind, *Dokl. Akad. Nauk SSSR* **102**, 1197 (1955).
102. A. J. Rivkind, *Dokl. Akad. Nauk SSSR*, **112**, 239 (1957).
103. P. A. Zagorets, V. J. Ermakov, and A. P. Granau, *Zh. Fiz. Khim.* **39**, 4 (1965).
104. R. E. Connick and E. D. Stover, *J. Phys. Chem.* **65**, 2075 (1961).
105. M. Eigen, *Z. Elektrochem.* **64**, 115 (1960).
106. R. A. Bernheim, T. H. Brown, H. S. Gutowsky, and D. E. Woersner, *J. Chem. Phys.* **30**, 950 (1959).
107. T. H. Brown, R. A. Bernheim, and H. S. Gutowsky, *J. Chem. Phys.* **33**, 1593 (1960).
108. S. Broersma, *J. Chem. Phys.* **53**, 1593 (1960).
109. G. D. Newman and G. E. Blomgren, *J. Chem. Phys.* **43**, 2744 (1965).
110. Z. Luz and S. Meiboom, *J. Chem. Phys.* **40**, 2686 (1964).
111. S. Nakamura and S. Meiboom, *J. Amer. Chem. Soc.* **89**, 1765 (1967).
112. R. G. Pearson and M. M. Anderson, *Angew. Chem. Int. Ed. Engl.* **4**, 281 (1965).
113. H. Taube, Mechanism of redox reactions of simple chemistry. *Advan. Inorg. Chem. Radiochem.* **1**, 1–53 (1959).
114. J. Halpern and L. E. Orgel, *Discuss. Faraday Soc.* **29**, 7, 32 (1960).
115. R. A. Marcus, *J. Chem. Phys.* **24**, 966 (1956).
116. F. Bloch, W. W. Hansen, and M. Packard, *Phys. Rev.* **70**, 474 (1946).
117. J. O. Wear, C. V. McNully, and E. S. Amis, *J. Inorg. Nucl. Chem.* **18**, 48 (1961).
118. J. O. Wear, C. V. McNully, and E. S. Amis, *J. Inorg. Nucl. Chem.* **19**, 278 (1961).
119. J. O. Wear, C. V. McNully, and E. S. Amis, *J. Inorg. Nucl. Chem.* **20**, 100 (1961).
120. G. W. Brady, *J. Chem. Phys.* **26**, 1371 (1958).
121. G. W. Brady, *Inorg. Chem.* **3**, 1168 (1964).
122. J. F. Hinton, E. S. Amis, and W. Mettetal, *Spectrochim. Acta Part A* **25**, 119 (1969).
123. O. Ya. Samoilov, *Discuss. Faraday Soc.* **24**, 141, 216 (1957).
124. O. Ya. Samoilov, *Zh. Fiz. Khim.* **29**, 1582 (1955).
125. G. A. Krestov, *Zh. Strukt. Khim.* **3**, 125 (1962).

126. J. O'M. Bockris and P. P. S. Saluja, *J. Phys. Chem.* **76**, 2298 (1972).
127. J. D. Bernal and R. H. Fowler, *J. Chem. Phys.* **1**, 515 (1933).
128. D. D. Eley and M. G. Evans, *Trans. Faraday Soc.* **34**, 1093 (1938).
129. H. S. Frank and W. Y. Wen, *Discuss. Faraday Soc.* **24**, 133 (1957).
130. J. O'M. Bockris and A. K. N. Reddy, "Modern Electrochemistry," Vol. I. Plenum, New York, 1970.
131. J. E. Desnoyers and C. Jolicoeur, *in* "Modern Aspects of Electrochemistry," (B. E. Conway and J. O'M. Bockris, eds.), Vol. 5, Chapter 1. Plenum, New York, 1969.
132. B. E. Conway, R. E. Verrall, and J. E. Desnoyers, *Z. Phys. Chem. Falkenhagen Festschr.* **230**, 157 (1965).
133. B. E. Conway, *in* "Physical Chemistry, An Advanced Treatise," (H. Eyring, ed.), Vol. IXA. Academic Press, New York, 1970.

Chapter 3

SOLVATION OF IONS

I. Introduction

From soon after the inception of theory of electrolytes until the present, the chemical literature is replete with articles dealing with the measurement, magnitude, and influence on chemical phenomena in solution of the solvation of ions. A recent extensive summary of such data has been compiled. The purpose of this chapter is to assemble and compare data on solvation.

II. General Concepts

This review includes various types of solvation—positive, negative outer sphere, and inner sphere—determined in pure and mixed solvents. To obtain actual solvation numbers for single ions, the solvation number of some reference ion has to be chosen arbitrarily. Even though the solvation number of a reference ion is chosen, the actual value of the solvation number of an individual ion depends on the method used in its determination.

III. Methods of Measurement

A. TRANSFERENCE

1. Theory and Experiment

The transference or transport number of an ion is the fraction of the total current carried by that ion. If the equivalent concentration and the speed of the cation are c_+ and u_+, respectively, and the corresponding quantities for the anion are c_- and u_-, respectively, the transference numbers for the cation and anion are, respectively,

$$t_+ = \frac{c_+ u_+}{c_+ u_+ + c_- u_-} \tag{3.1}$$

and

$$t_- = \frac{c_- u_-}{c_+ u_+ + c_- u_-} \tag{3.2}$$

For symmetric electrolytes in which the two ions have valences of the same magnitude, $c_+ = c_-$, Eqs. (3.1) and (3.2) become, respectively,

$$t_+ = \frac{u_+}{u_+ + u_-} \tag{3.3}$$

and

$$t_- = \frac{u_-}{u_+ + u_-} \tag{3.4}$$

Since the velocities of the two ions of an electrolyte are not equal, different concentration changes will occur in the regions of the cathode and anode, respectively, and these changes may be used to evaluate transference numbers [1, 2].

In general for an ith-type ion, the transport number is given by the expression

$$t_i = n_i u_i e_i / \sum n_i u_i e_i \tag{3.5}$$

where n_i is the number of ith-type ions per cubic centimeter, and u_i and e_i are the velocity and charge, respectively, of this type of ion. Since $e_i = z_i \varepsilon_i$, where z_i is the valence and ε_i the electronic charge, then

$$t_i = n_i u_i z_i / \sum n_i u_i z_i \tag{3.6}$$

where the summation in Eqs. (3.5) and (3.6) are taken over all the ions in solution.

Hittorf's method consists of chemically analyzing the contents of various portions of the solution of an electrolytic cell, and of asertaining the faradays which flow through cell using a silver or other type coulometer, or other current measuring device in series with the cell.

If a coulometer is included in the circuit, then according to Faraday's law, the same number of gram equivalents of material irrespective of its nature will be affected as in the transference cell; therefore,

$$t_+ = \frac{\text{Number of gram equivalents of electrolyte lost from the anode compartment}}{\text{Number of equivalents of metal deposited in the coulometer}}$$

(3.7)

$$t_- = \frac{\text{Number of gram equivalents of electrolyte lost from the cathode compartment}}{\text{Number of equivalents of metal deposited in the coulometer}}$$

(3.8)

Washburn [3] has given a formula for the calculation of transference number t of an ion formulated as follows. Let N_0 and N_F be the number of equivalents of an ion associated initially and finally with a given weight of solvent. Let N_E be the number of equivalents of this ion added to the sovlent by the electrode reaction, and tN_F be the number of equivalents of this ion lost to the solvent by migration. Then

$$N_F - N_0 = N_E - N_E t \qquad (3.9)$$

Wherefore,

$$t = (N_0 - N_F + N_E)/N_E \qquad (3.10)$$

The moving boundary method of transference numbers is based on the fact that when a solution of one electrolyte is placed above a solution of another electrolyte in a tube and a direct current is passed from bottom to top, the boundary between the two solutions will become sharp and will move up the tube.

Let V be the volume in liters swept out per faraday of current passed, and VC be the equivalents of selected ion constituent passing per faraday a fixed plane in the tube, where C is the equivalents per liter of the selected ion. The transference number t of the selected ion is, therefore,

$$t = VC \qquad (3.11)$$

For a smaller number of coulombs f passing through the solution, a smaller volume v will be swept out by the boundary between the electrolytes. The relationship between the two volumes and the numbers of coulombs becomes

$$v/V = f/F \qquad (3.12)$$

and Eqs. (3.10) and (3.11) yield

$$t = vCF/f \qquad (3.13)$$

But for a constant current I flowing for τ sec, $f = \tau I$, and hence

$$t = vCE/\tau I \qquad (3.14)$$

The moving boundary method can be used under certain conditions to measure the transference number of ion constituents in mixtures of electrolytes.

MacInnes [4] discusses in detail methods of forming the boundary and of making the measurements.

2. Solvation of Ions: "True" Transference Numbers

Transference number measurements may be used to determine the solvation of ions in solution. In the case of water solvent, solvation is termed hydration.

Thus for the cation carrying more solvent into the cathode portion of a transference cell, be it a Hittorf or moving boundary type cell, than the anion carries out, there will be a dilution of the solution around the cathode. In such a case the measured Hittorf transference number of the cation will be less than if the ions were not solvated and moved at the same relative velocity. For the anion carrying more solvent from the cathode compartment than the cation carries in, the effect on the Hittorf transference number will be opposite to that described above.

"True" transference numbers are those obtained by a method which is not influenced by the movement of solvents of solvation. Among the early workers in the field of solvation using the Hittorf method were Nernst [5], Buchböck [6], Washburn [7, 8], and their associates. A second solute, e.g., sucrose or raffinose was included in the solution in the Hittorf cell. The changes in salt and solvent concentrations were referred to the second solute. The second substance being composed of neutral molecules was uninfluenced by the passage of the electric current. However, the ratio of the reference substance to the water was changed in the electrode regions of the cell if solvent were carried along by the moving ions. By determining accurately the concentration of the reference substance, the "true" transference number of the ion constituent and Δn, the increase or decrease of the number of moles of water, in a given electrode portion per faraday of electricity passed was calculated. A polarimeter was used by Washburn [7, 8] to determine the concentration of reference substance. Recent investigators [9–16] have also studied solvation of ions using the polarimeter to determine the concentration of the optically active reference substance.

Taking Δn as the net effect of the solvent carrying by all the ions present, for the cathode portion in the case of a binary electrolyte Δn becomes,

$$\Delta n = \tau_c n_s^c - \tau_a n_s^a \qquad (3.15)$$

where τ_c and τ_a are the "true" transference numbers of the cation and anion, respectively; and $n_s{}^c$ and $n_s{}^a$ are the number of moles of solvent carried per mole by cation and anion, respectively. This approach gives only the differences in the extent of solvation of the cation and anion. These differences in terms of the moles of solvent carried per mole of cation as a function of the moles of solvent carried per mole of anion can be written from Eq. (3.15) as

$$n_s{}^c = \frac{\Delta n}{\tau_c} + \frac{\tau_a}{\tau_c} n_s{}^a \tag{3.16}$$

Mixed solvents are now used extensively, and since true transference numbers are used in electrochemical calculations, methods of obtaining true transference numbers and the solvation numbers of ions in mixed solvents have been devised [17].

Using Washburn's equation for the difference between the true transference number τ and the Hittorf transference number t namely,

$$\tau - t = \Delta n_{sol}^F n_s / n_{sol} \tag{3.17}$$

the above mentioned quantities were obtained. In Eq. (3.17) Δn_{sol}^F is the number of moles transferred per faraday, and n_s is the equivalents of solute in n_{sol} moles of solvent. Introducing the relations

$$\Delta n_{sol}^F = \Delta g_{sol}^F / M_{sol} \tag{3.18}$$

and

$$n_{sol} = g_{sol} / M_{sol} \tag{3.19}$$

where Δg_{sol}^F is the grams of solvent transferred per faraday, g_{sol} the grams of solvent, and M_{sol} the molecular weight of solvent, the following equation can be obtained:

$$\tau - t = \Delta g_{sol}^F n_s / g_{sol} \tag{3.20}$$

from which true transference numbers can be found from Hittorf values without a knowledge of the molecular weight of the solvent mixture.

With true transference, the following two relations can be written for a solvent containing the two components A and B:

$$\Delta g_{(sol) A}^F = M_{(sol) A} n_{(sol) A}^c \tau^c - M_{(sol) A} n_{(sol) A}^a \tau^a \tag{3.21}$$

$$\Delta g_{sol}^F = [M_{(sol) A} n_{(sol) A}^c + M_{(sol) B} n_{(sol) B}^c] \tau^c$$
$$- [M_{(sol) A} n_{(sol) A}^a + M_{(sol) B} n_{(sol) B}^a] \tau^a \tag{3.22}$$

In these equations $\Delta g_{(sol) A}^F$ and Δg_{sol}^F are grams of solvent A and total grams of solvent, transferred from the anode to the cathode; $M_{(sol) A}$ and $M_{(sol) B}$ are the molecular weights of solvent components A and B; $n_{(sol) A}^c$ and $n_{(sol) B}^c$

are the moles of solvent component A and B solvating the cation; $n^a_{(sol) A}$ and $n^a_{(sol) B}$ are the moles of solvent components A and B solvating the anion; and τ^c and τ^a are the true transference numbers of the cation and anion.

Dividing Eq. (3.21) by $M_{(sol) A}$ gives Eq. (3.15) for a pure solvent. Equation 15 has an infinite number of solutions, but these can be readily reduced by making the assumptions that $n_{(sol) A}$ and $n_{(sol) B}$ must be integers and that they have upper and lower limits.

A reasonable lower limit would be zero for the values of $n_{(sol) A}$ and $n_{(sol) B}$ in the case of a system about which nothing is known. The maximum upper limit for $n^c_{(sol) A}$ would be the moles of solvent A per mole of cation in the solution assuming the anion is not solvated, and the maximum upper limit of $n^a_{(sol) A}$ would be the moles of solvent per mole of anion in the solution assuming the cation is not solvated, and similarly for $n^c_{(sol) B}$ and $n^a_{(sol) B}$. The total number of moles of solvent A solvating the ions must not exceed the moles of solvent A in the solution. The relation

$$n^c_{(sol) A} + n^a_{(sol) A} \leqslant n_{(sol) A}/n_s \tag{3.23}$$

can be stated where $n_{(sol) A}$ is the total moles of solvent A in the solution and n_s is the equivalents of solute. Equation (3.23) can be programmed and solved for all possible values of the integers, provided the necessary transference and solution data are available.

Equation (3.22) can be expanded and Eq. (3.21) substituted into it to give

$$\frac{\Delta g^F_{sol} - \Delta g^F_{(sol) A}}{M_{(sol) B}} = n^c_{(sol) B} \tau_c - n^c_{(sol) B} \tau^a \tag{3.24}$$

Equation (3.24) can be solved in the same manner as Eq. (3.23).

Remy and co-worker [18] determined differences in ionic hydration by measuring the volume change in the two halves of a solution separated from each other by a parchment paper membrane in a cell through which current was passed. Changes in the positions of menisci in capillary tubes inserted in the stoppers of the two halves of the cell made possible the measurement of the change of volume. Change in volume due to electroosmosis of the solvent alone was negligably small in 1.0 normal electrolyte solutions at which concentrations their measurements were made. See also the work of Baborousky [19].

3. Solvation Numbers

MacInnes [4] quoted the data of Washburn and Millard [8] and calculated by employing Eq. (3.16) the number of moles of water carried per mole of cation for different assumed hydration of the chloride ion. His data are included in Table 3.1.

Table 3.1

Solvation of Various Ions by Different Electrolytic Transference Methods

Solvent	Reference ion	Assumed solvation number n of reference ion	Solvated ion	Solvation number n of ion	Concentration of electrolyte	Temperature (°C)	Reference
H_2O	Cl^-	0	H^+	0.3	1.2 N	25	[4][a]
	Cl^-	4	H^+	1.0	1.2 N	25	[4][a]
	Cl^-	8	H^+	1.8	1.2 N	25	[4][a]
	Cl^-	0	Cs^+	0.7	1.2 N	25	[4][a]
	Cl^-	4	Cs^+	4.7	1.2 N	25	[4][a]
	Cl^-	8	Cs^+	8.9	1.2 N	25	[4][a]
	Cl^-	0	K^+	1.3	1.2 N	25	[4][a]
	Cl^-	4	K^+	5.4	1.2 N	25	[4][a]
	Cl^-	8	K^+	9.5	1.2 N	25	[4][a]
	Cl^-	0	Na^+	2.0	1.2 N	25	[4][a]
	Cl^-	4	Na^+	8.4	1.2 N	25	[4][a]
	Cl^-	8	Na^+	14.9	1.2 N	25	[4][a]
	Cl^-	0	Li^+	4.7	1.2 N	25	[4][a]
	Cl^-	4	Li^+	14.0	1.2 N	25	[4][a]
	Cl^-	8	Li^+	23.0	1.2 N	25	[4][a]
	Cl^-	4	H^+	1.2	1.0 N	25	[4][b]
	Cl^-	4	K^+	5.0	1.0 N	25	[4][b]
	Cl^-	4	Na^+	9.8	1.0 N	25	[4][b]
	Cl^-	4	Li^+	14.3	1.0 N	25	[4][b]
	Cl^-	4	H^+	1.0	1.2 N	25	[8]
	Cl^-	4	Cs^+	4.7	1.2 N	25	[8]
	Cl^-	4	K^+	5.4	1.2 N	25	[8]
	Cl^-	4	Na^+	8.4	1.2 N	25	[8]
	Cl^-	4	Li^+	14.0	1.2 N	25	[8]
58%EtOH	Cl^-	18(max.H_2O)	Li^+	50(max.H_2O)	0.361 M	25	[17]
	Cl^-	0(min.H_2O)	Li^+	11(min.H_2O)	0.361 M	25	[17]
	Cl^-	13(max.EtOH)	Li^+	22(max.EtOH)	0.361 M	25	[17]
	Cl^-	2(min.EtOH)	Li^+	0(min.EtOH)	0.361 M	25	[17]
80%EtOH	Cl^-	4(max.H_2O)	Li^+	28(max.H_2O)	0.386 M	25	[17]
	Cl^-	0(min.H_2O)	Li^+	15(min.H_2O)	0.386 M	25	[17]
	Cl^-	17(max.EtOH)	Li^+	29(max.EtOH)	0.386 M	25	[17]
	Cl^-	3(min.EtOH)	Li^+	0(min.EtOH)	0.386 M	25	[17]
90%EtOH	Cl^-	10(max.H_2O)	Li^+	8(max.H_2O)	0.333 M	25	[17]
	Cl^-	5(min.H_2O)	Li^+	0(min.H_2O)	0.333 M	25	[17]
	Cl^-	17(max.EtOH)	Li^+	43(max.EtOH)	0.333 M	25	[17]
	Cl^-	0(min.EtOH)	Li^+	7(min.EtOH)	0.333 M	25	[17]
H_2O	Cl^-	144(max.H_2O)	Na^+	224(max.H_2O)	0.153 M	25	[17]
	Cl^-	2(min.H_2O)	Na^+	1(min.H_2O)	0.153 M	25	[17]
	Cl^-	161(max.H_2O)	Na^+	181(min.H_2O)	0.188 M	25	[17]
	Cl^-	1(min.H_2O)	Na^+	3(min.H_2O)	0.188 M	25	[17]

Table 3.1—continued

Solvent	Reference ion	Assumed solvation number n of reference ion	Solvated ion	Solvation number n of ion	Concentration of electrolyte	Temperature (°C)	Reference
H$_2$O	Cl$^-$	5	Cs$^+$	6	0.3 N	—	[58]
	Cl$^-$	5	Mg$_2{}^+$	36	0.3 N	—	[58]
	Cl$^-$	5	Ca$_2{}^+$	29	0.3 N	—	[58]
	Cl$^-$	5	Sr$_2{}^+$	29	0.3 N	—	[58]
	Cl$^-$	5	Ba^{2+}	28	0.3 N	—	[58]
	Cl$^-$	5	Cu^{2+}	34	0.3 N	—	[58]
	Cl$^-$	5	Zn^{2+}	44	0.3 N	—	[58]
	Cl$^-$	5	Cd^{2+}	39	0.3 N	—	[58]
	Na$^+$	13	Br$^-$	5	0.3 N	—	[58]
	Cl$^-$	5	F$^-$	7	0.3 N	—	[58]
	Cl$^-$	5	NO$_3{}^-$	6	0.3 N	—	[58]
	Na$^+$	13	CH$_3$COO$^-$	11	0.3 N	—	[58]
	Cl$^-$	5	SO$_4^{2-}$	12	0.3 N	—	[58]
NMA	R$_4$N$^+$	0	Li$^+$	5.1	—	40	[65]
	R$_4$N$^+$	0	Na$^+$	3.5	—	40	[65]
	R$_4$N$^+$	0	K$^+$	3.3	—	40	[65]
	R$_4$N$^+$	0	Cs$^+$	2.6	—	40	[65]
	R$_4$N$^+$	0	NH$_4{}^+$	2.7	—	40	[65]
	R$_4$N$^+$	0	Cl$^-$	2.1	—	40	[65]
	R$_4$N$^+$	0	Br$^-$	1.7	—	40	[65]
	R$_4$N$^+$	0	I$^-$	1.5	—	40	[65]
	R$_4$N$^+$	0	CN$_5$	1.3	—	40	[65]
	R$_4$N$^+$	0	NO$_3{}^-$	1.5	—	40	[65]
	R$_4$N$^+$	0	Ba$_2{}^+$	9.0	—	40	[65]
	R$_4$N$^+$	0	Sr^{2+}	8.6	—	40	[65]
	R$_4$N$^+$	0	Ca^{2+}	8.6	—	40	[65]
	R$_4$N$^+$	0	Mg^{2+}	10.3	—	40	[65]

^{a} Inert reference substance.
^{b} Parchment paper.

In Remy and Reisner's [18] work the moving boundary method was used effectively with the parchment paper serving as the reference boundary with respect to which volume changes in the electrode regions were measured. MacInnes [4] showed that in the case of the indifferent reference substance method, the number of moles of water transferred per faraday, Δn of Eq. (3.16), in various electrolytes in 1.0 normal solutions were identical to the moles of water carried per faraday in 1.3 normal solutions except for hydrochloric acid which shows only a small variation between the values of Δn in the two concentrations. (See MacInnes [4, Chapter 4, Tables X and XI].) He also

Table 3.1—continued

Solvent	Reference ion	Assumed solvation number n of reference ion	Solvated ion	Solvation number n of ion	Concentration of electrolyte
H_2O	Cl^-	61 (max.H_2O)	Na^+	102 (max.H_2O)	0.351
	Cl^-	0 (min.H_2O)	Na^+	3 (min.H_2O)	0.351
	Cl^-	42 (max.H_2O)	Na^+	73 (max.H_2O)	0.504 M
	Cl^-	0 (min.H_2O)	Na^+	4 (min.H_2O)	0.504 M
	Cl^-	271 (max.H_2O)	K^+	283 (max.H_2O)	0.101 M
	Cl^-	5 (min.H_2O)	K^+	0 (min.H_2O)	0.101 M
	Cl^-	181 (max.H_2O)	K^+	188 (max.H_2O)	0.158 M
	Cl^-	2 (min.H_2O)	K^+	0 (min.H_2O)	0.158 M
	Cl^-	69 (max.H_2O)	K^+	74 (max.H_2O)	0.373 M
	Cl^-	0 (min.H_2O)	K^+	3 (min.H_2O)	0.373 M
	Cl^-	52 (max.H_2O)	K^+	62 (max.H_2O)	0.505 M
	Cl^-	0 (min.H_2O)	K^+	2 (min.H_2O)	0.505 M
	Cl^-	4	Na^+	8 to 9	$1 N$
	Cl^-	4	K^+	5	$1 N$
	Cl^-	4	Li^+	13 to 14	$1 N$
	Cl^-	4	Br^-	3 to 4	$1 N$
	Cl^-	4	I^-	2	$1 N$
	Cl^-	4	H^+	1	$1 N$
	Cl^-	4	Mg_2^+	20	$1 N$
	Cl^-	4	Ca_2^+	17 to 16	$1 N$
	Cl^-	4	Sr_2^+	16	$1 N$
	Cl^-	4	Ba_2^+	11	$1 N$
	Cl^-	4	H^+	1.06	$1 N$
	Cl^-	4	SO_4^{2-}	2.8	$1 N$
	Cl^-	4	SO_4^{2-}	1.4	$1 N$
	Cl^-	4	K^+	5	$1 N$
	Cl^-	3.0	Na^+	7.4	$1 N$
	SO_4^{2-}	10.7	Cu_2^+	10.7	$1 N$
	Cu_2^+	8.5	$CuCl_4^-$	19.7	$1 N$
	Na^+	5	SO_4^{2-}	1.35	$1 N$
	K^+	5	NO_3^-	4.58	$1 N$
	Na^+	5	NO_3^-	4.54	$1 N$
	Li^+	7	H^+	0.5	—
	Li^+	7	Na^+	5	—
	Li^+	7	K^+	2	—
	Li^+	7	Cs^+	1.5	—
	Li^+	7	Ba_2^+	9.5	—
	Li^+	7	Cd_2^+	11	—
	Li^+	7	Cl^-	1	—
	Cl^-	5	Li^+	22	$0.3 N$
	Cl^-	5	Na^+	13	$0.3 N$
	Cl^-	5	K^+	7	$0.3 N$

compares Δn values for the indifferent reference substance and the parchment paper methods (see [4, Chapter 4, Table IX]). Since the Δn values at the two concentrations of electrolytes are almost identical using the indifferent reference substance approach, it can be assumed that the "true" and Hittorf transference numbers each have the same values for the same electrolyte at the two concentrations. The values of n_s^c were calculated from Eq. (16) for the cations of HCl, KCl, NaCl and LiCl for the parchment paper data using $n_s^a = 4$ for the chloride ion and the results are recorded in Table 3.1. It can be seen that the n_s^c for the various cations in the case of the indifferent reference substance and in the case of parchment paper approach each other.

Washburn [7] reported Hittorf transference measurement of H^+, Li^+, Na^+, Rb^+, and Cs^+ ions at 25°C in 1.25 N solution for $n_s^{Cl^-}$ equal to 4 and 9. The data for $n_s^{Cl^-}$ equal to 9 are included in Table 3.1. These data are quoted by Glasstone [20, 21].

From Hittorf data [12, 13] at 25°C using an inert reference Wear and Amis [17] calculated the maximum and minimum values for the moles of water and of ethanol solvating the cations and anions of LiCl at various weight % of the ethanol component of the solvent. In these calculations they employed Eqs. (3.20) and (3.21) as they did in calculating the solvation in multiples of 4 in the same solvent system at the same temperature for the cation and anion of LiCl. Similar calculations were made using Eq. (3.21) for the cations and anions of NaCl and KCl in water. The maximum and minimum number of moles of water and of ethanol transported at 25°C by Li^+ and Cl^- ions at different weight % of ethanol in water solvents are recorded in Table 3.1. Similar data are reported in the table for Na^+ and Cl^- ions and for K^+ and Cl^- ions for water solvent only. For lithium chloride the trends indicated in the table are what is expected, except for the maximum number of moles of water around the anion. It was felt by the authors that either the value at 80 weight % ethanol is too low or the value at 90 weight % ethanol is too high. Wear and Amis [17] in order to eliminate some of the possible calculations, assumed that solvation, especially in water occurs in multiples of four. They felt this to be reasonable since the primary solvation shell of lithium ion is four molecules in both water [20] and ethanol [22] and the primary solvation shell of chloride ion [23] is eight molecules of water. In addition hydrogen ion combines with four waters to form the pyramidal $H_9O_4^+$ complex stable up to 100°C [24], and it has been found that much hydration is a buildup of tetrahedrons [25]. The most common buildup appears to be a dodecahedron, which is five tetrahedrons. This is probably sound reasoning since the structure of water is tetrahedral and secondary solvation is probably greatly influenced by water structure as well as statistics, electrostatics, and chemical affinity.

A discrepancy between the numerical transport of water and the transport values of the cations found from cathodic or anodic determinations was observed [26] in the case of the chlorides, bromides, and iodides of the alkali metals, the alkaline earth metals, and hydrogen. These discrepancies were traced to the swelling of the parchment membranes used in the experiments caused by the electrolytes with a consequent decrease in mobility of the anions.

A distinction has been made [27–29] between close or chemical hydration and the wider or physical hydration of ions. In chemical hydration complex, ions are considered to be formed. Physical hydration is due to electrostatic attraction of the solvent dipoles in the field of an ion. It was felt that no method existed for the determination of chemical hydration alone. The electrolytic transfer of solvents is regarded as a hydrodynamic phenomena.

A comparison of the solvation numbers in normal solutions of several salts determined by the parchment paper, salting-out, and diffusion methods has been made [27]. Assuming the chloride ion to have a solvation number of 4 the solvation numbers of the other ions have the following ranges: Na^+, 8–17; K^+, 5–14; Li^+, 13–15.5; Br^-, 3–7.5, and I^-, 2–5. A comparison of the solvation of certain ions at different concentrations show [29] that the solvation numbers of the ions from 1 normal to 0.1 normal solution vary as follows: Li^+, 13–14 to 62; K^+, 5–29.3; Cl^-, 4–26.6; I^-, 2–31.4. The first solvation number(s) for each ion refers to the 1 normal and the last to 0.1 normal solution.

Solvation numbers of other ions have been measured using the parchment paper method [30, 31]. These data are listed in Table 3.1.

The solvation numbers listed in Baborousky [28, Table I] for K^+, Na^+, Li^+, Cl^- and Br^- ions were checked [32] and the transport number of the hydrogen ion determined. This value is listed in Table 3.1. This paper lists the hydration numbers as determined by Schreiner [33] of the previously-mentioned ions using their activities.

The degree of solvation of the ions of LiCl were studied [34] using camphor as the inert substance in Hittorf transference experiments using methyl, ethyl, and *n*-propyl alcohols as solvents. There was a net transfer of solvent by the lithium ions from the anode to the cathode. In a given alcohol, the degree of solvation of the lithium ion decreased with increasing concentration of the LiCl.

It was observed [35] through the hydration of ions derived from the transport of water that Kohlrausch's law of independent migration of ions does not hold exactly in fairly concentrated solutions. It was also found that for concentrated solutions electroosmotic transport does not interfere with electrolytic transport in the determination of hydration.

The hydration of the SO_4^{2-} ion was determined [36] using three parchment paper diaphragms in a four-compartment cell. The value for normal solutions is reported in Table 3.1.

In a review of the literature on the hydration of ions [37], it was concluded that all methods used to date were open to objections, but the most promising method involved using ultrasonic vibrations.

The agreement in transference number values with neutral solutions between the values obtained by the analysis of the anode and cathode layers suggests [38] that the high hydration values that have been ascribed [39, 40] to alkali metal cations in dilute solution as a result of transference measurements with or without a reference substance, are spurious.

The hydration of sulfate ions using the semipermeable membranes in electrolytic transport experiments has been measured [41] and is recorded in Table 3.1.

Ionic water transport using a parchment paper diaphragm was found [42] to be independent of the specific effects of the paper diaphragm at concentrations of 0.8 N and greater. In more dilute solutions the electroendosmotic effect is important. Data on solvation numbers are listed in Table 3.1.

The electroosmotic and electrolytic transport of water through parchment membrane in contact with $CuCl_2$ and $CuSO_4$ has been determined [43]. The electroosmotic transport was very much greater for $CuSO_4$. Valency alone could not account for the great effect of the sulfate ion. Evidence indicated [44] that the electrolytic transport of water per faraday does not increase with decreasing concentration of $CuSO_4$ from 0.4 N. Over the range from 0.0005 to 0.2 N there is a composite effect of a constant electrolytic water transport and of an electroendosmotic action by the parchment diaphragm. A diaphragm of powdered glass has been used [45] for measurements of liquid transport for a constant applied electromotive force using solutions of $CuSO_4$. For solutions greater than 0.0005 N, the liquid transport per faraday is related to the dilution by a linear law and decreases to a negligible value at 0.1-N solution.

The transport of water to the anode or cathode through benzaldehyde or carbon tetrachloride has been measured [46]. The moles of water transported per mole of NH_4^+ ion was found to be 892.

Hydration of ions has been investigated [47] using a three parchment membrane set up. The hydration numbers of the different ions are listed in Table 3.1.

MacDonald [48] using the mobility of hydrogen ion as about 65 estimated the hydrated ion to be H_3O^+. He points out that Darmois' method [49] would lead to negative values of hydration for chloride and potassium ions, and would give 0.3, 1, and 5, respectively, for the hydration numbers of H^+, Na^+, and Li^+ ions.

The apparent hydration numbers of migrating ions have been calculated [50] using Washburn's theory of ionic mobility.

The concept of a fixed number of water molecules bound to an ion and of a definite Stokes radius has been discussed [51]. The total hydration energy of an ion consisting of two parts. The first part is due to the nearest surrounding molecules, which may either increase or decrease the energy of the system ion-water, and the second part as due to the sum of the actions of the more remote water molecules, which always results in a decrease of that energy. The activation energy of viscosity of the water molecules nearest to the ion is increased by hydrated ions, but decreased by nonhydrated ions. In the first case, the mean time of stay of water molecules around the ion is increased while in the second case the mean time of stay is decreased. This latter case corresponds to negative hydration. Darmois' interpretation [52, 53] of the experimental fact that the temperature coefficient of the mobility of hydrated ions is greater than the temperature coefficient of the viscosity of water, by progressive dehydration with increasing temperature and the corresponding decrease in the Stokes radii of the ions, was criticized because it failed to recognize the difference between the relations existing between the temperature coefficients of the mobilities of hydrated and unhydrated ions and the temperature coefficient of the viscosity of the solvent.

It has been assumed [54–57] that a hydration flux exists and thus makes it necessary to consider a hydrodynamic reflux, whose effect modifies the transport number of ions. The experimental ionic mobilities are taken as the sum of the mobilities the ions would have in an immobile medium and the speed of entrainment which is that of the hydrodynamic reflux. This new viewpoint on the intervention of the hydration of ions helps explain the inversion of the Hittorf phenomenon.

Haase [58] using electrolytic water transport measurements obtained the solvation numbers listed in Table 3.1.

Three cellophane membranes were used [59] in a four-compartment vessel filled with 0.3-N electrolyte solution. The solvent was D_2O in compartments 1 and 2 and H_2O in compartments 3 and 4. The solvation numbers obtained are listed in Table 3.1.

The effect of the hydration of ions on the process of orientation in solutions of electrolytes has been formulated [60] in terms of a modified Kohlrausch rule.

It was found [61] in the range 2–11 N LiCl at 20°C, that the average number of molecules of water carried by Li^+ ion in its displacement by the action of the electric field decreased with increasing concentration. This was termed the kinetic hydration number of Li^+ ions.

Selective solvations of ions in solvent mixtures have been studied [62–64]. A transference technique was used to study the selective solvation of ions in various mixed solvents. The results are listed in Table 3.2. In the case of $CaCl_2$ in the water–methanol mixtures, a change in the composition of the solvent mixtures due to hydrolysis was determined by measuring the densities and refractive indices of the solutions.

Table 3.2

Selective Solvation of Ions

Solvent	Salt	Ion selectively solvated	Solvent component selectively solvating the ion	Ref.
H_2O-CH_3CN	$AgNO_3$	Ag^+	CH_3CN	[61]
	$AgNO_3$	NO_3^-	H_2O	[61]
	$CaCl_2$	Ca^{2+}	H_2O	[62]
	$CaCl_2$	Cl^-	H_2O	[62]
$H_2O-Hydrazine$	$ZnCl_2$	Zn^{2+}	Hydrazine	[63]
	$ZnCl_2$	Cl^-	H_2O	[63]
$H_2O-Acetonitrile$	$ZnCl_2$	Zn^{2+}	H_2O	[63]
	$ZnCl_2$	Cl^-	H_2O	[63]

The solvation number of Li^+ ion in Li_2SO_4 was found to decrease [65] in aqueous solutions from 9 for a 2 N solution to 5 for 5.5 N solution.

Appreciable ion–solvent interaction was observed [66] in potassium bromide solutions in N-methylacetamide (NMA). The solvation number of various ions in NMA are listed in Table 3.1. The improved method of Robinson and Stokes [67] as proposed by Nightingale [68] was used in calculating the solvation numbers. In these calculations the tetraalkyl-ammonium ions R_4N^+ except the $(CH_3)_4N^+$ were assumed to be unsolvated.

Solvation of the ions of hydrochloric and of p-toluenesulfonic acids in formic acid solutions were studied [69] by electrolysis in 0.8–0.2-N hydrochloric acid and 1.0–0.2-N p-toluenesulfonic acid solution using $H^{14}CO_2H$ as a tracer. It was assumed that the mobility of the anion and of the cation were equal. More $H^{14}CO_2H$ was transported into the cathode space than into the anode space. The solvation number of the Cl^- and $MeC_6H_4SO_3^-$ was approximately the same and was independent of the concentration. The solvation number of the cation tended to increase with decreasing concentration of the electrolyte.

In a electrical transport investigation of magnesium sulfate, magnesium chloride and magnesium nitrate solutions by the moving boundary method, it was found [70] that the unsymmetric nature of the electrolyte affected the hydration of the ions. For magnesium sulfate solutions there was a substantial transfer of solvent toward the cathode which amounted to 12 water molecules at a concentration of 3 N and 8.7 molecules at a concentration of 5 N. In the cases of $MgCl_2$ and $Mg(NO_3)_2$, there was an insignificant transfer of solvent in the opposite direction, that is, toward the anode.

The hydration number of H^+, Li^+, Na^+, and K^+ ions in highly concentrated solutions of chlorides have been determined [71] by the moving boundary method using indicator electrolytes at 20°C. The hydration numbers n of the ions were, with the exception of H^+, arranged in reversible dependence on their radii. Thus $n_{Li^+} > n_{Na^+} > n_{K^+}$. At low concentrations of HCl, the motion of the H^+ ions is mainly of a "relay" character, with a maximum mobility and a minimum hydration of the H^+ ions. In higher concentrations of HCl the H^+ ions are hydrated and form H_3O^+ ions. The "relay" mobility and the negative mobility of the H^+ ion decreased in importance as the HCl increases in concentration. At maximum HCl concentrations the mobility of the H^+ ions is determined by the H_3O^+.

Water transport in cation exchange membranes has been reported [72–74] to depend on current density in 0.01-N solution of an electrolyte. Two cation exchange membranes, cross-linked phenolsulfonic acid (PSA) and poly-ethylene–styrene graft copolymer supplied by the American Machine and Foundry Company (AMF) were used. The number, l_w, of moles of water transported per faraday of electricity passed in high water content membrane were high at low current densities and low at high current densities for all the alkali metal ions. Similar behavior was noted for the membrane of low water content for Li^+, Na^+, and K^+ ions, whereas l_w, was not significantly affected by current density in the case of Rb^+ and Cs^+ ions. Table 3.3 contains the data for the two membranes [72].

Table 3.3

Number of Moles of Water (l_w) Transported across Cation Exchange Membranes in Contact with 0.01 N Solutions of Alkali Metal Chlorides as a Function of Current Density at 25°C

Current density (mA/cm²)	Li^+	Na^+	K^+	Rb^+	Cs^+
	PSA membrane				
0.32	72.6	62.1	55.5	45.5	46.9
1.58	42.6	39.3	34.2	33.4	30.8
3.15	36.9	32.9	27.7	26.2	25.8
15.75	27.6	24.7	20.3	19.5	19.6
	AMF membrane				
0.32	12.1	9.6	7.1	4.9	4.8
1.58	10.3	7.0	4.7	4.9	4.9
3.15	10.3	7.0	4.5	4.5	4.9
15.75	10.1	6.9	4.2	4.6	5.0

The data with PSA membrane confirms earlier results [75–82] about the current density dependence of l_w in 0.01 N solutions. The data resulting from comparatively low water content AMF membranes support both the dependence and independence of l_w on current density. It is stated that the two factors that determine the behavior of l_w with respect to current density are membrane water content (or pore size) and the size of the ion. Some of the conditions existing both at the membrane-solution interface and within the membrane influence the variation of l_w with current density. Concentration polarization, which contributes to the interfacial phenomena involved, can be eliminated as in the experimental conditions under which the l_w values of Table 3.3 were measured, by efficient stirring of solutions and using low currents.

Dependence of electroosmotic water transport on current density across a polystyrenesulfonic acid cation-exchange membrane in contact with various solutions of sodium chloride (0.1, 1.0, and 4.0 M) has been reported [83].

Generally solutions at rest in the cell have been used [84–87] to measure electroosmotic water transport across cation-exchange membranes. Data obtained by this method have been substantiated by independent measurements of membrane potentials [88]. It was felt that considerable concentration polarization might occur in very dilute solutions. It was also observed that there was no significant liquid film resistance in concentrated solutions [89]. It was stated [73] that stirring did not eliminate the current dependence of electroosmotic water transport, but that the two major influences on this phenomenon for a given cation appear to be (1) the internal concentration of the membrane pore solution, and (2) the current density.

In answer to the above observation on the lack of the effect of stirring on electromosis it was pointed out [74] that the type of current dependence of l_w concerned with is confined to current density ranges in which counterion transport number remained unaffected and that l_+ is always unity. This is a membrane transport phenomenon which is not well understood, and is an effect in which there is an almost asymtotic rise of l_w with decrease in current density. This Bethe–Toropoff [90, 91] effect is due to concentration polarization at the region of the solution-membrane interface, ultimately leading to membrane polarization. This effect can be eliminated by stirring the solutions on either side of the membrane.

It can be seen from a perusal of Table 3.1 that there is variation among the reported solvation numbers of a given ion, but that there are certain trends which are consistent. The order of the solvation numbers of the alkali metal ions is $Li^+ > Na^+ > K^+ > Rb^+ > Cs^+$. Also, the more dilute the solution the greater the solvation of a given ion. The alkaline earth metal ions are more highly solvated than the alkali metal ions. The order of the solvation number of the alkaline earth ions is not so well established in all instances but tends to be in the direction $Mg^{2+} > Ca^{2+} > Sr^{2+} > Ba^{2+}$. The halogen ions are solvated in the order $F^- > Cl^- > Br^- > I^-$. Thus it would appear from electrolytic transport methods that the smaller the simple ion and the greater its charge, the more highly it is solvated. Cd^{2+}, Zn^{2+}, and Cu^{2+} ions are highly hydrated. There are a few data for hydration of complex anions. Their hydration numbers show no unusual tendencies, except that for the SO_4^{2-} ion, which tends to be strongly dependent on concentration. The date in Table 3.2 show the same order of water transport with respect to the alkali metal ions. Table 3.2 indicates that for the mixed solvent systems shown, cations are sometimes selectively solvated by the nonaqueous component, while anions are always selectively solvated by the water component of the solventsy stems. The transport methods are of the dynamic type and, it is believed [92], the only type of experiments in which hydrodynamical solvation is relevant. Transport methods as is to be expected [93–95a,b,c] should yield higher solvation numbers than the generally accepted [24, 92, 96] values since these methods measure the hydrodynamical solvation type in addition to the primary and secondary solvation types. The only exceptions to the above rule were hydrogen and hydroxyl ions which were found to yield smaller hydration numbers (e.g., $H^+ = 1$–5) by transport experiments than by other methods (e.g., $H^+ > 13$). The Grotthus method of conduction of these ions when the transport methods are used would result in low solvation numbers.

B. CONDUCTANCE

1. Theory

One approach to the calculation of solvation numbers of ions from electrolytic conductance data is to calculate the Stokes ionic radii r_s from the limiting ionic conductances at infinite dilution λ_0 which is obtained from the limiting conductance of electrolytes and the limiting conductance of the standard reference perchlorate ion [97, 98]. The procedure is as follows. For the tetraalkylammonium salts in benzene and sulfolane [97, 99], the λ_0 increase from ethyl to methyl substituent was lower than the expected value. Using the behavior of nitrobenzene solutions as an analogy, the small increase of the limiting conductance in sulfolane has been explained as an interaction between the tetramethylammonium charge and solvent molecule dipoles.

Using this approach, the Stokes ionic radii r_s were calculated from the λ_0 values of unsolvated Et_4N^+, Pr_4N^+, and Bu_4N^+ applying the equation

$$r_s = \frac{0.82|z|}{\lambda_0 \eta_0} \tag{3.25}$$

The corresponding Robinson and Stokes correction factor r_0/r_s was calculated using the crystallographic radii r_0 reported by these authors [100].

The linear plots of r_0/r_s versus r_s were used to calculate the correction factor of the alkali metal ions. This was permissible since the corresponding r_s values are between or near the r_s values of the tetraalkylammonium ions. From the plot the correct radii r_{cor} of the solvated ions were found, and the volume of the solvation shell surrounding the ions were obtained from the equation

$$V = \tfrac{4}{3}\pi(r_{cor}^3 - r_0{}^3) \qquad (3.26)$$

Assuming the contraction because of electrostriction of the solvent sulfolane next to the ions to be negligible, the number of sulfolane molecules in the solvodynamic unit has been calculated accepting the molecular volume of sulfolane to be 158 Å^3.

It was pointed out that the lowest values for the solvation numbers of cations are found in solvents having the greatest molecular volumes as sulfolane and nitrobenzene.

Nightingale [101] pointed out that the same procedure yielded an effective hydrated radius for those ionic species with a Stokes radius greater than 2.5 Å. The procedure was modified to apply to ions of smaller Stokes radii by assuming that all tetraalkylammonium ions, except the tetramethyl-ammonium ion, were unhydrated and by defining the effective hydrated radius r_H of an ion by a calibration curve which possessed a finite limit as the Stokes radii r_s went to zero. The calibration curve was prepared by plotting the crystal radii, r_0 as ordinate versus r_s as abscissa for the tetraalkylammonium ions, except for the tetramethylammonium ion. The use of this curve makes possible the evaluation of the deviations from Stokes's law radius and the necessary corrections thereto. This method has been used to calculate the hydration numbers of the lanthanide ions from conductance data [102, 103].

Earlier, Ulich [104] made a calculation of solvation numbers of ions from Stokes's law radii calculated from conductance and the radius proper of the ions in a manner similar to that illustrated by Hittorf [1, 2].

2. Experimental Techniques

In conductance methods for ion solvation, the conductance of electrolytes at different concentrations are obtained at a given temperature using standard procedures. The equivalent conductance as a function of concentration is mathematically or graphically extrapolated to yield the equivalent conductance of the electrolyte Λ_0 at infinite dilution. This Λ_0 is split into the equivalent conductances at the given temperature of the ions, λ_0^{+n} and λ_0^{-m}, using trans-ference data extrapolated to infinite dilution or by assuming a limiting con-ductance for a standard reference ion [97, 98]. In the above symbols n and m

are the valences of the positive and negative ions, respectively. Using the limiting ionic conductances, the crystalline radii, and the molecular volumes of the solvents, the solvation numbers of the ions are obtained from calculations presented above in the theoretical section.

3. Solvation Numbers Determined

In Table 3.4 solvation numbers of ions occurring in the literature [97, 102–104] are presented. The solvents and temperatures are specified.

Accepting that electrolytic conduction in solution is due mainly to electrolytic transport and that electroosmotic effects contribute only a negligible amount to the whole, the absolute ionic hydration of the ions listed in Table 3.1 were calculated from the electrolytic transport of water by alkali chloride and bromide solutions [105]. Both static and dynamic solvation were postulated.

It was stated [106] that data on conductance of alkali and alkaline earth metals indicated a relatively large increase of equivalent conductance with dilution corresponds to a greater degree of hydration. The greater the density of the electric charge of an ion, the greater is its hydration and viscosity effect.

In methyl ethyl ketone at 25°C, the electrical conductances of mono-, di-, tri-, and tetraalkylammonium picrates, iodides, bromides, chlorides, nitrates, and perchlorates of the picrates of Li, Na, K, Ag, and Cd, of the iodides of Na, K, Cd, and Hg, and of $HgCl_2$ were measured [107]. The conductance of NEt_4-picrate was also determined at 0° and 50°C. At 25°C in acetone the conductivities of tetra-n-propyl-, tetra-n-butyl-, and tetraisoamylammonium picrates and Li picrate were also determined. Using the square-root law, the limiting equivalent conductances were found. The mobilities and solvation numbers were calculated from the data. Extensive conductivity work has also been done in CH_3CN solvent [108] and the solvation numbers obtained. In Table 3.4 solvation numbers of ions obtained by these workers are listed.

The changes in composition of $CrCl_3$ in H_2O represented by

$$[Cr(H_2O)_4]Cl_3(I) \rightleftarrows [Cr(H_2O)_5Cl]Cl_2(II) \rightleftarrows [Cr(H_2O)_6]Cl_3(III)$$

was found to take place more slowly in 98.5% D_2O [109].

Conductance measurements have shown [110] that Li^+ is much more highly hydrated than Na^+ and K^+.

In a study of the partition of uranyl nitrate between water and organic solvents, conductance data on the organic phase showed, in general, an average of $4.0 H_2O$ per $UO_2(NO_3)_2$ molecules in the cases of ethers, esters, and ketones. Alcohols gave more complex data [111].

Ion hydration has been found to be high even at a temperature of 340°C.

Table 3.4

Solvation Numbers of Ions from Electrical Conductivity

Solvent	Temperature (°C)	Ion	Solvation number	Ref.
Sulfolane	30	Li^+	1.4	[97]
	30	Na^+	2.0	[97]
	30	K^+	1.5	[97]
	30	Pb^+	1.4	[97]
	30	Cs^+	1.3	[97]
	30	NH_4^+	0.9	[97]
	30	Cl^-	~0	[97]
	30	Br	~0	[97]
	30	I^-	~0	[97]
	30	ClO_4^-	~0	[97]
Water	25.0	La^{+3} to Nd^{+3}	12.8 ± 0.1	[102, 103]
	25.0	Dy^{+3} to Yb^{+3}	13.9 ± 0.1	[102, 103]
	25.0	Sm^{+3}	13.1	[102]
	25.0	Eu^{+3}	13.3	[102]
	25.0	Gd^{+3}	13.4	[104]
	25	Li^+	6–7	[104]
	25	Na^+	2–4	[104]
Methanol	25	Li^+	7	[104]
	25	Na^+	5–6	[104]
	25	K^+	4	[104]
	25	Cl^-	4	[104]
	25	Br^-	2–3	[104]
	25	I^-	0–3	[104]
Ethanol	25	Li^+	6	[104]
	25	Na^+	4–5	[104]
	25	K^+	3–4	[104]
	25	Cl^-	4–5	[104]
	25	Br^-	4	[104]
	25	I^-	2–3	[104]
Acetone	25	Li^+	4	[104]
	25	Na^+	4–5	[104]
	25	K^+	4	[104]
	25	Cl^-	2	[104]
	25	Br^-	1	[104]
	25	I^-	0–1	[104]
Acetonitrile	25	Na^+	5	[104]
	25	K^+	3–4	[104]
	25	Br^-	1–3	[104]
	25	I^-	0–2	[104]
Furfural	25	Na^+	5	[104]
	25	K^+	4	[104]
	25	I^-	0–1	[104]
Pyridine	25	Na^+	4	[104]

Table 3.4—continued

Solvent	Temperature (°C)	Ion	Solvation number	Ref.
Pyridine	25	K^+	2–3	[104]
	25	I^-	0–1	[104]
Water	—	Li^+	14	[105]
	—	Na^+	9	[105]
	—	K^+	5	[105]
	—	Cl^-	4	[105]
	—	Br^-	3	[105]
Methanol	25	Li^+	7	[107]
	25	Na^+	5–6	[107]
	25	K^+	4	[107]
	25	Cl^-	4	[107]
	25	Br^-	3	[107]
	25	I^-	1	[107]
	25	Pi^-	0	[107]
Ethanol	25	Li^+	6	[107]
	25	Na^+	4	[107]
	25	K^+	3–4	[107]
	25	Cl^-	4	[107]
	25	Br^-	4	[107]
	25	I^-	2	[107]
	25	Pi^-	0	[107]
Acetone	25	Li^+	5	[107]
	25	Na^+	4–5	[107]
	25	K^+	4	[107]
	25	Cl^-	2	[107]
	25	Br^-	1	[107]
	25	I^-	0–1	[107]
	25	Pi^-	0	[107]
Methyl ethyl ketone	25	Li^+	4	[107]
	25	Na^+	3	[107]
	25	K^+	2	[107]
	25	Cl^-	2	[107]
	25	Br^-	1	[107]
	25	I^-	0–1	[107]
	25	Pi^-	0	[107]
Acetonitrile	25	Li^+	9	[107]
	25	Na^+	6	[107]
	25	K^+	3	[107]
	25	Cl^-	2	[107]
	25	Br^-	1–2	[107]
	25	I^-	0–1	[107]
	25	Pi^-	0	[107]
Methylamine	25	Cl^-	1	[107]
	25	I^-	1	[107]
	25	Pi^-	0	[107]

Table 3.4—continued

Solvent	Temperature (°C)	Ion	Solvation number	Ref.
Ethylene chloride	25	Cl^-	2	[107]
	25	I^-	0–1	[107]
	25	Pi^-	0	[107]
Water	—	Li^+	21	[120]
	—	Na^+	10	[120]
	—	K^+	7	[120]
Formamide	25	Li^+	5.4	[132]
	25	Na^+	4.0	[132]
	25	K^+	2.5	[132]
	25	Rb^+	2.3	[132]
	25	Cs^+	1.9	[132]
	25	H^+	3.5	[132]
Water	25	Li^+	7.0	[133]
	25	Na^+	5.0	[133]

as shown by conductivity measurements of concentrated aqueous solutions of LiCl, NaCl, and KCl [112].

Conductivity studies at 25°C on lithium nitrate from 0.01 M up to saturation in 30, 70, and 100 weight % ethanol showed that the ion-size parameter in angstroms increased progressively in going from pure water to pure ethanol [113]. This was attributed to the change in solvation of the lithium ion from water molecules to alcohol molecules. It was concluded that nitrate as well as lithium ion is solvated to some extent at least in alcohol.

From conductivity measurements on the alkali chlorides in the $H_2O-H_2O_2$ system, it was concluded [114] that in pure H_2O_2 it appeared that the solvation of the alkali metal cations decreases in the order Li > Na > K > Rb > Cs.

The graphs of electrical conductance of electrolyte solutions against temperature were found to show breaks at the temperatures corresponding to the changes in the states of hydration [115]. The salts, the number of water moles per mole of salt below the temperature, the number of water moles per mole of salt above the temperature, and the temperature were thus given for each salt studied. Thus, $CoCl_2$, 6, 2, 48°C; Na_2CO_3, 7, 1, 35°C; $Sr(NO_3)_2$, 4, 0, 31°C; Na_2SO_4, 7, 0, 34°C; NaBr, 2, 0, 50.5°C; Na_2CrO_4, 4, 0, 67.5°C. In the case of phase changes, no breaks in the curves were found as when $NiSO_4 \cdot 6H_2O$ at 53.3°C changed from tetragonal to monoclinic structure.

Calculations of the degrees of ionization of various valence type electrolytes from boiling point, freezing point, and conductivity data gave differences in the apparent degrees of ionization which were explained in terms of, among other things, primary ion hydration [116].

Data showing the electrical conductances and viscosities of solutions of mixed electrolytes were lower than the means of the single salt values at the same total molarity have been discussed in terms of ion hydration and dielectric constant [117].

It has been observed that the electrical conductivity versus temperature plots for saturated aqueous solutions of electrolytes gave two straight lines, the extrapolation of which intersected at the transition temperature of the two hydrates [118]. Such plots for $CoSO_4 \cdot 6H_2O$ saturated at 60°C and for $MgSO_4$ were linear and without any breaks at the transition temperature. This was attributed to the metastable existence of the higher hydrates of these salts.

The limiting conductance of LiBr and KBr in SO_2 at 0.22°C indicates that Li^+ and K^+ ions are impeded to a comparable extent in their association with bromide ion as a result of solvation of the cations [119]. This was explained by assuming that the anions penetrate the solvation sheaths of the Li^+ and K^+ ions.

Countercurrent migration was used to determine the hydration numbers of lithium, sodium, and potassium ions [120]. These numbers are listed in Table 3.1.

From conductance data on $AgNO_3$ and KI between 15° and 40°C in ethanol and water, it was found that while in ethanol strong solvation results with large ions obeying Walden's rule, in water deviations arose from the change in the size of the ions with temperature due to difference in solvation [121].

From conductance data on magnesium chlorides and bromides in dimethylformamide (DMFA) and in DMFA–H_2O mixtures, the following solvates were found or inferred: $MgCl_2 \cdot 6DMFA$, $MgBr_2 \cdot 6DMFA$, $MgCl_2 \cdot 3DMFA$, $MgCl_2 \cdot 4DMFA \cdot 2H_2O$, $MgCl_2 \cdot 3DMFA \cdot 3H_2O$, and $MgCl_2 \cdot 2DMFA \cdot 4H_2O$ [122].

From conductance data at 35°C on potassium sulfate and sodium chlorides in dioxane–water mixtures, it was found that the solvation of K^+ and Na^+ vary in the same way in 10 and 20% dioxane solutions. In 30% dioxane solutions, the polarizing powers for dioxane and those for water molecules in the solvation sphere differ, and the preference of a particular ion for one species of solvent becomes evident [123, 124].

From conductivity data, the solvation properties of UO_2^{2+} ion were studied in solvents ranging in dielectric constant from 7 to 109 [125]. In several solvents solvation was isoergic and water and organic molecules competed in the primary solvation shell. In strongly polar solvents, water in the shell was completely replaced by the solvent which formed a complex with the UO_2^{2+}.

From the measurement of the surface conduction of NaCl it was found that for samples having low conductance, the conductance was through adsorbed semihydrated ions, since hexanol detergent did not decrease the conductance [126]. Samples having higher conductance were affected by hexanol, supposedly due to conduction in grooves. From the mass action law for the formation of adsorbed semihydrated ions, the average number of H_2O molecules of semihydrated cations and anions was found to be 3 at low humidities and about 10 at high humidities. The 10 molecules corresponded to a second hydration shell.

Conductance data on NaBr in liquid SO_2 at 0.02°C indicated a smaller degree of solvation of free Na^+ in SO_2 than in several other solvents [127]. In mobility behavior in SO_2, Na^+ resembles K^+ more than Li^+.

Selective solvation in asymmetric solvents has been studied using conductivity measurements [128].

A comparative study at 25°C of conductance data in protic and aprotic solvents indicated the role of the protons in enhancing the solvation of anions, but the very secondary effect of protons in influencing the solvation of cations which are solvated by ion–dipolar interaction [129].

From electrical conductance measurements on aqueous electrolytic solutions under pressure, it was found that the application of hydrostatic pressure reduced the radii of the hydrated ions, until above 2000 kg/m^2 the hydration atmospheres were stripped down to the innermost solvation sheaths of strongly bound waters [130].

Conductance measurements at 25°C in tetrahydrofuran (THF) of tetraphenylborides of Li^+, Na^+, K^+, Cs^+, Bu_4N^+, and $(isoamyl)_3BuN^+$ from 10^{-6} to 2×10^{-4} M were made and the limiting conductances obtained from the extrapolation of Fuoss plots [131]. The limiting conductance of the cations were found by assuming $\lambda_0^+[(Isoamyl)_3BuN]^+ = \lambda_0^-(BPh_4)^-$. The lowest value of λ_0^+ was found for Li^+, which was assumed to be most solvated, and the largest λ_0^+ was found for Cs^+, which was assumed to be unsolvated. The tetraalkyl ammonium ions were found to be unsolvated.

Transference numbers at 25°C were measured for the ions of KCl in the range from 0.01 to 0.1 N using formamide and formamide and water as the solvents [132]. The limiting transference numbers of the ions obtained from these moving boundary experiments were combined with equivalent conductance data from the literature to give individual limiting ionic conductances in formamide. From these ionic solvation numbers were calculated using the Robinson and Stokes [133] modification of Stokes's law and ignoring any electrostrictive contraction of the solvent molecules next to the ions. These data together with some of the corresponding hydration numbers of two ions (Li^+ and Na^+) are presented in Table 3.4.

Conductance measurements at 15°, 25°, and 35°C of aqueous solutions of KCl and LiCl mixed with glycine or urea showed that the mechanism and state of hydration of the ions and solutes were the same in mixed solutions as in water [134].

Agreement was found among four different methods of estimating the total hydration atmosphere or water-structure-enhanced regions surrounding the alkali metal cations in aqueous solutions [135]. Three of the methods involved conductance theory. For Na^+ at 5°, 20°, and 50°C, the number of water molecules in the total hydration atmosphere approximates closely the number of water molecules in the Frank–Wen cluster in pure water at these same temperatures.

4. Significance

Conductance data adds considerably to the varieties of solvents and ions investigated with respect to solvation. The variation of solvation number for a given ion with solvent is evident. Thus, at 25°C the solvation number of K^+ varies from 1.5 in sulfolane to 7 in formamide. The data indicate that in all solvents the solvation of the alkali metals is in the order $Li^+ > Na^+ > K^+ > Rb^+ > Cs^+$, except in acetone where the solvation numbers of Li^+, Na^+, and K^+ are about the same, at least as presented by Ulich [104]. The order of solvation of the halogen ions in all solvents is, in general, $Cl^- > Br^- > I^-$, though in methylamine at 25°C, Cl^- and I^- ions have the same solvation numbers.

In general solvation numbers from conductance data are the limiting values of these numbers since limiting values of the equivalent conductance of the ions are used in the calculations. These numbers are relative to some chosen ion since the limiting equivalent conductances of ions are determined relative to some standard reference ion as perchlorate ion.

Using this method, the results of different workers under similar conditions seem consistent. For example, reference is made to solvation numbers in methanol and ethanol presented in Table 3.4 from Ulich [104] and Walden and Birr [107].

C. ELECTROMOTIVE FORCE

1. Theory

Electromotive force data were used to find the solvation number of HCl [136]; and provided the solvation number of one ion was assumed or taken from some other type of measurement, the solvation number of the other ion would thus be readily available. The Hudson–Saville [137] simple approach to the problem of ionic solvation in liquid mixtures was used. The treatment was similar to that of Robinson and Stokes [138] and of Glueckauf [139] in the case of concentrated aqueous electrolyte solutions.

The Born equation, which depends on an equation for the potential of a charged sphere in a uniform dielectric, cannot be expected to apply accurately for the case of an ion in solution since the ion is comparable in dimensions to the solvent molecules. Conditions close to dielectric saturation prevails close to the ion. The first layer of solvent molecules around the ion can be considered as completely orientated and hence treated as a firmly bound solvation shell, the formation of which as the ion enters the solution from the gas phase will be accompanied by a loss of free energy by the coordinated water molecules. An equation of the Born type could be used to calculate the remaining free energy change which was assumed to be relatively small especially in solutions of high dielectric constants. Also, in all solutions considered, the ions were assumed to be preferentially solvated by the more polar molecules, water in this case. This would be particularly true when the water content was high.

For the coordination of n water molecules when one mole of HCl, as ions dissolve from the gas state to a standard state in aqueous solution, was written [1]

$$(H+Cl)_{gas} + nW \rightleftarrows (H+Cl), nW \tag{3.27}$$

where H^+ and Cl^- ions are written without their charges and W stands for water. The free energy change for the above process was

$$\Delta G^W = \mu^W(H+Cl), nW - \mu(H+Cl) - n\mu^W W \tag{3.28}$$

For the same process in an organic solvent–water mixture, the free energy change was

$$\Delta G^S = \mu^S(H+Cl), nW - \mu(H+Cl) - n\mu^S W \tag{3.29}$$

Consider the difference

$$\Delta G = \mu^S(H+Cl), nW - \mu^W(H+Cl), nW \tag{3.30}$$

If the interaction of the solvated ions with the solvent were assumed to be negligible, the difference in the partial molal free energies of the solvated ions in the two solvents depended only on the difference in the concentrations of the solvated ions in the two solvents. This term was taken as zero when the standard state was either the mole fraction or the molar one. The selection of either scale probably did not make the assumption exactly true.

Thus,

$$\mu^S(H+Cl), nW = \mu^W(H+Cl), nW \tag{3.31}$$

Hence,

$$\Delta G^S - \Delta G^W = n(\mu^W W - \mu^S W) = -F(^S E^0 - {}^W E^0) \tag{3.32}$$

For the water in the aqueous mixture, the partial molal free energy was expressed in terms of mole fraction or of volume fraction statistics, and for simplicity, expressions were developed for ideal mixtures. If N_W is the mole fraction and ϕ_W the volume fraction of water, then

$$\mu^W W - \mu^S W = -RT \ln N_W = -RT \ln \phi_W \tag{3.33}$$

Now using $E_N{}^0$ for the mole fraction and $E_\phi{}^0$ for the volume fraction models, respectively,

$$^S E_N{}^0 = {}^W E_N{}^0 + n\frac{RT}{F} \ln N_W \tag{3.34}$$

and

$$^S E_c{}^0 = {}^W E_c{}^0 + n\frac{RT}{F} \ln \phi_W \tag{3.35}$$

From Eqs. (3.34) and (3.35),

$$^S E_m{}^0 = {}^W E_m{}^0 + nk \log w + k(n-2) \log \frac{M_{XY}}{M_Y} \tag{3.36}$$

$$^S E_m{}^0 = {}^W E_m{}^0 + nk \log w + k(n-2) \log \rho \tag{3.37}$$

where w is the weight fraction of water and ρ is the density. For solutions rich in water, $\log \rho$ and $\log(M_{XY}/M_Y)$ are small, and if n is also close to 2, both Eqs. (3.36) and (3.37) approximate to

$$^S E_m{}^0 = {}^W E_m{}^0 + nk \ln w \tag{3.38}$$

Eqs. (3.34) and (3.35) can be used to obtain the solvation number by plotting $^S E_c{}^0$ versus $\ln \phi_W$ and plotting $^S E_N{}^0$ versus $\ln N_W$. The plot of $^S E_N{}^0$ versus $\ln N_W$ was shown to not achieve any striking correlation and was not applied to data [140].

2. Experimental Technique

From potential measurements in pure and in mixed solvents on galvanic cells at various concentrations of electrolyte and at fixed temperatures, the standard potential of the cell in each solvent and at each temperature is calculated using standard procedures [140]. The volume fractions of the components of the solvent are determined from weight and density measurements or from volume measurements. Plots of $^S E_c{}^0$ versus $\ln \phi_W$ are made; and from the slope of the resulting straight line, the solvation number n is obtained.

3. Solvation Numbers Determined

In the case of the cell [136]

$$\text{Pt, } H_2(1 \text{ atm})|HCl|AgCl-Ag$$

the solvation number of HCl was determined to be 2.2. If the solvation number of one ion could be determined, that of the other ion would be fixed.

From the electromotive force of the proper cell in mixtures of $H_2O-MeCN$

of various compositions, the dependence of the free energy of solvation of Ag^+ ion on the composition of the H_2O–MeCN solvent was determined [141].

Potentiometric measurements were used to determine the stability constants of the 1:1 and 2:1 complexes of Ag-1,10-phenanthroline [142]. Other metal complexes of 1,10-phenanthroline were also studied.

By following the change of pH with time, hydration and polymerization of various vanadate ions were followed [143]. It was found that VO_4^{-3} was protonated in steps much as PO_4^{-3}. The H_3VO_4 formed in solution partially hydrated to an equilibrium ratio $V(OH)_5/H_3VO_4 \sim 3$.

The effects of ion hydration and the dielectric constant on standard potentials were investigated in H_2O–MeOH, H_2O–EtOH, and in H_2O–dioxane mixed solvents [144]. For certain dielectric constant range E^0 was linear with $1/D$. The range broadened with increasing molecular weight of the nonaqueous solvent. Deviations from linearity occurred in all cases where the content of water in the mixed solvents attained 17 moles/1000 g of mixed solvent. The deviations were, therefore, connected with the loss of coordination water by the ions. Oversolvation of the ions beyond their coordination numbers did not cause a deviation from linearity of the E^0 versus $1/D$ plot. E^0 represents standard potential and D represents the solvent dielectric constant.

Ion solvation in nonaqueous solvents and their aqueous mixtures have been studied using electromotive force techniques; and the molar free energies of transfer of some halogen acids and alkali chlorides, from water to 10 and to 43.12 weight% methanol–water mixtures, were separated into values for the ion constituents by extra thermodynamic assumptions [145, 146]. A structural theory of ion solvation was used to explain the ionic free energy of transfer values so obtained.

Equations were given for the dependence of the normal potential (E_N^0, E_m^0, and E_c^0) of electrolytic cells without transference on the dielectric properties of a mixed solvent, primary hydration of an electrolyte, and the equilibria of the cation and anion [147]. From electromotive force the constants are obtained for the transsolvation ($H_2O \rightarrow MeOH$) equilibria in water for Li^+, Na^+, K^+, and the primary hydration number of these cations. These numbers are listed in Table 3.5.

Electromotive force studies on the proper galvanic cells showed that in molten salt solutions containing water, the water molecules could be considered as ligands competing with bromide ions for the displacement of nitrate ions from the coordination spheres of cadmium ions [148]. It was also observed that bromide ions displace nitrate ions from the coordination sphere of cadmium ions more readily than they displace water molecules since the association constant for the reaction $Cd^{2+} + Br^- \rightarrow CdBr^+$ increased with decreasing water content. The observed increase was a measure of the tendency of the Cd^{2+} ions to become hydrated.

Table 3.5

*Solvation Numbers for Ions from Electromotive
Force Data[a]*

Solvated ion	Solvation number of reference ion	Ref.
Li^+	4 (primary)	[149]
Na^+	4 (primary)	[149]
K^+	4 (primary)	[149]
H^+	1 (primary)	[149]

[a] The solvent is water–methanol; the reference ion is Cl^-, solvation number 0. From [149].

A further study by the application of electromotive force measurements has been made on ion solvation in nonaqueous solvents and their aqueous mixtures, but specific solvation numbers were not listed [149].

4. Significance

The value of $n = 2.2$ for HCl using the cell potential measurement is lower than that found from other methods with the exception of diffusion [150] ($n = 2.1$). From activity coefficients, Stokes and Robinson [151] found $n = 8$ and Glueckaut [139] $n = 4.7$. The $^S E_N{}^0$ versus $\ln N_W$ plots give somewhat higher than 2.2, namely 2.7–5.0.

The electromotive force approach to determination of solvation numbers seems to result in primary shell solvation. From Table 3.5 it would appear that, as detected by this method, Li, Na, and K ions are equally hydrated in their primary shells.

D. THERMODYNAMICS

1. Theory, Experiment, and Data

Solvation from solvation energies must depend on a knowledge of individual ion solvation energies. The estimation of these individual solvation energies are based on some nonthermodynamic, generally theoretical, principle for a division of the observed thermodynamic function for an electrolyte consisting of two ions. The principles used involve the dependence of the energy quantity on some function of the reciprocal of the ionic radii.

One of the earliest attempt's at finding individual ionic energies divided the total heat of solvation of K^+F^-, namely, $\Delta H_{KF}^0 = -191$ kcal/mole into equal parts so that $\Delta H^0_{K^+} = \Delta H^0_{F^-} = -95.5$ kcal/mole. This value was adjusted to $\Delta H^0_{K^+} = -94$ kcal/mole and $\Delta H^0_{F^-} = -97$ kcal/mole to account for the different spatial distribution of the water molecules about the K^+ and F^- ions owing to the noncentral location of the dipole in the water molecule. Using these values, $\Delta H^0_{H^+}$ was found [152] to be -276 cal/mole. It has been suggested [153–156] that this division of enthalpies of solvation was over-simplified and not in keeping with Bernal and Fowler's more complex expression for the calculations of enthalpies and free energies of hydration.

Another method involved the use of the Born equation with empirically corrected radii to represent the free energies of the solvation of ions [157]. This method is believed [156] to be even less satisfactory than that of Bernal and Fowler [152].

The calculations of Eley and Evans [158], from a comparative study of the methods proposed up to 1953, have been selected [153, 154] as the most acceptable basis for the determination of the heats of hydration of individual ions. The values of $\Delta H^0_{K^+}$ and $\Delta H^0_{F^-}$ was given as -90 and -91 kcal/mole, respectively, Verwey [159] split the value for ΔH^0_{KF} into $\Delta H_{K^+} = -75$ and $\Delta H^0_{F^-} = -122$ kcal/mole.

Buckingham [160] proposed a more complete calculation of hydration enthalpy which included terms dealing with ion–dipole, dipole–dipole, nonquadripole interactions, and effects of induced moments.

Instead of using a model for the hydration of the individual ions to make a division of salt solvation energy, Halliwell and Nyburg [161] used an approach for obtaining $\Delta H^0_{H^+}$ involving the difference of conventional hydration energies of opposite charged pairs of ions of the same radii. The relations between absolute and conventional standard enthalpies of hydration for a cation M of valence z^+ and for an anion A of valence z^- are given by

$$\Delta \overline{H}^0_{M^{z+}} = \Delta H^0_{M^{z+}} - z^+ \Delta H^0_{H^+} \tag{3.39}$$

$$\Delta \overline{H}^0_{A^{z-}} = \Delta H^0_{A^{z-}} + z^- \Delta H^0_{H^+} \tag{3.40}$$

Where the $\Delta \overline{H}^0$ are the conventional relative enthalpies and the ΔH^0 are the absolute ionic enthalpies of the indicated species. Subtracting Eq. (3.40) from Eq. (3.39) yields for univalent electrolytes.

$$\Delta \overline{H}^0_{M^{z+}} - \Delta \overline{H}^0_{A^{z-}} = [\Delta H^0_{M^{z+}} - \Delta H^0_{A^{z-}}] - 2 \Delta H^0_{H^+} \tag{3.41}$$

Experimental values of the conventional heats of hydration have been compiled [162]. These experimental values of $\Delta H^0_{M^{z+}}$ and $\Delta H^0_{A^{z-}}$ can, respectively, be plotted versus some function of the radius [163–165] of cations and anions, and from these curves a single curve can be plotted for the left-hand side of Eq. (3.41) [41] as the same function of ionic radius. The radius used in this latter plot is not necessarily that for the radii of any particular ions since few cation-anion pairs have identical radii. Then, if the theoretical relationship between the conventional enthalpies of hydration and the inverse function of the radius against which they have been plotted is known, the extrapolation of the left hand side of Eq. (3.43) [41] to infinite radius should ideally yield $2\,\Delta H^0_{H^+}$ as the intercept on the $\Delta H^0_{M^{z+}} - \Delta H^0_{A^{z-}}$ axis.

It has been pointed out [156] that usually no single-powered function of the reciprocal of the hydrated radius, i.e., of the sum of the radius of the ion and the radius of the water molecule, represents uniquely the heats of hydration of ions. This arises from the different dependencies on the radius of the primary shell [153–155, 166] and outer region contributions. The former depends principally on r_h^{2-} where r_h is the hydrated radius; the latter depends on $(r_i + 2r_{H_2O})^{-1}$, where r_i is the radius of the ion and r_{H_2O} that of the water molecule. To obviate this difficulty Halliwell and Nyburg plotted the best, self-consistent values of the conventional relative enthalpies, $\Delta\bar{H}^0$, for cations and anions versus $(r_i + r_{H_2O})^{-3}$, where the value of r_{H_2O} was taken as 1.38 Å and the values of r_i were taken from Pauling [167] and Ahrens [168] except for a few special cases. It was concluded, after considering "hard-sphere" and "soft-sphere" ion-solvent contact models and also accounting for a possible coordination number of 4 or 6 in the primary shell, that the best value for $\Delta H^0_{H^+}$ is -260.7 ± 2.5 kcal/mole. The form of the function for large r_h has been criticized [156], however, the plot of the left-hand side of Eq. (3.43) versus $(r_i + r_{H_2O})^{-3}$ has been approved [156] since for cations and anions of the same radii, the supposed differences in the heats of solvation may be regarded [160] as arising mainly from the ion-quadripole interaction terms. Verwey's [159] orientation of water at anions and cations may cause some error in difference plot. Halliwell and Nyburg's difference plot can be used to obtain a maximum of $\Delta H^0_{H^+}$ of about 267 kcal/mole. High values [169] such as -280 or -292 kcal/mole for free energy [170, 171] or, in like manner, -302 kcal/mole for the enthalpy of hydration of the H^+ is considered unlikely [156]. Other data in the literature are: free energy, -259 [172] and enthalpy -265 [173] and -263 [174] kcal/mole. It is believed [166] that the method discussed above now limits the uncertainity in $\Delta H^0_{H^+}$ to no more than 7 kcal/mole.

Much work has been done [156, 175–177] on the heat and standard free energy of formation of $H_3O^+_{(g)}$ from $H^+_{(g)}$ and $H_2O_{(g)}$. The subscripts (g), refer to the gaseous state. Thermodynamic cycles involving H_3O^+, ClO_4^- and NH_4^+, ClO_4^-, which were assumed to have the same crystal lattice energies were used since they involve ions of the same charge. Proton affinity of water was assumed to be identical with the enthalpy change for the protonation of water at room temperature. The validity of this assumption has been discussed [156]. The value of 298°C of $\Delta H^0_{H_2O}(-P_{H_2O})$, $\Delta S^0_{H_2O}$, and $\Delta F^0_{H_2O}$ are recorded [156] as -170, kcal/mole, -27 e.u., and -162 kcal/mole.

a. Equilibrium Methods   The hydration of ions in the gas phase have been studied [178–182] Hydrogen ions have been investigated [182] in irradiated water vapor from 0.1 to 6 Torr and temperatures from 5 to 600°. The equilibrium constants $K_{(n-1),n}$ were calculated from the concentration of the species $A^+ \cdot nS$, where the ion A^+ is produced in the gas phase by some form of ionizing radiation or thermal means, and mass spectrometric measurements were made of the relative concentrations of the ionic species A^+S_n, A^+S_{n-1}, B^-S_n, B^-S_{n-1} where A^+ and B^- are positive and negative ions and S is a solvent molecule. In the equilibrium expression

$$K_{(n-1),n} = \frac{P_{A^+ \cdot nS}}{P_{A^+ \cdot (n-1)S} P_S} \tag{3.42}$$

the equilibrium pressures of the different species were substituted for by the ratio of the ion mass-spectrometric intensities of the corresponding ions. From the equilibrium constants the stepwise free energies $\Delta G^0_{n-1,n} \Delta H_{n-1,n}$ and $\Delta S^0_{n-1,n}$ was determined for the processes:

$$A^+ + S \rightleftarrows A^+ \cdot S \quad (0,1) \tag{3.43}$$

$$A^+ \cdot S + S \rightleftarrows A^+ \cdot 2S \quad (1,2) \tag{3.44}$$

$$A^+ \cdot (n-1)S + S \rightleftarrows A^+ \cdot nS \quad ((n-1),n) \tag{3.45}$$

In the case of hydrogen ion studies H_2O^+ and OH^+ ions were produced by irridiation of water vapor by electrons, protons and α-particles. These ions react rapidly [183–185] with water vapor to produce H_3O^+, which in turn reacts with water vapor to produce the hydrates shown in Table 3.6. The hydration reactions are exothermic and a third body would be necessary in Eqs. 3.44 and 3.45.

Enthalpies $\Delta H_{n-1,n}$, which are the enthalpies for the hydration of the $H^+ \cdot (n-1)H_2O$ ion to $H^+ \cdot nH_2O$ ion have been compared with values of $Na^+ \cdot nH_2O$ which have been recorded [186, 187] in the literature. It was found, assuming the radius of the central H_3O^+ ion was similar to Na^+ ion, that $\Delta H_{0,4}$ and $\Delta H_{0,6}$ for Na^+ ion which were -104 and -114, respectively, compared favorably with $\Delta H_{1,5}$ and $\Delta H_{1,7}$ for H_3O^+ which were -91 and -115. Using the heat of solvation of the proton in liquid water [166, 172], namely, -261 kcal/mole, and using the enthalpy

$$\Delta H_1 = \Delta H_{0,8} + 8 \Delta H_{evap}(H_2O)$$

which is -213 kcal/mole, the heat of solvation of $H^+(H_2O)_8(g)$ into liquid water (-48 kcal/mole) was found from the difference. Thus these data would verify the solvation numbers for Na^+ given in Table 3.6.

Table 3.6

Solvation of Various Ions by Different Thermodynamic Methods

Method (solvent)	Reference ion	Assumed solvation number of reference ion	Solvated ion	Solvation number of ion	Concentration of electrolyte (moles)	Phase	Temperature (°C)	Reference
Equilibrium (water)	—	—	H^+	1	—	gas	15–600°C	[182]
	—	—	H^+	2	—	gas	15–600°C	[182]
	—	—	H^+	3	—	gas	15–600°C	[182]
	—	—	H^+	4	—	gas	15–600°C	[182]
	—	—	H^+	5	—	gas	15–600°C	[182]
	—	—	H^+	6	—	gas	15–600°C	[182]
	—	—	H^+	7	—	gas	15–600°C	[182]
	—	—	H^+	8	—	gas	15–600°C	[182]
	—	—	Na^+	4	—	gas	15–600°C	[182]
	—	—	Na^+	6	—	gas	15–600°C	[182]
Integral heats (water)	—	—	Cs^+	0.031	—	solution	—	[200]
	—	—	NH_4^+	0.042	—	solution	—	[200]
	—	—	Li^+	0.065	—	solution	—	[210]
	—	—	Na^+	0.053	—	solution	—	[210]
Activity coefficient (water)	Cs^+	0	Rb^+	0	1	—	—	[217]
	Cs^+	0	NH_4^+	0.2	1	—	—	[217]
	Cs^+	0	K^+	0.6	1	—	—	[217]
	Cs^+	0	Li^+	3.4	1	—	—	[217]
	Cs^+	0	H^+	3.9	1	—	—	[217]
	Cs^+	0	Ba^{2+}	3.0	0.7	—	—	[217]
	Cs^+	0	Sr^{2+}	3.7	0.7	—	—	[217]
	Cs^+	0	Ca^{2+}	4.3	0.7	—	—	[217]

Ion A		Ion B	Value	Conc. range		Temp.	Ref.
Cs⁺	0	Mg²⁺	5.1	0.7	—	—	[217]
Cs⁺	0	Zn²⁺	5.3	0.7	—	—	[217]
Cs⁺	0	UO₂²⁺	7.35	0.7	—	—	[217]
Cs⁺	0	La³⁺	7.5	0.7	—	—	[217]
Cs⁺	0	Al³⁺	11.9	0.7	—	—	[217]
Cs⁺	0	NO₃⁻	0	1	—	—	[217]
Cs⁺	0	ClO₄⁻	0.3	—	—	—	[217]
Cs⁺	0	Cl⁻	0.9	—	—	—	[217]
Cs⁺	0	Br⁻	0.9	—	—	—	[217]
Cs⁺	0	I⁻	0.9	—	—	—	[217]
Cs⁺	0	RSO₃⁻	0.9	—	—	—	[217]
Cs⁺	0	F⁻	1.8	—	—	—	[217]
Cs⁺	0	C₂H₃O₂⁻	2.6	—	—	—	[217]
Cs⁺	0	OH⁻	4.0	—	—	—	[217]
Cl⁻	0	H⁺	8	0.01–1.0	—	25	[216]
Br⁻	0	H⁺	8.6	0.1–1.0	—	25	[216]
I⁻	0	H⁺	10.6	0.1–0.7	—	25	[216]
ClO₄⁻	0	H⁺	7.4	0.1–2.0	—	25	[216]
Cl⁻	0	Li⁺	7.1	0.1–1.0	—	25	[216]
Br⁻	0	Li⁺	7.6	0.1–1.5	—	25	[216]
I⁻	0	Li⁺	9.0	0.1–1.0	—	25	[216]
ClO₄⁻	0	Li⁺	8.7	0.2–1.0	—	25	[216]
Cl⁻	0	Na⁺	3.5	0.1–5.0	—	25	[216]
Br⁻	0	Na⁺	4.2	0.1–4.0	—	25	[216]
I⁻	0	Na⁺	5.5	0.1–1.5	—	25	[216]
ClO₄⁻	0	Na⁺	2.1	0.2–4.0	—	25	[216]
Cl⁻	0	K⁺	1.9	0.1–4.0	—	25	[216]
Br⁻	0	K⁺	2.1	0.1–4.0	—	25	[216]
I⁻	0	K⁺	2.5	0.1–4.0	—	25	[216]
Cl⁻	0	Rb⁺	1.2	0.1–1.5	—	25	[216]
Br⁻	0	Rb⁺	0.9	0.1–1.5	—	25	[216]

Table 3.6—continued

Method (solvent)	Reference ion	Assumed solvation number of reference ion	Solvated ion	Solvation number of ion	Concentration of electrolyte (moles)	Phase	Temperature (°C)	Reference
Activity coefficient (water)	I^-	0	Rb^+	0.6	0.1–1.5	—	25	[216]
	Cl^-	0	Mg^{2+}	13.7	0.1–1.4	—	25	[216]
	Br^-	0	Mg^{2+}	17.0	0.1–1.0	—	25	[216]
	I^-	0	Mg^{2+}	19.0	0.1–0.7	—	25	[216]
	Cl^-	0	Ca^{2+}	12.0	0.1–1.4	—	25	[216]
	Br^-	0	Ca^{2+}	14.6	0.1–1.0	—	25	[216]
	I^-	0	Ca^{2+}	17.0	0.1–0.7	—	25	[216]
	Cl^-	0	Sr^{2+}	10.7	0.1–1.8	—	25	]216]
	Br^-	0	Sr^{2+}	12.7	0.1–1.4	—	25	[216]
	I^-	0	Sr^{2+}	15.5	0.1–1.0	—	25	[216]
	Cl^-	0	Ba^{2+}	7.7	0.1–1.8	—	25	[216]
	Br^-	0	Ba^{2+}	10.7	0.1–1.5	—	25	[216]
	I^-	0	Ba^{2+}	15.0	0.1–1.0	—	25	[216]
	Cl^-	0	Mn^{2+}	11.0	0.1–1.4	—	25	[216]
	Cl^-	0	Fe^{2+}	12.0	0.1–1.4	—	25	[216]
	Cl^-	0	Co^{2+}	13.0	0.1–1.0	—	25	[216]
	Cl^-	0	Ni^{2+}	13.0	0.1–1.4	—	25	[216]
	ClO_4^-	0	Zn^{2+}	20.0	0.1–0.7	—	25	[216]
Solubility (water–alcohol)	—	—	Ag^+	2	—	—	—	[222]
	—	—	Tl^+	4	—	—	—	[222]
	—	—	Br^-	3	—	—	—	[222]
	—	—	Cl^-	0	—	—	—	[222]
Acidity functions (water)	—	—	ArH_3N^+	3	—	—	—	[223]
	—	—	$Ar_2H_2N^+$	2	—	—	—	[223]

Colligative properties (sulfuric acid)

Anion		Cation					Ref.
HSO_4^-	0	Li^+	2.3	—	—	—	[225]
HSO_4^-	0	Na^+	3.0	—	—	—	[225]
HSO_4^-	0	K^+	2.1	—	—	—	[225]
HSO_4^-	0	NH_4^+	1.2	—	—	—	[225]
HSO_4^-	0	Ag^+	2.1	—	—	—	[225]
HSO_4^-	0	H_3O^+	1.8	—	—	—	[225]
HSO_4^-	0	Ba^{2+}	6.5	—	—	—	[225]

Colligative properties (water)

Anion		Cation					Ref.
Cl^-	4	NH_4^+	2.2	1	—	—	[228]
Cl^-	4	NH_4^+	3	0.5	—	—	[228]
Cl^-	4	K^+	4.6	1	—	—	[228]
Cl^-	4	K^+	6.2	0.5	—	—	[228]
Cl^-	4	Ba^{2+}	22.1	0.5	—	—	[230]
Cl^-	4	Ba^{2+}	25.1	0.25	—	—	[230]
Cl^-	4	Ni^{2+}	24.5	0.5	—	—	[231]
Cl^-	4	Ni^{2+}	29.1	0.25	—	—	[231]
K^+	0	Li^+	10.3	—	—	—	[232]
K^+	0	Na^+	3	—	—	—	[232]
K^+	0	NH_4^+	0	—	—	—	[232]
K^+	0	H^+	9	—	—	—	[232]
K^+	0	Br^-	1	—	—	—	[232]
K^+	0	I^-	2	—	—	—	[232]
Cl^-	4	Li^+	11	1	—	—	[233]
Cl^-	4	Li^+	13.8	0.5	—	—	[233]
K^+	4.6	Br^-	3.5	1	—	—	[234]
K^+	6.2	Br^-	3.4	0.5	—	—	[234]
Cl^-	4	Ca^{2+}	23	0.5	—	—	[235]
Cl^-	4	Ca^{2+}	26.6	0.25	—	—	[235]
Cl^-	4	Sr^{2+}	22.7	0.5	—	—	[236]
Cl^-	4	Sr^{2+}	26.3	0.25	—	—	[236]
Cl^-	4	Mg^{2+}	24.1	0.5	—	—	[237]
Cl^-	4	Mg^{2+}	27.6	0.25	—	—	[237]

Table 3.6—continued

Method (solvent)	Reference ion	Assumed solvation number of reference ion	Solvated ion	Solvation number of ion	Concentration of electrolyte (moles)	Phase	Temperature (°C)	Reference
Colligative properties (water)	Br^-	3.5	Na^+	9.4	1	—	—	[238]
	Br^-	3.5	Na^+	11.8	0.5	—	—	[248]
	K^+	6.2	I^-	2.4	1	—	—	[249]
	Cl^-	4	H^+	10.5	0.5	—	—	[250]
Entropy (water)	—	—	H^+	5	—	—	—	[251, 252]
	—	—	H^+	5	—	—	—	[253]
	—	—	Li^+	5	—	—	—	[189]
	—	—	Na^+	4	—	—	—	[189]
	—	—	K^+	2	—	—	—	[189]
	—	—	Rb^+	2	—	—	—	[189]
	—	—	F^-	5	—	—	—	[189]
	—	—	Cl^-	2	—	—	—	[189]
	—	—	I^-	0.5	—	—	—	[189]

From solubility [188] it was concluded that NaCl solutions consist of NaCl·2H$_2$O aggregates separated by water. Added HCl effectively removes 10 water molecules per HCl molecule and correspondingly reduces the solubility of NaCl. If the number of water molecules is reduced so that they are not numerous enough to separate the aggregates the aggregates combine and precipitate.

The entropy of neutralization of HCl can be calculated on the assumption that the ten water molecules surrounding each HCl molecule is composed of two layers of five water molecules each, and that two of the water molecules in the outside layer are replaced by two NaOH molecules. Thus in the systems of ten water molecules, there is the replacement of two adjacent molecules among five. The replacement can be accomplished in ten ways giving an entropy

$$\Delta S = 10R \ln 10 = 46 \quad \text{e.u.} \tag{3.46}$$

which agrees well with the value calculated from the heat of neutralization at 18°C which is 13,800 cal/mole. This gives, neglecting the change in heat content and the corresponding entropy change, 13,800/291 = 47 e.u.

The methods discussed in the preceding two paragraphs deal with solvation numbers of the electrolyte as a whole and not with the solvation numbers of individual ions.

It has been pointed out [189] that solvation numbers can be found from the entropy decrease which takes place when gaseous ions are dissolved in infinitely dilute solutions; and that entropy decrease, in turn, can be calculated from the observed heat effects. The decrease in entropy is caused by the free solvent molecules entering into tightly held solvation sheaths and thus losing some of their entropy. If it is assumed that this decrease in entropy is the same as the change in entropy when free water becomes bound as water of crystallization, in solid hydrates, the entropy of dissolution of an ion can be used to calculate its hydration number.

It is reasoned [190] that multiply charged ions or small, singly charged ions increase the viscosity of water and decrease the translational motion of the water molecules nearest them, while large singly charged ions decrease the viscosity of water and increase the translational motion of water molecules nearest them. This decrease in viscosity is termed negative viscosity. Since in this case the water molecules in the vicinity of the ions are more mobile than in pure water, and since these molecules exchange more frequently than water molecules in the neighborhood of other molecules in water, this phenomenon is termed "negative hydration" by analogy with negative viscosity. The ions which exhibit negative hydration are those which lower the structural temperature of water.

It was pointed out [190] that there is a qualitative dependence of the activity coefficient of water on ion hydration, as is shown by the experimental data that cations with negative hydration are associated with activity co-efficients of water that are greater than unity over a considerable concentration of electrolyte. On the other hand, the activity coefficients of water in solutions with hydrated cations are less than unity except for a relatively small region at lower concentrations.

   b. Thermochemical Method for Coordination Numbers   A thermochemical method of estimating the coordination number of an ion in dilute aqueous solutions, i.e., of obtaining the average number of constantly exchanging water molecules in the solution which form the immediate surroundings of the ion has been devised [191–195]. This method consists of measuring the integral heat of solution of salts in water and in aqueous solutions of acids of various concentrations. The integral heat of solution of a constant concentration of salt in different concentrations of acid is determined and used in calculating the coordination numbers of the ions in dilute solutions. The heats of solution of anhydrous and of hydrated salts were measured and the difference in the two heats were taken as the heats of hydration of the anhydrous salts.

   The protons from the acid and the proton charge were considered to be statistically distributed over the water molecules so that each water molecule became an ion with the average electronic charge of ε/e and the average protons per ion of ε/e, where ε is the average positive charge per ion and e is the electronic charge. Thus, the formula of the ions from water was written $H_{2+\varepsilon/e}O^{(\varepsilon/e)^+}$. The Coulombic interaction energies between these ions and cations and anions are, respectively,

$$E_k = -n_k \varepsilon e / R_k \tag{3.47}$$

and

$$E_a = n_a \varepsilon e / R_a \tag{3.48}$$

where R_k and R_a are the distances between the charges of the water ions in the first coordination spheres and the centers of the cations and anions, respectively, and n_k and n_a are the coordination numbers of the cations and anions, respectively. The minus sign corresponds to repulsion (reduction of exothermic heat of solution) between positive ions. If R_k is taken as $r_k + 1.38$ Å and R_a as $r_a + 1.38$ Å, where r is the crystal radius of an ion; and if ε is taken as proportional to the molality of the acid, that is $\varepsilon = km$, then Eqs. (3.47) and (3.48) can be written, respectively, as

$$E_k = -\frac{n_k kme}{r_k + 1.38} \cdot 10^8 \tag{3.49}$$

and

$$E_a = \frac{n_a kme}{r_a + 1.38} \cdot 10^8 \tag{3.50}$$

in which the units of the r's are now centimeters and the energies are in ergs per ion. To convert to calories per mole of salt, we take the sum of the two energies multiplied by Avogadro's number and by the factor for converting ergs to calories (2.389×10^{-8}); then setting $K = 2.389 Nek$ we can write

$$E = -K\left(\frac{n_k}{r_k + 1.38} - \frac{n_a}{r_a + 1.38}\right)m \tag{3.51}$$

where E is in calories per mole of salt and is greater than zero.

The integral heat of solution L of the salt in m molal HCl solutions is

$$L = L_0 - K\left(\frac{n_k}{r_k + 1.38} - \frac{n_a}{r_a + 1.38}\right)m \tag{3.52}$$

where L_0 is integral heat of solution of salt to the same concentration in pure water. Therefore,

$$\beta = -\frac{\Delta L}{\Delta m} = K\left(\frac{n_k}{r_k + 1.38} - \frac{n_a}{r_a + 1.38}\right) \tag{3.53}$$

The difference between two salts with the same anion but different cations [Eq. (3.53)] yields

$$\beta_1 - \beta_2 = K\left(\frac{n_{k_1}}{r_{k_1} + 1.38} - \frac{n_{k_2}}{r_{k_2} + 1.38}\right) \tag{3.54}$$

For a bi-univalent salt, a factor two would have to be introduced into Eqs. (3.53) and (3.54) and a factor $\beta' = \beta/2$ would be used. The values of β are approximately additive for salts and can be written $\beta = \delta_k - \delta_a$, where

$$\delta_k = K\frac{n_k}{r_k + 1.38}; \qquad \delta_a = K\frac{n_a}{r_a + 1.38} \tag{3.55}$$

and these relations do not change for various valence types of salts if β' is used for β. If K can be found, the coordination numbers of ions can be found from Eqs. (3.53) and (3.54). With this in view, the relation of coordination number to hydration can be considered.

Coordination numbers are not directly related to individual ion hydration numbers. The density of the molecular arrangement of water around the ions must be accounted for. The surface density ρ' of packing of water molecules in the first coordination layer of ions is directly associated with hydration. If n is the coordination number of the ion r its radius, and $4\pi(r + 1.38)^2$ is the area of the first coordination sphere (in Å^2), then

$$\rho' = \frac{n}{4\pi(r + 1.38)^2} \tag{3.56}$$

This can be compared with ρ' for the water molecule in water:

$$\rho'_{H_2O} = \frac{n_{H_2O}}{4\pi R^2} \tag{3.57}$$

where n_{H_2O} is the average coordination number of water molecules in water, and R is the distance between centers of neighboring water molecules in water. Morgan and Warren [196] using X rays have found n_{H_2O} for water at 25°C to be 4.6 and R to be 2.90 Å. Hence $\rho'_{H_2O} = 0.044$ Å^2.

Hydrated ions, such as Na$^+$, $\Delta E > 0$, cause the frequency of activated jumps in their vicinity to be less than that in pure water since they weaken the translational motion of the nearest water molecules. For such ions

$$\rho' > \rho'_{H_2O} \tag{3.58}$$

Ions, such as K$^+$, Cl$^-$, Br$^-$, I$^-$, showing negative hydration have $\Delta E < 0$, and cause the frequency of activated jumps of neighboring molecules to be greater than in pure water. For these ions

$$\rho' < \rho'_{H_2O} \tag{3.59}$$

Inequalities 58 and 59 permit the estimation of the lower limits for the coordination numbers of hydrated ions and the upper limits for ions which show negative hydration. Thus, $n_{Na^+} > 3.05$; $n_{K^+} < 4.02$; $n_{Cl^-} < 5.57$. Using the distinction between hydrated and nonhydrated ions (negative hydration), limiting values for coordination numbers can be estimated.

The lower limit for K in Eqs. (3.53) and (3.54) was found [190] from the β-values for NaCl and NaBr (NaBr·2H$_2$O). From Eq. (3.53) for NaCl, the corresponding equation for NaBr was subtracted, giving

$$K\left(\frac{n_{Br^-}}{3.34} - \frac{n_{Cl^-}}{3.19}\right) = 68 \tag{3.60}$$

where $r_{Br^-} + 1.38$ Å $= 3.34$, $r_{Cl^-} + 1.38$ Å $= 3.19$, and $\beta_{NaCl} - \beta_{NaBr} = 210 - 142 = 68$. From Eq. (3.60) $n_{Br^-} > n_{Cl^-}$ since $K > 0$. However, from heats of hydration of Cl$^-$, Br$^-$, and I$^-$ ions and also from certain properties of chlorides, bromides and iodides, it is reasoned that negative hydration increases in this sequence; and that, therefore, $\rho'_{Br^-} < \rho'_{Cl^-}$. Hence,

$$\frac{n_{Br^-}}{4\pi(3.34)^2} < \frac{n_{Cl^-}}{4\pi(3.19)^2} \tag{3.61}$$

and $n_{Br^-} < 1.096 n_{Cl^-}$. Substituting this value of n_{Br^-} into Eq. (3.60) yields

$$K\left(\frac{1.096 n_{Cl^-}}{3.34} - \frac{n_{Cl^-}}{3.19}\right) > 68 \tag{3.62}$$

or

$$K > \frac{68}{0.0147 n_{Cl^-}} \tag{3.63}$$

Substituting the upper limit for n_{Cl^-} into Eq. (3.63), gives the lower limit of K. In this calculation, the equation for KCl

$$K\left(\frac{n_{K^+}}{2.71} + \frac{n_{Cl^-}}{3.19}\right) = 11 \tag{3.64}$$

was used. Here $r_{K^+} + 1.38 = 2.71$, $r_{Cl^-} + 1.38 = 3.19$ and 11 is the experimental value of β for KCl. Since β and K are positive

$$\frac{n_{K^+}}{2.71} - \frac{n_{Cl^-}}{3.19} > 0 \tag{3.65}$$

It was shown that $n_{K^+} < 4.02$ and hence $n_{Cl^-} < 4.74$. This is smaller than the value ($n_{Cl^-} < 5.57$) obtained from the inequality $\rho'_{Cl^-} < \rho'_{H_2O}$. Substituting $n_{Cl^-} = 4.74$ in Eq. (3.63) yields $K > 976$.

Knowing the lower limit for K, the lower limit for the coordination number of K^+, n_{K^+}, can be estimated. Equation (3.54) for NaCl and KCl can be written

$$K\left(\frac{n_{Na^+}}{2.36} - \frac{n_{K^+}}{2.71}\right) = 199 \tag{3.66}$$

In this equation, $r_{Na^+} + 1.38$ Å $= 2.36$, the quantity 2.71 was identified in Eq. (3.64) and $\beta_{NaCl} - \beta_{KCl} = 210 - 11 = 199$. Solving Eq. (3.66) for n_{K^+} gives

$$n_{K^+} = 2.71\left(\frac{n_{Na^+}}{2.36} - \frac{199}{K}\right) \tag{3.67}$$

Since $n_{Na^+} > 3.05$ and $K > 976$, then $n_{K^+} > 2.95$ from Eq. (3.67). From Eq. (3.53) the upper limit of n_{Li^+} can be found and is 4.23.

A plot of the differential heat of solution ΔH, for HCl versus m gives a slope

$$\beta = \Delta H/\Delta m = 233 \quad \text{cal/mole/[m]} \tag{3.68}$$

This slope can be used to narrow the limits of the n_{K^+} values and to estimate K.

The interaction of $H_{2+\varepsilon/e}O^{\varepsilon/e^+}$ with each other and with chloride ions determine the difference in heat content of a mole of HCl in infinitely dilute and in m molal solutions. The ΔH versus m plot indicated that an equation similar to Eq. (3.53) is satisfied. Thus

$$K\left(\frac{n_{H^+}}{R} - \frac{n_{Cl^-}}{3.19}\right) = 233 \tag{3.69}$$

where n_{H^+} is the coordination number of water molecules in water and has the value of 4.6 at 25°C; R is the average distance of separation of neighboring molecules in water and from X-rays is 2.90 Å. n_{H^+} also denotes the coordination number of the $H_{2+\varepsilon/e}O^{\varepsilon/e^+}$ ions which are not in the vicinity of chloride ions.

If Eq. (3.53) for KCl is subtracted from Eq. (3.69),

$$K\left(\frac{4.6}{2.90} - \frac{n_{K^+}}{2.71}\right) = 222 \qquad (3.70)$$

where $222 = \beta_{HCl} - \beta_{KCl} = 233 - 11$. But $K > 976$, and therefore from Eq. (3.70) $n_K > 3.68$. Let $n_{K^+} = \frac{1}{2}(3.68 + 4.02) = 3.8$. Substituting this value of n_{K^+} into Eq. (3.70) gives 1.17×10^3 for K.

Using these values of n_{K^+} and K in Eqs. (3.53) and (3.54), the values for the coordination numbers n for various ions were obtained and are listed in Table 3.7.

Table 3.7

Coordination Numbers (n) for Various Ions in Dilute Aqueous Solutions at 25°C

Ion	n	Ref.	Ion	n	Ref.	Ion	n	Ref.
Li^+	3.8	[191]	Cs^+	3.6	[200]	NO_3^-	6	[211]
Na^+	3.7	[191]	NH_4^+	4.2	[200]	Li^+	4	[213]
K^+	3.8	[191]	ClO_4^-	6.4	[200]	Na^+	6	[213]
Rb^+	3.8	[191]	Rb^+	3.8	[203]	K^+	8	[213]
Cs^+	3.8	[191]	Cs^+	3.5	[203]	Rb^+	8	[213]
Cl^-	4.4	[191]	K^+	3.8	[204]	Cs^+	8	[213]
Br^-	4.8	[191]	HSO_4^-	4.5	[204]	F^-	6	[213]
I^-	5.3	[119]	Li^+	3.4	[205]	Cl^-	8	[213]
Be^{2+}	3.3	[191]	Na^+	3.6	[205]	Br^-	8	[213]
Mg^{2+}	3.8	[191]	K^+	4.0	[205]	I^-	8	[213]
Ca^{2+}	4.0	[191]	NH_4^+	4.8	[205]	Na^+	4	[214]
Sr^{2+}	4.4	[191]	Rb^+	4.2	[205]	ClO_4^-	8	[214]
Ba^{2+}	4.8	[191]	Cs^+	4.2	[205]			
			Cl^-	4.7	[205]			
			Br^-	4.9	[205]			
			I^-	5.2	[205]			

From the table it can be seen that the coordination number of the monatomic ions in dilute aqueous solutions are close in magnitude to the average coordination number (4.6) at 25°C of the water molecule in water. It is concluded [190, 197, 198] that the water structure controls the structure of dilute aqueous solutions. The state of the ions studied in dilute aqueous solutions correspond to the envelopment of the ions by water molecules with the smallest possible change of water structure [198].

From the integral heats of solution of CsI, NH_4Cl, and $NaClO_4$ in HCl and $NaBr \cdot 2H_2O$ in HBr and $HClO_4$ the coordination numbers were calculated [199] for the ions listed in Table 3.7. The coordination number of ClO_4^- given was that for which water molecules were considered to be perpendicular to the walls of the perchlorate tetrahedron. If the water molecules were considered on the Cl–O line the value of the coordination number is given [200] as 7.3. Solvation numbers for Cs^+ and NH_4^+ ions are given and are listed in Table 3.6.

Criteria for negative hydration of ions has been derived [201]. It has been concluded [202, 203] applying the thermochemical method that Rb^+ and Cs^+ ions belong to the class of ions characterized by negative hydration. The coordination numbers of these ions are listed in Table 3.7. Coordination numbers for the K^+ and HSO_4^- ions have been found [204] and are also listed.

From a study of the temperature dependence of the heats of solution of various electrolytes, the coordination numbers recorded in Table 3.7 were found [205]. Integral heat of solution data [206] on the coordination number of UO_2^{2+} ion seems to confirm Sutton's [207] view of $6H_2O$ molecules surrounding UO_2^{2+} ions in water solutions. The changes with temperature of coordination numbers of Li^+, Na^+, K^+, Rb^+, Cs^+, and Cl^- ions were found [208] to be respectively, -0, 44, 0.004, 0.10, 0.25, 0.49 and -0.074 per 10°C. The heats of solution of K_2SO_4 in aqueous solutions of HCl was measured [209] in aqueous solutions of HCl at 25° in an isothermal calorimeter. The coordination numbers of the ions were obtained. Coordination numbers identical to those listed in Table 3.7 from Samoilow [190] are given [210] together with the solvation numbers listed in Table 3.6 by Samoilov [210].

By thermochemical methods the coordination number (see Table 3.2) of NO_3^- ion was determined and its arrangement shown [211]. The effect of coordination number on thermochemical and thermodynamic properties of solutions also has been studied [212].

Heats of hydration of salts have been selected [213] on the basis of the heats of solution at infinite or very high dilution and values of lattice energies. To find the ionic heats, equal heats were ascribed to the Cs^+ and I^- ions. In calculating the coordination numbers of the ions from the radii of the ions and the radius r_w of water, the latter value was taken as 1.93 Å corresponding to liquid water rather than 1.38 Å on the assumption of a "frozen" hydrate envelope. The coordination numbers obtained are listed in Table 3.7. The most probable coordination numbers of Na^+ and ClO_4^- ions in acetone were determined [214] and are recorded in Table 3.7. The heat of proton solvation has been studied [215].

c. Activity Coefficient Methods   The application of volume statistics to a model similar to that of Stokes and Robinson [216] has resulted in the derivation of a new equation for osmatic and activity coefficients [217]. It is stated that the "hydration numbers" obtained do not show the anomalies exhibited by the Stokes–Robinson hydration parameters. These workers neglecting the covolume effect in dealing with ionic hydration parameters have obtained n values which are greater than any acceptable figure for the ionic hydration number, and which vary considerably according to the anion and in a direction opposite to that expected [217].

Using volume fraction instead of the customary mole fraction statistics the two factors, hydration and covolume effects, can be combined into a single theory. The use of a mean hydration number h (or the Stokes–Robinson parameter n) need not imply a sharp difference between "free" and "bound" water. For statistical purposes the weak association of a large number of water molecules with an ion [218] can be considered as a strong adsorption of a small fixed number [216]. Using these assumptions and applying the Gibb's free energy for the solution of a hydrated electrolyte in volume fraction statistics, the equation for the mean activity coefficient $\gamma_\pm$ for an electrolyte was obtained and can be written

$$\log \gamma_\pm = \log \gamma_\pm^{el} + \frac{0.018mr(r+h-v)}{23v(1+0.018mr)} + \frac{h-v}{v}$$

$$+ \log(1+0.018mr) - \frac{h}{v}\log(1-0.018mh) \qquad (3.71)$$

where h is the hydration number of the electrolyte, v is the number of ions per molecule of electrolyte, $r = \phi_v/v_w{}^0$, ϕ_v is the apparent molar volume of the unhydrated electrolyte, $v_w{}^0$ is the mole volume of pure water, m is the molality, and $\gamma_\pm^{el}$ is the electrostatic contribution to the mean activity coefficient.

Equation (3.71) can be used to calculate hydration numbers from recorded data on activity coefficients [219, 220]. In the calculations [216], the r used was constant and applied to a concentration $m = 1$ for 1:1 electrolytes and $m = 0.7$ for 2:1 electrolytes since at these concentrations Eq. (3.71) applies to all electrolytes. To find the hydration numbers of the individual ions it was assumed [217] that the halogen ions have a hydration value of 0.9 and cesium has a hydration value of zero. The hydration values of the individual ions are listed in Table 3.6.

Bernal and Fowler assumed that the large anions were unhydrated. Stokes and Robinson [216] accepted the idea that it is the cations rather than the anions that are hydrated. Their hydration number of cations listed in Table 3.6 are substantially greater than those based on the treatment of Bernal and Fowler [152] of apparent molal volumes in dilute solution or of those given by Gleuckauf [217]. The increase in hydration number of a given cation with increasing anion size was discussed in relation to the binding of water molecules not simply by the fields of isolated ions, but rather by the resultant field of an ion and its neighbors which naturally depends on the dimensions of the ions.

The anomalies of large hydration numbers of cations and of their increase with increasing anion size could tend to be removed by consideration of the volumes of hydrated ions if these volumes could be adequately dealt with [221].

Solubility and activities of electrolytes were measured [222] in mixed solvents, and the hydration numbers of the electrolytes and of ions and trans-solvation constants calculated. The hydration numbers of the ions are listed in Table 3.6.

Seven arylmethanols were used [223] to establish the acidity function H_0 in 44–64% H_2SO_4. The acidity function H_0 was obtained in 60 to 75% H_2SO_4 using three amines. Two other amines were also used to establish H_0. The data was used to establish the number of water molecules solvating the cations. Solvation numbers given are listed in Table 3.6.

d. Colligative Properties   Accurate freezing point depressions of solutions of metal hydrogen sulfates in sulfuric acid have been made [224] in order to determine the deviations from ideality of these solutions. Osmotic co-efficients ϕ were calculated using the equation

$$\phi = \theta(1+0.002\theta)/6.12 \sum m_{ij} \tag{3.72}$$

where θ is the molal freezing point depression measured from the freezing point of the hypothetical undissociated sulfuric acid and $\sum m_{ij}$ the molality summed over the various ion types in the solution. The standard state used in the calculations of ϕ was the hypothetical undissociated acid. Pure sulfuric acid has a minimum ionic strength μ of 0.0357 and a ϕ of 0.98. Its dielectric constant is 120 at the freezing point. This high dielectric constant does not necessarily mean that solutions in this solvent will be ideal.

The variation of the osmotic coefficient with molality is given by the equation

$$\phi = 1 + \phi_{el} + b \sum m_i \tag{3.73}$$

where ϕ_{el} is the electrostatic interionic contribution to the osmotic co-efficient, $\sum m_i$ is the sum of the molalities of the ionic species and b is a parameter related to the solvation numbers by the equation

$$b = [(r+s)^2/40.8] - (r/20.4) \tag{3.74}$$

where r is the ratio of the apparent molar volume of the electrolyte to the molar volume of the solute, 54 cm^3. The Debye–Huckel theory is used to calculate ϕ_{el} using an appropriate distance of closest approach a for the dielectric constant $\varepsilon = 120$ for sulfuric acid. Both b and a are therefore adjustable parameters. Thus Eq. (3.74) can be used to calculate solvation numbers for ions from freezing point data. Solvation numbers of ions have been calculated [225] and are listed in Table 3.6. The solvation numbers of the cations were referred to the common anion $HSO_4{}^-$ as unsolvated.

Ionic substances in sulfuric acid solutions using proton magnetic resonance measurements have been studied [225] and show that the cation solvation numbers are in the order $Na^+ > NH_4^+ \approx K^+ > Tl^+$.

Much work [226–243] has been done on the ebuilloscopic and cryoscopic determination of the total hydration of ions and the hydration of formula weights of salts. Assuming that the hydration number of one ion (reference ion) is known, the results of these studies are given in Table 3.6.

A method has been devised [232] for obtaining solvation and association from colligative properties by graphical means. These were applied to the total solvation of $GaCl_3$ in CH_3Cl, proving the existence [233] of $CH_3ClGaCl_3$ and to the AlI_3–CH_3I and $AlBr_3$–C_6H_6 systems [244, 245]. The method confirmed that a 1:1 complex existed in the AlI_3–CH_3I system but not in the $AlBr_3$–C_6H_6 system.

e. Entropy of Hydration   Several authors [246–249] have discussed entropy of hydration in conjunction with entropies of aqueous ions. The standard state chosen for the calculation of ionic entripies of ions in aqueous solution is the hypothetical one molal solution, i.e., a 1 molal solution of the ion obeying the perfect solution laws and the ions possessing the same partial molal heat content that they have at infinite dilution. Relative values of the ionic entropies are recorded. These relative ionic entropies are referred to $S^0_{298.1}$ of H^+ taken as zero.

The most direct method of obtaining ionic entropies in solution is to sum the ΔS of solution and the entropy of the solid salt. The ΔS of solution is calculated from the free energy and enthalpy of solution using the equation

$$\Delta G = \Delta H - T\,\Delta S \qquad (3.75)$$

where ΔG and ΔH are the free energy and enthalpy of solution, respectively. In recording the relative entropies referred to H^+ taken a zero, rather than these sums are frequently given.

It has been pointed out [246, 248] that the so-called entropy of hydration is a function of the size and charge of the ion:

$$\Delta S_{\text{hydration}} = f(e^2/r) \qquad (3.76)$$

where e is the charge and r the radius of the ion. The entropy of hydration is defined as the difference between the partial molal entropy of the ion in solution and the entropy of the ion in the gaseous state. This latter entropy, provided it is all translation, can be calculated from the Sackur equation

$$S_{298.1} = \tfrac{3}{2}R \ln M + 26.03 \qquad (3.77)$$

where R is the gas constant in calories/mole/degree, M is the molecular weight, and 26.03 is the constant for the gas at 1 atm and 298.1°K.

From straight-line plots of ionic entropies versus the reciprocals of the ionic radii as given by Pauling [250], it was concluded [248] that specific hydration effects are small compared to the electrostatic action of the charge on the water dipoles. In fact it was concluded earlier [246] that specific hydration effects do not exist, and that the chemical properties of these solutions of ions are those simply of a charged sphere of a given size in a medium of a certain dielectric constant.

By ionic entropy measurements [251, 252] the hydration number of hydrogen ion was found to be that given in Table 3.6. Values of the hydration number of various ions determined from ionic entropies have been listed [190, 253] and are given in Table 3.6.

From the heats and free energies of several metal ions, the entropies of ammoniation have been calculated and the results discussed [254] in relation to the hydration numbers of ions. The idea [192] that when a water molecule becomes bound to a cation the average entropy change will be approximately equal to the heat of fusion of ice at 298°, or 6 kcal/mole, and that the hydration number can be obtained by dividing the absolute entropy of the gaseous ion by 6 was considered [254] an oversimplification, because it does not take into consideration the different strengths with which water molecules may be bound to a cation. In like manner the average contribution of 9.4 kcal/mole to the entropy of a hydrate per water molecule [255] is not adequate. Rather electrostatic effects arising from orientation and restriction of translational freedom of water molecules due to ion–dipole interaction, and restriction due to the formation of true covalent bonds between a donating center of solvent and the ion must be considered. When the one and when the other effect predominates was considered. The results suggests that each of the ions carries no more water molecules covalently bound than ammonia. These results, recorded in Table 3.8, were obtained by assuming that in the calculation of entropies of ammoniation, the entropy change for each molecule of

Table 3.8

Ion	No. of ligands	$-\Delta S$	
		Calculated	Experimental
H	1	3	−0.5
Ag	2	6	12.4
Cu	4	12	16
Ni	6	18	27
Zn	4	12	14
Cd	6	18	27
Hg	4(2)	6(12)	8
Li	3	9	12
Mg	4	18	19

water replaced by a molecule of ammonia will equal the difference in the
entropies of fusion of ammonia and of water at 298°C (approximately 3
kcal/mole) and by assuming that the number of water molecules in the aquo-
cation is equal to the number of ammonia molecules in the ammine. It was
argued that if the ion carried more water than ammonia, the calculated entropy
change would be more positive, whereas actually, except for the hydrogen ion,
the measured results are more negative. The larger decrease in entropy on
ammoniation it was suggested might partly be due to the shortening of the
linkages in the ammines accompanied by correspondingly more configura-
tional restrictions.

Different methods of calculating the approximate values of heat, entropy,
and energy of hydration of 15 lanthanides resulted in close agreement with
known experimental values [256].

Comments have been made [257] on electrostatic volume and entropies
of solvation. It was found that the compression of the ion has an unimportant
effect on the free energy, entropy or enthalpy of solvation, but the thermal
expansion of the cavity containing the ion contributed significantly to the
electrostatic entropy of solvation.

The entropy of ion exchange with negative hydration has been studied [258].
It was observed that the principal contribution of the summary entropy of
exchange of the ions considered is due to the entropy of the structural changes
of water in the hydration of these ions.

f. Heat Capacity   From heat capacity measurements and calculations
[259] it was concluded that sodium ion was more highly hydrated than
potassium ion in aqueous solutions, and that the hydrogen ion formed
hydronium ion H_3O^+.

From heat capacity data over a range of temperature on LiOH and
$LiOH \cdot H_2O$ and from the entropy change at 25°C for the reaction
$LiOH \cdot H_2O(cryst.) \rightarrow LiOH(cryst.) + H_2O(gas)$ obtained from other thermodynamic data, it was confirmed [260] that the third law could be applied to
the calorimetric entropies of both LiOH and $LiOH \cdot H_2O$.

Specific heats and hydration of ions have been studied [261, 262]. From
the data it was concluded that the hydrated ions contained 10–15 molecules
of water in two energy layers, of which the second, more loosely held layer
was dissociated at the higher temperatures used in the investigations (140°C).

The temperature dependence of the entropy of hydration ΔS_h; energy of
hydration ΔE_h; and enthalpy of hydration ΔH_h, for some Li, Na, NH_4, and
Mg halides and hydroxides were calculated [263] from known heat capacity
$\bar{C}_p{}^0$ data of ions in solution and known data of entropy of ions in solution at
25°C, S_{soln}^{298}.

Specific heats have been measured [264] for solutions of $LaCl_3$, $NdCl_3$,

$DyCl_3$, $ErCl_3$, and $YbCl_3$ at 25°C over the concentration range from 0.1 *M* to saturation. From calculated apparent molal heat capacities and the concentration dependence of these for each solution, partial molal heat capacities of solute and solvent were obtained. The data indicated that the heavy rare earths have a lower coordination number than the light rare earths. In nearly saturated solutions as in crystals [265] it is assumed that complex ions of type $[Cl_2Gd(OH_2)_6]^+$ exist.

Earlier work [266] on specific heats indicated that solvation occured with chlorides, bromides and hydroxides of the alkali metals as well as with HCl and HBr.

g. Solubility and Dilution   From the difference in the heats of solution in dilute and concentrated solutions of KF and of $KF \cdot 2H_2O$ at 25°C the heat of solvation of the anhydrous to the dihydrated salt was found [267] to be 5912 cal/mole.

The solubility of CO_2 in methyl alcohol and ethyl alcohol and in solutions of lithium and sodium chlorides, bromides, and iodides in these alcohols has been determined [268] at 15° and 20°C. It was calculated that in concentrated solutions these salts combine with 3–5 moles of alcohol per mole of salt. In Table 3.9 the solvation in infinitely dilute solution of three lithium

Table 3.9

Solvation in Infinitely Dilute Solution of Lithium Halides and of Sodium Iodide in Methanol and Ethanol

	Solvent			
	Methanol		Ethanol	
Halide	Ulich	Lange and Eichler	Ulich	Lange and Eichler
LiCl	11	8	10–11	7
LiBr	9–11	8	10	8–9
NaI	5–9	8–9	6–8	7
LiI	7–10	9	8–9	10

halides and of sodium iodide obtained in methyl and ethyl alcohol are compared with the solvation obtained by Ulich [269] for these salts in these solvents. It can be seen that the two sets of measurements are in fair agreement. It might also be pointed out that if some manner of apportioning the solvation between the positive and negative ions could be devised, the solvation numbers of the individual ions could be estimated.

The solvation numbers, n for the salts in the two alcohols (Table 3.9) were calculated as follows. The solubilities in gram moles of $CO_2/100$ g moles of alcohol in the salt solutions and in the pure alcohols were represented, respectively, by M''_{CO_2} and M'_{CO_2}. The amount of salt in gram moles of salt/100 g moles of alcohol is given as M salt. Hence

$$n = 100\frac{1-(M''_{CO_2}/M'_{CO_2})}{M_{salt}} = 100\frac{M'_{CO_2}-M''_{CO_2}}{M'_{CO_2}M_{salt}} \qquad (3.78)$$

If ΔM_{CO_2} is defined as

$$\Delta M_{CO_2} = 100\frac{M'_{CO_2}-M''_{CO_2}}{M'_{CO_2}} \qquad (3.79)$$

then

$$n = \Delta M_{CO_2}/M_{salt} \qquad (3.80)$$

The solubilities of some strong electrolytes in alcohols were calculated [270] from their solubilities in water using a formula that involved only the charges and radii of the ions and the dielectric constant of the solvent. The K^+, Rb^+, Cs^+, Cl^-, Br^-, and ClO_4^- ions were found to occupy the same volumes in solution as in the crystalline salts, which led to the conclusions that there was no enrichment of either the water or alcohol around the ions, and that these ions were not solvated (hydrated). The abnormal solubility curve of NaCl led to the admission that the Na^+ ion was hydrated.

The temperature coefficients of the solubilities of salts were studied [271, 272]. It was found, among other things that the temperature coefficient of solubility for similar salts increased with the degree of hydration of the "dominant ion."

The solubility of isoamyl alcohol in aqueous alcohol-mixtures in the absence and in the presence of sodium and potassium halides was discussed [273], and the results attributed to hydration of the ions and their action on the alcohols. The action of the anions decreased in the order $Cl^- > Br^- > I^-$, while the cations caused an inversion of the lyotropic series.

The different solubilities of KNO_3 as affected by the presence of KNCS and KBr in aqueous solutions was discussed [274] in the light of the effects of these salts on the NCS^- and Br^- ions. The two systems were chosen because KNCS and KBr differ in ion structure but have the same hydration energy, namely, 72 kcal/g ion [212].

The solubility of silver bromate in sucrose–water and in tetrahydrofuron–water has been studied [275, 276]. The solvation radius of $AgBrO_3$ was calculated theoretically to be about 1 Å from 20° to 30°C.

The influence of the interaction of large univalent ions with water structure have been investigated by the solubility approach [277]. If the order of the orderliness in the original tetrahedral water cell was reduced by the intercalation of the ions in the cell, the effect of the nonhydrating large univalent ions was characterized as their "negative hydration" [278]. The negative hydration decreased in the order $NO_3^- < Br^- < SCN^-$.

From studies of the solubility of silver bromate in water–glycerol solvents, ranging from 0–90.713% by weight glycerol, at 15°, 20°, 25°, and 30°C, the solvation radius of the $AgBrO_3$ was calculated [279] to be 3–4 Å. From similar data in water–propyl alcohol solvents ranging from 0–73.777% by weight glycerol at 20°, 25°, and 30°C, the solvation radius of lead sulfate was found [280] to be 1–6 Å.

Solubilities of $AgBrO_3$ in water–urea and water–methanol solvents have been investigated [281, 282] at several temperatures. The solvation radii for $AgBrO_3$ were computed to be 5.5–7.6 Å in the former solvent and 0.5–1.3 Å in the latter solvent.

The influence of the character of the solvent and of the solvation of uranyl nitrate on its solubility in water–bis(2-chloroethyl) ether and in water–diethyl ether at 25°C has been studied [283].

The polytherm of the NaI–KBr–H_2O system between $-35.0°$ and 35.0°C, and the polytherm of the NaI–KBr–H_2O system between $-34°$ and 35°C were investigated. The hydrates NaBr, $2H_2O$, NaBr·$5H_2O$, NaI·$2H_2O$, and NaI·$5H_2O$ were observed [284, 285].

The solubility of rubidium and cesium nitrates in H_2O_2–H_2O solvents over the full range of H_2O_2 concentrations were measured [286] at 0°, 15°, and 25°C. In solid phases in the $RbNO_3$–H_2O–H_2O_2 system the solvates $RbNO_3 \cdot \frac{3}{7}H_2O_2$ and $RbNO_3 \cdot \frac{1}{2}H_2O_2$ were found. The formation of hydroxyperoxidates by KNO_3 and $RbNO_3$ indicates that these ions may be more highly solvated in H_2O_2 solutions than in H_2O. The solubilities of the alkali nitrates in 100% H_2O_2 increased as the diameter of the cation increased until cesium was reached when the smaller field strength of the ion connteracts the tendency of H_2O_2 to solvate the larger ion and thus decreases the solubility.

Investigations were made of double salts and complexes in saturated solutions of inorganic salts [287–292]. The concept of total hydration numbers [293] was used to describe solubility isotherms of ternary systems containing water as one component. It was found that the true total hydration numbers in saturated aqueous solutions of two inorganic salts are constantly independent of the analytical composition of the solution [291].

The effects of salts on the solubility of benzoic acid in mixed solvents has been determined [294]. The effects of salts on the solubility of the acid in 50% aqueous solutions of ethanol, methanol, isopropanol, and dioxane at 20° and 40°C was studied. The different salting-out effects are attributed to the solubility of some compounds in organic solvents, the molecules of which are used for solvation. In the case of the effects of salts on the solubility of naphthalene in 50% methanol [295], lithium salts depressed the solubility of naphthalene much more than did sodium compounds. Since the lithium salts are soluble in alcohol, they are not only hydrated but also solvated with methanol molecules and thus do not increase the relative concentration of alcohol in the solution.

The papers on solubility discussed above give solvation of salts rather than solvation of ions. They also indicate in mixed solvents when solvation by other than the aqueous component of the solvent occurs and the extent of solvation by nonaqueous solvents. If some method of apportioning the solvation data in Table 3.9, for example, between anion and cation could be devised, the solvation numbers of the individual ions could be obtained.

h. Structure and Ionic Heats and Entropies From the structure of the ionic solution for the divalent cation, the ionic heats of hydration of divalent ions in aqueous solutions at 25°C have been found [296] to depend on four terms, the major contribution being that of the catonium sheath. The most probable permanent solvation number was determined to be 4 for the metals

Table 3.10

Values of the Hydration Number for the Hydrogen Ion in Aqueous Solution at Room Temperature[a]

Year	Author and method	Hydration	Ref.
1930	Ulich (entropy)	4	[251]
1954	Azzam (theoretical)	3.9	[298]
1954	Wicke et al. (heat capacity)	4	[304]
1955	Glueckauf (activity coefficient)	4	[217]
1955	Glueckauf (uptake by ion exchange)	4	[218]
1957	Bascombe et al. (activity function)	4	[300]
1957	Ackermann (heat capacity, infrared)	4	[301]
1957	Van Eck et al. (X-ray, solution)	4	[302]
1957	Eigen	4	[303]
1957	Falk and Fignere	4	[304]
1958	Beckey	4	[305]
1920	Bjerrum (acitivity)	9	[306]
1936	Burion and Rouyer (cryoscopic)	10	[307]
1947	Van Ruyen (vapor pressure)	13	[308]
1948	Hasted et al. (dielectric)	10	[309]
1941	Darmois (density)	0.3	[310]
1909	Washburn (Hittorf transference) (1 N)	1	[311]
1927	Remy (Parchment paper) (1 N)	1	[312]
1938	Baborouvsky (1 N)	1	[313]
1938	Baborouvsky (0.1 N)	5	[313]

[a] From Koepp et al. [141].

of Group II A, 6 for metals of Group II B, and 4 and 6 for transition divalent ions. The nature of the permanent-type linkage is given, and it is concluded that the interaction energy of solvation is electrostatic. A discussion of the conflict of opinion on the nature of solvation bonding has been given [297].

Heats of hydration and hydration numbers for the hydrogen and hydroxide ions at 25°C have been examined theoretically [298]. Concentration variation of solvation is qualitatively predicted. The formulas of H^+ and OH^- ions in aqueous solutions, in the ideal case, could be given as $(H_3O^+)\cdot(H_2O)_3(H_2O)_9$ and $(OH^-)\cdot(H_2O)_{4.7}(H_2O)_{17}$. Simple models were developed for the arrangements of the H_2O molecules around the H_3O^+ and OH^- ions. Table 3.10 contains the various values that have been listed [299] for the hydration number of hydrogen ion found by different investigators using different methods.

Table 3.11

Hydration Numbers of Hydroxide Ions

Year	Authors and methods	Solvation number	Ref.
1920	Bjerrum (activity)	7.6	[306]
1941	Darmois (density)	0	[310]
1947	Van Ruyen (vapor pressure)	8	[308]
1955	Glueckauf (activity)	4	[217, 218]
1957	Van Eck *et. al.* (X-ray, solution)	4	[302]
1957	Ackermann (heat capacity)	6	[301]
1957	Ackermann (infrared)	3	[301]
1960	Azzam (theoretical)	4.7	[298]

Table 3.11 gives values for the hydration numbers of the hydroxyl ion by various workers employing different experimental techniques.

Positive and negative hydration of ions have been discussed with respect to structural changes in water and with respect to positive and negative entropies of hydration [314–316]. Table 3.12 lists ions showing positive and ions showing negative hydration.

i. Polarography   Polarographic data on some complex ions of Cu, Zn, Cd, and Pb were used to calculate [317] the approximate molecular weights of the ions using the Riecke formula [318]. The molecular weights thus obtained permitted the calculation of the degree of hydration of Cu, Zn, and Cd ions; Pb did not hydrate.

From polarographic data on the halfwave potential of Cd^{+2} and Tl^+ ions in various concentrations of sulfuric acid, it was concluded [319] in ^{17}F sulfuric acid Tl^+ binds 2.3 times as many water molecules as does Cd^{2+} ions.

Table 3.12

Ions Showing Positive and Negative Hydration

Positive hydration
  OH^-, $HCOO^-$, HF_2^-, HCO_3^-, BrO_3^-, HS^-, CO_3^{2-}, SO_3^{2-}, SO_4^{2-}, CrO_4^{2-}, SeO_4^{2-},
  MoO_4^{2-}, SiF_6^{2-}, PO_4^{3-}, AsO_4^{3-}

Negative hydration
  NH_4^+, OCN^-, CN^-, NO_3^-, ClO_3^-, HSO_4^-, HSe^-, ReO_4^-, ClO_4^-, MnO_4^-,
  IO_4^-

Polarography has been used [320] to study the hydration of certain bivalent ions in $MeOH-H_2O$ solvents of different compositions. Each ion showed a regular decrease in halfwave potential with increasing water concentration. Table 3.13 contains the values for the ions studied of the stability constants in NH_4ClO_4 solutions for steps of the type

$$M^{2+}A_n(H_2O)_m + H_2O \rightarrow M^{2+}A_{n-1}(H_2O)_{m+1} + A.$$

It was found that Tl^+ was not hydrated in these solutions.

Table 3.13

Stability Constants for Various Bivalent Ions in
NH_4ClO_4 Solutions

NH_4ClO_4 (M)	Ion	K_1	K_2	K_3
0.01	Cd^{2+}	0.36	2.00	2.7
0.05	Cd^{2+}	0.59	2.00	3.6
0.1	Cd^{2+}	0.60	1.65	—
0.05	Pb^{2+}	0.83	2.10	—
0.10	Pb^{2+}	0.46	—	—
0.01	Zn^{2+}	0.50	1.67	2.30

From the values obtained polarographically for the formation constants of the oxalate complexes of Cu(II) and Cd(II) in light and heavy water at $25.00 \pm 0.05°C$ and an ionic strength of 1.00 with $NaNO_3$, it was concluded [321] that light water is more strongly solvating than heavy water.

The solvations of bivalent metal ions in organic component–water and in mixed organic solvent systems have been investigated [322, 323]. In the case of acetone–methanol, acetone–ethanol, and acetone–water systems, polarographic investigations showed the general type of complexes to be $[M(ROH)_i(Me_2CO)_{n-i}]^{2+}$, where n is the coordination number, M = Cd or

Pb, and R = H, Me, or Et. Cd formed 6 hydrates and 4 solvates in MeOH and none in EtOH. Pb formed 5 hydrates and 2 solvates with MeOH and 3 with EtOH. In the acetone–alcohol solvents the method of DeFord and Hume [324] was used in calculating the composition and formation constants of the complexes.

Table 3.14

Solvation Numbers for Cd²⁺, Pb²⁺, and Zn²⁺ in Formamide–Methanol, Formamide–Ethanol, and in Formamide–Water Solvents

Ion		Methanol solvation number	Ethanol solvation number	Formamide solvation number	Water solvation number
Zn^{2+}	$CHONH_2-MeOH$	2		2	
Pb^{2+}	$CHONH_2-MeOH$	2		2	
Zn^{2+}	$CHONH_2-EtOH$		4	2	
Pb^{2+}	$CHONH_2-EtOH$		4	2	
Zn^{2+}	$CHONH_2-H_2O$			0	0
Pb^{2+}	$CH_2ONH_2-H_2O$			0	0
Cd^{2+}	$CH_2ONH_2-H_2O$			0	0

For Cd^{2+}, Pb^{2+}, and Zn^{2+} ions in formanide–methanol, in formanide–ethanol and in formanide–water solvents the solvation numbers are given in Table 3.14. For Cd^{2+} the coordination numbers in MeOH were 1 and 2 and for EtOH 1, 2, and 3.

j. Addenda to Thermodynamic Studies Some general studies of ion hydration have been made using the thermodynamic approach, Noyce found [325, 326] that the singly charged d^{10} ions, Cu^+, Ag^+, Au^+ exhibited extreme solvation effects not observed for any other ions, including the isoelectronic species Zn^{2+}, Cd^{2+}, and Hg^{2+}. He also found that the thermodynamic properties changes more during the hydration of anions than for cations of the same size, and that the few available data do not exhibit the monotonic variation with ionic size that is observed with cations.

The simple electrostatic model, ion-solvent interaction, and bi-bivalent sulfates in water have been discussed [327]. It was pointed out to understand the associational behavior of KI in a variety of solvents and of the bi-bivalent sulfates in water it is necessary to take into account the molecular nature of the solvent. The paper was a survey of ion association and solvation.

It has been concluded [328], in a discussion of the alternative to the primitive model, thermodynamics of solvation, and ionic mobilities that something radically different from the primitive theories is needed for the interpretation of ionization and solvation phenomena.

Enthalpies of solution of calcium, strontium, and barium chlorates and bromates in formamide, *N*-methylformamide, and *N,N*-dimethylformamide have been reported [329]. Solvation energies were evaluated and compared with those calculated from a Born model. It is concluded that if hydrogen bonding in the solvent is preferential to the formation of ion–solvent interactions, then the least "structured" solvents will have the greatest ion-solvating influence.

Experimental and absolute ionic enthalpies have been evaluated [330] for the alkali and halide ions in formamide. For these ions the absolute enthalpies of solvation were interpreted in terms of the interaction of the ion with six coordinated formamide molecules. The polarization energy of the solvent beyond the first coordination shell was taken into account.

The experimental real free energies of solvation of ions have been measured in several nonaqueous and mixed solvents [331]. The experimental real free energy of species i in solvent s, α_i^s, is defined as the free energy change in the process when an ion in field free space is inserted into a large quantity of solution s which carries no net electrical charge. For the Debye–Hückel region of activity a_i

$$\alpha_i^s = \alpha_i^{s,\theta} + RT \ln a_i \tag{3.81}$$

Where $\alpha_i^{s,\theta}$ is the standard real free energy which may be obtained by extrapolation. The standard real free energy was divided into bulk and surface contributions. The former contribution is identified with the "chemical" solvation energy which is the quantity desired in attempting to separate the free energy of solvation of a salt into its ionic components. Much data was included together with discussion and conclusions.

A discussion has been presented [332] on ionic solvation in mixed aqueous solvents and the changes in free energy that accompany the transfer of electrolytes from water to mixed aqueous solvents.

The thermodynamic treatment of a mixed fluid in electrostatic field is applied [333] to preferential solvation at infinite dilution in mixed solvents. Partial molar free energies are used to interpret the preferential solvation which is compared with literature data.

A method has been proposed [334] for calculating the individual and average activity coefficients of ions in solution based on the simultaneous consideration of the electrostatic interaction of the ions and of their solvation. It was found that a change in the enthalpy of solvation of the ions with change in concentration is a cause of the change in the activity coefficients.

The solvation of hydrogen ion by water molecules in the gas phase has been investigated [335]. Equilibrium constants and the thermodynamic functions for the reactions $H^+(H_2O)_{n-1} + H_2O \rightarrow H^+(H_2O)_n$ were determined. It was found that the structures of the hydrates $H^+(H_2O)_n$ changes quite continuously and that no single structure showed dominant stability. The findings suggested that in the lower hydrates ($n = 2$ to 4 or even 6) all water molecules are equivalent. Thus the notation $H_3O^+(H_2O)_n$ is inappropriate. The data also showed that beyond $n = 4$, a new shell is started or crowding of the first "shell" occurs. The solvation of the sodium ion is also discussed.

Thermodynamic functions of solvation of ions have been calculated theoretically [336] using a modification of the Born approach.

The conditions for positive and for negative hydration of ions have been discussed theoretically [337] in terms of the potential which a water molecule must overcome in passing from the nearest ion to an ion in the equilibrium position and in terms of the potential barrier which is overcome in exchange of the nearest molecule in water.

Negative hydration of ions has been found [338] to occur in the interval from the freezing point up to some limiting temperature. The effect of temperature on far hydration is insignificant. The temperature effect on near hydration is greater since most of the entropy characteristics in the variation of the structure of water are closeup effects. Near hydration may be negative as well as positive, owing to the existence of a destroying and stabilizing action of the ion in this region.

Solvent activity coefficients have been found to corroberate the qualitative observation that small densly charged anions are more strongly solvated by protic solvents than by dipolar aprotic solvents, whereas large polarizable anions are more solvated by dipolar aprotic than by protic solvents. Most anions are more solvated by water than by methanol; however, very large and polarizable anions are more solvated by methanol than by water. Cations are more solvated by DMF, DMSO, and HMPT than by water.

The statistical thermodynamics of solvation has been used [340] to derive an equation relating the change in volume in solution, the free energy of solvation, and the isothermal compressibility of the solvent. The relation was applied to a model of solid spheres in structureless water.

Compositions of solvates of alkali metal methylates with methanol were determined [341] by thermal analysis. Coordination numbers of 4, 6, 6, 8, and 8 for Li^+, Na^+, K^+, Rb^+, and Cs, respectively, were found.

It has been found [342] that the variation of the entropies of hydration across the lanthanide series support the suggestion of a change in the hydration number of the lanthanide ions somewhere in the middle of the series, form Nd to Tb. Also it was stated that the magnitude and trends of the thermodynamic parameters of hydration suggest that the variation in thermodynamic properties observed in complexation reactions often result from dehydration of the ions rather than of the ligation. Values of the hydration numbers of La and

Lu ions have been calculated [343] to be 7.5 and 8.7, respectively. The ratio of these values is 1.2, which is the ratio found [342] for the entropies of hydration of Lu and La ions. This, it is felt, might imply that the hydration sphere is 20% greater in the heavier lathenides.

Extrathermodynamic assumptions for the estimation of single-ion medium effects have been summarized and evaluated [344]. These assumptions are divided into two groups: (1) those which presume the liquid-junction potential at an aqueous-nonaqueous boundary can be suppressed by interposition of a salt bridge and (2) those based on proposed quantitative relationships between the size of ions or molecules and their energy of solvation.

In group (1) belong the work of Bjerrum and Larson [345] and of Parker and Alexander [346]. However, a derivation [347] showing that the interface liquid-junction potential E_j mainly depends on the medium effects for the single ions transported across the interface obviates the possibility of suppressing E_j by means of a salt bridge. It is recognized that the medium effect of an ion is comprised of electrostatic component obtainable from the Born equation and neutral component equated to the medium effect of an uncharged species similar in size and structure to the ion. The components correspond to standard free energies ΔG_{el}^0 and $\Delta G_n{}^0$, respectively. It was assumed that inert gas of corresponding sizes could be used as neutral analogs of small inorganic ions. For large ions, e.g., benzoate ion, their unionized counterparts, e.g., benzoic acid was taken as the neutral analog.

Group (2) methods relate to the Born equation, the simplest expression for the electrostatic standard free energy G_{el}^0 of the ion in terms of its crystallographic radius and the bulk dielectric constant of the media.

Because of the electrostatic nature of the equation it would seem that calculations based on the modifications of the Born equation [348–351] designed to correct for factors such as increase in ionic radius due to solvation, dielectric saturation near the ion, any specific solute–solvent interactions, and $G_n{}^0$, would be valid for only narrow ranges of conditions where interactions ignored by the model would tend to cancel.

Extrapolation to infinite radius of the varying ion thermodynamically allowable combinations (sums and differences) of ionic G^0's in which one ion was kept constant and the other varied was used [352–356] to find single ion contributions to medium effects. The extrapolated value being taken as G^0 for the invarient ion. In these estimations $G_n{}^0$ was neglected assuming that it canceled in the sums and differences or was only 1–2% of the total G^0. $G_n{}^0$ is proportioned to r^2 and would approach infinity at $1/r = 0$ and if its contribution were not negligible would cause additional errors in the extrapolation. It was suggested [356] that extrapolation of ΔG^0 to $1/r = 0$ is valid only after subtracting $\Delta G_n{}^0$. In methanol water mixtures [347–357] using this method gave positive medium effects for the halide and negative medium effects for the alkali–metal ions and protons. Other methods yielded positive values for both the anions mentioned above and cations. In a third method, oversimplification was made [347, 357, 358] by assuming ΔG^0 to be linear with $1/r$. Semiquantitative agreement between results were presented by DeLigny and Alfenaar [357] and Feakins and Watson [358].

The three methods of extrapolation described above led to large discrepancies, and caused serious doubts concerning the validity of determining an intercept by extrapolation from three to four distant points. Also theoretical objections may be raised to extrapolations versus functions of crystallographic (unsolvated) radii [359]. There is no general acceptance of the inert-gas assumptions [360] and there may be overcorrection in neglecting G_c^0. The determination of the G^0's of hydration and solvation by extrapolation in each solvent [352–356] and combining the results to obtain medium effects gave fair agreement with other literature values, and hence these G^0's may be reasonable for media of not too low a dielectric constant. The equation for combining extrapolated results is

$$_sG_i^0 - {_wG_i^0} = RT \ln {_m\gamma_i} \tag{3.82}$$

where the subscripts s and w denote nonaqueous and aqueous standard states, respectively, and $_m\gamma_i$ is the medium effect in terms of the activity coefficient of solute species i.

The assumption that any ion can undergo negligible change in G^0 on being transferred from water to any solvent is only a rough starting approximation on which more acceptable methods can be built.

The assumption that two cations of a redox system have equal medium effects in a given solvent has disadvantages which were discussed with respect to the $2+/3+$ charge type [348]. Thus for equal radii the $2+/3+$ couple will have a free energy difference five times as large as that for a $1+/0$ couple, and specific solvation effects are generally enhanced with ions of high-charge type. Complexes such as ferrocene and ferrion containing large organic ligands may be appropriate for comparison among similar nonaqueous solvents, but not for comparison with water in which exists at present incompletely understood specific interactions with such ligands. It is emphasized [348] that it is impossible at the existing state of knowledge to prove the validity of any split of solvation energies in any solvent.

The assumption that an uncharged species and the corresponding ion of unit charge have equal medium effects in a given medium in inherent in the use of the ferrocene–ferricenium and cobaltocene–cobaltocinium redox couples [344, 361], the Hammett acidity functions, the iodine–triiodide couple, and the reactant–transition state pair in aromatic nucleophile substitution [344, 346]. Several reviews have led to a possibly balanced perspective with reference to the first two couples. Based on the difference in their sizes and on the expected solvent–solute interactions for both components [344, 346], the iodine–triodide couple is probably inferior to the first two couples. The assumption of equal medium effects of an uncharged species and the corresponding unit charge ion, for which the medium effects approximately cancel, is superior to the assumption of a zero medium effect in all solvents of a reference ion or that of equal medium effects in a given solvent of two cations of a redox system. Also the ΔG_{el}^0 of the large unit charged ion is assumed to be small. However each ion has a presistent electrostatic component of the medium effect, and most of the above couples are subject to specific interactions with solvents. In some media the Hammett H_0 function properly corrected for ΔG_{el}^0 [344, 362] may be useful. The most promising approach to medium effects for single ions is based on the assumption of equal medium effects in a given solvent of a reference anion and a reference cation from an electrolyte ideally composed of large symmetrical ions of equal size and solvation properties. To minimize surface charge density and specific solvent interactions, the central atom and the charge of such ions should be shielded by large organic substituents. Electrolytes chosen to yield medium effects of single ions have been tetraphenyl phosphonium tetraphenyl borate [363], triisoamyl-*n*-butylammonium tetraphenyl borate [364], and tetraphenylarsonium tetraphenyl borate [346, 365]. Assumptions of equal medium effects of reference counterions are weak in that there is insufficient evidence of the equality of their ionic radii [364, 366] and there is the possibility of formation of crystal solvates [364, 367]. Experimental measurements, rather than calculations from models or Stokes's law, are needed for ion sizes. The determination of a medium effect of an univalent electrolyte from its solubility product constants K_{sp} by

$$\log_m \gamma_+ + \log_m \gamma_- = \log(_w K_{sp}/_s K_{sp}) \tag{3.83}$$

assume the solid phases in equilibrium with the saturated solutions are identical for both solvents. This might not be true if crystal solvates are formed. The energy difference for crystal solvates would perhaps be small for the reference ion where ΔG_n^0 would dominate the medium effects. Such solvates, if formed, could be corrected for. In dipolar aprotic solvents which differentiate strongly between anions and cations, the splitting of solvation energy of an electrolyte between its counterions might not be feasible [344, 348]. Only experiment will tell whether solvation effects based on large organic reference ions can be extrapolated to the behavior of small inorganic ions [344]. This approach was used [344] to express standard electrode potentials and ion activities in different solvents on a single aqueous scale and to evaluate liquid-junction potentials at aqueous–nonaqueous interphases. The reference electrolyte used was triisoamyl-*n*-butyl ammonium tetraphenyl borate.

The determination of absolute partial molal volumes was recently reviewed

[366–369]. A new method [368] is based upon the concepts of van der Waals volume, V_w and packing density d of the solute in solution, the former being calculated from a geometrical model [370] in which atoms are treated as overlapping spheres with volumes determined by covalent bond lengths, hydrogen bond lengths, and van der Waals radii; with the C–N covalent bond length being uncertain, ranging from 1.465 to 1.52 Å. Group contributions to V_w have been tabulated [370]. The equation

$$d = V_w/\overline{V} \tag{3.84}$$

was used to calculate d where $\overline{V}$ is the partial molal volume of the solute. As the size of the molecule increases d becomes uniform at about 0.654 for van der Waals volumes greater than about 50 ml/mole.

To determine the partial molar volume of the hydrogen ion, the standard partial molar volume of hydrohalic acids are subtracted from those of the alkylammonium halides, thus cancelling the effect of the anion:

$$\overline{V}(RA\,mm\,X) - \overline{V}(HX) = \overline{V}(RA\,mm^+) - \overline{V}(H^+) \tag{3.85}$$

Substituting $\overline{V}(RA\,mm^+)$ from Eq. (3.84) into Eq. (3.85)

$$\overline{V}(RA\,mm\,X) - \overline{V}(HX) = (V_w/d)(RA\,mm^+) - \overline{V}(H^+) \tag{3.86}$$

Since d is constant for large cations, the plot of the left-hand side of Eq. (3.86) versus the first term on the right-hand side of the equation should yield a linear plot with an intercept equal to $-\overline{V}(H^+)$.

Using 22 salts involving 13 different cations and three different halide anions, and taking the length of the C–N covalent bond as 1.465 Å, $\overline{V}(H^+)$ was found to be -4.9 ± 0.7 ml/mole compared to a reported [371] value of -5.4 ml/mole.

E. DIFFUSION

1. Theory

Diffusion coefficients are generally calculated [372–374] using the Onsager–Fuoss equation. The equation for calculating the diffusion coefficient, D_{calc}, of a binary electrolyte is

$$D_{calc} = 2000RT\frac{\overline{M}}{C}\left(1 + C\frac{\partial \ln y_\pm}{\partial C}\right) \tag{3.87}$$

where R is the molar gas constant in ergs per degree per mole; T is the absolute temperature; $\overline{M}$ is a function in diffusion theory and depends on the valences and number of ions in the formula of the electrolyte, the distance of nearest approach of the ions, the dielectric constant, the absolute temperature, and concentration of the solution and the limiting conductances of the electrolyte and the ions; C is the molar concentration of the solution; and $y_{\pm}$ is the mean activity coefficient of an electrolyte.

From the calculated and observed values of the diffusion coefficient, the solvation number n of the electrolyte can be found from the equation,

$$D_{obs} = D_{calc}(1 - 0.018n) \tag{3.88}$$

2. Experimental Technique

Diffusion coefficients can be obtained from conductivity measurements as a function of time [375, 376] using an accurately machined parallelopiped cells with electrodes at top and bottom positions. These positions may be suitably determined by theory. If K_B and K_T are the reciprocal resistances measured at the bottom and top of the cell, then

$$\ln(K_B - K_T) = (t/\tau) + \text{constant} \tag{3.89}$$

where t is the time and τ is defined by the equation

$$1/\tau = \pi^2 D/a^2 \tag{3.90}$$

in which a is the height of the cell. Also τ is the slope of the line obtained when $\ln(K_B - K_T)$ is plotted versus t. From the slope of the line, therefore, D can be calculated from experimental data using the expression

$$D = (a^2/\pi^2) \cdot (1/\tau) \tag{3.91}$$

This is the value of D_{obsd} given in Eq. (3.88).

There are other methods for obtaining diffusion coefficients such as the porous diaphragm approach [377, 378], but the above conductance method will suffice for illustrating one experimental approach to obtaining diffusion information.

3. Solvation Numbers Determined

From the measurement of the diffusion coefficient in calcium chloride in aqueous solution, and using Eq. (2.88), the hydration number of $CaCl_4$ was found to be approximately 24 which is not in agreement with the value of 11.9 found [379] from activity studies.

From diffusion coefficient measurements, degrees of hydration of metallic ions were obtained that agreed with those obtained by reliable methods [380]. For some bivalent ions, the hydration was found to correspond to the formulas of solid hydrates. The hydration numbers at infinite dilution of elementary ions are listed in Table 3.1.

It was found [381] from diffusion and mobility studies that the number of water molecules m in the inner coordination sphere of an ion was given by the expression

$$m = (n' - 1)/3 \qquad (3.92)$$

where n' is the number of atoms composing the ion. The equation applied to complex ions containing one metallic atom.

The hydration of ions as determined [382] in 0.1 M solutions using parchment membranes are listed in Table 3.15. With increasing dilution the hydration number of anions seems to tend toward a common limiting value, indicating a physical nature for hydration in dilute solutions, and a chemical nature in concentrated solutions.

Table 3.15

Solvation Numbers of Ions from Diffusion Measurements

Solvent	Temperature (°C)	Reference ion	Solvation number of reference ion	Ion	Solvation number of ion	Reference
Water	25	—	—	Li^+	5	[380]
	25	—	—	Na^+	3	[380]
	25	—	—	K^+	1	[380]
	25	—	—	Rb^+	1	[380]
	25	—	—	Cs^+	1	[380]
	25	—	—	Tl^{1+}	1	[380]
	25	—	—	Ag^+	2	[380]
	25	—	—	Cl^-	1	[380]
	25	—	—	Br^-	1	[380]
	25	—	—	I^-	1	[380]
	25	—	—	Mg^{2+}	9	[380]
	25	—	—	Ca^{2+}	9	[380]
	25	—	—	Sr^{2+}	9	[380]
	25	—	—	Ba^{2+}	8	[380]
	25	—	—	Cu^{2+}	11	[380]
	25	—	—	Zn^{2+}	11	[380]
	25	—	—	Cd^{2+}	11	[380]
	25	—	—	Co^{2+}	13	[380]
	25	—	—	Fe^{2+}	12	[380]
	25	—	—	Mn^{2+}	12	[380]
	—	—	—	Li^+	62	[382]
	—	—	—	Na^+	44.5	[382]
	—	—	—	K^+	29.3	[382]
	—	—	—	H^+	5	[382]
	—	—	—	Cl^-	26.6	[382]
	—	—	—	Br^-	29.6	[382]
	—	—	—	I^-	21.4	[382]

Table 3.15—continued

Solvent	Temperature (°C)	Reference ion	Solvation number of reference ion	Ion	Solvation number of ion	Reference
Water	—	—	—	Cu^{2+}	7–8	[383]
	—	—	—	Ni^{2+}	10	[383]
	10	Tl^+	0	K^+	9.5	[383]
	10	Tl^+	0	Cl^-	3.5	[384]
	10	Tl^+	0	I^-	10	[384]
	10	Ba^{2+}	0	Be^{2+}	10	[384]
	10	Ba^{2+}	0	Mg^{2+}	9	[384]
	10	Ba^{2+}	0	Zn^{2+}	9	[384]
	10	Ba^{2+}	0	Cu^{2+}	7.4	[384]
	10	Ba^{2+}	0	Ni^{2+}	8.5	[384]
	10	Bi^{3+}	0	Al^{3+}	13	[384]
	10	Bi^{3+}	0	Y^{3+}	11	[384]
	10	Bi^{3+}	0	Ce^{3+}	6.7	[384]
	10	Bi^{3+}	0	Cr^{3+}	16.5	[384]
	10	Bi^{3+}	0	Fe^{3+}	10.5	[384]
Water (H_2O or D_2O)	—	—	—	Li^+	22	[391]
	—	—	—	Na^+	13	[394]
	—	—	—	K^+	7	[394]
	—	—	—	Cs^+	6	[394]
	—	—	—	Cl^-	5	[394]
	—	—	—	Br^-	5	[394]
	—	—	—	F^-	7	[394]
	—	—	—	No_3^-	6	[394]
	—	—	—	CH_3COO^-	11	[394]

From diffusion measurements using a membraneless cell [383] copper and nickel ions from 0.01 to 0.1 M appear to have the numbers of waters per ion as indicated in Table 3.15.

It was concluded that, of dialysis and diffusion methods, only diffusion measurements give a reliable degree of hydration [384]. Choosing a reference ion, assumed to have zero hydration, for each valence group (Tl^+, Ba^{2+}, Bi^{3+}, Th^{4+}), the degree of hydration of other ions were given as listed in Table 3.15. Other standards of reference for hydration, for example Ulich's [385] value of 2 for Tl^{1+}, yields other hydration values for the other ions.

In methods of diffusion the reference substance, for example allyl alcohol, is carried by the electrolyte [386]. In some instances the electrolyte was found to carry more reference substance than water [387]. Pyridine as a reference substance was observed to combine in large proportions with silver ion in the electrolysis of silver nitrate solution containing pyridine [388].

The theoretical treatment of diffusion, self-diffusion of electrolytes and hydration effects were discussed [389]. The additional effect of counter-diffusion of the solvent was pointed out, and an alternative theory based on the absolute rate theory of diffusion was presented.

A potential barrier was assumed over which ions in water jump either alone or together with their hydration shells [390]. Using the mobility and the coefficient of self-diffusion of the ion and the absolute temperature, the change in the potential barrier (energy of activation) was calculated and found to be positive for Li^+, Na^+, Mg^{2+}, and Ca^{2+} and negative for K^+, Cs^+, Cl^-, Br^-, and I^-. Hence for these latter ions, water molecules remain in their vicinity a shorter time than in the vicinity of other molecules. The activation energy for the solvent between solvent bound by the ion and free solvent was considered to be the main quantity characteristic of the solvation of ions [391]. When this was negative, as in the case indicated above, the water near the ions became more mobile than in pure water. This was called negative hydration.

Experimental diffusion coefficients in several solutions of $LiNO_3$ and $LiCl$ agreed well with the theoretical values calculated using the equation deduced from the assumptions that only molecules of water attached directly to Li^+ moved with the ion and that other water molecules were not affected [392].

From self-diffusion studies by means of nuclear magnetic resonance spin-echo techniques, hydration numbers z were deduced using self-diffusion coefficients D and concentrations C and assuming rapid exchange between "free" and "hydrated" water [393]. Thus,

$$\frac{D_{solution}}{D_{pure\ water}} = 1 - z\frac{C_{ion}}{C_{water}} \tag{3.93}$$

Hydration values found were: In LiCl, $z = 4$, with Li^+ the hydrated species since $z < 1$ in HCl; in $Th(ClO_4)_4$, $z = 30$; and in $Be(ClO_4)_2$, $z = 6$ for low concentrations and $Z \cong 16$ for higher concentrations.

A potential of 400 V applied across a membrane separating a 0.3 N electrolyte solution in H_2O from a 0.3 N electrolyte solution in D_2O caused a current of 0.1 A to flow accompanied by the passage of H_2O or D_2O through the membrane depending on the direction of flow of the current [394]. From the amounts of H_2O and D_2O the relative solvation numbers were calculated. Absolute solvation numbers were obtained from the relative ones by conductivity measurements using a cell closed at one end by cation-permeable and at the other end by cellophane membranes. Table 3.15 contains the solvation (hydration) numbers found.

The coefficient of self-diffusion for water in aqueous 0.5 M solutions of KCl, NaCl, and KI were measured using a nuclear magnetic spin-echo relaxometer at 21–23°C [395]. The hydration of Na^+ was found to be positive and the solvation numbers of K^+, I^-, and Cl^- ions were found to be negative.

Using radioisotopes as tracers and a capillary tube measurement at 25°C of self-diffusion coefficients of ions and solvents from 0.1 M to saturated solutions of LiCl, NaCl, KCl, and $CaCl_2$ were made [396]. The hydration of Cl^- was found to be weak and constant which suggested a hydrogen bond with water. The hydration number of cations diminished with increased concentration.

The negative hydration of ClO_4^- and NH_4^+ ions was observed by determining the concentration dependence of self-diffusion and the macroscopic viscosity in aqueous solutions of various perchlorate and ammonium salts [397]. The negative hydrated ClO_4^- ion was observed to be weaker than that of Cl^-, Br^-, or I^-, and the negative hydrated NH_4^+ was observed to be stronger than K^+ and weaker than Cs^+.

4. Significance

The method of diffusion depending on conditions can yield widely divergent results. Thus lithium ion may have a solvation number of 5, 22, or 62. The value of 62 obtained using a parchment paper diaphragm no doubt measures, not only the primary, but also several secondary solvation sheaths of the lithium ion. The value of 22, measured by imposing a relatively high potential across the diffusion cell also involves at least two or three solvation shells of the lithium ion. The hydration number of 5 no doubt is limited to the primary solvation shell.

There is not, in general, a consistent order in the degree of solvation of the ions within a family, though the order in the alkai metal family of ions is ordinarily $Li^+ > Na^+ > K^+$. There is a tendency of the halogen family of ions to show the order of hydration $I^- > Br^- > Cl^-$, though in some cases $I^- = Br^- = Cl^-$, also in some cases $K^+ = Rb^+ = Cs^+$ and $Mg^{2+} = Sr^{2+} = Ca^{2+} \cong Ba^+$. The orders of degree of solvation in some cases, e.g., halogen ions, are opposite to those found by other methods.

In the authors' opinion, diffusion is one of the less dependable and more inconsistent methods of measuring solvation, and the data from this method are most difficult to rationalize and interpret.

F. ISOTOPIC EQUILIBRIUM

1. Theory

It was proposed that ionic hydrates $A(H_2O)_n^z$ be considered to exist if, and only if, characteristic internal vibrations can be detected by spectral or other effects [398]. The purpose of the investigation was to show the connection between the appreciable fractionation by ions [399] of the oxygen isotopes in the water in which they are dissolved and the ionic hydration.

The hydration of ions in the sense mentioned above will produce group vibrations of bound isotopic water molecules which will differ in their zero-point vibrational energy levels because of the difference in their masses. The effect of such difference was to decrease the activity of the heavier species relative to the lighter ones, and this was indicated by the equation

$$\alpha \times 10^3 = 1 - R/R_0 = \sum_A m_A n_A (K_A - 1)/55.51 \qquad (3.94)$$

where α was the enrichment factor, R was the ratio of the activity of the heavy to light species in the solution, R_0 was the ratio of the activity of the heavy to light species in the absence of solute, m_A was the molality of ion A, n_A was the hydration number of ion A, K_A was the intensity factor K_1/n_0, and K_1 was the equilibrium constant for the reaction

$$A(H_2O)_n + H_2O^* \rightleftarrows A(H_2O)_{n-1}(H_2O^*) + H_2O \qquad (3.95)$$

where the asterisk indicates the heavier species. All species are in solution.

In the derivation of Eq. (3.88), the following assumptions were made: (a) Heavy and light water species formed an ideal solution; (b) the addition of solute did not change the gross isotopic composition of water; (c) in the water replacement reactions, the successive steps differed only by a statistical factor; i.e.,

$$A(H_2O)_n + H_2O^* \rightleftarrows A(H_2O)_{n-1}(H_2O)^* + H_2O; \ K_1 = n_A K_A \qquad (3.96)$$

$$A(H_2O)_{n-1}(H_2O^*) + H_2O^* \rightleftarrows A(H_2O)_{n-2}(H_2O^*)_2 + H_2O;$$

$$K_2 = \frac{(n_A - 1)K_A}{2}, \quad \text{etc.} \qquad (3.97)$$

2. Experiment

One method of determining the enrichment and intensity factors, and thus ultimately the hydration number of ion A, n_A, was to vigorously dry solutes to the anhydrous states and dissolve the solute with minimum exposure to air in distilled water which had been stored in pyrex vessels. About 10 g of the solution were degassed on the vacuum line by repeated cycles of freezing, pumping, and thawing, and about 0.2 millimole of carbon dioxide admitted to the equilibration vessel. After equilibration from 5 to 30 days at a controlled temperature, the gas was sampled, dried, and measured for isotopic ratio using a mass spectrometer. The isotopic ratio (mass 46/mass 44) is a measurement if a very large molar ratio of water to carbon dioxide is maintained, which faithfully reflects the relative activity of the water species via the equilibrium

$$H_2{}^{18}O(soln) + C{}^{16}O_2(gas) \rightleftarrows H_2{}^{16}O(soln) + C{}^{18}O_2(gas) \qquad (3.98)$$

Pure water, for which all the conditions, preparation, equilibration, sampling, and measurement were made as nearly identical as possible, was the standard against which each solution was compared. The standard deviation of a single measurement of α was ± 0.02. The concentrations of solutes ranged from 0.4 to 2.84 M in the single experiments reported at 25°C.

3. Solvation Numbers Determined

In the method involving the determination of the enrichment factor, it was found [398] that a reasonable and consistent distribution of the separate effects due to cation and anion could be made only if in the sense defined, i.e., by detection of characteristic internal vibrations, the ions ClO_4^-, Cl^-, I^-, Na^+, and $Co(en)_3^{3+}$ were not detectably hydrated. Table 3.16 contains the hydration numbers of the other ions investigated.

The results of equilibrium data of HCl solutions up to 1.7 molal containing radioactive alkali metal chlorides and of LiCl and KCl solutions containing radioactive Na with synthetic ion exchange resins at 25°C were discussed [400] in terms of the Harned rule and the hydration theory of Stokes and Robinson. Competitive hydration of cations was indicated in mixed electrolyte solutions.

In the study of the exchange of chromium between Cr(II) and Cr(III) using radioactive techniques, the effect of ion hydration and of chloride complexing on the rate of exchange is discussed [401]. Such ions as $CrCl_2(H_2O)_4^{1+}$ and $Cr(H_2O)_6^{3+}$ are mentioned.

The use of oxygen-isotope in the study of the hydration of ions was discussed in an article which was largely a review of Laube's work in the field [402]. A less direct method, but one which could have great power, for studying solvation was suggested. This was an equilibrium method which involves the study of the measurement of the effect that salts exert on the activity ratio of $H_2{}^{18}O$ compared to $H_2{}^{16}O$ in the liquid.

Isotopic exchange equilibrium data of the $H_2{}^{18}O$ and $H_2{}^{16}O$ concentrations in the body of the liquid (K_l) and in the immediate vicinity of the ions (K_h) and of the energy involved (ΔE) were used to formulate the criteria for negative hydration of ions [403]. These criteria were $K = (K_l/K_h) < 1$ and $\Delta E < 0$.

A method for rapid mixing and sampling in the application of the isotope dilution technique to the study of ionic hydration gave the results recorded in Table 3.16 [404].

In a kinetic study of the isotopic exchange between $[Lu(EDTA)(H_2O)]^-$ and Lu^{3+}, a simple displacement reaction was suggested for the exchange of the Lu of the chelated compound with the uncomplexed Lu [405].

It was shown that solvent isotope effects for H_2O–D_2O mixtures and for

Table 3.16

Solvation Numbers of Ions Determined by the Isotopic Equilibrium Method [a]

Ion	Solvation number of ion	Temperature (°C)	Reference ion	Solvation number of reference ion
Li^+	1.0	25	ClO_4^-, etc.	0
Li^+	2.0	4	ClO_4^-, etc.	0
Ag^+	0.7	25	ClO_4^-, etc.	0
Ag^+	0.7	4	ClO_4^-, etc.	0
H^+	4.6	25	ClO_4^-, etc.	0
H^+	2.0	4	ClO_4^-, etc.	0
Mg^{2+}	6.2	25	ClO_4^-, etc.	0
Mg^{2+}	7.1	4	ClO_4^-, etc.	0
Cr^{3+}	19 ± 10	25	ClO_4^-, etc.	0
Al^{3+}	6.0 ± 0.5	25	Cr^{3+}	6.0 ± 0.5

[a] Data from [398], except for Al^{3+}, which is from [404]. The solvent is water in each case.

ionic hydration equilibrium in such mixtures can be calculated from the structure differences between D_2O and H_2O and that between HDO and H_2O and the relative amounts of H_2O, D_2O, and HDO [406].

The flow method of isotope dilution was used [407] in the study of hydration, polymerization, oxidation, and reduction of aquocations.

In the study of the exchange of water between oxygen-18-labeled solvent and aquorhodium(III) ion using the isotopic dilution procedure, the hydrated ions $Rh(H_2O)_6^{3+}$, $Rh(H_2O)_5OH^{2+}$, $Rh(H_2O)_5^{3+}$, and $Rh(H_2O)_4OH^{2+}$ were used in explaining the exchange rate [408].

The solvation of Cr(III) ion in acidic water–methanol mixed solvents was investigated using ion–exchange procedures and an isotopic dilution procedure [409]. At 60°C the composition of the Cr(III) ion varied from $Cr(H_2O)_{5.831}(MeOH)_{0.169}^{3+}$ to $Cr(H_2O)_{2.40}(MeOH)_{3.60}^{3+}$ when the H_2O–MeOH solvent ranged in composition from 0.154 mole fraction of MeOH to 0.982 mole fraction of MeOH, respectively. The data indicate that with respect to first shell coordination, Cr(III) ion discriminates in favor of H_2O over MeOH.

In water–ethanol mixed solvents, separation by ion-exchange methods of individual differently solvated species, $Cr(OH_2)_{6-n}(OHC_2H_5)_n^{3+}$ ($n = 0, 1, 2,$ and 3), present in equilibrated solutions were made [410]. Values of $\bar{n}$, the average number of ethanol molecules bound per Cr(III), are the same within experimental error at 50° and 75°C, indicating that the enthalpy change associated with the solvent replacement reaction is small (0 ± 0.5 kcal/mole).

4. Significance

No general trend of hydration number with temperature is evident, though the data would indicate that cations other than hydrogen have the same or greater hydration numbers at lower than at higher temperatures. It would be expected that ions be more highly solvated at lower than at higher temperatures since at lower temperatures, the decreased thermal agitation would permit a stronger attraction between ions and solvent molecules. That hydrogen ion is less solvated at lower temperature might be due to stronger hydrogen bonding among water molecules at lower temperature which prevents the breaking down of the water structure by protons and the consequent formation of hydronium and other hydrated hydrogen ions. The hydration number exhibits a variation with ionic charge, radius, and type which is qualitatively reasonable [411].

Some ions found unhydrated by this approach, namely ClO_4^-, Cl^-, I^-, and Na^+, generally evidence hydration by other methods. The cation with a completed inner sphere, $Co(en)_3^{3+}$, shows zero hydration probably because of the identity of the hydration number and the number of water molecules in the inner sphere.

As do several other methods, this procedure shows, in general, increasing hydration with increasing ionic valence. The lithium ion by this method has a relatively low hydration number, but in contrast with NMR procedures does show solvation rather than the breaking of solvent structure.

G. SPECTROSCOPIC AND OTHER OPTICAL METHODS

1. Theory

The coordination and solvation numbers of ions and complexes can be obtained from the displacement of absorption bands and a change in their intensity. The distribution of intensity along the water band can arise from the hydration of ions in the case of acids in aqueous solutions. Change in the shape of hypersensitive and normal absorption bands with concentration of electrolyte and temperature can be used to determine the degree of solvation and change of solvation or coordination number. The appearance of new bands and their change in intensity with temperature, electrolyte concentration, and solvent composition can be interpreted in terms of solvation of the ions. The use of rotation of polarized light and fluorescence in the determination of solvation is common.

These optical methods depend among other things on the change in the absorption bands of valence vibrations, and the change in rotational bands of solvent and complex molecules with the nature and concentration of electrolyte, nature and composition of solvent, and temperature. The mathematical treatments of these phenomena are many and varied, and the reader is referred to the references which follow for the details of these theories and calculations.

2. Experiment and Results

The displacement of absorption bands and a change in their intensity caused by solvation process has been discussed [412]. It was also pointed out that a very sensitive criterion for solvation exists in the investigation of fluorescence.

It has been found that the distribution of intensity along the water band are due, among other things, to the hydration of the ions in the case of acids in aqueous solutions [413].

On the assumption of negligible hydration of $NaNO_3$ in aqueous solution occurs, the study of the Raman band of water in electrolyte solutions showed HCl changes trihydral into dihydral water upon the formation of hydrates with two molecules of water [414].

In a Raman study of aqueous solutions of salts and acids it has been shown that the change in water equilibrium is due to the anions, and changes in band structure due to hydration are caused by the cations [415].

Evidence from electrical conductance and viscosity measurements indicate that Mg^{2+} and Li^+ ions are relatively small and hydrated, or assuming all ions are hydrated, then these two ions are more hydrated than is usual for ions [416]. Spectral data concerning solvation was not definitive.

The effects of the ions, Cl^-, Br^-, I^-, NO_3^-, ClO_3^-, ClO_4^-, IO_4^-, CO_3^{2-}, and SO_4^{2-} on the structure of water have been studied using Raman spectra [417]. It was found that the first maximum of liquid water (~ 3200/cm) is, with the exception of ClO_4^-, SO_4^{2-}, CrO_3^{2-}, and IO_3^-, considerably weakened. In contrast, the second band at ~ 3400/cm is very much strengthened. In the case of KCl the 3400/cm maximum is doubled and the 3200/cm maximum invisible. An additional maximum occurs at 3600/cm with perchlorates. The effects of electrolytes and of temperature on water are quite different, and the previously considered similarity of these two effects must be discarded. The experimental data are discussed with relation to hydration, viscosity, and the mobilities of the ions.

From the rotation in water and 2 M solution of KCl of the optically active complex salt of molybdic acid and malic acid of the type $[4MoO_3 \cdot 2C_4O_5H_4](NH_4)_4$ it was found that a mole of KCl binds 14 moles of water [418]. At a concentration of 10 g/100 cm^3 the rotation $[\alpha]_c$ approached a limiting value, $[\alpha]_{lim}$. In the ion equilibrium

$$[4MoO_3 \cdot 2C_4O_5H_4]^{4-} + H_2O \rightleftarrows [4MoO_3O]^{2-} + 2C_4O_5H_5^-$$

the ion $C_4O_5H_5^-$ has very little effect on the rotation of light. From the ratio $[\alpha]_c/[\alpha]_{lim}$ in water and in KCl the extent of hydration of KCl was found.

Using the method described by Darmois [419] the solvation of ions in methanol and in acetophenone were found [420] and are listed in Table 3.17. In the method of the Darmois, Stokes Law dealing with the mobilities of ions is used to calculation the radius of the ion and from this its real volume. Another part obtained from the densities of the solutions is assumed to be additive in obtaining the apparent volumes of the ions. When extrapolated to infinite dilution, the apparent volume at most equals to the real volume. In case they are not equal, the apparent volume is adjusted by the addition to it of the volume of *n*- molecules of solvent so as to equal the real volume. Thus

Table 3.17

Solvation of Ions Determined by Optical Methods

Solvent	Solvated ion	Solvent particles per ion	Ref.
Optical rotation			
Methanol	Li^+	4	[420]
Methanol	Na^+	2.2	[420]
Methanol	K^+	1.2	[420]
Methanol	Cl^-	1	[420]
Methanol	I^-	0	[420]
Acetophenone	Li^+	1	[420]
Acetophenone	Na^+	0.5	[420]
Acetophenone	K^+	0.5	[420]
Acetophenone	Cl^-	0	[420]
Acetophenone	I^-	0	[420]
Absorption spectroscopy			
Water	Cu^{2+}	4	[426]
Water	Nd^{3+}	3	[426]
Water	Co^{2+}	6	[426]
Water	VO^{2+}	5	[443]
Water	Nd^{3+}	6	[463]
Water–MeOH	Nd^{3+}	$4H_2O$–2MeOH	[463]
Water–MeOH	Nd^{3+}	$2H_2O$–4MeOH	[463]
Methanol	Nd^{3+}	6	[463]
Water	Nd^{3+}	9	[466]
Water	Nd^{3+}	8	[466]
Spectrophotometry			
Water	Cu^{2+}	6	[444]
Water–acetone	Cu^{2+}	$4H_2O$–2$(CH_3)_2O$	[444]
Water–EtOH	Cu^{2+}	$4H_2O$–2EtOH	[444]

one obtains the molecules of solvation n of the solvent associated with the ion. The values of n were obtained from data involving the rotation of light in the previous reference.

Microwave absorption studies were made of water–salt equilibrium [421]. In a solution of NaCl in 10% water maintained at $-10°C$ the rate of absorption decreases with time and ceases when the reaction NaCl + solution → $NaCl \cdot 2H_2O$ is completed.

Light absorption by a single ion is shown to depend critically on the configuration of the adjoining polarized medium [422]. The role of this configuration in electronic transitions involving one or a pair of ions is treated in a general way based on Debye's "diffusion" theory for ion collision rates and a nonadibatic type of potential curve. It was found that rates of chemical processes depend decisively on the interaction of two ions at separations where the simple Caulombic law is not valied.

Specific refraction was used to study solvation [423]. An equation was formulated relating specific refraction r of a solution to concentration c of the dissolved substance as functions of the specific refrations R of dissolved substance and R' of the solvent. In case solvation exists, and m moles of solvent solvates 1 mole of dissolved substances, mc is substituted for c for the solvated molecule. Then $1 - c - mc$ = concentration of free solvent. Using the Lorentz formula, R' and r are calculated from densimetric and refractometric measurements. Combining terms and solving the resulting equations, it was observed that r did not depend on c, and that r almost equals R'. Four mathematical identities followed from this, and it was shown that if r is known for any two neighboring concentrations, the values R, R'', and finally m for any solution can be calculated. R'' is the specific refration of dissolved ions.

Adsorption spectra of alcohol–water solutions of 0.0784 M $CuCl_2$ were used in the determination of the number of water molecules in the mixed water–ethanol envelope of the Cu^{2+} ion [424]. The $CuCl_2$ was dissolved in alcohol and some water added for the absorption spectra study. The number of water molecules entering the ionic envelopes were calculated from derived transcendental equations. The equations became linear at a high proportion of water to copper ions. Two water molecules were found to be present in the mixed $EtOH–H_2O$ envelope of Cu^{2+} ions in the 0.0784 M $CuCl_2$.

The solvation of anhydrous neodymium chloride in ethanol to which water was added in various amounts was investigated by absorption spectrometry [425]. When there was insufficient water to hydrate all of the Nd^{3+} ions, the ions were solvated simultaneously by EtOH and H_2O. The complete displacement of EtOH molecules of solvation occured when there were 20–30 molecules of water per Nd^{3+} ion. The slightly greater solvation energy with EtOH than with H_2O caused the large excess of water over ethyl alcohol necessary for complete hydration.

The adsorption spectrum data of the pure anhydrous chlorides of Co^{2+}, Cu^{2+}, and Nd^{3+} ions in EtOH–H_2O were used to determine the hydration of these ions [426]. The hydration numbers found for the ions are listed in Table 3.17.

In a quantative spectral measurement of a liquid phase above its normal boiling point, the observations suggested an effect of temperature on the solvation sphere of NpO_2^+ and a concomitant interaction of NO_3^- ion present with the NpO_2^+-solvate system [427].

In methyl alcohol solutions of HCl a line of 710–13/cm appeared which was apparently related to the formation of the chemical compound $MeOH \cdot HCl$ and therefore to the oscillation of the $MeOH \cdot HCl$ bond [428]. This line was observable at very low dilutions.

Shifts of the ultraviolet absorption band of the solvated iodide ion at 2200 Å were observed for changing environment brought about by using various pure and mixed solvents, adding salts which did not absorb strongly in the $220\bar{0}$ Å region and by changing the temperature [429, 430]. The shifts of the $220\bar{0}$ Å band covered a range of more than 20 kcal. It was suggested that these large and characteristic shifts are a common feature for all negative ions absorbing by a charge-transfer-to-solvent mechanism.

From infrared spectroscopic observations of water in nonaqueous solutions of uranyl nitrate it was determined that two moles of water of hydration in $UO_2(NO_3)_2 \cdot 6H_2O$ are in a state of marked deformation due to the stable bonds between the water molecules and the $UO_2(NO_3)_2$ [431]. The nature of the solvent used determines the degree of deformation. In organic solvents, the remaining moles of water present in $UO_2(NO_3)_2 \cdot 6H_2O$ are deformed to a much lesser degree and are less tightly bound to the $UO_3(NO_3)_2$.

The values of the Raman active vibration frequency v_1 in dilute solutions of $HgCl_2$, when measured in various solvents were all less than that for the gas [432]. The relative change in frequency was roughly linear with the dielectric constant of the solvent as found for other bonds with partial ionic character. The ionic character of the Hg–Cl bond was about 28%, in reasonable agreement with the value obtained from the bond dipole moment.

In further studies of the effect of environmental changes on the ultraviolet spectrum of iodide ion in a variety of pure and mixed solvents, the first electronic band at 200–50 mμ of solvated I^- at less than 0.01 M was observed over a range of temperatures [433]. The data on band maxima (mμ), molar extinction coefficients, and temperature coefficients were in accord with a simple square-well model of the excited state. No specific complex formation between I^- and solvent was indicated in mixed solvents. The most suitable model was that in which the excited electrons were confined to the first layer of solvent molecules.

From refractive index and density measurements on strong electrolytes in aqueous solutions the total hydration numbers of electrolytes and the polarizability of the water molecules were obtained [434]. These total hydration numbers were divided into total primary and total secondary hydration numbers and are listed in Table 3.18.

Table 3.18

Primary and Secondary Overall Hydration Numbers[a]
for Some Univalent Electrolytes

Electrolyte	Overall primary hydration number	Overall secondary hydration number	Reference
NaCl	6	28	[434]
NaBr	6	26	[434]
KCl	5	34	[434]
KBr	5	22	[434]
KI	5	19	[434]

[a] These were found to be nearly independent of temperature.

From observing the shifts and broadening of the absorption bands for Cu^{2+} and Co^{2+} salts in alcohol solutions upon the addition of water, it was concluded that ions exist which are simultaneously solvated by molecules of water and of alcohol [435]. The maximum values for the coordination number of hydration were 6.

In an investigation of the effect of the solvent on the electron spectra of phthalimides, it was found that the degree of action of the solvent on the electron-vibration levels of dissolved molecules increases with increasing complexity and with the intensification of intermolecular interaction in the medium [436].

By means of absorption spectra it was observed that in alcohol–water solutions Cu^{2+} ion shows a strong tendency to be solvated with alcohol rather than with water [437].

Using optical absorption spectra to study the solvation of ions in water, methanol, and ethanol solutions it was stated that all the solvates formed in the cases of $Cu(NO_3)_2 \cdot 3H_2O$ and $CuCl_2 \cdot 2H_2O$ had similar structure, which changed strongly with concentration only in solutions of $CuCl_2$ in water and methanol [438]. A higher intensity of field was created by alcohol than by water molecules as was shown by the difference in forces of oscillators in the alcohol and water solutions. The ligands distribute themselves around the Cu^{2+} ion with the symmetry of a tetragonal bipyramid, with four of the ligands forming a square in the center of which is the Cu^{2+} ion, and the other two are on a perpendicular to the square at large distances from it.

Absorption spectra studies of Ti^{3+} ion in aqueous solutions of 0.01 to 1.0 N $TiCl_3$ in the 200 to 1500 mμ range indicated that in dilute solutions 0.01 N in Ti^{3+}, the hydrates formed had a D_{4h} tetragonal bipyramidal and not tetrahedral symmetry [439]. The changes in spectra at high Ti^{3+} concentrations were due to penetration of the Cl^- ion into the Ti^{3+} shell.

Unchanged spectra indicated that $FeCl_3$ and $CuCl_2$ solvates in dimethylsulfoxide solutions were not affected by excess alkali chloride [440]. However, excess alkali chloride causes $CoCl_2$ to form $CoCl_4^{2-}$, which could be changed to the solvate by chloride acceptors $ZnCl_2$ and $HgCl_2$ in 2:1 concentration, but completely restored by KCl when the ratio $CoCl_2 : MCl_2 : KCl$ became 1:2:4.

Optical absorption spectra in the 220–2000 mμ region were used to study the solvation of Co^{2+} and Cu^{2+} ions in the form of chloride and nitrate salts in solutions of methanol, ethanol, acetone, and methyl ethyl ketone [441]. The Co^{3+} concentration ranged from 0.02 to 2.5 N in alcohols and from 0.001 to 0.3 N in other solvents. In all solvents with no chloride ion present the Co^{2+} ion had octahedral symmetry and in methyl alcohol it had this symmetry with Cl^- present but low in concentration. In methanol with appreciable Cl^- concentration and in ethanol with any Cl^- ion, two chloride ions replaced two solvent molecules in the *cis* position giving a complex with rhombic symmetry. Methanol solvated Co^{2+} ion more strongly than did ethanol. In acetone and methyl ethyl ketone the solvated Co^{2+} ion showed tetragonal symmetry at all Cl^- concentrations. The complex possessed either one or two Cl^- ions in the trans position. The solvated Cu^{2+} complex showed tetragonal bipyramidal symmetry in all solutions with no Cl^- and in alcohol with a low Cl^- concentration. The Cu^{2+} complex had the same form in alcohol solutions containing an appreciable concentration of Cl^- but consisted of a Cl^- and two solvent molecules in the trans position. In acetone and methyl ethyl ketone solutions containing any Cl^-, the complex had rhombic symmetry with two chlorides in the cis position.

The onset of the absorption band of I^- ion (260–70 mμ) was investigated using fifteen different anions of various types and valencies [442]. It was found that the spectrographic effect of the added anions depend mainly on their power to dehydrate the test ion, I^-, Complex ions such as citrate, tartrate, and phosphate ions show much larger effects than halide ions. Thus at like concentrations, citrate ion shows ten times the effect of bromide ion.

Optical spectra in the 200–1500 mμ range was used to investigate the hydration of the vanadyl ion in aqueous solutions as functions of $VOCl_2$ and of Cl^- ion concentrations [443]. The concentration of $VOCl_2$ varied from 0.009 to 2.3 M. The concentration of Cl^- ion was varied by adding $CaCl_2$. The presence of two bands indicated that the symmetry around V^{4+} was lower than the octahedral or tetrahedral. In the presence of an excess of Cl^- ion,

$VOCl_5^{3-}$ was formed otherwise $VO(H_2O)_5^{2+}$ was produced. The hydrates and the solvates both had the same C_{4v} symmetry. The solvation of VO^{2+} differed from those of Ti^{3+} and Cu^{2+} because of the presence of the V:O bond.

Spectrophotometric measurements have been made of $Cu(NO_3)_2$ and $Cu(ClO_4)_2$ in acetone–water and ethanol–water solutions [444]. The marked increase in the visible region of the extinction coefficients with decrease in water concentration is interpreted as being due to the successive replacement of the two water molecules of the Cu^{2+} hydration sphere by two acetone or alcohol molecules. The equilibrium constants for the first replacement are 3.7 and 1.8 for acetone and ethanol, respectively, and for the second replacement 3.1×10^{-3} and 3.1×10^{-2} for acetone and ethanol, respectively. Ligand field strength can account for the small deviation of the first constants from statistical expectation, but cannot account for the very small values of the second constants, which are apparently depressed by a general solvent effect.

In an ultraviolet absorption spectrum investigation of iodides in dioxane–water mixtures the shift in the wave length of maximum absorption for Bu_4NI was attributed to the replacement of water molecules around the I^- ion by dioxane as the mole fraction of dioxane increases, and is regarded to be equivalent to an increase of an excited electronic orbital [445]. This is due to the presence or absence of solvent molecules between ions of the ion pair.

Equilibrium between violet and green CrK alums as a function of time was studied using visible absorption spectra [446]. At equilibrium both gave identical adsorption spectra with peaks at 410 and 580 mμ. As a result it was supposed that two processes, hydration-dehydration and hydrolysis, came to equilibrium in solutions of these alums.

The copper(II) and cobalt(II) chlorides and nitrates have been further examined using optical absorption spectra in the solvents water, methanol, ethanol, acetone, and methyl ethyl ketone [447]. Much of the same material is presented as in the work of Baborousky [30]. However, the order of bonds strengths were included and were in the orders Cu^{2+}–Cl^- = Cu^{2+}–solvent; Co^{2+}–H_2O > Co^{2+}–Cl^- = Co^{2+}–ROH > Co^{2+}–ketone.

Nitrates of copper, zinc, mercury, and indium; sulfates of copper, magnesium, zinc, gallium, indium, and thallium; and perchlorates of copper, mercury, indium, and magnesium have been studied by Raman spectroscopy in nearly saturated aqueous solutions [448]. All bivalent ions except copper gave lines in the narrow region 360–400 cm^{-1} indicating the formation of hexacoordinated aquo complexes. Higher intensities and frequencies for the gallium and indium aquo complexes than for the other metal complexes indicate stronger bonding of the hydration sheaths due to increased ionic charge. The comparatively greater viscosities of the gallium and indium solutions confirm the greater solvation [449–450].

A comparison has been made of the infrared spectra from 4000 to 70 cm^{-1}

of a hydrate with a full coordination shell of water, of lower hydrates, and of anhydrous salt [451]. Differences in the spectra were discussed.

Salt effects on aqueous solutions based on the infrared bands of water have been used to show that there is considerable secondary hydration of ions, up to several hundred water molecules per pair of ions [452]. In the case of these large secondary hydration numbers, there are specific differences between ions, analogous to the lyotropic sequence of the ions.

The stepwise displacement of two water molecules of the solvation sphere by acetone and by ethanol in Co^{2+} and Ni^{2+} ions were determined from visible absorption spectra of cobalt(II) perchlorate and nickel(II) perchlorate measured in the widest accessible range of acetone–water and ethanol–water solutions [453]. The values of the equilibrium quotients are given for the two displacements in the hydration spheres of the two ions by the two solvents, and the magnitudes of these quotients are discussed in relation to previously determined values for other metal ions, especially Cu^{2+}. It is concluded that the low values of the quotients may arise from both a thermodynamic trans effect and a general solvent effect.

Bound water was determined in electrolyte solutions by absorption spectrometry in the near infrared [454]. From the decrease in the absorption of the 1.45 μ (O–H overtone stretching vibration in water) due to added electrolyte, it was concluded that the electrolyte kept OH groups in bonds that prevent the OH stretching vibration. In salts forming hydrates, bound water corresponds to water of crystallization; however, bound water was also found in non-hydrated salts.

Absorption spectra of the water–organic-solvent-salt ternary system have been investigated [455]. When lithium, magnesium, and aluminum perchlorates were added to methyl ethyl ketone, there resulted only a shift of the lower frequency component of the triplet which occurs in the region of 1000–1200 cm^{-1}. When water or deuterium oxided was added, the cation becomes hydrated and the $v(ClO_4)$ frequency approaches that of the $NaClO_4$. When the molar ratio of methyl ethyl ketone to water approaches 1:4, the anion is hydrated and the band shifts to 1090–1095 cm^{-1}.

The Raman and infrared studies of aqueous calcium nitrate solution as well as mixtures involving high Ca^{2+} to NO_3^- or NO_3^- to Ca^{2+} concentration ratios have been observed [456]. The removal of the degeneracy of the E′ modes, even in dilute solution, and the activity of the A′ modes led to the suggestion that the symmetry of the NO_3^- ion had been lowered by solvation with water. The perturbation was advanced by ionic interaction with hydrated Ca^{2+} ions. A model was proposed and its implications discussed.

The effect of various ions on the infrared spectra of water in crystal hydrates and solutions, and also the effect of water on the infrared absorption of ions containing oxygen have been investigated [457]. Water of crystallization form hydrogen bonds with anions containing oxygen. The bonds become remarkably weaker on going from crystal hydrates to solutions. Al^{3+}, Cr^{3+}, Be^{2+}, Zn^{2+}, Li^+, SO_4^{2-}, OH^{-1} and F^- increased the strength of the hydrogen bonds, while Ca^{2+}, K^+, ReO_4^-, ClO_4^-, I^-, NO_3^- and SCN^- weakened the strength of the bonds.

The ratio of water to acetate ion in the solvation shell of europium(III) and of terbium(III) as a function of acetate concentration has been examined using the fluorescence intensity and the fluorescence lifetimes of Eu^{3+} and Tb^{3+} in water and in deuterium oxide solutions [458]. For the ratio of water to acetate ion found to agree with those found by Sonessen [459, 460] by potentiometric measurements, the ratio of the hydration number in pure water to the number of water molecules replaced by each acetate ion must be taken as 6.

From the infrared spectra of the Al^{3+} and Sc^{3+} salts of polystyrenesulfonic acid it was shown that for 5 μm thick foils of the polyelectrolytes under quasi-liquid conditions the molecules of water of hydration are incorporated directly between cation and adjacent anion [461]. First two and then more layers of molecules of water are added between cation and anion at higher degrees of hydration.

New thorium(IV) complexes with 4,4,4-trifluoro-1-(2-thienyl)-1,3-butane-dione (HTTA), orthophenanthioline (phen), and 1,1'-dipyridyl, namely, Th(TTA)$_4$(phen) and Th(TTA)$_4$(1,1'-dipyridyl), have been characterized by spectroscopy and both have a coordination number of ten [462].

The compositions and stabilities of neodymium solvates in water–methyl alcohol solutions have been studied. The solvated ions identified are listed in Table 3.17. The equilibrium constants for the formation of the complexes were recorded [463]. The higher the MeOH concentration in the mixed solvent, the greater the number of methanol molecules substituting for water in the ion solvation sheath.

Up to 540°C, the spectra over the wavelength range of 200–2400 mm of uranium(IV) and uranium(III) in molten fluoride solvents compared to their spectra in other molten salt systems suggested that the coordination number of the uranium species in the molten fluoride media studied is possibly 8 or 9 [464].

Hydration and the reaction of ions with the nearest water molecules in an organic medium was investigated in a spectroscopic study of solutions of the perchlorates of Na^+, Li^+, Mg^{2+}, and Co^{2+} and iodides of Na^+, Li^+, and $(C_4H_9)_4N^+$ in acetone or methyl cyanide containing semiheavy water $(H_2O + D_2O)$ [465]. From the changes in the absorption bands of valence vibrations, the interactions of the various ions with the nearest water molecules could be studied in the absence of hydrogen bonds between water molecules themselves. A new band of lower frequency in addition to the 3575-cm^{-1} band appeared in the absorption spectrum due to the hydroxyl group in methyl cyanide solutions. The intensity of the new band increased and that of the 3575-cm^{-1} band decreased with increasing salt concentration. The magnitude of the new band shifted in the order $NaClO_3 < LiClO_3 < Mg(ClO_3)_2 < Co(ClO_3)_2$. It was hazarded that the new band was due to hydrated complexes of the type $M^{n+} \cdots O(D)H \cdots N:CMe$. A third band of even lower frequency appeared in iodide solutions, and may have arisen from complexes of the type $M^{n+} \cdots O(D)H \cdots I^-$. A similar study was made of duteroxyl absorption spectra in methyl cyanide and acetone solutions of the same salts, and the influence of the various cations presented.

Optical properties of rare earth complexes including in detail their fluorescence and luminescence properties were discussed and significant experimental proofs compiled for the existence of coordination numbers greater than 6 in rare earth complexes [466]. Details of proof were given for the high coordination number of lanthanides in chelates with polyamino-carboxylic acids and in β-diketone complexes.

From the change in shape of hypersensitive and normal absorption bands in the spectra of aqueous Nd^{3+} as the concentration increased from 0 to 12 M for the electrolytes HCl, LiCl, and $HClO_4$ or as the temperature increased in concentrated LiCl solutions, the coordination number of Nd^{3+} in dilute and concentrated electrolyte solutions was proposed. By comparison with absorption spectra of 9- and 8-coordinated Nd^{3+}, the shape change in aqueous Nd^{3+} spectra were interpreted as a change in coordination number of the aquo Nd^{3+} from 9 in dilute solutions to 8 in concentrated solutions. It was assumed that other normally 9-coordinated lanthanide ions undergo a similar change. It was believed in concentrated chloride that all lanthanide ions are 8-coordinated, possibly $Ln(H_2O)_8^{3+}$ ions. In strong chloride solutions, trivalent actinide ions were thought to be chlor complexes. In Table 3.17, the coordination numbers 9 and 8 refer to dilute and concentrated chloride solutions, respectively.

Correlations were investigated between the solvation numbers of ions and the characteristics of the rotational Brownian movement of the molecules of the solvent, namely, relaxation time of reorientation τ and the coefficient of the rotational diffusion D_r in nonaqueous solutions of electrolytes, including LiCl, NaCl, CsCl, NaBr, and NaI in methanol and LiCl in alcohols [467]. A complete correlation was found between the change in the solvation number and the change in the total interaction between the ions and molecules of the solvent and the time of relaxation of reorientation τ as well as D_r. It was found that the smallest ions as Li^+ have the greatest solvation numbers and exert the greatest inhibiting effect on methanol molecules. Of several alcohols, Li^+ and Cl^- ions exerted the greatest inhibiting effect on methanol molecules and the least on amyl alcohol molecules.

3. Significance

The significance of the optical approach to the study of solvation and coordination in general involves several facets. The optical methods are accurated, comparatively rapid, and do not involve alteration of the sample. Some of them, especially spectroscopy, allow observations on the individual components of a sample rather than on the composite sample since the bond and rotational frequencies are characteristic parameters for each component. In addition, structural features of the sample are determinable using spectroscopic and other light measuring approaches. Spectral studies are not definitive in all cases. The spectroscopic approach has been able to show that the effects of electrolytes and temperature on water, which was assumed to be the same, are in fact quite different.

H. Sound Velocity—Compressibility

1. Theoretical

The adiabatic compressibility β_a of a liquid is related to the velocity v of sound in cm/sec for a liquid of density, ρ g/ml, by the equation [468, 469]

$$\beta_a = 10^6 v^{-2} \rho^{-1} \quad \text{bar}^{-1} \tag{3.99}$$

From β_a the isothermal compressibility β can be calculated using the equation

$$\beta = \beta_a C_p / C_v = \beta_a + \alpha^2 T / J C_p \rho \tag{3.100}$$

where C_p and C_v are the heat capacities at constant pressure and constant volume, respectively, and α is the coefficient of volume expansion. Substituting $-1/\rho \cdot d\rho/dT$ for α in Eq. (3.100) yields,

$$\beta = \beta_a + \{[(d\rho/dT)_p]^2 T\}/\rho^3 J C_p \tag{3.101}$$

The term $(d\rho/dT)_p$ may be approximated for by the expression $1/\delta T(\rho T + \delta T - \rho T)$, and then at $t = 25°C$,

$$\beta = \beta_a + 0.07125(\rho 30° - \rho 20°)/(\rho 25°)^3 C_p \tag{3.102}$$

Passynski [470] tried to explain the decrease in compressibility of electrolyte solutions with increasing electrolyte concentration by assuming that the solvent molecules solvating the electrolyte ions were fully compressed by the electrical forces of the ions. Thus the solvent molecules in the primary solvent shells of the ions were rendered incompressible. Passynski related the adiabatic compressibilities of solution β_a and that of solvent $\beta_{0,a}$ by the equation,

$$\beta_a + \beta_{0,a} = (1 - Sn_2/n_1) \tag{3.103}$$

where S is the primary solvation number of the electrolyte and n_1 and n_2 are the numbers of moles of solvent and solute present, respectively. By extrapolation of the graph of S versus n_2 to $n_2 = 0$, "true" solvation numbers S_0 may be obtained.

Allam and Lee [468] expressed the compressibilities of various electrolytes in water, methanol, and ethanol by the equation

$$\beta_a = \beta_{0,a}[1 + n_2(S_0 + An_2 + Bn_2{}^2)/n_1] - 1 \qquad (3.104)$$

where A and B are constants, and the remaining terms have already been defined. The above equation reproduced all the data observed. Expanding Eq. (3.104) binomially, the first two terms correspond to Eq. (3.103). The remaining terms represent contributions to the compressibility of the solution from the interactions of the solvated ions and solvent and of the solvated ions and solvated ions, assuming that the solvated species retain part of the solvent compressibility. From Eqs. (3.103) and (3.104)

$$\text{Limit}\,(d\beta/dn_2)_{n_2 \to 0} = -\beta_0 S_0/(n_1)_0 \qquad (3.105)$$

and thus the limiting slope of the compressibility versus concentration plot is proportional to the limiting solvation number, S_0. Equation (3.104) was used to obtain the limiting solvation numbers of various electrolytes at 25°C in water, methanol, and ethanol.

Assuming an incompressible solvation shell model, an entirely different method has been adopted [470–474] to give a parameter of hydration. The model assumes that the hydration shells, together with the central solute ions, are incompressible, and gives the volume of water V_h bound to one mole of solute in solution reduced to one atmosphere of pressure P_0 to be

$$V_h = (n_1 \beta_1^{(P_0)} V_1^{(P_0)} - \beta^{(P_0)} V^{(P_0)})/\beta_1^{(P_0)} = -K_2/\beta_1^{(P_0)} \qquad (3.106)$$

where n_1 is the number of moles of water in the solution, β_1 is the isothermal compressibility of the water, V_1 is the molar volume of water, P_0 is one bar standard pressure, β is the isothermal compressibility of the solution, the superscript (p_0) indicates the properties under standard pressure of one bar, and K_2 is the apparent molal compressibility of the solute.

Using compressibility and density data, the value of V_h may be calculated. Also the average hydration number h of a pair of solute ions can be calculated using the equation

$$h = V_1/V_h^{(po)} \qquad (3.107)$$

Several authors [470–476] have attempted to evaluate V_h and h.

2. Experimental Technique

Several workers have used compressibility for obtaining solvation numbers of ions [468–477].

Two types of ultrasonic interferometer, the Hubbard and Loomis type and the Debye and Sears type, have been used to measure the wavelengths of ultrasonic waves in pure solvents and in solutions of electrolytes in these solvents. From the wavelengths the ultrasonic velocities in these solvents and solutions have been calculated, and from these and measured densities the adiabatic and isothermal compressibilities obtained for the solvents and solutions. From the compressibilities of the pure solvent and solutions the hydration numbers of the electrolytes were obtained as explained in the theory section above [468–478]. Then making an assumption concerning the solvation number of some particular ion, the solvation numbers of various ions have been calculated. Thus, in general, the determination of hydration numbers of ions can be resolved into the measurement of the wavelengths of ultrasonic waves in solvents and solutions and the measurements of the densitities of the solvents and solutions.

3. Solvation Numbers Found

In Table 3.19 are presented the limiting solvation numbers of electrolytes at infinite dilution calculated from Eq. (3.104); in Table 3.20 are listed the solvation numbers of individual ions assuming that in water solvent the hydration numbers of K^+ and Cl^- ions in KCl are equal, and the solvation numbers of individual ions assuming that in water, methanol, and ethanol the nitrate ion, owing to its configuration and size, has two solvent molecules of solvation.

In Table 3.19 are also presented average hydration numbers for electrolytes (ion pairs). Multiple temperatures and multiple values of the average hydration numbers at these different temperatures are enclosed in parenthesis. The average hydration numbers are sequential with the temperatures.

Tables 3.19 and 3.20 contain [510], respectively, hydration numbers at infinite dilution of some acids, alkalis and tetraalkylammonium salts, and the limiting hydration numbers of the ions involved assuming the planar nitrate ion can hold two water molecules in its primary hydration shell [478–479]. The data are for 25°C.

The solvation of ions has been found by determining the compressibility of solutions from measurements of the velocity of ultrasonic waves in the solutions [480]. The mathematical theory was developed relating the degree of solvation of an electrolyte to the compressibility of its solution. The values of the solvation numbers found for various ions are listed in Table 3.20.

Table 3.19

Solvation Numbers of Electrolytes

Solvent and electrolyte	Temperature (°C)	Concentration	Solvation number	Reference
Adiabatic compressibility method				
Water				
NaOH	25	0	9.9	[468]
KOH	25	0	9.2	[468]
LiOH	25	0	8.7	[468]
NaCl	25	0	7.1	[468]
KCl	25	0	6.4	[468]
LiCl	25	0	6.0	[468]
NH_4Cl	25	0	4.2	[468]
$MgCl_2$	25	0	12.5	[468]
$MgSO_4$	25	0	14.8	[468]
$NaNO_3$	25	0	6.9	[468]
KNO_3	25	0	6.1	[468]
$AgNO_3$	25	0	6.0	[468]
$LiNO_3$	25	0	5.6	[468]
NH_4NO_3	25	0	4.1	[468]
NaBr	25	0	6.5	[468]
KBr	25	0	5.9	[468]
LiBr	25	0	5.4	[468]
NH_4Br	25	0	3.7	[468]
NaI	25	0	6.0	[468]
KI	25	0	5.4	[468]
LiI	25	0	4.7	[468]
NH_4I	25	0	3.2	[468]
Methanol				
NaI	25	0	6.2	[468]
KI	25	0	6.0	[468]
LiI	25	0	5.6	[468]
NH_4I	25	0	5.5	[468]
$NaNO_3$	25	0	5.9	[468]
$LiNO_3$	25	0	5.3	[468]
NH_4NO_3	25	0	5.2	[468]
NaBr	25	0	5.6	[468]
KBr	25	0	5.2	[468]
LiBr	25	0	5.0	[468]
NH_4Br	25	0	5.0	[468]
NaCl	25	0	4.7	[468]
LiCl	25	0	4.2	[468]
NH_4Cl	25	0	4.9	[528]
Ethanol				
LiI	25	0	3.7	[468]
NaI	25	0	3.2	[468]
NH_4I	25	0	3.3	[468]

Table 3.19—continued

Solvent and electrolyte	Temperature (°C)	Concentration	Solvation number	Reference
$LiNO_3$	25	0	3.4	[468]
NH_4NO_3	25	0	2.9	[468]
LiBr	25	0	3.4	[468]
NaBr	25	0	2.9	[468]
NH_4Br	25	0	2.5	[468]
LiCl	25	0	2.7	[468]
Water				
LiCl	25	—	4.24	[534]
$LiNO_3$	25	—	2.98	[534]
NaCl	20(25)	—	5.96(4.98)	[534]
NaBr	20	—	4.82	[534]
NaI	20	—	3.37	[534]
NaOH	25	—	8.40	[534]
KCl	20(25)	—	5.17(4.41)	[534]
KBr	20	—	3.97	[534]
KI	20	—	2.92	[534]
$(C_2H_5)_4NBr$	25	—	10.4	[478]
NH_4Br	25	—	3.7	[478]
NH_4Cl	25	—	1.50	[474]
$MgCl_2$	20(25)	—	11.5(10.4)	[474]
$Mg(NO_3)_2$	25	—	8.38	[474]
$CaCl_2$	20	—	11.3	[474]
$Ca(NO_3)_2$	25	—	7.60	[474]
$BaCl_2$	20(30)(40)	—	12.6(11.4)(11.1)	[474]
Na_2CO_3	20(25)	—	19.1(15.7)	[474]
Na_2SO_4	20	—	18.2	[474]
K_2SO_4	25	—	14.4	[474]
$(NH_4)_2SO_4$	25	—	9.34	[474]
$MgSO_4$	25	—	14.7	[474]
$CuSO_4$	25	—	13.2	[474]
NaOH	25	—	9.9	[478]
KOH	25	—	9.2	[478]
LiOH	25	—	8.7	[478]
HCl	25	—	2.1	[478]
HNO_3	25	—	1.8	[478]
CH_3COOH	25	—	1.8	[478]
$(CH_3)_4NBr$	25	—	6.7	[478]

Isothermal compressibility method

Water				
NaCl	25	—	5.70	[517]
KCl	25	—	5.0	[517]
NaBr	25	—	4.3	[517]
KBr	25	—	3.8	[517]
NaI	25	—	3.4	[517]

Table 3.19—continued

Solvent and electrolyte	Temperature (°C)	Concentration	Solvation number	Reference
KI	25	—	2.6	[517]
$MgCl_2$	25	—	11	[517]
$SrCl_2$	25	—	13	[517]
Na_2SO_4	25	—	18	[517]
$MgSO_4$	25	—	17	[517]

Table 3.20

Solvation Numbers of Ions

Solvent and reference ion	Solvation number of reference ion	Concentration	Temperature (°C)	Solvated ion	Solvation number of solvated ion	Reference
Water						
K^+ or Cl^-	3.2	0	25	Na^+	3.9	[468]
K^+ or Cl^-	3.2	0	25	Li^+	2.7	[468]
K^+ or Cl^-	3.2	0	25	NH_4^+	1.1	[468]
K^+ or Cl^-	3.2	0	25	Ag^+	3.1	[469]
K^+ or Cl^-	3.2	0	25	Mg^{2+}	7.0	[468]
K^+ or Cl^-	3.2	0	25	NO_3^-	2.9	[468]
K^+ or Cl^-	3.2	0	25	Br^-	2.6	[468]
K^+ or Cl^-	3.2	0	25	I^-	2.1	[468]
K^+ or Cl^-	3.2	0	25	OH^-	6.1	[468]
K^+ or Cl^-	3.2	0	25	SO_4^{2-}	8.8	[468]
NO_3^-	2.0	0	25	Na^+	4.8	[468]
NO_3^-	2.0	0	25	K^+	4.1	[468]
NO_3^-	2.0	0	25	Li^+	3.6	[468]
NO_3^-	2.0	0	25	NH_4^+	2.0	[468]
NO_3^-	2.0	0	25	Ag^+	4.0	[468]
NO_3^-	2.0	0	25	Mg^{2+}	7.9	[468]
NO_3^-	2.0	0	25	Cl^-	2.3	[468]
NO_3^-	2.0	0	25	Br^-	1.7	[468]
NO_3^-	2.0	0	25	I^-	1.2	[468]
NO_3^-	2.0	0	25	OH^-	5.2	[468]
NO_3^-	2.0	0	25	SO_4^{2-}	7.9	[468]
Methanol						
NO_3^-	2.0	0	25	Na^+	3.9	[468]
NO_3^-	2.0	0	25	K^+	3.6	[468]
NO_3^-	2.0	0	25	Li^+	3.3	[468]
NO_3^-	2.0	0	25	NH_4^+	3.3	[468]
NO_3^-	2.0	0	25	Br^-	1.7	[468]
NO_3^-	2.0	0	25	Cl^-	0.9	[468]
NO_3^-	2.0	0	25	I^-	2.4	[468]

Table 3.20—continued

Solvent and reference ion	Solvation number of reference ion	Concentration	Temperature (°C)	Solvated ion	Solvation number of solvated ion	Reference
Ethanol						
NO_3^-	2.0	0	25	Na^+	0.9	[468]
NO_3^-	2.0	0	25	Li^+	1.4	[468]
NO_3^-	2.0	0	25	NH_4^+	0.9	[468]
NO_3^-	2.0	0	25	I^-	2.3	[468]
NO_3^-	2.0	0	25	Br^-	2.0	[468]
NO_3^-	2.0	0	25	Cl^-	1.3	[468]
Water						
NO_3^-	2.0	0	25	Na^+	4.8	[478]
NO_3^-	2.0	0	25	K^+	4.1	[478]
NO_3^-	2	0	25	Li^+	3.6	[478]
NO_3^-	2	0	25	Cl^-	2.3	[478]
NO_3^-	2	0	25	Br^-	1.7	[478]
NO_3^-	2	0	25	OH^-	5.2	[478]
NO_3^-	2	0	25	$(CH_3)_4N^+$	5.0	[478]
NO_3^-.	2	0	25	$^bC_2H_5)_4N^+$	8.7	[478]
NO_3^-	2	0	25	NH_4^+	2.0	[478]
—	—	—	—	H^+	1–2	[480]
—	—	—	—	Li^+	5–6	[480]
—	—	—	—	Na^+	6–7	[480]
—	—	—	—	K^+	6–7	[480]
—	—	—	—	Mg^{2+}	16	[480]
—	—	—	—	Ba^{2+}	16	[480]
—	—	—	—	Be^{3+}	8	[480]
—	—	—	—	Al^{3+}	31	[480]
—	—	—	—	F^-	2	[480]
—	—	—	—	Cl^-	0–1	[480]
—	—	—	—	Br^-	0	[480]
—	—	—	—	I^-	0	[480]
—	—	—	—	Li^+	2	[493]
—	—	—	—	Na^+	3	[493]
—	—	0	—	K^+	2	[493]
—	—	0	—	Cs^+	1	[493]
—	—	0	—	NH_4^+	0	[493]
—	—	0	—	Mg^{2+}	8	[493]
—	—	0	—	Ca^{2+}	8	[493]
—	—	0	—	Ba^{2+}	8	[493]
—	—	0	—	Cd^{2+}	5	[493]
—	—	0	—	Cu^{2+}	5	[493]
—	—	0	—	Zn^{2+}	6	[493]
—	—	0	—	Be^{2+}	1	[493]
—	—	0	—	Ce^{3+}	13	[493]
—	—	0	—	OH^-	8	[493]

Table 3.20—continued

Solvent and reference ion	Solvation number of reference ion	Concentration	Temperature (°C)	Solvated ion	Solvation number of solvated ion	Reference
Water						
—	—	0	—	Cl^-	3	[493]
—	—	0	—	Br^-	2	[493]
—	—	0	—	I^-	0	[493]
—	—	0	—	CH_3COO^-	4	[493]
—	—	0	—	NO_3^-	1	[493]
—	—	0	—	CNS^-	0	[493]
—	—	0	—	HCO_3^-	2	[493]
—	—	0	—	CrO_4^-	13	[493]
—	—	0	—	CO_3^{2-}	15	[493]
—	—	0	—	SO_4^{2-}	11	[493]

Adiabatic compressibility studies of aqueous electrolyte solutions showed that the liquid structure of water became more highly coordinated and compacted with the introduction of ions [481].

Measurements with an ultrasonic interferometer on aqueous solutions of electrolytes permitted the calculation of the volumes and the amount of hydration of various ions [482–485].

The absorption of sound by aqueous solutions of electrolytes were investigated for frequency, concentration, and temperature dependence of their absorption of sound in the frequency range 8 kHz to 15 MHz [486]. Uni-univalent and bi-bivalent electrolytes showed no measurable absorption except for higher concentrations of Na_2SO_4 and $MgCl_2$. The bi-bivalent sulfates showed very great absorption which from the slope of the frequency response was interpreted to be due to the relaxation process. For $MgSO_4$ from 0.002 to 0.2 M, the dependence on concentration proved linear; this indicates a constant absorption effect or absorption cross section per mole. The further increase above 5 MHz was attributed to being due possibly to another relaxation process caused by a change in the hydration shells of the reacting ions.

Adiabatic compressibility of gelatin sol [487], and aqueous solutions of sugars [488] and of dicarboxylic and hydroxycarboxylic acids [489] have been made. It was found that a molecule of monosacchride combined with about four molecules of water as would be expected if each OH group combined by hydrogen bond with a molecule of water. A polysaccharide molecule also combined with approximately four molecules of water. It was thought that some of the OH groups in the molecule hydrogen bonded with each other, and were thus not available to hydrogen bond with water. The adiabatic compressibility data for maleic anhydride and of malonic, succinic, glutaric, oxalic, tartaric, citric, and malic acids were interpreted in terms of hydration effect of radicals. The negativity of the coefficient for the methylene radical was attributed to its hydrophobic nature.

Ultrasonic absorption studies on electrolyte solutions have been made and for bi-bivalent electrolytes a relaxation maximum was found [490], the frequency of which is independent of concentration, but depended on the metal ion, and increased with temperature. Hydration, association, dissociation, and hydrolysis mechanisms were used in an attempt to explain the data. Only the hydrolysis mechanism explained all the observed facts.

By measuring the speed of propagation of ultrasonic waves in solutions of sucrose, galactose, arbinose, maltose, and lactose, two moles of water were found to be linked to each OH group in the carbohydrate molecule [491].

Hydration studies by means of a ultrasonic interferometer indicated the mean pressure of hydration to be 4000 atm if it was assumed that water of volume V is compressed by ΔV owing to hydration [492].

From compressibilities solvation numbers of ions at infinite dilution were determined and are listed in Table 3.20 [493]. In calculating the solvation numbers it was assumed that the solvation layer had negligible compressibility.

Studies of hydration–dehydration in polyelectrolyte solutions have been made using ultrasonic technique [494]. The amount of hydration was determined for methacrylic and polyacrylic solutions which were progressively neutralized by NaOH and Bu_4NOH.

From ultrasonic velocity measurements in nitrate solutions, it was found that the larger the ionic radius the greater the hydration [495]. It was also found that the relation of the compressibility of aqueous solutions of electrolytes with common anions to the ionic size of the cation can be explained by a dependence on both the hydration and charge of the ions [496].

Ultrasonic absorptions of aqueous zinc acetate as a function of temperature were found to pass through a maximum [497]. The observations were explained on the basis of chemical relaxation and formation of an activated complex with an activation energy of 4.38 kcal/mole. With solutions above 0.05 N the reaction is assumed to be

$$Zn(OAc)_2 + nH_2O \rightarrow Zn(OAc)_2(H_2O)_n \qquad (3.108)$$

This activated complex may dissociate as follows:

$$Zn(OAc)_2(H_2O)_n \rightleftarrows Zn(H_2O)_n^{2+} + 2(OAc^-) \qquad (3.109)$$

or

$$Zn(OAc)_2(H_2O)_n \rightleftarrows Zn^{2+} + nOH^- + nH^+ + 2(OAc^-) \qquad (3.110)$$

From densitometric determinations on aqueous solutions containing two electrolytes, the densities of packing of the systems $KCl-NaCl-H_2O$ and $KCl-KF-H_2O$ were calculated [498]. NaCl added to KCl solution caused a tighter packing of hydrated ions as compared to that in pure KCl solution, but KF lossened the packing greatly. The presence of H^+ decreases the volume gap among hydrated ions.

From ultrasonic studies of aqueous solutions of the nitrates of Li, Ag, Be, Ca, Cd, UO_2^{2+}, Al, and La, the hydration numbers were established for all the electrolytes and compared with data compiled by other methods [499].

The hydration number and other ultrasonic parameters of the nitrates of Li, Ca, Cu, Mg, Mn, Ni, Zn, Fe, Al, and Cr were determined from ultrasonic velocity data [500]. Ultrasonic studies of the hydration of NaCl, KCl, K_2CO_3, and $BaCl_2$ in mixtures of ethanol–water also have been made [501].

The hydration numbers of $CaI_2 \cdot 4H_2O$ and $Th(NO_3)_4 \cdot 4H_2O$ were determined from the ultrasonic behavior of the salts at different dilutions [502].

Total hydration numbers were determined from ultrasonic velocities and compressibilities as functions of concentration in aqueous solutions of $Na_2MoO_4 \cdot 2H_2O$, $Na_2CrO_4 \cdot 10H_2O$, K_3PO_4, $KHCO_3$, $Ca(HCO_3)_2 \cdot H_2O$ and CaI_2 [503]. In general, the hydration number decreased with increasing cationic radius except in the cases of LiOH, $Be(NO_3)_2$, and $Ca(NO_3)_2 \cdot H_2SO_4$ gave a very high hydration number, HNO_3 and HCl have very low hydration numbers, and NH_4OH gave a negative value.

Experimental data on the apparent volumes and compressibilities of ions in aqueous solutions showed that the assumption of equal hydration of anions and cations is incorrect [504]. The apparent incompressible volumes of ions increased with decreased ionic radius, especially for anions.

Ultrasonic velocities were obtained at 1.4 MHz at 28°C in aqueous solutions of the sulfates of Li, Mg, Cd, Co, Ni, Fe, Cr, Mn, and Al [505]. From these velocities hydration numbers and other properties were calculated. The variation of adiabatic compressibility with concentration was inversely dependent on the hydration number.

The solvation approach has been used to interpret ion–solvent interaction [506]. Solvated ions in solution were assumed to be hard incompressible spheres.

By the method of ultrasonic interferometry the hydration numbers for NaCl, RbCl, LiCl, KCl, CsCl, $SrCl_2$, $NiCl_2$, $BaCl_2$, $CoCl_2$, $MgCl_2$, $MnCl_2$, $CaCl_2$, $ZnCl_2$ and $CuCl_2$ in aqueous solutions and their dependence on temperature in the range 20°–40°C were determined [507]. The relations between hydration and compressibility of aqueous electrolytes were discussed. Further evidence was obtained for the existence of the clathrates as hydration shells.

The spin-lattice relaxation time for protons for solutions of $CoCl_2$, $CuCl_2$

and LiCl were measured as a function of m, the number of moles of water per mole of solvent in aqueous methanol as water was added to the starting solution [508]. As greater proportions of water were included in the solvent, there was a basic structural change in the first coordination sphere as the alcohol solvate shell was completely replaced by the hydrate solvate shell. The filling of the sphere of influence of the cation and anion occurred in a set sequence in relation with their solvation energies.

Ultrasonic measurements permitted the calculation of the hydration numbers and other properties of aqueous solutions of the chlorides of Li, Al, Cu, Cd, Mn, Mg, Ca, Ni, Co, and Sr [509].

Ultrasonic velocities and viscosities of aqueous solutions of mixtures of KCl and NaCl at high concentrations were measured, and the hydration numbers calculated. It was found that hydration decreased with total molality [510]. With increasing concentration, there was either not enough water to go around to all the ions or preferentially ion-pairing occurs. It was concluded that there is some doubt whether "relative association" is a measure of hydration.

Ultrasonic absorption studies by the Carstensen method were made on 0.1 M MnSO$_4$ in water and on 0.05 M MnSO$_4$ in 25 weight% dioxane in water [511]. The data do not bear out the observations of Atkinson and Kor [512] or of Smithson and Litovitz [518] that a relaxation peak existed at 35 MHz. This brought into question their assignment of relaxation peaks to the steps in the following association processes:

$$M^{2+} + A^{2-} \underset{\longleftarrow}{\overset{a}{\longrightarrow}} M^{2+}OH_2OH_2A^{2-} \underset{\longleftarrow}{\overset{b}{\longrightarrow}} M^{2+}OH_2A^{2-} \underset{\longleftarrow}{\overset{c}{\longrightarrow}} M^{2+}A^{2-}$$
$$(3.111)$$

On theoretical grounds, step [a] is not expected to yield a separate relaxation peak [514]. Hence it was suggested [511] that the peak at 200 MHz ought not to be assigned to step [a] but rather to step [b]. In the above dissociation mechanism only the water molecule interposed between the aquated ion–pair members are shown.

Ultrasonic absorption of aqueous solutions of (R$_4$N$^+$) halides, where R was ethyl, propyl, and butyl, showed that marked relaxation occurred at 1.5 to 300 MHz [515]. The relaxations were believed to be associated with the special hydration properties of the alkylammonium cations.

In a study of the influence of ultrasound on electrode processes, the differences in potential of discharge of the chloride anion in the different alkali salts were correlated with, among other things, the degrees of hydration of the respective cations [516].

In a study of the velocity of sound in aqueous solutions of NaCl, KCl, NaBr, KBr, NaI, KI, MgCl$_2$, SrCl$_2$, Na$_2$SO$_4$, and MgSO$_4$, the concentration ranges up to 0.0625 M at 25°C were investigated using a frequency of 4.04 MHz [517]. From the data, the hydration numbers of the electrolytes were calculated and compared with those obtained from compressibility measurements at higher concentrations. The values obtained represented more truly the primary hydration numbers and conform more nearly to the values determined empirically by Robinson and Stokes [518] to fit the experimental activity coefficient values. These solvation numbers for electrolytes are presented in Table 3.19. It is again pertinent to point out that were some choice made for dividing the solvation numbers between the ions, e.g., selecting the hydration number for one ion, the electrolyte solvation numbers could be resolved into individual ion solvation numbers.

The molar quantity of movement MV was defined by the equation

$$MV = a + bM \tag{3.112}$$

where M was the molecular weight of the examined liquid and V the velocity of sound in the examined medium [519]. For solutions, formulas were derived taking into account dissociation and hydration of solute molecules. The degrees of hydration were estimated for different concentrations of NaCl, KCl, $NaNO_3$, KNO_3, urea, α-alinine, and glycine.

Ultrasonic absorption in aqueous solutions of the sulfates of Mg, Li, Na, and K was measured [520]. For a given concentration the ultrasonic absorption was found to decrease in the order $MgSO_4 > Li_2SO_4 > Na_2SO_4 > K_2SO_4$. In terms of size the Mg ion is between Li and Na, therefore it was believed that the double charge of Mg ion must be considered. It was thought that these data confirm the hypothesis that the larger the number of molecules bound by ions of large ionic radius, the lower the value of the ultrasonic absorption.

Measurements of the densities and sound velocities, the latter at 1 MHz, in liquid SO_2 of KI, MePyrI, Me_4NI, Et_4NI, Et_4NBr, Et_4NCl, Et_4NPic, Et_4NClO_4, $MePyrClO_4$, and Me_4NClO_4, where MePyr is methylpyridinium and Pic is picrate, were made [521]. The volume change accompanying the solution of an electrolyte at zero concentration was found to be much smaller for KI and Et_4NI in SO_2 than in water. This was explained as arising from the slight association of SO_2 in the liquid state and from its possible ease of compression in the vicinity of the ionic charge. In addition the SO_2 molecule is larger than the water molecule and the interaction of the SO_2 with a large ion such as Me_4N^+ might be predominant.

The velocity of ultrasonic wave propagation, densities, and adiabatic compressibilities of methanolic solutions of LiCl, NaCl, NaBr, NaI, and CsCl were measured [522]. It was found that the solvation numbers decreased in the order LiCl > NaCl > CsCl. The solvation numbers were found to be practically independent of the anion present.

Ultrasonic interferometry and a method for the quantitative determination of combined water were used to obtain complete saturation boundary values for the alkali halides, and for some sulfates and nitrates of Na, K, Li, Ba, Cd, and Pb in aqueous solutions [523]. For alkali halides, a crystal lattice energy value greater than the solvation energy in solution ($\Delta H > 0$) indicated a complete solvation boundary concentration greater than the saturated concentration. A crystal lattice energy value less than the solvation energy in solution ($\Delta H < 0$) implied a complete solvation boundary concentration less than the saturated concentration.

If the hydration capability of cations is less than that of anions, the complete solvation boundary concentration lies lower than the saturated concentration and vice versa. It was found in general that the complete saturation boundary concentration relative to the saturated concentration depends on the corresponding hydration entropy of the ions composing the electrolyte with the hydration capability of the cation a large factor in the complete solvation boundary values.

I. EFFECTIVE VOLUME

1. Theory and Experiment

Darmois [524] showed that density measurements could be used to calculate primary solvation numbers. Consider a volume of solution V containing a total number of water molecules n_1 of which n are attached to each of the n_2 salt molecules present in the total volume of solution as primary hydration water. If v_1 is the volume of one free water molecule, then $(n_1 - n_2 n) v_1$ is the volume of free water in the volume V. Let V_s be the Stokes volume of one of the hydrated ions in the volume V, then the primary hydration number n can be calculated from the equation

$$(n_1 - n_2 n) v_1 + n_2 V_s = V \qquad (3.113)$$

However, it has been shown that this method has a basic weakness in that it depends upon the Stokes volume which is often too small thereby yielding hydration numbers that may also be too small [525].

Conway and Bockris and co-workers [525, 526] suggest that a better method would be to express the Stokes volume in terms of the crystallographic volumes of the ions v_i and of the compressed water molecules, $n v_1^h$. Therefore

$$V_i = n v_1^h + v_1 \qquad (3.114)$$

or

$$V = (n v_1^h + v_i) n_2 + (n_1 - n_2 n) v_i \qquad (3.115)$$

Introducing the apparent molal volume ϕ and solving for n, one obtains

$$n = \frac{\phi - v_i}{v_1^h - v_1} \tag{3.116}$$

The $v_1^h - v_1$ term was evaluated from data for the pressure created by the coulombic ionic field in a surrounding dielectric at a mean distance, $r_i + r_w$, from the center of the ion [527] and from data for the compressibility of water at these pressures [528]. Using known literature values for the apparent molal volume [529–532], the primary hydration numbers for the following ions were obtained: Li^+ ($n = 2.5$), Na^+ ($n = 4.8$), K^+ ($n = 1.0$), F^- ($n = 4.3$) and Cl^- ($n = 0$).

Goto [533] has determined the effective volumes, the apparent volumes, and the magnitudes of electrostriction for a number of 1-1 electrolytes in aqueous solution from an analysis of the densities of solutions at different concentrations. From these data the number of molecules of water of hydration associated with the various electrolytes was obtained. The method of Goto may be outlined as follows. The reciprocal of the density (d_m) of a solution containing m g of solute in 100 g of the solution corresponds to the volume V_m of 1 g of the solution. Consequently, the volume V_c of a solution containing 100 g of water and c g of solute may be written as

$$V_c = \left(\frac{100}{d_m}\right)\left(\frac{100}{100 - m}\right) \tag{3.117}$$

or

$$V_c = c\bar{v} + \frac{100}{d_0} - V_c^e \tag{3.118}$$

where $\bar{v}$ is the effective specific volume of salt in the solution, d_0 is the density of pure water, V_c^e is the volume of water decreased by electrostriction and c is given by

$$c = 100m/(100 - m) \tag{3.119}$$

If the concentration of the solute c is changed from c_i to c_j then

$$\bar{v}_{i,j} = \frac{V_{c_i} - V_{c_j}}{c_i - c_j} = \frac{dV}{dc} \tag{3.120}$$

$$\bar{v}_{i,j} = \bar{v} - \frac{V_{c_i}^e - V_{c_j}^e}{c_i - c_j} = \bar{v} - \frac{dV^e}{dc} \tag{3.121}$$

where $\bar{v}_{i,j}$ is the apparent specific volume of the salt for a given concentration range. If $dV^e/dc = 0$ or if $V_{c_i}^e$ becomes equal to $V_{c_j}^e$ as the concentration is increased, then $\bar{v}_{i,j}$ will correspond to the effective specific volume of the salt in solution since all the solvent water will be attracted by ions at a high concentration. Under such conditions, the volume of water (100 g) may approach a constant $(100/d_0 - V^e_{max})$ with an increase in the concentration of the solute. The correlation between V_c and the salt concentration was found to be represented empirically by

$$V_c = c\bar{v} + (100/d_0) - (1 - r^c)V^e_{max} \qquad (3.122)$$

The parameters, V^e_{max}, r, and $\bar{v}$ were determined by the method of least squares. Also the difference between $\bar{v}_{i,j}$ at infinite dilution or the apparent specific volume $\bar{v}_0$ of a salt, and its effective specific volume $\bar{v}$ corresponds to the magnitude of electrostriction v^e/g of salt at infinite dilution or

$$v^e = \bar{v} - \bar{v}_0 = kV^e_{max} \qquad (3.123)$$

Since the volume of the solution does not increase linearly with concentration as the result of the electrostriction of the solvent, two curves were drawn for the volume of the solution (V_c) and of the solvent (V_s) under conditions: $dV_c/dc > 0$, $d^2V_c/dc^2 > 0$ and $dV_s/dc < 0$, $d^2V_s/dc^2 > 0$. These curves were analyzed using Eq. (3.116) for solutions of 1-1 electrolytes from which it was determined that the effective specific volumes $\bar{v}$ of a salt in aqueous solution and the maximum volumes V^e_{max} of water decreased by the electrostriction. Then the apparent specific volumes $\bar{v}_0$ and the magnitudes of the electriction v^e were calculated from Eq. (3.117). It was then shown that the number of molecules of water of hydration per molecule of solute is given by the expression

$$N = (100v^e/V^e_{max})\,(\text{molecular weight of solute})/18 \qquad (3.124)$$

The values obtained are given in Table 3.21. The large values for the hydration numbers were discussed in terms of the probability of clathrates around the ions. The state of water molecules around ions and in particular the existence of clathrates around the ions was discussed.

The previous discussion has shown that the volume contraction produced by the interaction between ions and water molecules can be used as a measure of the magnitude of hydration. The volume contraction calculated from compressibility data has been shown to be proportional to the amount of hydration water; the proportionality constant being independent of the temperature and the nature of the electrolyte [534–536].

Tamura and Sasaki [537] have developed a method based upon the empirical equation of Tait [538], which gives the compressibility of water under pressure, from which a quantity is obtained that the authors [537] propose as a common parameter of hydration.

Table 3.21

Hydration Numbers of 1–1 *Electrolytes at 20°C*

Salt	N (mol. H_2O/mole solute)
LiCl	30
LiBr	51
LiI	66
NaCl	22
NaBr	37
NaI	42
KCl	34
KBr	46
KI	47
RbCl	49
RbBr	38
RbI	28
CsCl	45
CsBr	87
CsI	96

The Tait equation may be written

$$\beta^{(p)}V_1^{(p)} \equiv (\partial V_1^{(p)}/\partial P)_T = 0.4343C/(B+P) \tag{3.125}$$

or in the integrated form

$$V_1^{(p_0)} - V_1^{(p)} = C \log[(B+P)/(B+P_0)] \tag{3.126}$$

where β is the isothermal compressibility, V the molar volume of water, B and C constants dependent upon the temperature and nature of the liquid, P the external pressure and P_0 the standard pressure of 1 bar. The dissolution of an electrolyte in water produces a state of electrostriction [539] that causes water to behave like a substance placed under a constant effective pressure which is greater than atmospheric pressure by an amount P_e [540]. Equation (3.126) can now be written as

$$V_1^{(p)} - \Phi_1^{(p)} = C \log[(B+P+P_e)/(B+P_0)] \tag{3.127}$$

where $\Phi^{(p)}$ is the volume occupied by one mole of water in an electrolyte solution under an external pressure P which is, therefore, the molar volume of pure water under the pressure $(P+P_e)$. A solution containing n moles of water and one mole of solute has the volume

$$V^{(p)} = n\Phi_1^{(p)} + \Phi_2^{(p)} \tag{3.128}$$

where $\Phi_2^{(p)}$ is the contribution to the volume of the solution made by one mole of solute. Therefore, if $P = P_0$

$$V^{(p)} = nV_1^{(p_0)} - nC \log[(B+P_0+P_e)/(B+P_0)] + \Phi_2^{(p_0)} \qquad (3.129)$$

or

$$\beta(V)^p = 0.4343Cn/(B+P+P_e) - (\partial\Phi_2^{(p)}/\partial P)_T \qquad (3.130)$$

Assuming the last term to be negligible at moderate pressures and concentrations [540], one can obtain the expression

$$\Phi_2^{(p_0)} = V^{(p_0)} - n_1V_1^{(p_0)} + nC \log(n\beta^{(p_0)}V_1^{(p_0)}/\beta^{(p_0)}V^{(p_0)}) \qquad (3.131)$$

Equation (3.131) was used to calculate $\Phi_2^{(p_0)}$ from the isothermal compressibilities, obtained by sound velocity measurement [541], and densities of both pure water and the solutions. Values of $\Phi_2^{(p_0)}$ were obtained for a number of electrolytes and compared with the molar volumes of the same electrolytes in the super-cooled liquid state (the molar volume of a molten electrolyte at a high temperature was calculated and extrapolated to the temperature corresponding to those at which $\Phi_2^{(p_0)}$ were calculated). Satisfactory agreement was found between the two values obtained for each electrolyte, which supports the view that the solute in a solution behaves as a super-cooled liquid with respect to its contribution to the volume of the solution.

In obtaining a parameter of hydration a method was adopted based on the incompressible hydration shell model [542, 543]. Assuming that the hydration shells together with the hydrated ions are incompressible, then the volume of water V_h bound to 1 mole of solute in solution, reduced to atmospheric pressure P_0 is

$$V_h = (n\beta^{(p_0)}V_1^{(p_0)} - \beta^{(p_0)}V^{(p_0)})/\beta_1^{(p_0)} \qquad (3.132)$$

$$V_h = -K_2/\beta_1^{(p_0)} \qquad (3.133)$$

where K_2 is the apparent molal compressibility of the solute. The values of V_h may be determined from compressibility and density data. The average hydration number of water molecules for a pair of solutes is then defined by

$$h = V_h/V_1^{(p_0)} \qquad (3.134)$$

Values of h obtained [537] for a number of electrolytes are shown in Table 3.22. It was also shown that h is a common hydration parameter to the concentration coefficient of electrostrictive pressure, the volume contraction due to the dissolution of 1 mole of solute and the volume of hydration water per mole of solute.

In the development of an equation for the molar volume of ions in aqueous solution which includes all types of ions to temperatures up to 200°C, Glueckauf [544] calculated the number of water molecules in a number of

Table 3.22

Salt	Hydration number	Temperature (°C)
LiCl	4.24	25
LiNO$_3$	2.98	25
NaCl	5.96	20
	4.98	25
NaBr	4.82	20
NaI	3.37	20
NaOH	8.40	25
KCl	5.17	20
	4.44	25
KBr	3.97	20
KI	2.92	20
NH$_4$Cl	1.5	25
MgCl$_2$	11.5	20
	10.4	25
Mg(NO$_3$)$_2$	8.38	25
CaCl$_2$	11.3	20
Ca(NO$_3$)$_2$	7.6	25
BaCl$_2$	12.6	20
	11.4	30
	11.1	40
Na$_2$CO$_3$	19.1	20
	15.7	25
Na$_2$SO$_4$	18.2	20
K$_2$SO$_4$	14.4	25
(NH$_4$)$_2$SO$_4$	9.34	25
MgSO$_4$	14.7	25
CuSO$_4$	13.2	25

layers around the ions. The apparent molar volume ϕ of ions in water at infinite dilution was expressed in terms of the intrinsic volume V^0 of the ion from which was subtracted the volume change arising as the result of forces exerted by the ion on the water dipoles. Assuming that water at its normal density surrounds a spherical ion, then because of the open structure of water, there will be a certain amount of dead space that has to be included in the intrinsic volume. The intrinsic volume can then be expressed by

$$V^0 = \tfrac{4}{3}\pi(r_0+a)^3 N \qquad (3.135)$$

where

$$a = (3v_w/4\pi N)^{\frac{1}{3}} - r_w \qquad (3.136)$$

where r_0 is the radius of the ion, r_w is the radius of the water molecule (1.38 Å), v_w is the molar volume of water at a given temperature [545]. If one envisions each ion originally surrounded by water molecules in such a way that the density is everywhere that of water, then the first layer will have its center at $r_1 = r_0 + r_w = \bar{r}$, and for subsequent layers it is assumed that the distance of each layer increases by 2.76 Å. It is noted that this distance is only correct for ions equal in size to the water molecule. However, this difference is not important at these relatively large distances from the ion. The number of water molecules χ_n in the nth layer surrounding an ion was calculated in the following manner. Each sphere of radius r_n contains approximately $\frac{1}{2}\chi_n$ water molecules with the center at r_n plus all the water molecules at $r < r_n$, plus the intrinsic volume of the ion. Therefore

$$\chi_1 = \tfrac{8}{3}(\pi/v_w)r_1^3 - 2(V^0/v_w) \tag{3.137}$$

or

$$\chi_n = \tfrac{8}{3}(\pi/v_w)r_n^3 - 2(V^0/v_w) - 2\sum_{a=1}^{a=n-1}\chi_a \tag{3.138}$$

The number of water molecules in the nth layer was calculated for the ions shown in Table 3.23.

Table 3.23

Number of Water Molecules in the nth Layer at
25°C in the Uncompressed State

Ion	Number of water molecules			
	$n=1$	$n=2$	$n=3$	$n=4$
I^-	7.3	22	35	60
Cs^+	5.1	18	30	53
K^+	3.8	16	27	48
Li^+	2.0	11	22	41

Benson and Copeland [546] have shown that the Mukerjee [547, 548] hypothesis of correlating partial molar volumes of ions with the continuum model of Born can be understood in terms of an isomorphic replacement of water molecules in a simple cubic lattice by ions whose sizes range from smaller to a little larger than water. From the arguments developed, optimum values for coordination numbers of 6–8 were calculated for ions of radius between 1 and 2 Å [546].

Partial molal volume data was used by Padova [549] to determine the solvation numbers of the lanthanides in aqueous solution. The solvation numbers were obtained with the equation [550]

$$\bar{V}_h = \bar{V}_2^0 + n^0\bar{V}^0 \tag{3.139}$$

where $\bar{V}_h$ is the molar volume of the hydrated ion, $\bar{V}_2^0$ the partial molal volume at infinite dilution of the lanthanide salts in water, $\bar{V}^0$ the partial molal volume of water at infinite dilution and n^0 the solvation number. The values of $\bar{V}_h$ were obtained from the relationship between B coefficient of the Jones–Dole equation [551] and $\bar{V}_h$, $\bar{V}_h = B/2.5$ [552]. The values for B and $\bar{V}_2^{\,0}$ were taken from the literature [553, 554]. Values for the ionic solvation numbers were then obtained by using values for the chloride ion of $n^0 = 1$ and $\bar{V}_2^{\,0} = 18.60$ [555]. The following solvation number values were obtained for the lanthanides: La^{3+} (8.5), Pr^{3+} (9.5), Nd^{3+} (9), Sm^{3+} (9.5), Tb^{3+} (10), Dy^{3+} (11), Ho^{3+} (11) and Er^{3+} (11).

The apparent molar volumes of HCl, HNO_3, H_2SO_4, KOH and NaOH have been determined at 15°C [556]. Assuming that the hydration numbers of H^+ is 1 and that of OH^- is O, the following hydration numbers were determined from the data: SO_4^{2-} (2), Cl^- (3), NO_3^- (3), K^+ (4), and Na^+ (10).

Solvation numbers have been calculated from the apparent molar volumes of electrolytes in sulfuric acid [557, 558]. The values obtained (n) are shown in Table 3.24 based upon a value of 3 for the sodium ion [559]. The relationship between osmotic coefficient and apparent molar volume was also used to obtain [559] the solvation numbers of ions(s) in sulfuric acid as shown in Table 3.24.

Table 3.24

Solvation Numbers (n) of Ions in Sulfuric Acid

Ion	n	s
Ba^{2+}	6	6.5
Na^+	3	3.0
Li^+	3	2.3
K^+	2	2.1
NH_4^+	1	1.2
Rb^+	0.7	—

If one treats the solutions as if containing a single electrolyte species of molality m the variation of the osmotic coefficient with electrolyte concentration may be expressed as

$$\phi = 1 + \phi_{el} + b \sum m \tag{3.140}$$

where ϕ_{el} is the electrostatic interionic contribution to the osmotic coefficient, $\sum m$ the total concentration of ionic species and b the osmotic coefficient parameter which is related to the solvation number(s) of the electrolyte by

$$b = [(r+s)^2/40.8] - (r/20.4) \tag{3.141}$$

In Eq. (3.141), r is the ratio of the apparent molar volume of the electrolyte to the molar volume of the solute, 54 cc, and s is the solvation number of the electrolyte [559]. Values of b were determined from osmotic coefficients using freezing point data [560]. A comparison has been made of the solvation numbers derived from cryoscopy and density measurements [561].

J. DIELECTRIC PROPERTIES

1. Theory and Experiment

   The molar dielectric depression observed in aqueous solutions of ions has been discussed assuming that the first shell of water molecules is dielectrically saturated with respect to positive ions and unsaturated in the case of negative ions [562]. It was found from the results of dielectric constant measurements made at centimeter wavelengths and extrapolated to zero frequency that the dielectric constant depends on the salt concentration in the following manner

$$E_s = E_{H_2O} - \delta c \qquad (3.142)$$

where E_s is the dielectric constant of the electrolyte solution, E_{H_2O} the dielectric constant of pure water, c the electrolyte concentration, and δ given by

$$\delta = 1.5 \left[V_2 \frac{(E_{H_2O} - E_{\infty \, ions})}{1000} + V_{H_2O} \frac{(E_{H_2O} - E_{\infty, H_2O})}{1000} n \right] \qquad (3.143)$$

where values of $E_{\infty \, ions}$ and E_{∞, H_2O} are taken as 2 and 5.5, respectively, V_2 and V_{H_2O} are the molar volumes of the solute and water, respectively, and n is the primary hydration number. Equations (3.142) and (3.143) were used to calculate the primary hydration numbers for a number of salts [563] as shown in Table 3.25. To determine primary hydration numbers for individual ions one must assume the relative contributions of the two ions in a single case.

Table 3.25

Primary Hydration Numbers of Salts from Dielectric Constant Measurements

Salt	n
NaF	4 ± 1
NaCl	6 ± 1
NaBr	6 ± 1
NaI	6 ± 1
LiCl	6 ± 1
KCl	5 ± 1
RbCl	4 ± 1
NH$_4$Cl	4 ± 1
KF	5 ± 1

The parameter δ can then be written as $\delta = (\delta^+ + \delta^-)/2$. Using NaCl as a reference and taking into consideration that a small amount of water outside the first hydration sphere is also dielectrically saturated, the following minimum hydration numbers were obtained (Table 3.26 [562]).

Table 3.26

Minimum Hydration Numbers of Positive Ions

Ion	n
H^+	10
Li^+	6
Na^+	4
K^+	4
Rb^+	4
Mg^{2+}	14
Ba^{2+}	14
La^{3+}	22

Glueckauf [564] states that the assumption made that positive ions are surrounded by a dielectrically saturated first shell of water molecules, while negative ions have their first shell of water molecules completely unsaturated is incapable of giving a quantitative description of the decrease in dielectric constant with added electrolyte and that, in fact, the fields in the neighborhood of all monovalent ions with the exception of H^+ are such that the first shell is far from saturated. Glueckaut [564] determined the dielectric constants of aqueous electrolyte solutions by integrating over the spatial distribution of the local dielectric constants in the manner appropriate for disperse systems. Using the dielectric constant change of the water molecules as a rough measure of their immobilization a mean hydration number was determined from the equation

$$h = \sum_p \bar{n}_p \left(1 - \frac{D_{wp} - n^2}{D_0 - n^2} \right) \tag{3.144}$$

where $\bar{n}_p$ is the number of water molecules that can be fitted into the pth shell, D_{wp} is the mean dielectric constant of the water molecules in that layer and D_0 in the dielectric constant of water at zero field strength. A rough estimate of the mean hydration number h from D_{wi} the value of D for first layer water molecules is given in Table 3.27 [565].

The second layer hydration makes a significant contribution only for polyvalent ions and H^+ and Li^+. For H^+ and Li^+ the second layer was assumed to be spheres of diameter 3.1 Å closely packed into the opening of the first layer.

Table 3.27

Hydration Number h Determined from D_{wi}

Ion	h
H^+	4
Li^+	6
Na^+	8
K^+	10
Cs^+	12
F^-	10
Cl^-	13
I^-	16

K. X RAY

1. Theory and Experiment

There have been a number of attempts to use X rays to study the structure of ionic solutions. Stewart studied the variation in the structure of water in ionic solutions using liquid diffraction curves of water in these solutions [565].

Prinns [566] obtained qualitative evidence of structure in aqueous solutions of $Th(NO_3)_4$, $UO_2(NO_3)_2$, $AgNO_3$, $Pb(NO_3)_2$, $Ba(NO_3)_2$, LiI, RbBr, and LiBr. It was shown that the X-ray diffraction patterns could be associated with the different structural elements in the solution. The intensity was assumed to be the result of three distributions: the scattering due to the ions, the scattering due to the water structure and the scattering due to ion–water structure. If the distribution function characteristic of each of these terms were known, one would then be able to obtain information about the hydration number and interionic distance as a function of concentration. Other ionic solutions have been studied for which radial distribution functions of NaOH, HCl, and H_3PO_4 have been determined, however, peak resolution in the distribution functions was not sufficient to permit the attainment of direct quantitative information [567–570].

Brady and Krause performed X-ray diffraction studies on concentrated KOH and KCl solutions and from the radial distribution functions obtained, hydration numbers could be calculated [571]. The distribution functions of two KOH solutions were found to contain primary peaks with maxima at about 2.87 Å for a 18.8% solution and 2.92 Å for the 11.4% solution. Beyond this primary peak the curves indicate that there is a region of decreased electron density followed by a peak at 4.75 Å. The identification of the peaks was made from information pertaining to the structure of water and the ionic radii of the species involved. The 4.75 Å peak corresponds to the second nearest-neighbor distance in water. Water with its tetrahedral structure has its second nearest-neighbor distance as the length of the tetrahedral edge. The calculated value of this length based on a nearest-neighbor distance of 2.92 Å is 4.75 Å in good agreement with the observed value. The ionic radii of K^+ and OH^- are both 1.33 Å and H_2O has an effective radius of 1.38 Å. The nearest-neighbor peak for liquid H_2O at 30°C has its maximum at 2.94 Å. Since the radius of H_2O and the radii of the ions are very similar, one would expect only a slight change in peak position in KOH solution. The primary peaks exhibited only one maximum and it was concluded that the primary peak in the KOH distribution functions includes nearest-neighbor H_2O molecules, K^+–H_2O neighbors and OH^-–H_2O neighbors. There was also evidence that some of the nearest neighbors are constantly changing position with other molecules in solution. From an analysis of the peak area in terms of the number of molecules around OH^-, the number of molecules around K^+ and the number of nearest-neighbor H_2O molecules around any other H_2O molecule, a hydration number of 4 was obtained for the K^+ ion. A subsequent X-ray investigation of aqueous solutions containing K^+, Li^+, OH^- and Cl^- gave hydration numbers of 4, 4, 6, and 8, respectively [572]. Coordination numbers have also been determined for Li^+, Na^+, K^+, OH^- and Cl^- from the X-ray analysis of aqueous solutions of these ions by Skryshevskii [573]. Hydration numbers for K^+ and Na^+ of 4 and 5.5, respectively, have been determined from the X-ray analysis of sulfate solutions [574]. The X-ray diffraction of water and a number of electrolyte solutions has been measured from which radial distribution functions were obtained [575]. Evidence was obtained that suggests that the K^+ and Ca^{2+} ions are surrounded by six water molecules coordinated octahedrally. X-ray diffraction studies of aqueous solutions of magnesium tetrafluoroborate indicate that number of water molecules in the first hydration shell around the Mg^{2+} ion is six and that in dilute solutions the number of water molecules in the second hydration shell is ten to eleven [576]. Aqueous solutions of $ZnBr_2$ and $ZnCl_2$ have been studies using X-ray techniques and evidence obtained for a coordination number of 4 for Zn^{2+} [577, 578].

L. NUCLEAR MAGNETIC RESONANCE

1. Theory and Experimental Technique

Solvation numbers have been determined for a number of ions by various nuclear magnetic resonance (NMR) techniques [579]. In an electrolyte solution, the solvent molecules can exist in several environments. These environments may be arbitrarily divided into bulk solvent regions, where solvent molecules are effectively out of range of ionic influence, secondary solvation regions, and primary solvation regions [580]. If exchange of molecules between all of these environments were very slow, then a number of peaks would be expected in the NMR spectrum of the solvent nuclei corresponding to the different interactions. Generally, the exchange of solvent molecules between the different environments is very rapid; consequently, the separate resonance signals expected for each environment are time averaged to a single peak whose shift from the pure solvent resonance peak reflects the mean effect of the different environments. However, certain experimental techniques have been developed that permit the observation of bulk and bound solvent molecules.

Primary and secondary solvation numbers have been determined for a number of ions by various NMR techniques. Probably the most direct NMR approach involves the observation and peak area determination of resonance peaks associated with the bulk solvent and with the bound solvent (i.e., solvent molecules coordinated with the ions). The initial study of this type was that of Jackson *et al.*, [581] in which the $H_2{}^{17}O$ NMR spectra of several aqueous solutions revealed peaks attributed to bulk and bound solvent. The ^{17}O nucleas was used as a "probe" due to the magnitude of the chemical shifts in NMR spectra which had been observed [582].

The ion first chosen for study was $(NH_3)_5CoOH_2^{3+}$ since the number of water molecules attached to the Co^{3+} ion was known [583] and because of the long exchange time of the bound water with bulk water. The ^{17}O NMR spectra of the aqueous solution of $(NH_3)_5CoOH_2^{3+}$ ion showed a well-defined resonance peak for the bound water that had a diamagnetic shift of 1.3 G with respect to the bulk water ^{17}O resonance peak. In a study of the effect of gadolinium ions on the $H_2{}^{17}O$ NMR, Shulman and Wyluda [584] had previously reported a diamagnetic shift of the ^{17}O resonance that was linear with gadolinium concentration; however, only one resonance peak was observed in the system. Shulman [585] had also observed that paramagnetic ions can produce large shifts in the resonance absorption of other nuclei. When Co^{2+} ions were introduced into the $(NH_3)_5CoOH_2^{3+}$ system, Jackson *et al.* [581] observed that the resonance peak of ^{17}O in bulk water experienced a paragmagnetic shift of -3.3 G while the ^{17}O resonance peak of the water bound to the Co^{3+} ion was unaffected. The explanation for the difference in behavior lies within the fact that a great many more bulk water molecules come into contact with the Co^{2+} ions than the water molecules in the $(NH_3)_5CoOH_2^{3+}$ ion. Consequently, the bulk water resonance is shifted while no significant contribution is made by the Co^{2+} ion to the average field which the water molecule in the $(NH_3)_5CoOH_2^{3+}$ ion experiences.

The ^{17}O NMR spectra of aqueous solutions of Al^{3+}, Be^{2+}, and Ga^{3+} show only a single resonance peak for water; however, upon the addition of a small amount of Co^{2+} to the system, separate peaks appear associated with bulk solvent and those water molecules bound by the ions [581]. This method failed for solutions of Mg^{2+}, Ba^{2+}, Sn^{2+}, Hg^{2+}, and Bi^{3+}. The requirements for the detection of the two signals are that the residence time of a water molecule bound to the diamagnetic cation in the primary solvation sphere be 10^{-4} sec or longer and that the residence time of the water in the primary sphere of the paramagnetic ion be significantly less than 10^{-4} sec.

Since it is possible to separate the ^{17}O NMR signals of bound and bulk solvent water, the number of water molecules held in the primary sphere could be calculated in one of two ways—by comparing the areas under the two ^{17}O resonance peaks or by measuring the shift of the resonance peak produced by the paramagnetic ion in the presence and absence of the diamagnetic ion.

Although the precision of measurement (unfavorable signal-to-noise ratio) was not sufficiently high for Jackson et al. [581] to determine the number of water molecules associated with Al^{3+}, Be^{2+}, or Ga^{3+}, Connick and Fiat [586] were able to determine the coordination of Al^{3+} and Be^{2+} with the peak area comparison method by increasing the signal-to-noise ratio using water of greater ^{17}O enrichment and the sideband detection technique. Solutions of $AlCl_3$ and $BeCl_2$ to which Co^{2+} ions had been added gave an average solvation number of 5.9 and 4.2 for Al^{3+} and Be^{2+}, respectively.

Alei and Jackson [587] have determined the solvation numbers of Al^{3+}, Be^{2+}, and Cr^{3+} ions using the chemical shift method suggested by Jackson et al. [581]. In an aqueous solution of a paramagnetic ion in which there is a rapid exchange of water molecules between the first coordination sphere of the ion and the bulk of the solvent, a single $H_2^{17}O$ NMR signal is observed. This resonance signal is shifted from its normal position in pure water by an amount dependent on the ratio of the concentration of the paramagnetic ion to the total labile water present. The addition of a second ion for which the water in the first coordination sphere is nonlabile causes a decrease in the amount of water available to interact with the paramagnetic ion. Consequently, an increase in the shift of the labile $H_2^{17}O$ resonance is observed. In general the ^{17}O NMR shifts, observed in solutions containing a paramagnetic ion and a diamagnetic ion and in solutions containing only the paramagnetic ion, are related by the equation

$$\delta m'_{H_2O}/m_{Dy(III)} = \delta^0 m^0_{H_2O}/m^0_{Dy(III)} \qquad (3.145)$$

where m_x represents the millimoles of species x in solutions and the superscript 0 denotes the quantity in the solution containing only the paramagnetic ion Dy(III). The only unknown in this expression is m'_{H_2O}, the number of millimoles of labile solvent water in the solution containing the diamagnetic ion. The difference between the known total millimoles of water and m'_{H_2O} divided by the number of millimoles of the diamagnetic ion gives the solvation number of the diamagnetic ion. The Dy(III) ion was used as the paramagnetic ion because it produces a large shift with minimum broadening. The shift of the labile $H_2{}^{17}O$ resonance by $Dy(III)$ was shown to be proportional to the ratio of $Dy(III)$ to labile water and that the proportionality constant is independent of the concentrations of $Dy(III)$, H^+, and $ClO_4{}^-$ ions in the concentration range of interest. The molal shift thus determined for the $Dy(III)$ ion was 3.68 in agreement with a value of 3.81 ± 0.10 determined previously [588]. Solvation numbers of 5.9 (the average of 5 different measurements) and 3.8 (the average of three different measurements) were obtained for the Al^{3+} and Be^{2+} ions, respectively. The ratio of $Dy(III)$ or H^+ to diamagnetic ion had no significant effect on the solvation number in the region studied.

The chemical shift method of determining solvation numbers may also be applied to paramagnetic ions if the magnetic influence of the paramagnetic ion does not extend beyond the first solvation sphere [587]. However, this method was applied to the $Cr(III)$ ion whose magnetic influence does extend beyond the first solvation sphere and produces a shift in the labile $H_2{}^{17}O$ resonance by the following modification. Alei [589] has shown that in perchlorate solutions of $Cr(III)$ the shift of the labile $H_2{}^{17}O$ resonance is related to $Cr(III)$, $ClO_4{}^-$, and H_2O by the expression

$$1/\delta_{H_2O} = (0.0415)(m_{ClO_4^-}/m_{Cr(III)}) + (0.0346)(m_{H_2O}/m_{Cr(III)}) - 0.0113 \tag{3.146}$$

In solutions containing both $Dy(III)$ and $Cr(III)$, Eq. (3.146) represents the contribution by $Cr(III)$ to the labile $H_2{}^{17}O$ shift, the observed shift being the sum of the contributions attributed to $Cr(III)$ and $Dy(III)$. This permits the calculation of the shift of the labile $H_2{}^{17}O$ resonance by $Dy(III)$ in the $Cr(III)$–$Dy(III)$ solutions. Comparing this derived shift with that in a reference $Dy(III)$ solution, the solvation number of the $Cr(III)$ ion may be determined in the same manner as described for a diamagnetic ion.

A value of 6.8 (the average of three different measurements) was obtained for the solvation number of $Cr(III)$ by this method. This value is significantly greater than the value of 6 determined by the isotope dilution method [590]. This discrepancy and the validity of the assumptions made in acquiring the value, 6.8, were discussed by the authors.

The chemical shift method [587] has been used to determine the hydration numbers of certain organometallic cations [591, 592]. The hydration number of $(CH_3)_3Pt^+$ was found to be 3. In solutions containing $(NH_3)_2Pt^+$ ions, however, bulk and bound water resonance signals were observed from which a hydration number of 2 was obtained for the cation.

Cation hydration numbers in hydrate melts have also been obtained for melts of calcium nitrate tetrahydrate with anhydrous potassium nitrate, tetramethylammonia nitrate, and magnesium nitrate using the chemical shift method [593]. The proton chemical shifts of water indicated that in melts with nitrates of univalent cations the Ca^{2+} ion is selectively hydrated. In melts with $Mg(NO_3)_2$, the Mg^{2+} ion was found to be selectively hydrated at the expense of the Ca^{2+} ion.

A solvation number of 3.8 ± 0.2 has been obtained for the vanadyl(IV) ion by the chemical shift method [594]. Fiat and Connick [595] have determined the solvation number of gallium(III) from the ratio of the area of the bound to bulk water resonance signal, where one resonance signal is shifted relative to the other by the addition of Co^{2+} ions, and also by a chemical shift method which involved the measurement of the shift of the free water relative to reference pure water. The $H_2{}^{17}O$ NMR was studied in each method. The area ratio method gave a solvation number of 5.89 ± 0.20 while a value of 6.28 ± 0.26 was obtained from the chemical shift method.

The hydration numbers for several electrolytes have been determined from temperature effects on the proton shift of water [596–601]. The proton shift of water is strongly dependent upon temperature [602]. The observed proton shift is believed to be an average of the bonded and nonbonded environments of the proton; water vapor having few hydrogen bonds is characterized by a high field shift. Likewise, the proton chemical shift of liquid water would be expected to shift to high field with increasing temperature.

The chemical shift of pure water was measured relative to gaseous ethane over the temperature range 5°–95°C [596]. Ethane was used as the reference because its shift is independent of temperature [603]. Over the temperature range studied, the proton chemical shift was found to be a linear function of temperature as described by the equation

$$\delta_{H_2O} = 0.0102t - 4.38 \quad \text{ppm} \tag{3.147}$$

where t is the temperature in degrees centigrade. The value of the slope, 0.0102, is in reasonable agreement with the value 0.0095 ppm/°C [602] obtained using liquid cyclopentane as a reference whose shift varies slightly with temperature.

In an aqueous electrolyte solution, a single resonance signal is observed for water protons that is the weighted average of the different environments experienced by the protons (i.e., hydrated, nonhydrogen bonded and hydrogen bonded forms). Any given proton residing in the normal water structure has a shift δ_N that is identical to the shift, δ_{H_2O}, of pure water at the temperature of the observation. Protons of water molecules involved in the solvation of an ion have a resonance signal that is shifted relative to that of pure water. This solvation shift δ_S depends upon both the cation and anion. Since the average time a proton resides in any environment is proportional to its instantaneous mole fraction, the proton shift observed for an aqueous solution can be written as

$$\delta_{H_2O} = X_N \delta_N + X_S \delta_S \tag{3.148}$$

where X_N and X_S are the mole fractions of protons in the normal water and hydrated form. This assumes that water molecules beyond the first solvation sphere are unaffected by the ions. If h represents the total effective solvation number (i.e., moles of solvated water per stoichiometric moles of salt) and m represents the stoichiometric molality, then

$$\delta_{H_2O} = (hm/55.55)(\delta_S - \delta_N) + \delta_N \qquad (3.149)$$

Equation (3.149) predicts a linear relationship between shift and molality if h and δ_S remain independent of concentration. Experimentally, it was observed that at the various temperatures the shift-molality curves were linear at low concentrations but deviated from linearity at high concentrations. This deviation was thought to result from ion pairing and increased competition between cations and anions for water molecules. Equation (3.149) also indicates that at some temperature $\delta_{H_2O} = \delta_N$. Obviously, at this temperature $\delta_S = \delta_N$. A plot of δ_{H_2O} of various molal solutions of sodium chloride as a function of temperature showed that all the curves intersect the pure water line at approximately 110°C, giving a value for $\delta_S = 3.26$ ppm. Assuming that the shift of a proton in the solvated environment does not vary with temperature, the effective solvation number of the salt, NaCl, may be calculated from the equation

$$h = (55.55/m)[(\delta_{H_2O} - \delta_N)/(\delta_S - \delta_m)] \qquad (3.150)$$

Substitution of the experimentally determined values of the parameters into Eq. (3.150) gives a total effective hydration number of 4 for NaCl over the concentration and temperature range studied (see Table 3.28). The data indicate that within experimental error the total solvation number is independent of temperature and concentration. The effective solvation number of 4 obtained for NaCl is in general agreement with that determined by other techniques but differs widely from the value of 16.8 determined by another NMR method [604].

This method has also been applied to solutions of sodium perchlorate, hydrochloric acid, perchloric acid [605] and aluminum nitrate [600]. Total hydration numbers of 3.0 ± 0.2, 3.4 ± 0.4 and 2.6 ± 0.4 were obtained for $NaClO_4$, HCl, and HClO, respectively. Assignments of individual hydration numbers for $Na^+ = 2$, $^4Cl^- = 2$, and $ClO_4^- = 1$ were made by assuming that H^+ is coordinated to a single water molecule. A total hydration number of 13.4 was obtained for $Al(NO_3)_3$. Two explanations of this value were postulated:

1. The aluminum ion is hydrated to six water molecules and each nitrate ion is hydrated to one or three water molecules.

2. Each of the six water molecules in the primary layer of the aluminum ion are bound to two molecules of the secondary layer, with nitrate ions replacing some secondary water molecules.

Broersma [606] obtained a value of 16 for the aluminum ion from proton relaxation-time measurements; consequently, if the nitrate ions replace some of the secondary water molecules, the value 13.4 would seem to be plausible.

A study of the ^7Li NMR of aqueous lithium halide solutions revealed a downfield chemical shift in resonance position with an increase in concentration [607]. Plots of chemical shift versus mole fraction of salt were linear up to a mole fraction of 0.3. The shifts were also apparently temperature independent. The observed shifts were interpreted in terms of a progressive polarization of the lithium ion by the close approach of an increasing proportion of anions. The data suggest a solution model of a relatively random mixture of ions and molecules with no lasting interaction between them; furthermore, the data are not compatible with the concept of a tightly bound hydration shell or of a tightly bound ion-pair. The average number of sites available for competition around the lithium ion was calculated from $\delta = \delta^0 Xm$, where δ is the observed chemical, δ^0 is the limiting shift corresponding to the pure halide, X is the number of sites, and m is the mole fraction of salt. From a plot of δ/δ^0 versus m, a value of $X = 2$ was obtained for values of $m < 0.3$. It was suggested that this value of X does not so much represent the classical hydration number as the average number of sites accessible to halide substitution. The number might also represent the effectiveness of the water molecules to exclude anions from the cation.

Swinehart and Taube [608] determine the solvation number of the magnesium ion in methanol by an NMR technique.

Proton NMR spectra of solutions composed of 1 mole $Mg(ClO_4)_2$: 17.1 moles CH_3OH: 1.4 mole H_2O and 1 mole $Mg(ClO_4)_2$: 17.1 moles CH_3OH: 3.8 moles H_2O taken at $-75°C$ showed separated hydroxyl proton resonance peaks for methanol in the bound and bulk state. A solvation number of 5.7 ± 0.2 was determined for the magnesium ion from measurements of the resonance peak area associated with the bulk and bound hydroxyl protons of methanol and with the water proton. The deviation from a value of 6 for the solvation number was attributed to association of magnesium with the perchlorate ion. The separation of the absorption signals for the bulk and bound solvent environments was feasible because of the low temperature employed in recording the spectra. At $-75°C$ the frequency of exchange of nuclei between the two environments was small compared to the frequency difference between the chemical shifts. At $-85°C$ it was observed that a decrease in the amount of solvation water relative to the amount of solvation methanol occurred at this temperature indicating a reequilibration of water between the solvated cation and solvent.

A more extensive study of the $Mg(ClO_4)_2$–methanol system has been made by Nakamura and Meiboom [609]. At low temperature the OH protons of the methanol molecules bound to the Mg^{2+} ion gave a resonance peak well separated from the bulk OH peak. The symmetry of the OH quadruplet of

methanol molecules in the solvation shell suggests a regular octahedral configuration in the first solvation shell under the conditions studied. The exchange
of the OH protons of the methanol molecules in the solvation shell with bulk
solvent and for exchange within the solvent complex is very slow as indicated
by the well defined structure of the quadruplet. The solvation number of the
magnesium ion in anhydrous methanol was determined from the molar ratio
of $Mg(ClO_4)_2$ and methanol in the solution and the ratio of the area under
each resonance signal. The solvation number obtained in this manner was 6.
No appreciable "close" ion association was indicated by the data since the
solvation number determined remained constant even at high concentration
of $Mg(ClO_4)_2$ and at very low temperature. Penetration of the perchlorate
ion into the first solvation shell of the Mg^{2+} ion would have caused a decrease
in the calculated solvation number.

The addition of a small amount of $Cu(ClO_4)_2$ to the $Mg(ClO_4)_2-CH_3OH$
system was found to broaden the bulk OH and CH_3 proton resonance signals
while the proton signals of solvation shell methanol molecules are only
slightly affected. This technique [581] provides an excellent method for
determining whether the exchange of protons of the solvation shell is due to
protons only or to whole molecules. It was found that the CH_3 and OH proton
resonance signals of the solvation shell molecules broaden and disappear at
the same time as the temperature is increased demonstrating that the whole
molecule is exchanging in the solution.

Other NMR studies of the Mg^{2+} ion in aqueous acetone solution [610–612]
and in methanolic acetone solution [612] indicate that the solvation number
for this cation if 6. This value was obtained by the bulk-bound signal area
ratio method. Direct evidence of anion shifts in the hydroxyl proton resonance
spectrum of methanol solutions of magnesium perchlorate has been obtained
[613]. The solvation number of the Mg^{2+} ion in liquid ammonia has been
determined to be 5 [614].

A solvation number of 6 has been obtained for Co^{2+} in anhydrous methanol
[615], water [616], N,N-dimethylformamide [617] and acetonitrile [618],
using the bulk-bound signal area ratio method. In these solutions no consistent
trend of the solvation number was discernible with either temperature or
solution composition.

A solvation number of 6 has been obtained for Al^{3+} in dimethylformamide
[619], anydrous dimethylsulfoxide [620], aqueous dimethylsulfoxide [621],
and in water [622–627] by the bulk-bound signal area ratio method. It is also
of interest to note that separate resonance peaks corresponding to differently
hydrated aluminum ions have been observed in the spectrum of aqueous
acetonitrile solutions of $Al(ClO_4)_3$ [628]. Solvation numbers of 4 for Be^{2+} in
dimethylformamide [629] and in aqueous and aqueous–acetone solution
[622]; 6 for Ga^{3+} [629, 623, 630 a minimum of 4]; 6 for In^{3+} [622]; 3.9 for
Sc^{3+} [631, 632]; 2.4 for Y^{3+} [631, 632]; 2.9 for Th^{4+} [631, 632]; 2 for Sn^{2+}

[633]; and about 4 for UO_2^{2+} in aqueous mixed solvent systems [634, 635] have also been obtained using the area ratio method.

A modified NMR technique has been proposed for the determination of the solvation number of metal ions in aqueous solutions consisting of a mixture of a salt of the metal ion under investigation and a very soluble salt of a metal ion with which the coordinated water has been shown to be labile [636]. Bulk and bound water peaks are still used to determine solvation numbers. From the values of the solvation numbers obtained for Al^{3+} and Ga^{3+}, evidence for contact ion pairing was obtained.

A method has been developed to determine hydration numbers which utilizes integral measurements of resonance singals [637]. The method has proven to be accurate for determining hydration numbers from measurements of the total hydrogen content of materials dissolved in D_2O. The requirements are that the materials contain some nonexchangeable protons and no paramagnetic ions.

The solvation of several paramagnetic ions have been studied in aqueous alcohol mixtures using selective broadening of the NMR absorption peaks of ethyl alcohol [638]. The solvation studies were conducted in deuterated ethyl alcohol solutions of $Cu(ND_3)_2 \cdot 3H_2O$, $CuCl_2$, $MnCl_2$, and $Cr(NO_3)_3 \cdot 9H_2O$ with varying amounts of D_2O added. From the ratio of amplitudes for the protons resonance of the CH_2 and CH_3 groups as a function of D_2O concentration, selective solvation was observed. A hydration number of 120 was obtained for the Cu^{2+} ion.

Proton relaxation time values as a function of salt concentration have been used to determine the hydration numbers of several cations and anions in alkali halide solutions [639]. Assuming the hydration number of the K^+ ion to be 6, hydration numbers for the following ions were obtained: $Li = 1 \pm 1$; $Na^+ = 3.6 \pm 1$; $Rb^+ = 9.9 \pm 2$; $Cs^+ = 14.6 \pm 2$; $F^- = 9.9 \pm 2$; $Cl^- = 13.2 \pm 2$; $Br^- = 16.2 \pm 2$; and $I^- = 21.8 \pm 2$.

The correlation between relaxation time and salt concentration for methanol solutions of $CoCl_2$ and for aqueous solutions of HCl, NaCl, KCl, $MgCl_2$, and $CaCl_2$ has been used to calculate solvation numbers for the cations and anions present [640, 641]. The first shell solvation numbers determined for Co^{2+}, Cu^{3+}, and Cl^- ions in methanol are 4, 6, and 8, respectively.

Hydration numbers have been determined ($\pm 40\%$) for $Li^+ = 5$, $Na^+ = 3$, $K^+ = 1$, $Mg^{2+} = 6$, $Zn^{2+} = 12$, $Al^{3+} = 16$, $Cl^- = 2$, $OH^- = 4$, $SO_4^{2-} = 5$, ice $= -5$ (Broersma states that the value for ice is to be subtracted per monovalent ion) from proton relaxation time measurements using a modified Stokes equation in the calculation [642].

Hindman [643] has obtained effective hydration numbers for univalent cations and anions in aqueous solution from PMR chemical shifts. A solution model is adopted that subdivides the observed chemical shift produced by an ion into four terms: bond-breaking, structural, polarization, and electrostatic. In this interpretation the hydration of an ion is considered in terms of effective hydration numbers in which weaker interactions with a large number of water molecules are replaced by strong interaction with a limited number. Values of the effective hydration numbers of the following ions were obtained: $Li^+ = 4.0$, $Na^+ = 3.1$, $K^+ = 2.1$, $Rb^+ = 1.6$, $Cs^+ = 1.0$, $Ag^+ = 2$, $F^- = 1.6$ (Cl^-, Br^-, I^-, NO_3^-, $ClO_4^- = 0$). Hindman [643] concluded that the concept of a complete hydration sphere of tightly bound water is not compatible with the data, and that the effectiveness of these ions in coordinating strongly with water decreases with increasing ionic radius. A structure making effect is suggested by the data for the Li^+ ion while the larger halide ions tend to break down the water-structure. The data also indicate that of the halide ions, only the F^- ion forms a hydrate in the chemical sense.

Swift and Sayre [644] have developed a technique for the determination of the primary hydration numbers of cations in aqueous solutions. The primary hydration number is defined as the number of water molecules associated with the ion for a time that is long compared to the time of diffusion, the method being applicable to those ions for which the average lifetime of an associated water molecule in the primary hydration shell is less than about 10^{-4} sec. This technique is based upon the kinetically distinguishable water molecule exchange with hydrated manganous ion and the relationship between PMR relaxation time and this exchange.

In an aqueous electrolyte solution there are a number of different environments with which a water molecule might be associated: those associated with an anion; those in the primary hydration sphere of a cation; those associated only with other water molecules. Kinetically it might be possible to distinguish between these water molecules by adding to the solution another species which is capable of reacting with water molecules. If the water molecules are kinetically distinguishable, the reaction will proceed as the sum of parallel reactions characterized by different rate constants. A "probe" species must satisfy the following conditions if this method is to be successful; it must react directly with water molecules; the rate constant for the direct reaction between the "probe" species and the water in a primary hydration sphere must be considerably different from the rate of reaction for all other types of water molecules; the lifetime of the "probe" must be relatable to a precisely measurable experimental quantity.

The manganous ion was found to be a suitable probe and the pertinent exchange reactions involved are

$$H_2O_a + Mn(H_2O)_5(H_2O^*)^{2+} \xrightarrow{k_a} H_2O_a^* + Mn(H_2O)_6^{2+} \quad (3.151)$$

$$H_2O_b + Mn(H_2O)_5(H_2O^*)^{2+} \xrightarrow{k_b} H_2O_b^* + Mn(H_2O)_6^{2+} \quad (3.152)$$

The rate of disappearance of $Mn(H_2O)_5(H_2O^*)^{2+}$ is, therefore,

$$\frac{-d[Mn(H_2O)_5(H_2O^*)^{2+}]}{[Mn(H_2O)_5(H_2O^*)^{2+}]\,dt} = k_a[H_2O_a] + k_b[H_2O_b] = \frac{1}{\tau_{Mn}}$$

$$(3.153)$$

where τ_{Mn} is the lifetime, $[H_2O_a]$ and $[H_2O_b]$ are bulk water molecules or a water molecule associated with an anion and a water molecule in the primary hydration sphere of a cation, respectively. Assuming that k_a is much greater than k_b, the rate equation can be reduced to

$$\frac{-d[Mn(H_2O)_5(H_2O^*)^{2+}]}{[Mn(H_2O)_5(H_2O^*)^{2+}]} = \frac{1}{\tau_{Mn}} = k_a[H_2O_a] \qquad (3.154)$$

The dependence of the lifetime τ_{Mn} on the concentration of added cation can theoretically be determined from PMR linewidth measurements; however, ionic strength effects and anion effects prevented the formulation of an absolute emperical relationship between linewidths and the hydration number of a cation. A technique involving a comparison of linewidths obtained for two solutions, one containing a cation of known primary hydration number with the other solution containing the cation being studied was used. A common anion was present in each solution along with the "probe" manganous cation. An empirical formula was derived relating the ratio of proton linewidths to the primary hydration number of the cation studied

$$\frac{W_{AB} - W_A}{W_{AB'} - W_{A'}} = \frac{[H_2O]_{AB'} - N_{A'}[A']}{[H_2O]_{AB} - N_A[A]} \qquad (3.155)$$

where the subscript AB refers to a solution containing a standard cation of known hydration number, AB' to a solution containing the cation being studied, A and A' to solutions of the same composition as AB and AB' except containing no manganous ion. W_{AB}, W_A, $W_{AB'}$, and $W_{A'}$ are the linewidths determined in a single comparison, $[A]$ and $[A']$ are cation concentrations, $[H_2O]_{AB}$ and $[H_2O]_{AB'}$ are water concentrations, and N_A and $N_{A'}$ are the primary hydrations numbers of the two cations. Equation (3.155) was tested with respect to temperature, anion, and concentration using the ions, Al^{3+}, Be^{2+}, NH_4^+, and H^+, whose primary hydration numbers are known. An application of this technique gave primary hydration numbers for the following cations: $Mg^{2+} = 3.8$, $Ca^{2+} = 4.3$, $Sr^{2+} = 5.0$, $Ba^{2+} = 5.7$, $Zn^{2+} = 3.9$, $Cd^{2+} = 5.0$, $Hg^{2+} = 4.9$, and $Pb^{2+} = 5.7$. A direct correlation between primary hydration number and ionic radius was found for the doubly charged series. The relatively low primary hydration numbers obtained by this method are interpreted as a reflection of the effect of water structure in determining the structure of hydrated ions.

Swift [645] has also determined the hydration numbers of the Ga^{3+} and Th^{4+} ions by the method just described. A value of 6 was obtained for the Ga^{3+} ion. The hydration number determined for the Th^{4+} ion was 10 ± 0.2, which is especially interesting with respect to the recently prepared [646, 647] complexes of thorium with a coordination number of 10. The hydration numbers for both Ga^{3+} and Th^{4+} were found to be temperature independent. Linewidth measurements as a function of temperature for aqueous solutions of the Ni^{2+} ion were used by Swift and Weinberger to obtain a solvation number of 6 for this ion [648].

A discussion of the validity of the method developed by Swift [644] for the determination of primary solvation numbers of cations has been presented in the literature by Meiboom [649] as well as the rebuttal argument of Swift [650].

Proton chemical shift measurements of aqueous solutions of Co^{2+} and histidine have been used to estimate the number of first hydration sphere sites of Co^{2+} that are utilized by a ligand and, therefore, are not accessible to water [651]. Since the PMR of coordinated and free water is a single resonance signal at constant Co^{2+} concentration, displacement of water by another ligand, histidine, causes a shift in water PMR toward the cobalt-free resonance line. This shift is used to estimate the number of available sites.

PMR shifts in aqueous solutions of paramagnetic metal ions have also been used to determine the number of sites in the first hydration shell of Co^{2+} which are occupied by water [652]. The chemical shift produced by the presence of paramagnetic ions in solution can be related to the coordination number of the metal ion by the equation

$$\Delta\omega = pq\omega_1[S(S+1)g|B|/3ktv_1]\,A \qquad (3.156)$$

where q is the coordination number and $p = [M/H_2O]$. Introduction of a ligand into the metal ion solution causes the displacement of water from the first hydration sphere, thereby producing a difference in chemical shift as compared to the ligand-free solution. Plots of water proton shift versus $[Co^{2+}]$ for solutions with and without an excess ligand (NaH_2PO_4) added were linear. In the phosphate solution the slope was 0.83 times as great as that for the ligand-free solutions. Therefore, assuming the coordination number of CO^{2+} to be 6, one water site is then occupied by a phosphate group. Similar results were obtained for RNA Co^{2+} solutions; Co^{2+} binds to one phosphate of RNA.

Schaschel and Day [653] have studied ion solvation in nonaqueous solution using a PMR technique that involves measuring the proton chemical shifts of molecules such as tetrahydrofuran, diethyl ether, and triethylamine as they are added to a solution of sodium tetrabutyl aluminate in cyclohexane. Evidence for a four-coordinated sodium ion was obtained with tetrahydrofuran while the data indicate a 1:1 complex for sodium diethyl ether and sodium triethylamine.

The solvation number of the electron in liquid ammonia has recently been calculated to be in the range of 20–40 for a dilute solution of potassium in liquid ammonia [654]. From an analysis of the contribution to the proton line shape from N^{14} spin lattice relaxation the range of the solvation number was determined.

According to Samoilov [190], in aqueous electrolyte solutions, a water molecule not close to an ion remains in its equilibrium position for an average time t_1 which is approximately that for pure water. In the neighborhood of an ion, the water molecule would have a different average time t_2 for maintaining its equilibrium position depending upon the potential barrier opposing exchange of nearest water molecules under the influence of the ions. The change in potential barrier between the two possible equilibrium states of water molecules produced by ions may be defined as $t_2/t_1 = e^{\Delta E/RT}$. If water molecules are firmly held by the ions, the ratio t_2/t_1 is large ($t_2 > t_1$) and $\Delta E > 0$; if $t_2 < t_1$, then weakening of the ion–water bond is indicated and $\Delta E < 0$. In the latter case water molecules around an ion are more mobile than in pure water or the frequency of exchange of water molecules around the ions is greater than in regions of pure water. This phenomenon is termed "negative" hydration.

Since proton relaxation times in water molecules are a direct function of water mobility, PMR relaxation time measurements in various aqueous salt solutions should give some indication of "negative" and "positive" hydration. Zhernovoi and Yakovlev [655] have measured the water proton relaxation times in solutions containing KH_2PO_2 and NaH_2PO_2. Plots of water proton relaxation time versus salt concentration showed that in dilute solutions the mean relaxation time in potassium solutions was much greater than in sodium solutions. It was determined that the mean relaxation time in K^+ solutions (the mean of relaxation of protons close to the K^+ ion and those remote was greater than that for remote) was greater than that for remote water protons while in Na^+ solutions the mean was less than that of remote water. This was interpreted as indicating "negative" hydration for the K^+ ion, the water close to the ion being more mobile than remote water, therefore giving a mean relaxation time greater than just remote water. Positive hydration was, therefore, indicated for Na^+ ions.

Another way to investigate the mobility of water molecules in electrolyte solutions is by determination of the self-diffusion coefficient of water in these solutions. The self-diffusion coefficients of water molecules in NaCl, KCl, and KI solutions have been determined by nuclear magnetic spin-echo techniques [656]. The amplitudes of the spin-echo signals as a function of magnetic field gradient G is related to the self-diffusion coefficient D by the equation

$$E_1/E_2 = \exp[-\tfrac{3}{2}v^2 D\tau^3](G_2{}^2 - G_1{}^2) \qquad (3.157)$$

All parameters, except D, are experimentally determined or obtained from tables. Self-diffusion coefficients were determined as a function of salt concentration for the three solutions studies. Plots of D versus [NaCl] decreased with concentration and were less than pure water whereas the diffusion coefficients obtained for KCl and KI solutions were greater than in pure water. Since both cation and anion affect the diffusion of water, the coefficients obtained are therefore a mean of the two effects. It was assumed that the anion–water bond resembled a water–water bond more than a cation–water bond; and, therefore, the effect of cations on the state of water should predominate. The experimental results indicate that the Na^+ ion is positively hydrated and the K^+ ion is negatively hydrated. Since the diffusion coefficients obtained for the KI solutions are greater than those for the KCl solutions, I^- ions are thought to be more negatively hydrated or reinforce the mobility of water more than Cl^- ions. The curves for the KCl and KI solutions showed definite maxima which indicate that in relatively concentrated solutions, in which there is a disappearance in ordering to a large degree, negative hydration is not prominent.

Valiev [657] in studying the role of ion–ion interactions in the quadrupole relaxation of the nuclear spins of diamagnetic ions presents a theory for this relaxation process and analyzes the existing experimental data in light of the theory. From an analysis of the data of Hertz [658] on linewidth measurements of Br^{79} and I^{127} in aqueous solution of a series of bromides and iodides, the interaction energy of I^- with water was determined. The I^-–H_2O interaction was found to be weaker than that of H_2O–H_2O interactions; $\Delta E = E_{I-H_2O} - E_{H_2O-H_2O} = -1.6$ kcal/mole. Similar calculations were made for the Br^- ion and a value of $\Delta E = -1.02$ kcal/mole obtained. The quantity ΔE is the same as that used by Samoilov [190] in defining negative hydration. Therefore, negative hydration is indicated for the I^- and Br^- ions. The data of Richards and Yorke [659] on quadrupole relaxation of the Br^{81}, Br^{79}, and Na^{23} nuclei in solutions of NaBr, CaBr$_2$, and KBr are discussed in terms of positive and negative solvation. The cations, Na^+ and Ca^{2+}, are considered to be positively hydrated while the I^-, Br^-, and Cs^+ ions are negatively hydrated. Nuclear spin relaxation times T_1 and T_2 are used to deduce the degree of hydration of electrolytes, particularly the cases of negative hydration [660].

Spin-echo techniques have been used to determine the frequency dependence of the relaxation times of proton spins in aqueous solutions containing $[Cr(CN_6)]^{3-}$ complexes [661]. The relaxation was found to be due to dipole–dipole interactions between protons and the paramagnetic complex and excited by the translational difusion of water molecules in relation to these ions. From this the diffusion coefficient of water near the $[Cr(CN)_6]^{3-}$ complexes was determined and was found to be higher than the diffusion coefficient in pure water. It was, therefore, assumed that the $[Cr(CN)_6]^{3-}$ complex is negatively hydrated.

Proton relaxation studies of ionic solutions by Broersma [642] also indicate negative solvation by some ions although the data is discussed in terms of decreasing order of water; more water aggregates being broken up than are formed when ions are dissolved in the water.

The effective hydration number obtained by Hindman [643] of zero for Cl^-, Br^-, I^-, NO_3^-, and ClO_4^- could also be looked upon as an indication of negative solvation.

IV. Evaluation

The results presented in this review indicate that, as one referee commented, the field is in a state of confusion. To some extent the confusion is more imaginary than real because the solvation number determined for any ion depends upon the method of measurement. For example, mobility measurements indicate the number of molecules moving with the ion while dielectric constant measurements indicate the number of molecules in the first sheath. However, a real source of confusion resides in the fact that there is no standard reference ion; coosequently, in many instances the solvation number determined depends upon the reference ion and its assumed solvation number. The above statement is obvious since most methods measure the total solvation of the electrolyte and not that of the individual ions. One purpose of writing this review was to call attention to the state of the field at this time in hopes that it would have some heuristic value. There can be little hope of reconciling the results of different methods of measurement until there is definite knowledge of the structure of solutions and of the structure of solvations. That the possession of this knowledge is not imminent is testified to be the many models of the structure of solutions and solvated ions currently appearing in the literature.

With respect to the individual methods the following comments can be made. With mobility measurements one would obtain erroneous results for ions whose sizes as determined in solution are small compared to the particles of the medium, because of the limitation of Stokes's law. However, one must bear in mind that ion size as measured by X-ray diffraction on crystals is not necessarily the same as that of the ion in solution, and, therefore, there is no *a priori* way of determining whether Stokes's law applies from just a consideration of X-ray determined ion sizes. Bockris [525] indicates that the measurement of solvation by mobility measurements would be smaller than is the actual case because of the loss, arising from Brownian motion, of solvent molecules attached to the ion. Mobility measurements determine the total number of solvent molecules transferred through the solution by the ion. Solvation numbers determined by these methods will in general be greater than those determined by methods giving the number of molecules in the first solvation shell. Krishnan and Friedman [662] points out that while the study of transport coefficients of aqueous systems produce results that may be interpreted in terms of solution structure effects the molecular theory of transport coefficients is so difficult that it would seem impossible to use such data to distinguish among the various models proposed to explain these same structural effects.

It has also been shown that spectroscopic and diffraction methods which are so often used for structural problems prove to be indecisive in determining absolute solution structure because these methods yield structure data which are averages over the contributions of various structures. This difficulty results primarily from the short lifetimes, about 10^{-12} sec, that characterize the hydrogen-bonded structure in hydrogen-bonding solvents and their solutions. Raman and IR methods are further complicated by spectral band components arising from both ion–ion and ion–solvent interactions because

Table 3.28

Total Effective Hydration Numbers of Electrolytes Determined from Temperature Dependence of Proton Shifts

Compound	h	Reference	Compound	h	Reference
HCl	3.4	[605]	KCl	4.6	[603]
HNO_3	2.8	[602]	KI	4.1	[602]
$HClO_4$	2.6	[605]	RbOH	5.2	[602]
LiCl	3.2	[602, 603]	RbCl	4.0	[603]
LiBr	4.4	[602]	CsCl	3.9	[603]
NaOH	2.9	[602]	$MgCl_2$	8.2	[603]
NaCl	4.5	[603]	$Mg(ClO_4)_2$	8.0	[602]
NaBr	4.4	[603]	$CaCl_2$	9.5	[602, 603]
$NaNO_3$	3.3	[602]	$Ca(NO_3)_2$	7.8	[602]
$NaClO_4$	3.0	[603, 605]	$Ca(ClO_4)_2$	8.6	[602]
Na *p*-toluene sulfonate	2.8	[603]	$Al(NO_3)_3$	13.4	[605]
KF	3.8	[602]			

high concentrations are required for sufficient intensity. The interpretation of relaxation data also suffers from the same limitation as that described above for transport coefficients. Another weakness of methods is that, in general structural models are used that ignore the anion.

References

1. W. Hittorf, *Z. Phys. Chem. Stoechiom. Verwandschaftslehre* **39**, 612 (1901).
2. W. Hittorf, *Z. Phys. Chem. Stoechiom. Verwandschaftslehre* **43**, 49 (1903).
3. E. W. Washburn, "Principles of Physical Chemistry," p. 276. McGraw-Hill, New York, 1921.
4. D. A. MacInnes, "The Principles of Electrochemistry," Chapter 4. Van Nostrand-Reinhold, Princeton, New Jersey, 1939.
5. W. Nernst, C. C. Gerrard, and E. Oppermann, *Nachr. Kgl. Ges. Wiss. Goettingen Math. Phys. Kl.* **56**, 86 (1900).
6. G. Buchböck, *Z. Phys. Chem. Stoechiom. Verwandschaftslehre* **55**, 563 (1906).
7. E. W. Washburn, *J. Amer. Chem. Soc.* **31**, 322 (1909); *Z. Phys. Chem. Stoechiom. Verwandschaftslehre* **66**, J13 (1909).
8. E. W. Washburn and E. B. Millard, *J. Amer. Chem. Soc.* **37**, 694 (1915).
9. D. M. Mathews, J. O. Wear, and E. S. Amis, *J. Inorg. Nucl. Chem.* **13**, 298 (1960).
10. W. V. Childs and E. S. Amis, *J. Inorg. Nucl. Chem.* **16**, 114 (1960).
11. J. O. Wear, C. V. McNully, and E. S. Amis, *J. Inorg. Nucl. Chem.* **18**, 48 (1961).
12. J. O. Wear, C. V. McNully, and E. S. Amis, *J. Inorg. Nucl. Chem.* **19**, 278; **20**, 106 (1961).
13. J. O. Wear, J. T. Curtis, Jr., and E. S. Amis, *J. Inorg. Nucl. Chem.* **24**, 93 (1962).
14. J. A. Bard, J. O. Wear, R. G. Griffin, and E. S. Amis, *J. Electroanal. Chem.* **8**, 419 (1964).
15. R. G. Griffin, E. S. Amis, and J. O. Wear, *J. Inorg. Nucl. Chem.* **28**, 543 (1966).
16. J. R. Bard, E. S. Amis, and J. O. Wear, *J. Electroanal. Chem.* **11**, 296 (1966).
17. J. O. Wear and E. S. Amis, *J. Inorg. Nucl. Chem.* **24**, 903 (1962).
18. H. Remy, *Z. Phys. Chem. Stoechiom. Verwandschaftslehre* **118**, 161 (1925); **124**, 41, 394 (1926).
19. G. Baborousky, *Z. Phys. Chem. Stoechiom. Verwandschaftslehre* **129**, 129 (1927).
20. S. Glasstone, "Textbook of Physical Chemistry," 2nd ed. Van Nostrand-Reinhold, Princeton, New Jersey, 1946.
21. S. Glasstone, "An Introduction to Electrochemistry," p. 115. Van Nostrand-Reinhold, Princeton, New Jersey, 1942.
22. O. Ya. Samoilov, *Izv. Akad. Nauk SSSR Otd. Tekh. Nauk* p. 398 (1952).
23. Q. Osaka, *Bull. Chem. Soc. Jap.* **12**, 177 (1937).
24. K. P. Mischenko, *Zh. Fiz. Khim.* **26**, 1736 (1950).
25. M. Eigen and L. DeMaeyer, *in* "The Structure of Electrolytic Solutions" (W. J. Hamer, ed.), Chapter 5. Wiley, New York, 1959; *Chem. Eng. News* **39**, No. 7, 40 (1961).
26. J. Baborousky, *Chem. Listy* **26**, 414 (1932).
27. J. Baborousky, *Collect. Czech. Chem. Commun.* **6**, 283 (1934).
28. J. Baborousky, *Z. Phys. Chem. Abt. A* **168**, 135 (1934).
29. H. Ulich, *Z. Phys. Chem. Abt. A.* **168**, 141 (1934).

30. J. Baborousky, *Z. Phys. Chem. Stoechiom. Verwandschaftslehre* **129**, 129 (1927).
31. J. Baborousky and O. Viktorin, *Collect. Czech. Chem. Commun.* **5**, 518 (1933).
32. J. Baborousky, J. Vesilek, and A. Wagner, *J. Chim. Phys. Physicochim. Biol.* **25**, 452 (1928).
33. E. Schreiner, *Zh. Anorg. Allg. Chem.* **121**, 321 (1922); **135**, 333 (1924).
34. J. N. Pearce and W. G. Euirsole, *Proc. Iowa Acad. Sci.* **33**, 151 (1926).
35. J. Baborousky, *Collect. Czech. Chem. Commun.* **1**, 315 (1929).
36. J. Baborousky, *Trans. Electrochem. Soc.* **75**, 6 pp. (preprint) (1939).
37. J. Baborousky, *Collect. Czech. Chem. Commun.* **19**, 542 (1938).
38. L. Nickels and A. J. Allmand, *J. Phys. Chem.* **41**, 873 (1937).
39. H. Remy, *Z. Phys. Chem. Stoechiom. Verwandschaftslehre* **118**, 161 (1926).
40. J. Velisek, *Chem. Listy* **20**, 242 (1926).
41. J. Baborousky and G. Koudela, *Chem. Listy* **32**, 5 (1938).
42. H. C. Hepburn, *Phil. Mag.* **25**, 1074 (1938).
43. H. C. Hepburn, *Proc. Phys. Soc. London* **45**, 755 (1933).
44. H. C. Hepburn, *Proc. Phys. Soc. London* **43**, 524 (1931).
45. H. C. Hepburn, *Proc. Phys. Soc. London* **39**, 99 (1927).
46. C. C. Rainey, *Science* **89**, 435 (1939).
47. J. Baborovsky and M. Hrusovsky, *Chem. Listy* **34**, 191 (1940).
48. J. Y. MacDonald, *Trans. Faraday Soc.* **43**, 674 (1947).
49. E. Darmois, *C. R. Acad. Sci.* **221**, 290–292 (1954).
50. T. Watari, *J. Electrochem. Soc. Jap.* **18**, 183 (1950).
51. O. Ya. Samoilov, *Dokl. Akad. Nauk SSSR* **77**, 633 (1951).
52. E. Darmois, *Chem. Zentralbl.*, **II**, 259 (1942).
53. E. Darmois, *J. Chim. Phys. Physicochim. Biol.* **43**, 1 (1946).
54. L. H. Collet, *J. Chim. Phys. Physicochim. Biol.* **49**, No. 7/8, C69 (1952).
55. L. H. Collet, *J. Chim. Phys. Physicochim. Biol.* **49**, No. 7/8, C69 (1952).
56. L. H. Collet, *C. R. Acad. Sci.* **237**, 252 (1953).
57. L. H. Collet, *C. R. Acad. Sci.* **239**, 266 (1954).
58. R. Haase, *Z. Elektrochem.* **62**, 279 (1958).
59. A. J. Rutgers and Y. Hendrikx, *Trans. Faraday Soc.* **58**, 2184 (1962).
60. V. P. Troshim, *Zh. Prikl. Khim.* (*Leningrad*) **36**, 1342 (1963).
61. B. P. Konstantinov and V. P. Troshim, *Zh. Prikl. Khim.* (*Leningrad*) **36**, 449 (1963).
62. H. Strethlow and H. M. Koepf, *Z. Elektrochem.* **62**, 372 (1958).
63. H. Schneider and H. Strethlow, *Z. Elektrochem.* **66**, 309 (1962).
64. H. Schneider and H. Strethlow, *Ber. Bunsenges. Phys. Chem.* **69**, 674 (1965).
65. V. P. Troshin, *Elektrokhimiya* **2**, 232 (1966).
66. R. Gopal and O. N. Bhatnagar, *J. Phys. Chem.* **69**, 2382 (1965).
67. R. A. Robinson and R. H. Stokes, "Electrolyte Solutions," pp. 124–126. Butterworth, London, 1959.
68. E. R. Nightingale, Jr., *J. Phys. Chem.* **63**, 1381 (1959).
69. E. F. Ivanova, *Zh. Fiz. Khim.* **39**, 1446 (1965).
70. V. P. Troshim, *Elektrokhimiya* **2**, 500 (1966).
71. P. Konstantinov and V. P. Troshim, *Izv. Akad. Nauk. SSSR Ser. Khim.* p. 1907 (1966).
72. N. Lakshminarayanaiah and V. Subrahmanyan, *J. Phys. Chem* **72**, 1253 (1968).
73. A. S. Tombalakian, *J. Phys. Chem.* **72**, 3698 (1968).
74. N. Lakshminarayanaiah, *J. Phys. Chem.* **72**, 3699 (1968).
75. N. Lakshminarayanaiah, *Curr. Sci.* **28**, 321 (1959).
76. N. Lakshminarayanaiah, *Desalination* **3**, 97 (1967).
77. V. Subrahmanyan, Thesis, Madras Univ., Madras, India, 1961.

78. V. Subrahmanyan and N. Lakshminarayanaiah, *Curr. Sci.* **29**, 207 (1960).
79. V. Subrahmanyan and N. Lakshminarayanaiah, *Bull. Chem. Soc. Jap.* **34**, 587 (1961).
80. N. Lakshminarayanaiah, Thesis, London Univ., London, 1961.
81. N. Lakshminarayanaiah, *Proc. Indian Acad. Sci. Sect. A* **55**, 200 (1952).
82. N. Lakshminarayanaiah, *Chem. Rev.* **65**, 525 (1965).
83. A. S. Tombalakian, H. J. Barton, and W. F. Graydon, *J. Phys. Chem.* **72**, 1253 (1968).
84. A. G. Winger, R. Ferguson, and R. Kunin, *J. Phys. Chem.* **60**, 556 (1956).
85. J. W. Lorimer, E. I. Boterenbrood, and J. J. Hormans, *Discuss. Faraday Soc.* **21**, 141 (1956).
86. A. Despic and G. J. Hills, *Discuss. Faraday Soc.* **21**, 150 (1956).
87. R. Stewart and W. Graydon, *J. Phys. Chem.* **61**, 164 (1957).
88. A. S. Tombalakian and W. F. Graydon, *J. Phys. Chem.* **70**, 3711 (1966).
89. A. S. Tombalakian, C. Y. Yeh, and W. F. Graydon, *Can. J. Chem. Eng.* **42**, 61 (1964).
90. A. Bethe and T. Toropoff, *Z. Phys. Chem. Stoechiom. Verwandschaftslehre* **88**, 686 (1914).
91. A. Bethe and T. Toropoff, *Z. Phys. Chem. Stoechiom. Verwandschaftslehre* **89**, 597 (1915).
92. A. M. Azzam, *Z. Phys. Chem. N. F.* **32**, 309 (1962).
93. A. M. Azzam, Part 1. *Z. Elektrochem.* **58**, 889 (1954).
94. B. E. Conway and J. O'M. Bockris, "Modern Aspects of Electrochemistry," p. 47. Butterworth, London, 1954.
95. J. D. Bernal and R. H. Fowler, *J. Chem. Phys.* **1**, 515 (1933).
95a. A. M. Azzam, Part II. *Can. J. Chem.* **38**, 993 (1960).
95b. A. M. Azzam, Part III. *Can. J. Chem.* **38**, 2203 (1960).
95c. A. M. Azzam, Ph.D. Thesis, London, 1949.
96. Interactions in Ionic Solutions. *Discuss. Faraday Soc.* **24**, 216 (1957).
97. M. D. Monica and U. Lamanna, *J. Phys. Chem.* **72**, 4329 (1968). (12)
98. M. D. Monica, U. Lamanna, and L. Senatoro, *J. Phys. Chem.* **72**, 2124 (1968).
99. E. G. Taylor and C. A. Kraus, *J. Amer. Chem. Soc.* **69**, 1731 (1947).
100. R. A. Robinson and R. H. Stokes, "Electrolyte Solutions," pp. 124–125. Butterworth, London, 1959.
101. E. R. Nightingale, Jr., *J. Phys. Chem.* **63**, 1381 (1959).
102. G. R. Choppin and A. J. Graffeo, *Inorg. Chem.* **4**, 1254 (1965).
103. S. L. Bertha and G. R. Choppin, *Inorg. Chem.* **8**, 613 (1969).
104. H. Ulich, *Trans. Faraday Soc.* **23**, 388 (1927).
105. J. Baborovsky and J. Velisek, *Chem. Listy* **21**, 227 (1927).
106. N. R. Dher, *Z. Anorg. Allg. Chem.* **159**, 57 (1926).
107. P. Walden and E. J. Birr, *Z. Phys. Chem. Abt. A* **153**, 1 (1931).
108. P. Walden and E. J. Birr, *Z. Phys. Chem. Abt. A* **144**, 269 (1929).
109. G. Champetier, *C. R. Acad. Sci.* **201**, 1118 (1935).
110. M. Jacopetti, *Gazz. Chim. Ital.* **72**, 251 (1942).
111. H. A. C. McKay and A. R. Machieson, *Trans. Faraday Soc.* **47**, 428 (1951).
112. I. M. Rodnyanskii and I. S. Galinker, *Dokl. Akad. Nauk SSSR* **105**, 115 (1955).
113. A. N. Campbell and G. H. Debres, *Can. J. Chem.* **34**, 1232 (1956).
114. D. K. Thomas and D. Maass, *Can. J. Chem.* **36**, 449 (1958).
115. B. F. Markov, *Ukr. Khim. Zh.* **23**, 706 (1957).
116. B. H. van Ruyven, *Chem. Weekbl.* **54**, 636 (1958).
117. A. N. Campbell, E. M. Kartzmark, and A. G. Sherwood, *Can. J. Chem.* **36**, 1325 (1958).
118. E. M. Loebl and J. J. O'Neal, *J. Polym. Sci.* **45**, 538 (1960).

119. N. N. Lichtin and K. N. Rao, *J. Phys. Chem.* **64**, 945 (1960).
120. G. G. Devyatykh, Z. B. Kuznetsova, and A. L. Agafonova, *Tr. Khim. Khim. Tekhnol.* **1**, 75 (1958).
121. H. Brussett and M. Kikindal, *C. R. Acad. Sci.* **252**, 1777 (1961).
122. H. J. Bittrich, W. Gaube, and R. Landsberg, *Wiss. Z. Tech. Hochsch. Chem. "Carl Schorlemmer" Leuna-Merseburg* **2**, 443 (1959–1960).
123. P. B. Das, P. K. Das, and D. Patrick, *J. Indian Chem. Soc.* **36**, 761 (1959).
124. P. K. Das, *J. Indian Chem. Soc.* **36**, 613 (1959).
125. B. Jezowska-Trzeblatowska and M. Chmielowska, *J. Inorg. Nucl. Chem.* **20**, 106 (1961).
126. G. Simkovich, *J. Phys. Chem.* **67**, 1001 (1963).
127. N. N. Lichtin and H. Kliman, *J. Chem. Eng. Data* **8**, 178 (1963).
128. J. A. Powell, Univ. Microfilm, Ann Arbor, Michigan, Order No. 63-5715, 110 pp.; Dissertation Abstract **14**, 978–979 (1963).
129. R. Gopel and M. M. Hussain, *J. Indian Chem. Soc.* **40**, 981 (1963).
130. R. A. Thorne, *Nature (London)* **200**, 418 (1963).
131. D. N. Bhattacharyya, C. L. Les, J. Smid, and M. Swarc, *J. Phys. Chem.* **69**, 608 (1965).
132. J. M. Motley and M. Spiro, *J. Phys. Chem.* **70**, 1502 (1966).
133. R. A. Robinson and H. H. Stokes, "Electrolyte Solutions," 2nd ed., pp. 120–126, 461. Butterworth, London, 1959.
134. T. Tonomura and K. Okamoto, *Bull. Chem. Soc. Jap.* **39**, 1621 (1966).
135. R. A. Horn and J. D. Birkett, *Electrochim. Acta* **12**, 1153 (1967).
136. D. Feakings and C. M. French, *J. Chem. Soc. (London)* p. 2681 (1957).
137. R. F. Hudson and B. Saville, *J. Chem. Soc. (London)* p. 4144 (1955).
138. R. A. Robinson and R. H. Stokes, *J. Amer. Chem. Soc.* **70**, 1870 (1948).
139. E. Glueckauf, *Trans. Faraday Soc.*, **51**, 1235 (1955).
140. H. S. Harned and B. B. Owen, "The Physical Chemistry of Electrolyte Solutions," 2nd ed. Van Nostrand–Reinhold, Princeton, New Jersey, 1956.
141. H. M. Koepp, H. Wendt, and H. Strehlow, *Z. Elektrochem.* **64**, 483 (1960).
142. J. M. Dale and C. V. Banks, *Inorg. Chem.* **2**, 591 (1963).
143. G. Schwarzenbach and G. Geier, *Helv. Chim. Acta* **46**, 906 (1963).
144. A. M. Shkodin, *Ukr. Khim. Zh.* **29**, 400 (1963).
145. D. Feakins and P. Watson, *J. Chem. Soc. (London)* p. 4686 (1963).
146. D. Feakins and P. Watson, *J. Chem. Soc. (London)* p. 4734 (1963).
147. Ya. I. Tur'yan, *Zh. Fiz. Khim.* **38**, 1853 (1964).
148. P. C. Lammers and J. Braunstein, *J. Phys. Chem.* **71**, 2626 (1967).
149. D. Feakins and P. T. Tomkins, *J. Chem. Soc.* p. 1458 (1967).
150. R. A. Robinson and R. H. Stokes, "Electrolyte Solutions," p. 319. Butterworth, London, 1955.
151. R. H. Stokes and R. A. Robinson, *J. Amer. Chem. Soc.* **70**, 1870 (1948).
152. J. D. Bernal and R. H. Fowler, *J. Chem. Phys.* **1**, 515 (1933).
153. B. E. Conway and J. O'M. Bockris, "Modern Aspects of Electrochemistry," Vol. 1. Academic Press, New York, 1954.
154. B. E. Conway and J. O'M. Bockris, "Modern Aspects of Electrochemistry," Vol. 2. Academic Press, New York, 1954.
155. P. Mukerjee, *J. Phys. Chem.* **65**, 740, 744 (1961).
156. B. E. Conway and J. O'M. Bockris, "Modern Aspects of Electrochemistry," Vol. 3. Butterworth, London, 1964.
157. W. M. Latimer, K. S. Pitzer, and C. M. Slansky, *J. Chem. Phys.* **7**, 108 (1939).
158. D. D. Eley and M. G. Evans, *Trans. Faraday Soc.* **34**, 1093 (1938).

159. E. J. W. Verwey, *Rec. Trav. Chim. Pays-Bas* **61**, 127 (1942).
160. A. D. Buckingham, *Discuss. Faraday Soc.* **24**, 151 (1957).
161. B. F. Halliwell and S. C. Nyburg, *Trans. Faraday Soc.* **59**, 1126 (1963).
162. L. Benjamine and V. Gold, *Trans. Faraday Soc.* **50**, 797 (1954).
163. M. Boudart, *J. Amer. Chem. Soc.* **74**, 1531, 3556 (1952).
164. Ya. Zeldovich, *Acta Phys. Chim. USSR* **1**, 961 (1935).
165. J. Horvichi and G. Okamoto, *Sci. Pap. Inst. Phys. Chem. Res. Tokyo* **28**, 231 (1936).
166. J. O'M. Bockris, *Quart. Rev. Chem. Soc.* **3**, 173 (1949).
167. L. Pauling, "Nature of the Chemical Bond," 3rd ed. Cornell Univ. Press, Ithaca, New York, 1960.
168. L. H. Ahrens, *Geochim. Cosmochim. Acta* **2**, 155 (1952).
169. D. D. Eley and M. G. Evans, *Trans. Faraday Soc.* **34**, 1093 (1938).
170. K. J. Laidler and C. Pegis, *Proc. Roy. Soc. Ser. A* **241**, 80 (1957).
171. A. M. Couture and K. J. Laidler, *Can. J. Chem.* **35**, 202 (1957).
172. A. M. Noyes, *J. Amer. Chem. Soc.* **84**, 513 (1962).
173. V. P. Vasil'ev, E. K. Zolotarev, A. F. Kapustykoku, K. P. Mischenko, B. A. Podgornaya, and K. B. Yakimirskii, *Russ. J. Phys. Chem.* **34**, 1763 (1960).
174. L. Brewer, L. A. Bromley, P. W. Gilles, and N. L. Lafgren, *in* "Chemistry and Metallurgy of Miscellaneous Materials: Thermodynamics" (L. L. Quill), p. 165. McGraw-Hill, New York, 1950.
175. K. Fajans, *Ber. Phys. Ges.* **21**, 549, 709 (1919).
176. J. Sherman, *Chem. Rev.* **11**, 93 (1932).
177. H. G. Grimm, *Handb. Phys.* **27**, 518 (1927).
178. R. Kebarle and A. M. Hogg, *J. Chem. Phys.* **42**, 798 (1965).
179. A. M. Hogg and P. Kebarle, *J. Chem. Phys.* **43**, 449 (1965).
180. A. M. Hogg, R. M. Haynes, and P. Kebarle, *J. Amer. Chem. Soc.* **88**, 28 (1966).
181. P. Kebarle, Mass Spectrometric Study of Ion-solvent Molecule Interactions in the Gas Phase, *Advan. Chem. Ser.* **72**, p. 24 (1968).
182. P. Kabarle, S. K. Searles, A. Zoila, J. Scarborough, and A. Arshadi, *J. Amer. Chem. Soc.* **89**, 6393 (1967).
183. M. M. Mann, A. Hustralid, and J. T. Late, *Phys. Rev.* **58**, 340 (1940).
184. F. W. Lampe, F. W. Field, and J. L. Franklin, *J. Amer. Chem. Soc.* **79**, 6132 (1957).
185. P. Kebarle, R. M. Haynes, and S. K. Searles, *Advan. Chem. Ser.* **58**, 210 (1966).
186. F. J. Garrick, *Phil. Mag,* **9**, 131 (1930).
187. F. J. Garrick, *Phil. Mag.* **10**, 76 (1930).
188. T. G. Owe Berg, *Ann. N. Y. Acad. Sci.* **125**, 298 (1965).
189. G. Körtum and J. O'M. Bockris, "Textbook of Electrochemistry," Vol. 1, p. 136. Amer. Elsevier, New York, 1951.
190. O. Ya. Samoilow, "Structure of Aqueous Electrolyte Solutions and Hydration of Ions" (translated by D. J. G. Ives). Consultants Bureau, New York, 1965.
191. A. F. Kapustinski and O. J. Samoilow, *Zh. Fiz. Khim.* **26**, 918 (1952).
192. O. J. Samoilov, *Izv. Akad. Nauk SSSR* **3**, 398 (1952).
193. O. J. Samoilov, *Izv. Akad. Nauk SSSR* **4**, 627 (1952).
194. A. F. Kapustinski and O. J. Samoilov, *Izv. Akad. Nauk SSSR* **4**, 337 (1950).
195. A. F. Kapustinski, I. I. Lipilini, and O. J. Samoilov, *Zh. Fiz. Khim.* **30**, 896 (1956).
196. J. Morgan and B. E. Warren, *J. Chem. Phys.* **6**, 666 (1938).
197. D. Jelenkow and L. Gemov, *Ber. Bulg. Akad. Wiss.* **7**, 37 (1954).
198. D. Jelenkow and L. Gemow, *Bulg. Akad. Wiss. Nachr. Chem. Inst.* **3**, 7 (1955).
199. L. Genow, *Dokl. Bolg. Akad. Nauk* **9**, 23 (1956).
200. L. Genow, *Izv. Khim. Inst. Bulg. Akad. Nauk* **4**, 251 (1951).

201. O. Ya. Samoilov, *Dokl. Akad. Nauk SSSR* **102**, 173 (1956).
202. O. Ya. Samoilov, *Izv. Akad. Nauk SSSR Otd. Khim. Nauk* p. 1415 (1956).
203. O. Ya. Samoilov, *Bull. Acad. Sci. USSR Div. Chem. Sci.* p. 1449 (1956).
204. O. Ya. Samoilov, K.-Y. Hu, and T. A. Nosova, *Zh. Strukt. Khim.* **1**, 131 (1960).
205. M. N. Buslaeva and O. Ya. Samoilov, *Zh. Strukt. Khim.* **2**, 551 (1961); **4**, 682 (1963).
206. I. I. Lipilina and O. Ya. Samoilov, *Dokl. Akad. Nauk SSSR* **98**, 99 (1954).
207. J. Sutton, *Nature (London)* **160**, 235 (1952).
208. V. P. Il'insku and A. T. Uverskaya, *Sb. Tr. Inst. Prikl. Khim.* No. 41, 112 (1958).
209. O. Ya. Samoilov, K.-Y. Hu, and T. A. Nasova, *Zh. Strukt. Khim.* **1**, 404 (1960).
210. O. Ya. Samoilov, *Izv. Akad. Nauk SSSR Otd. Khim. Nauk* p. 398 (1952).
211. I. I. Lipilina, *Dokl. Akad. Nauk SSSR* **102**, 525 (1955).
212. B. P. Burylev, *Zh. Fiz. Khim.* **41**, 676 (1967).
213. K. P. Mishchenko, *Zh. Fiz. Khim.* **26**, 1336 (1952).
214. K. P. Mishchenko and V. V. Sokolov, *Zh. Strukt. Khim.* **4**, 184 (1963).
215. M. Depas, J. J. Leventhal, and L. Friedman, *J. Chem. Phys.* **49**, 5543 (1968).
216. R. H. Stokes and R. A. Robinson, *J. Amer. Chem. Soc.* **70**, 1870 (1948).
217. E. Glueckauf, *Trans. Faraday Soc.* **51**, 1235 (1955).
218. E. Glueckaus and G. P. Kitt, *Proc. Roy. Soc. Ser. A* **228**, 322 (1955).
219. R. A. Robinson and R. H. Stokes, *Trans. Faraday Soc.* **45**, 612 (1924).
220. R. H. Stokes, *Trans. Faraday Soc.* **44**, 295 (1948).
221. R. A. Robinson and R. H. Stokes, "Electrolyte Solutions," p. 254. Butterworth, London, 1959.
222. Ya. I. Tur'yan, *Zh. Fiz. Khim.* **38**, 1690 (1964).
223. R. W. Taft, Jr., *J. Amer. Chem. Soc.* **82**, 2965 (1960).
224. J. Bass, R. J. Gillespie, and J. V. Oubridge, *J. Chem. Soc.* p. 837 (1960).
225. W. H. Lee, Sulfuric Acid. In "The Chemistry of Non-Aqueous Solvents" (J. J. Lagowski, ed.). Academic Press, New York, 1967.
226. R. J. Gillespie and R. F. M. White, *Can. J. Chem.* **38**, 1371 (1960).
227. F. Burion and E. Rouyer, *C. R. Acad. Sci.* **188**, 626 (1929).
228. E. Royer and O. Hun, *C. R. Acad. Sci.* **196**, 1015 (1933).
229. F. Burion and E. Rouyer, *C. R. Acad. Sci.* **197**, 52 (1933).
230. F. Burion and E. Rouyer, *C. R. Acad. Sci.* **196**, 1111 (1933).
231. E. Royer, *C. R. Acad. Sci.* **198**, 742 (1934).
232. O. Hun, *C. R. Acad. Sci.* **198**, 740 (1934).
233. K. Jablezynski and A. Balczewski, *Rocz. Chem.* **12**, 880 (1932).
234. F. Bourion and E. Rouyer, *C. R. Acad. Sci.* **198**, 1490 (1934).
235. F. Bourion and O. Hun, *C. R. Acad. Sci.* **198**, 1921 (1934).
236. F. Bourion and E. Royer, *C. R. Acad. Sci.*, **198**, 1944 (1934).
237. E. Rouyer, *C. R. Acad. Sci.* **198**, 1156 (1934).
238. F. Bourion and E. Rouyer, *C. R. Acad. Sci.* **201**, 65 (1935).
239. O. Hun, *C. R. Acad. Sci.* **201**, 547 (1935).
240. O. Hun, *C. R. Acad. Sci.* **202**, 1779 (1936).
241. F. Burion and O. Hun, *C. R. Acad. Sci.* **202**, 2149 (1936).
242. G. B. Smith, C. A. Hollingsworth, and D. H. McDaniel, *J. Chem. Educ.* **38**, 489 (1961).
243. H. C. Brown, L. P. Eddy, and R. Wong, *J. Amer. Chem. Soc.* **75**, 6275 (1953).
244. H. C. Brown and W. J. Wallace, *J. Amer. Chem. Soc.* **75**, 6279 (1953).
245. H. C. Brown and W. J. Wallace, *J. Amer. Chem. Soc.* **75**, 6265 (1953).
246. W. L. Latimer and R. M. Buffington, *J. Amer. Chem. Soc.* **48**, 2297 (1926).
247. W. L. Latimer, P. W. Schutz, and J. F. G. Hicks, Jr., *J. Chem. Phys.* **2**, 82 (1934).
248. W. L. Latimer, *Chem. Rev.* **18**, 349 (1936).

249. J. Chanu, *J. Chim. Phys. Physicochim. Biol.* **55**, 733, 743 (1958).
250. L. Pauling, *J. Amer. Chem. Soc.* **49**, 765 (1927).
251. H. Ulich, *Z. Elektrochem. Abt. A* **36**, 497 (1930).
252. H. Ulich, *Z. Phys. Chem.* **168**, 141 (1934).
253. B. E. Conway, *in* "Modern Aspects of Electrochimestry" (J. O'M. Bockris and B. E. Conway, eds.), Chapter 2. Butterworth, London, 1964.
254. W. S. Fyfe, *J. Chem. Soc.* p. 2023 (1952).
255. W. M. Latimer, *J. Amer. Chem. Soc.* **73**, 1480 (1951).
256. E. K. Zolotarev and V. E. Kalinini, *Zh. Neorg. Khim.* **7**, 1225 (1962).
257. E. Whalley, *J. Chem. Phys.* **38**, 1400 (1963).
258. G. L. Starobinets, A. B. Chizhevskaya, and T. L. Dubovik, *Vestn. Akad. Nauk Belarusk. SSR Ser. Khim. Nauk.* p. 110 (1965).
259. M. Randall and F. D. Rossini, *J. Amer. Chem. Soc.* **51**, 323 (1929).
260. T. W. Bauer, H. L. Johnson, and E. C. Kerr, *J. Amer. Chem. Soc.* **72**, 5174 (1950).
261. A. Eucken, *Z. Elektrochem.* **52**, 6 (1948).
262. A. Eucken and M. Eigen, *Z. Elektrochem.* **55**, 343 (1951).
263. V. I. Lebed and V. V. Alksandrov, *Elektrokhimiya* **1**, 1359 (1965).
264. F. H. Spedding and K. C. Jones, *J. Phys. Chem.* **70**, 2450 (1966).
265. M. Marezio, H. A. Plettinger, and W. H. Zachariasen, *Acta Crystallog.* **14**, 234 (1961).
266. K. Bennewitz, *Z. Elektrochem. Angew. Phys. Chem.* **33**, 540 (1927).
267. E. Lange and A. Eichler, *Z. Phys. Chem. Stoechiom. Verwandshaftslehre* **129**, 285 (1927).
268. P. P. Kosakevich, *Z. Phys. Chem. Abt A* **143**, 216 (1929).
269. H. Ulich, *Trans. Faraday Soc.* **23**, 388 (1927).
270. R. Flatt and A. Jordan, *Helv. Chim. Acta* **16**, 37 (1933).
271. S. S. Chin, *Zh. Fiz. Khim.* **26**, 960 (1952).
272. S. S. Chin, *Zh. Fiz. Khim.* **26**, 1125 (1952).
273. E. Angelescu and D. Motoc, *Acad. Repub. Pop. Rom. Commun.* **3**, 267 (1953).
274. P. S. Bogoyovlenskii, *Dokl. Akad. Nauk SSSR* **101**, 865 (1955).
275. E. Koizumi and H. Miyamoto, *Nippon Kagaku Zasshi.* **77**, 193 (1956).
276. E. Koizumi and H. Miyamoto, *Bull. Chem. Soc. Jap.* **29**, 250 (1956).
277. P. S. Bogoyavlenskii, *Zh. Fiz. Khim.* **32**, 2035 (1958).
278. A. Ya. Samoilov, *Dokl. Akad. Nauk SSSR* **102**, 1173 (1955).
279. H. Miyamoto, *Nippon Kagaku Zasshi* **80**, 825 (1959).
280. H. Miyamoto, *Nippon Kagaku Zasshi* **80**, 4 (1959).
281. H. Miyamoto, *Nippon Kagaku Zasshi* **81**, 54 (1960).
282. H. Miyamoto, *Nippon Kagaku Zissha* **81**, 1376 (1960).
283. V. M. Vdovenko, I. G. Suglobova, and D. N. Suglobov, *Radiokhimiya* **1**, 637 (1959).
284. A. I. Agaev and A. I. Dzhabarov, *Tr. Azer. Gos. Univ. Ser. Khim.* p. 8 (1959).
285. A. I. Dzhabarov and A. I. Agaev, *Ach. Zap. Uzerb. Gas. Univ. Ser. Fiz. Mat. Khim. Nauk* p. 77 (1960).
286. M. B. Everhard and P. M. Gross, Jr., Office Tech. Serv. P. B. Rept. No. 145, 174, 4 pp. U.S. Dept. of Commerce, Washington, D. C., 1959.
287. S. Poczopko, *Rocz. Chem.* **34**, 1245 (1960).
288. S. Poczopko, *Rocz. Chem.* **34**, 1255 (1960).
289. S. Poczopko, *Rocz. Chem.* **36**, 103 (1962).
290. S. Poczopko, *Rocz. Chem.* **36**, 111 (1962).
291. S. Poczopko, *Rocz. Chem.* **36**, 295 (1962).
292. S. Poczopko, *Rocz. Chem.* **36**, 303 (1962).
293. A. Basinski and S. Poczopko, *Rocz. Chem.* **34**, 1061 (1960).

294. V. F. Sergeeva, *Izv. Vyssh. Ucheb. Zaved. Khim. Khim. Teknol.* **5**, 905 (1962).
295. V. F. Sergeeva, *Izv. Vyss. Ucheb. Zaved. Khim. Khim. Teknol.* **5**, 908 (1962).
296. A. M. Azzam, *Z. Phys. Chem. N.F.* **33**, 23 (1962).
297. H. J. Emeleus and J. S. Anderson, "Modern Aspects of Inorganic Chemistry." Van Nostrand–Reinhold, Princeton, New Jersey, 1943.
298. A. M. Azzam, *Z. Phys. Chem. N.F.* **32**, 309 (1962).
299. E. Wicke, M. Eigen, and T. Ackermann, *Z. Phys. Chem. N.F.* **1**, 340 (1954).
300. K. S. Bascombe and R. P. Bell, *Discuss. Faraday Soc.* **24**, 158 (1957); P. A. Wyatt, *Ibid.* **24**, 162 (1957); S. R. Gupta, G. J. Hills, and D. G. Ives, *Ibid.* **24**, 147 (1957).
301. T. Ackermann, *Discuss. Faraday Soc.* **24**, 180 (1957).
302. C. L. Van Eck, H. Mendel, and W. Boog, *Discuss. Faraday Soc.* **24**, 200 (1957).
303. M. Eigen, *Discuss. Faraday Soc.* **24**, 235 (1957).
304. M. Falk and P. A. Gignere, *Chem. Eng. News* p. 59 (1957).
305. H. D. Beckey, *Rep. Int. Congr. Electromicrosc. ,4th.* Springer-Verlag, Berlin and New York, 1958.
306. N. Bjerrium, *Z. Anorg. Allg. Chem.* **109**, 275 (1920); *Z. Phys. Chem. Stoechiom. Verwandschaftslehre* **106**, 231 (1923).
307. F. Burton and E. Rouyer, *C. R. Acad. Sci.* **202**, 214 (1936).
308. B. H. Van Ruyen, *Chem. Weekbl.* **53**, 461 (1947).
309. J. B. Hasted, C. H. Collie, and D. M. Riston, *J. Chim. Phys. Physicochim. Biol.* **16**, 1 (1948).
310. E. Darmois, *J. Phys. Radium* **2**, 2 (1941).
311. E. W. Washburn, *J. Amer. Chem. Soc.* **31**, 322 (1909).
312. H. Rèmy, *Z. Phys. Chem. Stoechiom. Verwandschaftslehre* **126**, 161 (1927).
313. J. Baborouvsky, *Collect. Czecho. Chem. Commun.* **11**, 542 (1938).
314. O. Ya. Samoilov and V. I. Yashkichev, *Zh. Strukt. Khim.* **3**, 334 (1962).
315. G. A. Krestov, *Zh. Strukt Khim.* **3**, 137 (1962).
316. G. A. Krestov, *Zh. Strukt. Khim.* **3**, 402 (1962); G. A. Krestov and V. I. Klophov, *Ibid.* **4**, 502 (1963).
317. Ya. P. Gokhshtein, *Tr. Kamissii Anal. Khim. Otd. Khim. Nauk Akad. Nauk SSSR* **2**, 5 (1949).
318. E. Riecke, *Z. Phys. Chem. Stoechiom. Verwandschaftslehre* **6**, 564 (1890).
319. A. A. Vleck, *Collect. Czech. Chem. Commun.* **16**, 230 (1951).
320. P. K. Migal and N. Kh. Grinberg, *Zh. Neorg. Khim.* **6**, 727 (1961).
321. D. L. McMasters, J. C. DiRaimondo, L. H. Jones, R. P. Lindley, and E. W. Zeltmann, *J. Phys. Chem.* **66**, 249 (1962).
322. P. K. Migal and N. Kh. Grinberg, *Zh. Neorg. Khim.* **7**, 527 (1962).
323. P. K. Migal and N. Kh. Grinberg, *Zh. Neorg. Khim.* **7**, 531 (1962).
324. D. A. DeFord and D. N. Hume, *J. Amer. Chem. Soc.* **73**, 5321 (1951).
325. R. M. Noyes, *J. Amer. Chem. Soc.* **84**, 513 (1962); **86**, 971 (1964).
326. R. M. Noyes, *J. Phys. Chem.* **69** 3181 (1965).
327. J. E. Prue, *Chem. Phys. Ionic Solutions Selec. Invited Pap. Discuss.* p. 163 (1964) (Pub. 1966).
328. H. S. Frank, *Chem. Phys. Ionic Solutions Selec. Invited Pap. Disucss.* p. 53 (1964) (Pub. 1966).
329. A. Finch, P. J. Gardner, and C. J. Steadman, *J. Phys. Chem.* **71**, 2996 (1967).
330. G. Somsen, *Rec. Trav. Chim. Pays-Bas* **85**, 317, 526 (1966).
331. B. Case and R. Parsons, *Trans. Faraday Soc.* **63**, 1224 (1967).
332. D. Feakins, *Phys. Chem. Processes Mixed Aqueous Solvents Lect. Symp.* p. 71 (1966) (Pub. 1967).

333. J. Padova, *J. Phys. Chem.* **72**, 796 (1968).
334. N. E. Khomutov, *Zh. Fiz. Khim.* **41**, 2258 (1967).
335. P. Kebarle, S. K. Searles, A. Zolla, J. Scarborough, and M. Arshadi, *J. Amer. Chem. Soc.* **89**, 6393 (1967).
336. W. A. Millen and D. W. Watts, *J. Amer. Chem. Soc.* **89**, 6051 (1967).
337. O. Ya. Samoilov and G. G. Malenkov, *Zh. Strukt. Khim.* **8**, 618 (1967).
338. G. A. Krestov and V. K. Abrosimov, *Zh. Strukt. Khim.* **8**, 522 (1967).
339. R. Alexander and A. J. Parker, *J. Amer. Chem. Soc.* **89**, 5549 (1967).
340. Yu. V. Gurikov, *Teor. Eksp. Khim.* **4**, 61 (1968).
341. S. I. Drakin, R. Kh. Kurmalieva, and M. Kh. Karapet'yants, *Reaks. Sposobnost Org. Soedin. Tattusk. Gos. Univ.* **2**, 267 (1965).
342. S. L. Bertha and G. R. Choppin, *Inorg. Chem.* **8**, 613 (1969).
343. E. Glueckauf, *Trans. Faraday Soc.* **51**, 1235 (1955).
344. O. Popovych and A. J. Dill, *Anal. Chem.* **41**, 456 (1969).
345. N. Bjerrum and E. Larson, *Z. Phys. Chem. Stoechiom. Verwandschaftslehre* **127**, 358 (1927).
346. A. J. Parker and R. Alexander, *J. Amer. Chem. Soc.* **90**, 3313 (1968).
347. M. Alfenaar, C. L. DiLigny, and A. G. Remijnse, *Rec. Trav. Chim. Pays-Bas* **86**, 929, 986 (1967).
348. J. F. Coetzee and J. J. Campion, *J. Amer. Chem. Soc.* **89**, 2513 (1967).
349. H. Streklow, in "The Chemistry of Non-Aqueous Solvents" (J. J. Lagowski, ed.), Chapter 2. Academic Press, New York, 1966.
350. A. J. Dill, L. Mitzkowitz, and O. Popovych, *J. Phys. Chem.* **72**, 4580 (1968).
351. H. Strethlow, *Z. Elektrochem.* **56**, 827 (1952).
352. N. A. Ismaylov, *Z. Elecktrochem.* **56**, 827 (1952).
353. N. A. Ismaylov, *Dokl. Akad. Nauk SSSR* **126**, 1033 (1959).
354. N. A. Ismaylov, *Dokl. Akad. Nauk SSSR* **127**, 104 (1959).
355. N. A. Ismaylov, *Dokl. Akad. Nauk SSSR* **149**, 884 (1963).
356. N. A. Ismaylov, *Dokl. Akad. Nauk SSSR* **149**, 1364 (1963).
357. C. L. DeLigny and M. Alfenaar, *Rec. Trav. Chim. Pays-Bas* **84**, 81 (1965).
358. D. Feakins and P. Watson, *J. Chem. Soc.* p. 4734 (1963).
359. B. E. Conway and E. Solomon, in "Chemical Physics of Ionic Solvations" (B. E. Conway and R. G. Barrada, eds.), Chapter 24. Wiley, New York, 1966.
360. R. M. Noyes, *J. Amer. Chem. Soc.* **84**, 513 (1962).
361. H. M. Koepp, H. Wendt, and H. Strehlow, *Z. Elektrochem.* **64**, 483 (1960).
362. H. Strehlow, in "The Chemistry of Non-Aqueous Solvents" (J. J. Lagowski, ed.), Chapter 4. Academic Press, New York, 1966.
363. E. Grundwald, G. Baughman, and G. Kohnstam, *J. Amer. Chem. Soc.* **82**, 5801 (1960).
364. O. Popovych, *Anal. Chem.* **38**, 558 (1966).
365. R. Alexander and A. J. Parker, *J. Amer. Chem. Soc.* **89**, 5549 (1967).
366. M. H. Paekhurst, *Rev. Pure Appl. Chem.* **19**, 45 (1969).
367. E. J. King, *J. Phys. Chem.* **73**, 1220 (1969).
368. E. J. King, *J. Phys. Chem.* **74**, 4590 (1971).
369. F. Millero, in "Structure and Transport in Water and Aqueous Solutions" (R. A. Horne, ed.). Wiley (Interscience), New York, to be published.
370. A. Bondi, *J. Phys. Chem.* **68**, 441 (1964).
371. R. Zana and E. Yeager, *J. Phys. Chem.* **71**, 521 (1967).
372. R. A. Robinson and C. L. Chia, *J. Amer. Chem. Soc.* **74**, 2776 (1952).
373. L. Onsager and R. M. Fuoss, *J. Phys. Chem.* **36**, 2689 (1932).

374. H. S. Harned and B. B. Owen, "The Physical Chemistry of Electrolyte Solutions," 2nd ed., pp. 86–90. Van Nostrand-Reinhold, Princeton, New Jersey, 1956.
375. H. S. Harned and J. French, *Ann. N. Y. Acad. Sci.* **44**, 267 (1945).
376. H. S. Harned and R. L. Nuttall, *J. Amer. Chem. Soc.* **69**, 736 (1947).
377. J. H. Northrop and M. L. Anson, *J. Gen. Physiol.* **12**, 543 (1929).
378. R. H. Stokes, *J. Amer. Chem. Soc.* **72**, 763 (1950).
379. R. H. Stokes and R. A. Robinson, *J. Amer. Chem. Soc.* **70**, 1870 (1948).
380. E. N. Gapon, *Z. Anorg. Allg. Chem.* **168**, 125 (1927).
381. E. N. Capon, *Zh. Russ. Fiz. Khim. Obshchest. Chast.* **60**, 237 (1928).
382. J. Baborovsky, O. Viktorin, and A. Wägner, *Collect. Czech. Chem. Commun.* **4**, 200 (1932).
383. G. Jander and H. Möhr, *Z. Phys. Chem. Abt. A* **190**, 81 (1942); *Chem. Zentralbl.* **1**, 2098 (1942).
384. H. Spandau and C. Spandau, *Z. Phys. Chem. (Leipzig)* **192**, 211 (1943).
385. H. Ulich, *Z. Elektrochem.* **36**, 505 (1930).
386. A. Hunyar, *J. Amer. Chem. Soc.* **71**, 3552 (1949).
387. E. Pogany, *Magy. Chem. Foly.* **48**, 85 (1942).
388. J. L. R. Morgan and C. W. Kanolt, *J. Amer. Chem. Soc.* **28**, 572 (1906).
389. A. W. Adamson, *J. Phys. Chem.* **58**, 514 (1954).
390. O. Ya. Samoilov, *Zh. Fiz. Khim.* **29**, 1582 (1955).
391. O. Ya. Samoilov, *Discuss. Faraday Soc.* **24**, 141, 151 (1957).
392. G. A. Andrew, *Dokl. Akad. Nauk SSSR* **145**, 358 (1962).
393. D. W. McCall, D. C. Douglass, and E. W. Anderson, *Ber. Bunsenges. Phys. Chem.* **67**, 336 (1963).
394. A. J. Rutgers, W. Rigole, and V. Hendrix, *Chem. Weekbl.* **59**, 33 (1963).
395. I. V. Matyash, A. I. Toryanik, and V. I. Vashkichev, *Zh. Strukt. Khim.* **5**, 777 (1964).
396. J. Salvinien and B. Brun, *C. R. Acad. Sci.* **259**, 565 (1964).
397. M. I. Emel'yanov and A. Sh. Agishev, *Zh. Strukt. Khim.* **6**, 909 (1965).
398. H. M. Feder and H. Taube, *J. Chem. Phys.* **20**, 1335 (1952).
399. J. P. Hunt and H. Taube, *J. Chem. Phys.* **19**, 602 (1951).
400. R. H. Betts and A. N. McKenzie, *Can. J. Chem.* **30**, 146 (1952).
401. H. van der Straaten and A. H. W. Aten, Jr., *Rec. Trav. Chim. Pays-Bas* **73**, 89 (1954).
402. H. Taube, *J. Phys. Chem.* **58**, 523 (1954).
403. A. Ya. Samoilov, *Dokl. Akad. Nauk SSSR*, **102**, 1173 (1956).
404. H. W. Baldwin and H. Taube, *J. Chem. Phys.* **33**, 206 (1960).
405. T. Asano, *Radioisotopes (Tokyo)* **14**, 363 (1965).
406. C. G. Swain and R. F. W. Bader, *Tetrahedron* **10**, 200 (1960).
407. L. B. Anderson, Univ. Microfilms, Ann Arbor, Michigan, Order No. 62-117, 125 pp.; Dissertation Abstract **22**, 3002 (1962).
408. W. Plumb and G. M. Harris, *Inorg. Chem.* **3**, 542 (1964).
409. J. C. Jayne and E. L. King, *J. Amer. Chem. Soc.* **86**, 3989 (1964).
410. D. W. Kemp and E. L. King, *J. Amer. Chem. Soc.* **89**, 3433 (1967).
411. J. D. Bernal and R. H. Fowler, *J. Chem. Phys.* **1**, 515 (1933).
412. W. Weyl, *Beih. Z. Ver. Deut. Chem.* **18** (1935); *Angew. Chem.* **48**, 573 (1935).
413. C. S. Rao, *Curr. Sci.* **4**, 649 (1936).
414. C. S. Rao, *Phil. Mag.* **24**, 87 (1937).
415. C. S. Rao, *Indian J. Phys.* **11**, 143 (1937).
416. C. H. Cartwright, *J. Chem. Phys.* **5**, 776 (1947).
417. Th. C. Kryumzelis, *Z. Phys.* **110**, 742 (1938).
418. E. Darmois, *J. Phys. Radium Suppl.* [8], **2**, 2–3 (1941); *Chem. Zentralbl.* p. 259 (1942).

419. M. E. Darmois, *J. Phys. Radium* **8**, 1 (1944).
420. M. Cardier, *C. R. Acad. Sci.* **214**, 707 (1942).
421. R. Freymann and R. Rohmer, *C. R. Acad. Sci.* **233**, 951 (1951).
422. R. Platzman and G. Branch, *Z. Phys.* **138**, 411 (1954).
423. R. Amiot, *C. R. Acad. Sci.* **243**, 1311 (1956).
424. I. G. Pominov, *Zh. Fiz. Khim.* **31**, 2184 (1957).
425. I. S. Pominov, *Zh. Fiz. Khim.* **31**, 1926 (1957).
426. K. P. Mishchenko and I. S. Pominov, *Zh. Fiz. Khim.* **31**, 2026 (1957).
427. W. C. Waggener, *J. Amer. Chem. Soc.* **80**, 3167 (1958).
428. G. S. Karetnikov, *Nauch. Dokl. Vyssh. Shk. Khim. Khim. Tekhnol.* p. 213 (1958).
429. M. Smith and M. C. R. Symons, *Trans. Faraday Soc.* **54**, 338 (1958).
430. M. Smith and M. C. R. Symons, *Trans. Faraday Soc.* p. 346 (1958).
431. Y. I. Ryskin, V. I. Zemlyanukhin, A. A. Solov'eva, and N. A. Derbeneva, *Zh. Neorg. Khim.* **4**, 393 (1959).
432. G. Allen and E. Warhurst, *Trans. Faraday Soc.* **54**, 1786 (1958).
433. T. R. Griffiths and M. C. R. Symons, *Trans. Faraday Soc.* **56**, 1125 (1960).
434. T. Sato and K. Hayashi, *J. Phys. Soc. Jap.* **15**, 1658 (1960).
435. I. S. Poninov, *Fiz. Sb. L'vov. Gos. Univ.* p. 213 (1967).
436. L. G. Pikulik, *Tr. Inst. Fiz. Mat. Akad. Nauk Beloruss. SSR* p. 167 (1959).
437. I. I. Antipova-Karataeva and E. E. Vainshetein, *Zh. Neorg. Khim.* **5**, 107 (1960).
438. E. E. Vainshtein and I. I. Antipova-Karataeva, *Termodin. Str. Rastvorov Tr. Soveshch.* p. 266 (1958) (Pub. 1959).
439. I. I. Antipova-Karataeva, E. E. Vainstein, and Yu. I. Kutsenko, *Zh. Neorg. Khim.* **6**, 816, 2329 (1961).
440. V. Guttmann and L. Huebner, *Monatsh. Chem.* **92**, 1262 (1961).
441. I. I. Antipova-Karataeva and E. E. Vainshtein, *Zh. Neorg Khim.* **6**, 1115 (1961).
442. D. Meyerstein and A. Treinin, *J. Phys. Chem.* **66**, 446 (1962).
443. I. I. Antipova-Karataeva, Yu. I. Kutsenko, and G. I. Yatsun, *Zh. Neorg. Khim.* **7**, 1913 (1962).
444. N. J. Friedman and R. A. Plane, *Inorg. Chem.* **2**, 11 (1963).
445. F. H. A. Rummens, *Rec. Trav. Chim. Pays-Bas* **81**, 758 (1962).
446. R. Bastrocchi and M. Costa, *Ann. Chim. (Rome)* **52**, 1285 (1962).
447. I. I. Antipova-Karataeva and E. E. Vainshtein, *Fiz. Probl. Spektrosk. Mater. 13th Soveshch.* pp. 1, 339–341 (1960) (Pub. 1962).
448. R. E. Hester and R. A. Plane, *Inorg. Chem.* **3**, 768 (1964).
449. G. Jones and M. Dole, *J. Amer. Chem. Soc.* **51**, 2950 (1929).
450. R. W. Gurney, "Ionic Processes in Solutions," Chapter 9. McGraw-Hill, New York, 1953.
451. J. R. Ferraro and A. Walker, *J. Chem. Phys.* **42**, 1278 (1965).
452. W. Luck, *Ber. Bunsenges. Phys. Chem.* **69**, 69 (1965).
453. R. F. Pasternack and R. A. Plane, *Inorg. Chem.* **4**, 1171 (1965).
454. A. Groskaufmania, A. Sakaline, and L. Liepina, *Latv. PSR Zinat. Akad. Vestis Kim. Ser.* p. 449 (1965).
455. A. V. Karyakin, A. V. Petrov, and Yu. B. Gerlit, *Dokl. Akad. Nauk SSSR* **168**, 588 (1966).
456. D. E. Irish and G. E. Walrafen, *J. Chem. Phys.* **46**, 378 (1967).
457. A. V. Karyakin, A. V. Petrov, Yu. E. Gerlit, and M. E. Zubrilina, *Gheor. Eksp. Khim.* **2**, 494 (1966).
458. J. L. Kropp and M. W. Windsor, *J. Phys. Chem.* **71**, 477 (1967).
459. A. Sonesson, *Acta Chem. Scand.* **12**, 165 (1958).

460. A. Sonesson, *Acta Chem. Scand.* 12, 1937 (1958).
461. G. Zundel and A. Murr, *Z. Phys. Chem. (Leipzig)* 233, 415 (1966).
462. E. Butter, *Z. Chem.* 7, 199 (1967).
463. N. A. Kostromina and T. V. Ternovaya, *Zh. Neorg. Khim.* 12, 700 (1967).
464. J. F. Young, *Inorg. Chem.* 6, 1486 (1967).
465. I. S. Perelygin and N. R. Saffiullina, *Zh. Strukt. Khim.* 8, 205 (1967).
466. E. Antonescu, *Stud. Cercet Chim.* 15, 645 (1967).
467. G. P. Roshchina, A. S. Kaurova, and I. D. Kosheleva, *Zh. Strukt. Khim.* 9, 3 (1968).
468. D. S. Allam and W. H. Lee, *J. Chem. Soc.* p. 5 (1966).
469. D. S. Allam and W. H. Lee, *J. Chem. Soc.* p. 6049 (1964).
470. A. G. Passynski, *Acta Physicochim. USSR* 8, 385 (1938).
471. Y. Wada, S. Shimbo, M. Oda, and J. Nagumo, *Oyo Butsuri* 17, 257 (1948).
472. T. Yasunaga, *Nippon Kagaku Zasshi* 72, 87 (1954).
473. T. Sassaki and T. Yasunaga, *Bull. Chem. Soc. Jap.* 28, 269 (1955).
474. K. Tamura and T. Sasaki, *Bull. Chem. Soc. Jap.* 36, 975 (1963).
475. T. Yasunaga and T. Sasaki, *Nippon Kagaku Zasshi* 72, 89 (1951).
476. T. Sasaki, T. Yasunaga, and H. Fujihara, *Nippon Kagaku Zasshi* 73, 181 (1952).
477. S. Barnartt, *Quart. Rev. Chem. Soc.* 7, 84 (1953).
478. D. S. Allam and W. H. Lee, *J. Chem. Soc.* p. 426 (1966).
479. M. F. C. Ladd and W. H. Lee, *J. Inorg. Nucl. Chem.* 13, 218 (1960).
480. A. Pasynskii, *Acta Physicochim. USSR* 8, 385 (1938); *J. Physchem. USSR* 11, 608 (1938).
481. V. B. Cory, *Phys. Rev.* 64, 350 (1943).
482. T. Yasunaga and T. Sasaki, *Nippon Kagaku Zasshi* 72, 366 (1951).
483. T. Yasunaga and T. Sasaki, *Nippon Kagaku Zasshi* 72, 87 (1951).
484. T. Sasaki, T. Yasunaga, and J. Fujiwara, *Nippon Kagaku Zasshi* 73, 181 (1952).
485. K. Tamm and G. Kurtze, *Nature (London)* 168, 346 (1951).
486. K. Tamm, *Int. Conf. Ultrasonics Gases Liquids Pap. Conf., 1951*, p. 214.
487. Y. Miyahara and H. Shiio, *Nippon Kagaku Zasshi* 73, 265 (1952).
488. Y. Miyahara and H. Shiio, *Nippon Kagaku Zasshi* 72, 876 (1951).
489. Y. Miyahara, *Bull. Chem. Soc. Jap.* 25, 326 (1952).
490. K. Tamm, *Nachr. Akad. Wiss. Goettingen Math. Phys. Kl. 2A Math. Phys. Chem. Abt.* p. 81 (1952).
491. Yu. K. Novodranov and S. N. Mal'tsman, *Uch. Zap. Leningrad. Gos. Univ.* im. A. A. Zhdanova, No. 150, *Ser. Khim. Nauk* No. 10, 163 (1951).
492. T. Sasaki and T. Yasunaga, *Bull. Chem. Soc. Jap.* 28, 269 (1955).
493. J. Padova, *Bull. Res. Counc. Isr. Sect. A. Chem.* 10, 63 (1961).
494. H. Asai, *J. Phys. Soc. Jap.* 16, 761 (1951).
495. S. V. Subrahmanyam and J. Bhimasenachar, *J. Acoust. Soc. Amer.* 32, 835 (1960).
496. M. Suryanarayana, *J. Acoust. Soc. Amer.* 33, 1245 (1961).
497. M. Pancholi and S. P. Singal, *Nuovo Cimento* 28, 292 (1963).
498. S. Lengyel and G. Felzer, *Magy. Kem. Foly.* 60, 183 (1963); *Acta Chim. (Budapest)* 37, 319 (1963).
499. M. G. S. Rao and B. R. Rao, *Indian J. Phys.* 36, 613 (1962).
500. M. S. Murty and Bh. Krishnamurty, *Indian J. Pure Appl. Phys.* 1, 332 (1963).
501. T. Yasunaga, Y. Hirata, Y. Kawano, and M. Miura, *Bull. Chem. Soc. Jap.* 37, 867 (1964).
502. T. Satyavati, *Indian J. Pure Appl. Phys.* 2, 201 (1964).
503. M. G. S. Rao and B. R. Rao, *Indian J. Pure Appl. Phys.* 1, 362 (1963).
504. I. V. Litvinenko, *Zh. Strukt. Khim.* 4, 830 (1963).

505. M. S. Murty and Bh. Krishnamurty, *Indian J. Phys.* **37**, 359 (1963).

506. J. Padova, *Isr. At. Energy Comm. IA* [*Rep.*] **IA-823**; **IA-830** (1963).

507. T. Isemura and S. Goto, *Bull. Chem. Soc. Jap.* **37**, 1690 (1964).

508. P. A. Zagorets, V. I. Ermakov, and A. P. Grunau, *Zh. Fiz. Khim.* **39**, 456 (1965).

509. M. S. Marty, *Indian J. Pure Appl. Phys.* **3**, 156 (1965).

510. J. Kuppuswamy, A. S. Lakshmanan, R. Laskshminarayanan, N. Rajaram, and C. V. Suryanarayana, *Bull. Chem Soc. Jap.* **38**, 1610 (1965).

511. L. G. Jackopin and E. Yeager, *J. Phys. Chem.* **70**, 313 (1966).

512. G. Atkinson and S. Kor, *J. Phys. Chem.* **69**, 128 (1965).

513. J. Smithson and T. Litovitz, *J. Acoust. Soc. Amer.* **28**, 462 (1956).

514. M. Eigen and K. Tamm, *Z. Elektrochem.* **66**, 107 (1962).

515. M. J. Blandamer, M. J. Foster, N. J. Hidden, and M. C. R. Symons, *Chem. Commun.* **3**, 62 (1966).

516. S. Lica and G. Tudorache, *Bul. Inst. Politch. Bucuresti* **25**, 35 (1963).

517. M. V. Kaulgud, *Z. Phys. Chem.* (*Frankfurt am Main*) **47**, 24 (1965).

518. R. A. Robinson and R. H. Stokes, "Electrolyte Solutions," p. 247. Butterworth, London, 1955.

519. Yu. A. Petrenko, L. A. Petrenko, and V. K. Tkach, *Biofiz. Radiobiol.* p. 3 (1966).

520. F. Fittipaldi, *Nature* (*London*) **210**, 5038 (1966).

521. Y. Kondo, T. Goto, I. Suo, and N. Tokura, *Bull. Chem. Soc. Jap.* **39**, 1230 (1966).

522. A. S. Kaurova and G. P. Roshchina, *Akust. Z.* **12**, 319 (1966).

523. L. A. Petrenko and Yu. A. Petrenko, *Zh. Strukt. Khim.* **8**, 212 (1967).

524. E. Darmois, *J. Phy. Radium* **2**, 2 (1941).

525. J. O'M. Bockris and B. E. Conway, "Modern Aspects of Electrochemistry," No. 1, p. 67. Butterworth, London, 1954.

526. B. E. Conway, R. E. Verrall, and J. E. Desnoyers, *Z. Phys. Chem.* (*Leipzig*) **230**, 157 (1965).

527. T. J. Webb, *J. Amer. Chem. Soc.* **48**, 2589 (1926).

528. P. W. Bridgman, *J. Chem. Phys.* **3**, 597 (1935).

529. J. D. Bernal and R. H. Fowler, *J. Chem. Phys.* **1**, 515 (1933).

530. D. A. McInnes and M. Dayoff, *J. Amer. Soc.* **74**, 1017 (1952).

531. G. W. Stewart, *J. Chem. Phys.* **7**, 381 (1939).

532. H. Lee, *J. Chin. Chem. Soc.* (*Peiping*) **9**, 46 (1942).

533. S. Goto, *Bull. Chem. Soc. Jap.* **37**, 1685 (1964).

534. K. Tamura and T. Sasaki, *Bull. Chem. Soc. Japan* **36**, 975 (1963).

535. T. Yasunaga and T. Sasaki, *Nippon Kagaku Zasshi* **72**, 89 (1951).

536. T. Sasaki, T. Yasunaga, and H. Fujihara, *Nippon Kagaku Zasshi* **73**, 181 (1952).

537. K. Tamura and T. Sasaki, *Bull. Chem. Soc. Jap.* **36**, 975 (1963).

538. P. G. Tait, "The Physics and Chemistry of the Voyage of H.M.S. Challenger," Part IV, 1888.

539. W. Nernst and P. Drude, *Z. Phys. Chem. Stoechiom. Verwandschaftslehre* **15**, 79 (1894).

540. G. Tammann, "Uber die Beziehungen Zwischen den Kräften and Eigenschaften der Läsungen," p. 36. Voss, Leipzig, 1907.

541. T. Yasunaga, *Nippon Kagaku Zasshi* **72**, 87 (1951).

542. A. Passynski, *Acta Physicochim. USSR* **8**, 385 (1938).

543. Y. Wada, S. Shimbo, M. Oda, and J. Nagumo, *Oyo Butsuri* **17**, 257 (1948).

544. E. Glueckauf, *Trans. Faraday Soc.* **64**, 2423 (1968).

545. E. Glueckauf, *Trans. Faraday Soc.* **61**, 914 (1965).

546. S. W. Benson and C. S. Copeland, *J. Phys. Chem.* **67**, 1194 (1963).

547. P. Mukerjee, *J. Phys. Chem.* **65**, 740 (1961).
548. P. Mukerjee, *J. Phys. Chem.* **65**, 744 (1961).
549. J. Padova, *J. Phys. Chem.* **71**, 2347 (1967).
550. J. Padova, *Bull. Res. Counc. Isr. Sect. A Chem.* **10**, 63 (1961).
551. G. Jones and M. Dole, *J. Amer. Chem. Soc.* **51**, 2959 (1929).
552. R. H. Stokes and R. Mills, "Viscosity of Electrolytes and Related Properties," 1st ed. Pergamon, Oxford, 1965.
553. F. H. Spedding and M. J. Pikal, *J. Phys. Chem.* **70**, 2430 (1966).
554. F. H. Spedding and M. J. Pikal, *J. Phys. Chem.* **70**, 2440 (1966).
555. J. Padova, *J. Chem. Phys.* **39**, 1552 (1963).
556. A. I. Kaz'min, *Zh. Fiz. Khim.* **11**, 585 (1938).
557. R. J. Gillespie and S. Wasif, *J. Chem. Soc.* p. 215 (1953).
558. R. H. Flowers, R. J. Gillespie, and E. A. Robinson, *J. Chem. Soc.* p. 845 (1960).
559. W. H. Lee, *in* "The Chemistry of Non-aqueous Solvents" (J. J. Lagowski, ed.), Vol. II. Academic Press, New York, 1967.
560. S. J. Bass, R. J. Gillespie, and J. V. Oubridge, *J. Chem. Soc.* p. 837 (1960).
561. R. H. Flowers, R. J. Gillespie, and E. A. Robinson, *J. Chem. Soc.* p. 4327 (1960).
562. J. B. Hasted, D. M. Ritson, and C. H. Collie, *J. Chem. Phys.* **16**, 1 (1948).
563. G. H. Haggis, J. B. Hasted, and T. J. Buchanan, *J. Chem. Phys.* **20**, 1452 (1952).
564. E. Glueckauf, *Trans. Faraday Soc.* **60**, 1637 (1964).
565. G. W. Stewart, *J. Chem. Phys.* **7**, 869 (1939).
566. J. A. Prinns, *J. Chem. Phy.* **3**, 72 (1935).
567. O. Bastiansen and C. Finbak, *Tidsskr. Kjemi Bergv. Met.* **4**, 50 (1944).
568. O. Bastiansen and C. Finbak, *Tidsskr. Kjemi Bergv. Met.* **3**, 98 (1943).
569. O. Bastiansen and C. Finbak, *Tidsskr. Kjemi Bergv. Met.* **4**, 40 (1944).
570. C. Finbak, *Tidsskr. Kjemi Bergv. Met.* **4**, 77 (1945).
571. G. W. Brady and J. T. Krause, *J. Chem. Phys.* **27**, 304 (1957).
572. G. W. Brady, *J. Chem. Phys.* **28**, 464 (1958).
573. A. F. Skryshevskii, *Str. Fiz. Svoistva Veshchestva v Zhidkom* Sostoyanii (Iieva: Izdatel. Univ.) **27** (1954); *Ref. Zh. Khim.* Abstract No. 32076 (1956) (Chem. Abstr. 52: 17910 f).
574. I. M. Shapovalov, I. V. Radchenko, and M. K. Lesovitskaya, *Zh. Strukt. Khim.* **4**, 10 (1963).
575. C. I. Von Panthaleon van Eck, H. Mendel, and W. Boog, *Discuss. Faraday Soc.* **24**, 200 (1957).
576. A. I. Ryss and I. V. Radchenko, *Zh. Strukt. Khim.* **6**, 449 (1965).
577. R. F. Kruh and C. L. Standley, *Inorg. Chem.* **1**, 941 (1962).
578. D. Wertz, R. M. Lawrence, and R. F. Kruh, *J. Chem. Phys.* **43**, 2163 (1965).
579. J. F. Hinton and E. S. Amis, *Chem. Rev.* **67**, 367 (1967).
580. J. Burgess and M. C. R. Symons, *Quant. Rev.* **3**, 276 (1968).
581. J. A. Jackson, J. T. Lemons, and H. Taube, *J. Chem. Phys.* **32**, 553 (1960).
582. H. Weaver, B. M. Tolbert, and R. C. LaForce, *J. Chem. Phys.* **23**, 1956 (1955).
583. A. C. Rutenberg and H. Taube, *J. Chem. Phys.* **20**, 825 (1952).
584. R. G. Shulman and B. J. Wyluda, *J. Chem. Phys.* **30**, 335 (1959).
585. R. G. Shulman, *J. Chem. Phys.* **29**, 945 (1958).
586. R. E. Connick and D. N. Fiat, *J. Chem. Phys.* **39**, 1349 (1963).
587. M. Alei and J. A. Jackson, *J. Chem. Phys.* **41**, 3402 (1964).
588. W. B. Lewis, J. A. Jackson, J. T. Lemons, and H. Taube, *J. Chem. Phys.* **36**, 694 (1962).
589. M. Alei, *Inorg. Chem.* **3**, 44 (1964).
590. J. P. Hunt and H. Taube, *J. Chem. Phys.* **18**, 757 (1950).

591. G. E. Glass and R. S. Tobias, *J. Amer. Chem. Soc.* **89**, 6371 (1967).
592. G. E. Glass, W. B. Schwabacher, and R. S. Tobias, *Inorg. Chem.* **7**, 2471 (1968).
593. C. T. Moynihan and A. Fratiello, *J. Amer. Chem. Soc.* **89**, 5546 (1967).
594. J. Reuben and D. Fiat, *Inorg. Chem.* **6**, 579 (1967).
595. D. Fiat and R. E. Connick, *J. Amer. Chem.* **88**, 4754 (1966).
596. E. R. Malinowski, R. S. Knapp, and B. Feuer, *J. Chem. Phys.* **45**, 4274.
597. F. J. Vogrin, P. S. Knapp, W. L. Flint, A. Anton, G. Highberger, and E. R. Malinowski, *J. Chem. Phys.* **54**, 178 (1971).
598. R. W. Creekmore and C. N. Reilley, *J. Phys. Chem.* **73**, 1563 (1969).
599. P. S. Knapp, R. O. White, and E. R. Malinowski, *J. Chem. Phys.* **49**, 5459 (1968).
600. E. R. Malinowski and P. S. Knapp, *J. Chem. Phys.* **48**, 4989 (1968).
601. E. R. Malinowski, P. S. Knapp, and B. Feuer, *J. Chem. Phys.* **47**, 347 (1967).
602. W. G. Schneider, H. J. Bernstein, and J. A. Pople, *J. Chem. Phys.* **28**, 601 (1958).
603. L. Petrokis and C. H. Sederholm, *J. Chem. Phys.* **35**, 1174 (1961).
604. B. P. Fabricand, S. S. Goldberg, R. Leiferaand, and S. G. Ungar, *Mol. Phys.* **7**, 425 (1964).
605. P. S. Knapp, R. O. White, and E. R. Malinowski, *J. Chem. Phys.* **49**, 5459 (1968).
606. S. Broersma, *J. Chem. Phys.* **28**, 1158 (1958).
607. J. W. Akitt and J. A. Downs, *Chem. Commun.* p. 222 (1966).
608. J. H. Swinehart and H. Taube, *J. Chem. Phys.* **37**, 1579 (1962).
609. S. Nakamura and S. Meiboom, *J. Amer. Chem. Soc.* **89**, 1765 (1967).
610. A. Fratiello, R. E. Lee, V. M. Nishida, and R. E. Schuster, *Chem. Commun.* p. 173 (1968).
611. R. G. Wawro and T. J. Swift, *J. Amer. Chem. Soc.* **90**, 2792 (1968).
612. N. A. Matwiyoff and H. Taube, *J. Amer. Chem. Soc.* **90**, 2796 (1968).
613. R. N. Butler, E. A. Phillpot, and M. C. R. Symons, *Chem. Commun.* p. 371 (1968).
614. T. J. Swift and H. H. Lo, *J. Amer. Chem. Soc.* **89**, 3987 (1967).
615. Z. Luz and S. Meiboom, *J. Chem. Phys.* **40**, 1058 (1964).
616. N. A. Matwiyoff and P. E. Darley, *J. Phys. Chem.* **72**, 2659 (1968).
617. N. A. Matwiyoff, *Inorg. Chem.* **5**, 788 (1966).
618. N. A. Matwiyoff and S. V. Hooker, *Inorg. Chem.* **6**, 1127 (1967).
619. W. G. Movius and N. A. Matwiyoff, *Inorg. Chem.* **6**, 847 (1967).
620. S. Thomas and W. L. Reynolds, *J. Chem. Phys.* **44**, 3148 (1966).
621. A. Fratiello and R. E. Schuster, *Tetrahedron Lett.* **46**, 4641 (1967).
622. A. Fratiello, R. E. Lee, V. M. Nishida, and R. E. Schuster, *J. Chem. Phys.* **48**, 3705 (1968).
623. R. E. Schuster and A. Fratiello, *J. Chem. Phys.* **47**, 1554 (1967).
624. A. Fratiello and R. E. Schuster, *J. Chem. Educ.* **45**, 91 (1968).
625. D. Fiat and R. E. Connick, *J. Amer. Chem. Soc.* **90**, 608 (1968).
626. N. A. Natwiyoff, P. E. Darley, and W. G. Movius, *Inorg. Chem.* **7**, 2173 (1968).
627. A. Fratiello, V. Kubo, R. E. Lee, S. Peak, and R. E. Schuster, *J. Inorg. Nucl. Chem.* **32**, 3114 (1970).
628. L. D. Supran and N. Sheppard, *Chem. Commun.* p. 832 (1967).
629. N. A. Matwiyoff and W. G. Movius, *J. Amer. Chem. Soc.* **89**, 6077 (1967).
630. A. Fratiello, R. E. Lee, and R. E. Schuster, *Inorg. Chem.* **9**, 82 (1970).
631. A. Fratiello, R. E. Lee, V. M. Mishida, and R. E. Shuster, *J. Chem. Phys.* **50**, 3624 (1969).
632. A. Fratiello, R. E. Lee, and R. E. Schuster, *Inorg. Chem.* **9**, 391 (1970).
633. A. Fratiello, D. D. Davis, S. Peak, and R. E. Schuster, *J. Phys. Chem.* **74**, 3730 (1970).
634. A. Fratiello, V. Kubo, R. E. Lee, and R. E. Schuster, *J. Phys. Chem.* **74**, 3726 (1907).

635. A. Fratiello, V. Kubo, and R. E. Schuster, *Inorg. Chem.* **10**, 744 (1971).

636. J. F. Stephens and G. K. Schweitzer, *Spectrosc. Lett.* **1**, 373 (1968); **3**, 11 (1970).

637. R. J. Kula, D. L. Rabenstein, and G. H. Reed, *Anal. Chem.* **37**, 1783 (1965).

638. Yu. Shamoyan and S. A. Yan, *Dokl. Akad. Nauk SSSR* **152**, 677 (1963).

639. B. P. Fabricand, S. S. Goldberg, R. Leiferaard, and S. G. Unger, *Mol. Phys.* **7**, 425 (1964).

640. P. A. Zagorets, V. I. Ermakov, and A. P. Granaw, *Zh. Fiz. Khim.* **37**, 1167 (1963).

641. P. A. Zagorets, V. I. Ermakov, and A. P. Granaw, *Zh. Fiz. Khim.* **39**, 4 (1965).

642. S. Broersma, *J. Chem. Phys.* **28**, 1158 (1958).

643. J. C. Hindman, *J. Chem. Phys.* **36**, 1000 (1962).

644. T. J. Swift and W. G. Sayre, *J. Chem. Phys.* **44**, 3567 (1966).

645. T. J. Swift, O. G. Fritz, and T. A. Stephenson, *J. Chem. Phys.* **46**, 406 (1967).

646. E. Butler, *Z. Chem.* **5**, 199 (1967).

647. E. L. Muetterties, *J. Amer. Chem. Soc.* **88**, 305 (1966).

648. T. J. Swift and G. P. Wienberger, *J. Amer. Chem. Soc.* **90**, 2023 (1968).

649. S. Meiboom, *J. Chem. Phys.* **46**, 410 (1967).

650. T. J. Swift and W. G. Sayre, *J. Chem. Phys.* **46**, 410 (1967).

651. C. C. McDonald and W. D. Phillips, *J. Amer. Chem. Soc.* **83**, 3736 (1963).

652. Z. Luz and R. G. Shulman, *J. Chem. Phys.* **43**, 3750 (1965).

653. E. Schaschel and M. C. Day, *J. Amer. Chem. Soc.* **90**, 503 (1968).

654. R. A. Pinkowitz and T. J. Swift, *J. Chem. Phys.* **54**, 2858 (1971).

655. A. I. Zhernovoi and G. I. Yakovlev, *Zh. Strukt. Khim.* **4**, 914 (1963).

656. I. V. Matyash, A. I. Toryanik, and U. I. Yashkichev, *Zh. Strukt. Khim.* **5**, 777 (1964).

657. K. A. Valiev, *Zh. Strukt. Khim.* **5**, 517 (1964).

658. H. G. Hertz, *Z. Elektrochem.* **65**, 20 (1961).

659. R. Richards and B. A. Yorke, *Mol. Phys.* **6**, 289 (1963).

660. K. A. Valiev and B. M. Khabibullin, *Zh. Fiz. Khim.* **35**, 2265 (1961).

661. K. Grunther and H. Pfeifer, *Zh. Strukt. Khim.* **5**, 193 (1964).

662. C. V. Krishnan and H. L. Friedman, *J. Phys. Chem.* **74**, 2357 (1970).

Chapter 4

STRUCTURAL ASPECTS OF MIXED AQUEOUS SOLVENT SYSTEMS

Because of its role in our environment, water and aqueous solutions have been the subject of countless investigations. For many years the search for a structural model that would account for the unique physical properties of pure water has intrigued, stimulated and, at times, probably frustrated the scientific community. Indeed, quite recently the existence of a new type of water called anomolous or polywater (sometimes referred to by less impressive titles by opponents) with physical properties more peculiar than "common" water has been debated in scientific literature and even in the news media. Although many model systems have been proposed for the structure of water, the final answer to its structure is yet to come.

As perplexing as the search for a structural definition of water is, the eccentric physicochemical properties of mixed aqueous solutions is at times even more baffling. Perhaps in the near future we may sit in the vaunted position of knowing how to describe all of the molecular interactions of liquid water and hence how to formulate its structure. Then we shall apply this knowledge along with that most important portion of acumen, hindsight, to those problems in solution chemistry that have plagued us and say, "It really wasn't so difficult after all." However, until that state of bliss is achieved it seems that unanswered questions increase almost exponentially with each problem solved in investigations attempting to characterize the molecular interactions and structure of mixed aqueous systems and mixed aqueous system containing electrolytes.

In this short chapter, we shall present a number of examples of the unusual properties of certain mixed aqueous systems with no attempt being made to review the existing literature. The examples presented were chosen so that a variety of experimental techniques used in obtaining information about solution structure could be covered.

In the study of the amount of solvent transported into the cathode region of a transference apparatus relative to the amount transported into the anode region using ethanol–water and other mixed solvents, it has been found in several instances [1–4] that there was a periodic variation of the relative amount of solvent, $\Delta G_{\text{sol}}^{\text{F}}$, transferred into the cathode portion as the weight percent of alcohol increased. Attempts have been made at explaining these periodic effects. Two sets of these data will be presented here together with the explanation of the phenomena. In Fig. 4.1 is depicted [2] the effect for LiCl in the ethanol–water system. The curve resembles a sine curve which increases in frequency and amplitude with increasing weight percent of ethanol. It has been suggested [5] that in $0.5\ M$ solution, the chloride ion has a hydration number of 8 including both primary and secondary solvation shells. According to MacInnes [6], a hydration number of 8 for the chloride ion would give a hydration number of 23 for the LiCl. Accepting this value at $0.5\ N$, and taking values of 89 and 19 for the hydration number of LiCl in

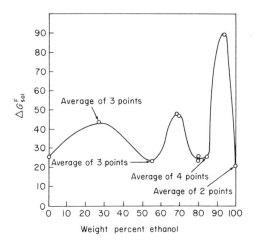

Fig. 4.1. Calculated grams of solvent transferred from anode to cathode per faraday of electricity interpolated to 1.60 weight% LiCl plotted against percent of ethanol in solvent.

0.1 and 1 N solutions, respectively [2], then the hydration number of LiCl in 0.383 molal LiCl extrapolates to a value of about 38 of which about 29 to 30 would be associated with Li^+ ion and 4 to 8 with chloride ion A 0.383-N.

The lithium ion has been shown to have a primary solvation number of 4 in water [7] and 4 in ethanol [8], and the chloride ion a primary solvation shell of 4 in water [5]. It has been found [9] that lithium ion is preferentially solvated by water in ethanol–water solvents. The above data for 0.383 N LiCl indicate that lithium ion probably has three shells of solvent of solvation, and chloride ion probably has two shells of solvent of solvation. Assuming that these assumptions are correct, we can possibly explain the periodicity of the curve in Fig. 4.1 in the following manner.

The outer solvation shell of the lithium ion is probably the most labile, and the relative mass of solvent transported into the cathode region of the cell by the cation at first increases due to the gradual exchange of the water for the higher molecular weight alcohol in the outer solvation shell of the lithium ion. At about 27 weight % ethanol in the solvent, a maximum in the curve is reached, and the curve then has a downward trend to about 55.0 weight % ethanol. This decrease in the relative weight of the solvent carried into the cathode region by the cation could arise from the gradual replacement of water by ethanol in the outer solvation shell of the chloride ion as the ethanol increase in weight percent. Of the various solvation shells of the two ions, this outer solvation shell of the chloride ion is perhaps next most labile after the outer shell of lithium ion. The second solvation shell of lithium would possibly be next most labile, the inner shell of chloride ion next most labile, and the inner shell of lithium ion least labile of all. This order of labialities would probably be true because of the relatively smaller size of the lithium ion and its relatively higher charge density compared to the chloride ion.

Thus, after the decrease in ΔG_{sol}^F out to about 55 weight % ethanol, ΔG_{sol}^F again increase out to about 70 weight % ethanol presumably because the alcohol is substituting in the second solvation shell of the lithium ion. About 70 weight % ethanol, the remaining most labile shell of solvent water in the two ions is the inner solvation shell of the chloride ion. In this shell exchange of water for ethanol becomes dominant and causes a decrease in value of ΔG_{sol}^F up to about 80 weight % ethanol, when the attack by ethanol of the inner shell of water in the lithium ion again causes ΔG_{sol}^F to increase. That the last remaining shell of water in the two ions is breached at 80 weight % ethanol should be especially noted since this rather rapid replacement of the last solvent–water shell beginning at this weight percent organic composition of the solvent is evidenced by several other phenomena. From the precipitous and large rise of the ΔG_{sol}^F–weight % ethanol, one would believe that this last solvent shell is replaced rapidly and rather completely. The final maximum occurs at about 94 weight % alcohol when presumably the last water is replaced by ethanol in the inner solvation shell of lithium ion.

Other information shows [10] that certain ions are preferentially solvated by water in ethanol–water mixtures. However, there is a region of composition of ethanol above which even the inner layers of water of these ions apparently become subject to replacement by ethanol as has been indicated by other considerations [10–13]. This is apparently the case with lithium ion even in its inner solvation layer when ethanol reaches the high weight percent represented by this final large maximum.

The decrease beyond this final maximum was caused presumably by the precipitation at the cathode of the reference material, α-methyl-D-glucoside, in this ethanol-rich solvent range. This precipitation could have resulted from the increased concentration of the sugar in the cathode portion due to its transport into that portion by the lithium ion. The sugar apparently could not compete with water in solvating the lithium ion in water rich solvents, however, once the inner water sheath of lithium ion was broken, the α-methyl-D-glucoside presumably could compete with ethanol as a solvating agent. The transport of the sugar into the cathode portion by the lithium ion resulted in a relative decrease of solvent concentration in that portion when the ethanol was greater than 94 weight % as compared to an ethanol concentration of 94 weight %. Longsworth [11b] and others have shown inert reference substances to not be completely inert. In the instance described above, the solubility of the sugar was exceeded only in the immediate region of the cathode even when the alcohol was greater than 94 weight %.

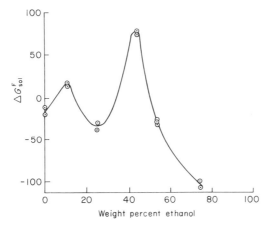

Fig. 4.2. Calculated grams of solvent transferred from anode to cathode per faraday of electricity at ∼0.15 moles KCl per 1000 gm of solution plotted against weight% ethanol.

In Fig. 4.2 is presented data for ΔG_{sol}^{F} versus weight % ethanol for 0.15 moles of KCl per 1000 g of solution [14]. This concentration of KCl could not be maintained beyond about 74 weight % methanol, and lower concentrations of KCl made the accurate determination of the relative amount of solvent transported from anode region to cathode region very difficult. Hence the whole range of solvent composition was not observed in these experiments. Over the region of solvent composition studied, this curve is similar to the one observed for LiCl. Apparently a minimum in the KCl curve could occur at 80 weight % ethanol as it did in the curve for LiCl. This would be the solvent composition at which the last water solvent sheath of the potassium ion is breached by alcohol if one uses the analogy with lithium ion. This region of solvent compositions seems significant with respect to several solution phenomena and with respect to several organic component–water mixtures.

The region of 10–30 weight % of organic component in a mixed organic component–water solvent in many instances shows significant variation in chemical phenomena occuring in mixed solvent media of which water is one component. This is the region in which the first solvent water layer, the most labile layer, of one of the ions involved in a chemical phenomenon is in the process of being replaced by organic solvent. For example it is the composition region in which apparently the outermost water solvent layer of lithium ion in LiCl is replaced by ethanol. Other instances of the effect on chemical phenomena of this composition of solvent will be presented later in this book.

The compositions of about 20 and 80 weight % of solvents are also critical regions with respect to the structure of solvents as will be illustrated later. Thus, the marked effects of these solvent compositions on chemical phenomena affecting dissolved solutes is perhaps a combination of critical solvation and solvent structure effects.

That the 10–30 and about 75–90 weight % regions of organic component in a mixed organic component–water media are critical with respect to replacement of water by organic component in the solvent sheaths of ions is strikingly evident from the following considerations.

Walden's rule can be obtained by balancing electrostatic and resistence forces. Mathematically the rule may be expressed as

$$\Lambda_0 = \frac{\varepsilon F}{1800\pi\eta}\left(\frac{Z^+}{r_0^{+}} + \frac{Z^-}{r_0^{-}}\right) \tag{4.1}$$

where Λ_0 is the equivalent conductance at infinite dilution of an electrolyte possessing positive ions of valence Z^+ and radius r^+ and negative ions of valence Z^- and radius r_0^-, ε is the electronic charge, F is the faraday, and η is the viscosity of the medium.

For a uniunivalent electrolyte $Z^+ = Z^- = 1$ and

$$\Lambda_0 = \frac{\varepsilon F}{1800\pi\eta}\left(\frac{1}{r_0{}^+} + \frac{1}{r_0{}^-}\right) \qquad (4.2)$$

If from Eq. (4.2) the value of $(1/r_0{}^+) + (1/r_0{}^-)$ for an electrolyte at different solvent compositions is calculated from experimentally found values of Λ_0 and η, and the reciprocals $r_0{}^+ r_0{}^- / (r_0{}^+ + r_0{}^-)$ (hereafter designated as the r function of this function) are obtained, these reciprocal quantities are of the order of magnitude of r_i an ionic radius. In Fig. 4.3 values of this r function

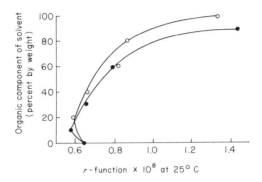

Fig. 4.3. The r function $\times 10^8$ at 25°C versus percent by weight organic component of the solvent. Open circle = water–methyl alcohol; solid circle = water–acetone.

versus solvent composition are plotted for conductance data on potassium chloride in water–methanol and water–acetone solvents [10]. The curves are striking in several respects. In the first place they are similar for both solvent systems. In the second place there are minimum values of the r function, at about 10 weight% acetone in the water–acetone solvent and at about 20 weight% methanol in the water methanol–solvent. This is the region where transference data indicated that water was being replaced in the first solvation shell of the potassium ion. Apparently the increased electrostatic force due to increasing amounts of the lower dielectric organic component of the solvent pulled the solvent layers in closer to the ions and caused a decrease in the r function in spite of the fact that the larger organic solvent molecules were replacing the smaller water molecules in the outer solvation layer of the potassium ion. Beyond the minimum values of the r functions, the replacement effect predominates over the increased electrostatic effect, and the r functions increase in an accelerated manner; these increases become precipitous at around 80 weight% organic component of the solvent, which is the composition of solvent at which, according to transference data, the inner layer of solvent water around the potassium ion begins to be abruptly displaced by the larger organic-solvent-component molecules. The authors are now convinced that it is the net effect on the r functions of the opposing replacement and increasing electrostatic effects, rather than selective solvation by water as such which causes the r function to be relatively constant up to about 40 weight% of the organic-solvent-component, though up to about 20 weight% of organic component, the solvent sheaths of both ions are principally water, and beyond about 80 weight% the solvent sheaths are almost completely organic solvent.

A striking example of the effectiveness of small amounts of water in breaking the hydrogen-bonded structure of the nonaqueous component of the solvent is the dependence at 25°C of the equivalent conductance at infinite dilution Λ_0 of perchloric acid on traces of water when methanol and ethanol are used as solvents [11]. When 0.3 weight% water was added to pure methanol Λ_0 dropped by about 22 units or about 10% and when 0.3 weight% water was added to pure ethanol, Λ_0 dropped about 28 units or about 30%. The data are shown in Fig. 4.4. In the pure alcohols the conductance of the hydrogen ion was of the Grotthus type along the hydrogen-bonded chains of the pure alcohols. The large decrease in Λ_0 values when small amounts of water are added to pure methanol and ethanol solutions of perchloric acid is probably

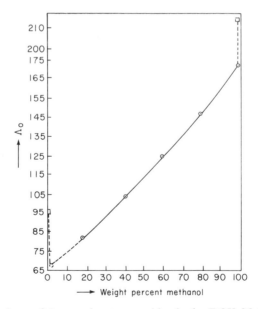

Fig. 4.4. Dependence of Λ_0 on solvent composition in the EtOH–MeOM–0.3 weight% H_2O at 25°C.

largely hydrogen–ion dependent and could be related to two effects. These are perhaps the disruption of alcohol chains with the consequent decreased effectiveness of Grotthus conduction along the shortened alcohol chains, and conductance by the movement of H_3O^+ and $H_9O_4^+$ ions as entities through the liquid media. The effect of the addition of small amounts of water to methanol is not as great as in the case of ethanol because the alkyl group is less prominent in methanol and it approaches water more nearly in composition and polar properties.

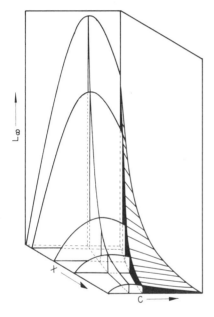

Fig. 4.5. Specific conductance of sodium perchlorate in water–NMA solvents at 50°C.

Conductivity studies of sodium perchlorate in aqueous *N*-methylacetamide (NMA) solutions also clearly indicate the marked effect small amounts of one solvent component can have on the structure of the other component (12). Figure 4.5 is a three-dimensional plot of the specific conductance of sodium perchlorate at infinite frequency L_∞ as a function of increasing mole fraction of NMA, X, and as a function of increasing salt concentration C from extremely dilute to almost saturated solutions. The curve with the highest maximum in the $L_\infty C$ plane at $X = 0$ represents pure water and the curve in the $L_\infty C$ plane for $X = 1$ represents pure NMA. The stripped area is the extension of the concentration curves beyond saturation. The frontal curve of the plane through the maxima of the $L_\infty C$ plots indicates the precipitous decrease in conductance with the first additions of NMA to the aqueous solution. At higher mole fractions of NMA, the rate of decrease of L_∞ with

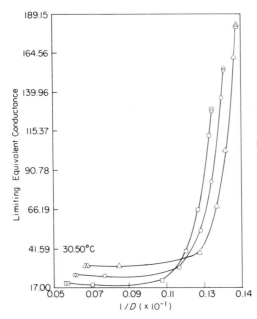

Fig. 4.6. $\Lambda_0(\infty)$ versus $1/D$ for sodium perchlorate in water–NMA mixtures. $\square = 30.00°C$; $\circ = 40.00°C$; $\triangle = 50.00°C$.

Fig. 4.7. Standard potentials at 25°C as functions of methanol–water compositions. $\bullet = H_2|HCl(m)|AgCl,Ag$ cell; $\blacksquare = H_2|HBr(m)|AgBr$, Agcell; $\blacktriangle = H_2|HI(m)|AgI,Ag$ cell.

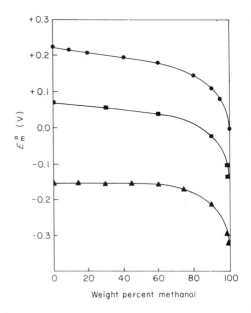

X falls dramatically. The $L_\infty C$ curve with the second lowest maximum is that of mole fraction of NMA at the concentration represented by the plane where the rate of change of L_∞ with X is making the sharpest transition from a very large to a very small rate of change. This mole fraction of NMA is in the region of maximum viscosity of the solvent system. However, viscosity is not the sole factor in controlling the changes of specific conductance with increasing mole fraction of NMA as indicated by the fact that the specific conductance does not again increase after the point of maximum viscosity is passed.

Nuclear magnetic resonance experiments with the aqueous-NMA system indicate the hydrogen-bonded structure of NMA "tightens' as the mole fraction of NMA increases (13). This "tighter" solvent structure in NMA rich solutions slows the motion of the ions through the solvent and hence decreases the specific conductance.

Another factor which would produce a consistent decrease in specific conductance with increasing mole fraction of NMA would be a continual replacement of water in the primary solvation sphere of the ions by NMA molecules thus increasing the effective size of the ions. The specific conductance would then become essentially constant at the point where the ions were solvated by only NMA. This phenomenon is demonstrated clearly in Fig. 4.6 where the limiting conductance is plotted as a function of the reciprocal of the dielectric constant. The break in the curves occurs in the region of maximum viscosity.

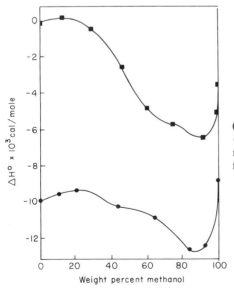

Fig. 4.8. Enthalpies for the Pt,H_2|HI (m)|AgI,Ag cell(III) and the Pt,H_2|HCl (m)|AgCl,Ag cell (I) as functions of methanol–water compositions. ■ = 35°C for cell III; ● = 35°C for cell I.

Electromotive force measurements in mixed solvent systems also exhibit structural changes in the solvent. The characteristic curves of the standard potentials E_m^0 at 25°C versus solvent composition in water methanol solvents are presented in Fig. 4.7 for the cells [14]

$$pt.H_2(1 \ atm)|HCl(m)|AgCl,Ag \qquad (I)$$

$$pt,H_2(1 \ atm)|HBr(m)|AgBr,Ag \qquad (II)$$

and

$$pt,H_2(1 \ atm)|HI(m)|AgI,Ag \qquad (III)$$

In Fig. 4.8 are presented plots at 35°C of the enthalpies for the reactions in the cells (I) and (III) versus solvent composition of solvent in water–methanol media. Also in Fig. 4.9 is given a plot at 35°C of the entropy for the reaction in cell (III) above versus solvent composition in the above mentioned media. In Fig. 4.10 are plotted at 25°C the ion-size parameter a and the salting-out coefficient B for the cells (I) and (II) [15]. For all the cells, the E_m^0 versus solvent composition curves show a marked change in slope in the region of 80 weight% methanol. The curves in Figs. 4.8 and 4.9 show maxima in the region of 20 weight% methanol, and marked inflections (in Fig. 4.8 minima) in the 80–90% methanol region. In Fig. 4.10 the a versus weight% methanol and B versus weight% methanol curves show pronounced maxima around 90 weight% methanol.

Fig. 4.9. Entropies for the $Pt,H_2|HI$ (m)|AgI,Ag cell as functions of methanol –water compositions. ■ = 35°C.

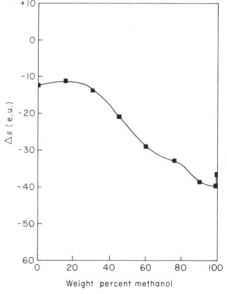

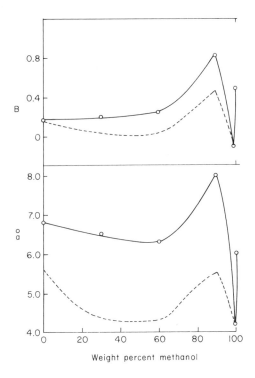

Fig. 4.10. Plots of ion-size parameter $\mathring{a}$ and salting-out coefficient B for HBr (solid line) and HCl (broken line) versus weight% MeOH at 25°C.

In emf data, too, there is the evidence of a marked change of the potential and derived phenomena in the organic rich solvent (80–90 weight% methanol) and for some of the phenomena in the 10–30% by weight organic component of the solvent. These are the regions of solvent composition in which data indicate that the outermost and innermost layers, respectively, of the most highly hydrated of the ions of an electrolyte are being replaced by the organic component of the solvent. The other hydration layers of both ion are replaced at intermediate solvent composition. Thus many chemical phenomena show marked changes when the outermost hydration layer is breached for any ion and when the innermost and final hydration layer for any ion is breached.

Kinetic data show striking examples of the effect on chemical phenomena of the solvent composition involving 10–30 and about 80 weight% organic component of the solvent. Figures 4.11 and 4.12 are for data on reaction between U(IV) and U(VI) ions in the water–ethanol solvent systems [16]. Figures 4.11 and 4.12 are for volume percent ethanol and present the orders of the various reactants and the logarithm of the specific velocity constant, respectively, as functions of solvent composition. The drastic changes of the orders of the various reactants and of $\log k'$ with increasing percent of alcohol in the solvent at high ethanol concentrations are striking. Like data are shown in Figs. 4.13 and 4.14 for the above described reaction in water–ethylene glycol solvents [17]. Here the marked effect of the solvent composition on both the order of the various reactants and the specific velocity constants are dramatically evident at both low and high percentages of ethylene glycol in the mixed solvents. Actually, in Fig. 4.12 the weight % ethanol at which the maximum and minimum for the orders with respect to U(IV), and U(VI), respectively, occur is 85.5. In Fig. 4.13 the weight % glycol at which the maxima for the orders with respect to U(IV) and acid, respectively, occur is 90.9, while at the low ethylene glycol ends of the curves, the inflection points of

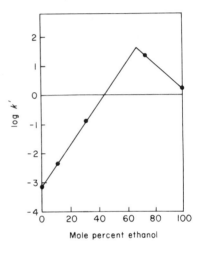

Fig. 4.11. Logarithm of the specific velocity constant versus the mole percent ethanol in the solvent.

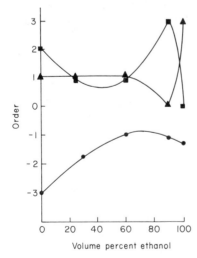

Fig. 4.12. Order of the various reactants versus the volume percent ethanol. ● = acid; ▲ = U(VI); ■ = U(IV).

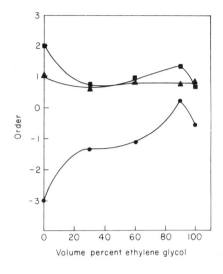

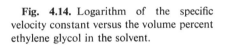

Fig. 4.13. Order of the various reactants versus the volume percent ethylene glycol. ● = acid; ▲ = U(VI); ■ = U(IV).

Fig. 4.14. Logarithm of the specific velocity constant versus the volume percent ethylene glycol in the solvent.

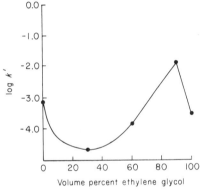

the curves occur at about 31.5 weight% glycol. The 85.5 weight% ethanol in water–ethanol, and the 90.9 weight% glycol in water–glycol correspond, respectively, to 70.0 and 74.3 mole% of the two organic components. Thus in any units the exact composition at which drastic effects occur depend to some extent on the nature of the organic component, which perhaps is to be expected since the places at which initial and final breaches in the water sheaths of the solute particles occur depend no doubt on the polarity, size, geometry, and other properties of the substituting reagent. However, these two critical regions do manifest themselves in many phenomena.

Figures 4.15 and 4.16, respectively, present the $\log k$ versus percent ethanol by weight for various molarities of hydrochloric acid and $\log k$ versus $\log [\mathrm{H}^+]$ for various weights percents of ethanol [18]. These data are for the electron exchange reaction between U(VI) and Sn(II) ions in ethanol–water media in which the hydrochloric acid varied from 1 to 6 M. In Fig. 4.15 the drastic changes in the rate constants at about 20 and 80 weight % ethanol are obvious. The actual magnitudes and even directions of the changes depend on the hydrochloric acid concentrations. Figure 4.16 reveals a periodic dependence of $\log k$ on $\log[\mathrm{H}^+]$ in all save the 80 weight % ethanol solvent. For this solvent there is no maximum in the curve at 2 M acid, and a much more vertical climb of the curve beyond 4 M acid. Thus once again the 20 and 80 weight % of the organic component in the solvents show remarkable effects on a chemical phenomenon.

Thermodynamic parameters are frequently used to obtain information about the structure and intermolecular interactions in aqueous mixed solvent systems. Although quite useful, these parameters have rarely provided immediate solutions to the description of the systems under investigation. There are a great many such examples in the literature, however, we shall describe only a few in this chapter.

Figure 4.17 shows a plot of the molal heat of vaporization as a function of the mole fraction of dioxane in aqueous dioxane solutions [19]. This plot of the total molal heat of vaporization for the isothermal–isobaric transformation of the liquid into the vapor phase certainly seems to indicate significant changes in intermolecular interactions and solution structure as the mole

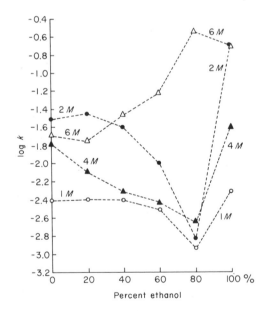

Fig. 4.15. Plot of log k versus percent ethanol in the solvent at different molarities of hydrochloric acid. Sn(II) = 1.0 M, U(VI) = 0.1 M, T = 35°C. ○ = 1 M HCl; ● = 2 M HCl; ▲ = 4 M HCl; △ = 6 M HCl.

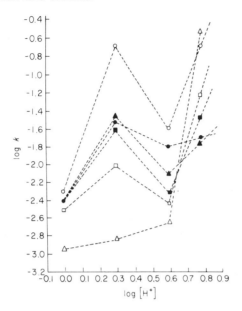

Fig. 4.16. Hydrogen ion dependence of the apparent second order rate constnat k at different ionic strengths. Sn(II) = 1.0 M, U(VI) = 0.1 M, T = 35°C. Percent EtOH: ● = 0%; ▲ = 20%; ■ = 40%; □ = 60%; △ = 80%; ○ = 99.9%.

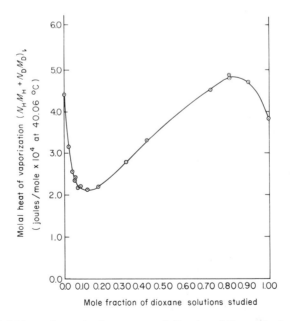

Fig. 4.17. Molal heat of vaporization versus mole fraction of dioxane in the solutions.

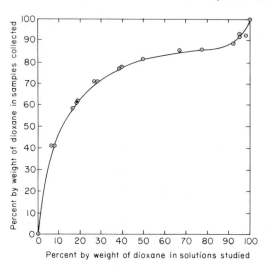

Fig. 4.18. Percent by weight of dioxane in vapor versus percent by weight of dioxane in solution.

fraction of solvent components is varied. The rapid decrease of the molal heat of vaporization upon the initial addition of dioxane to water was attributed to a decrease in hydrogen bonding and dipolar forces among the water molecules. After passing through a minimum, the molal heat of vaporization of the solution increases with increasing mole fraction of dioxane until a mole fraction of 0.8 dioxane is attained. It was suggested that this increase could be due to the rapid proportionate increase of water coming off as vapor as shown in Fig. 4.18. Beyond a mole fraction of 0.8 dioxane a decrease in molal heat of vaporization occurs which was ascribed to the predominance of weak induced dipole-induced dipole forces in the dioxane-rich solutions.

Glew and co-workers [20] have also investigated the thermodynamic and physical properties of aqueous dioxane solutions as well as other aqueous nonelectrolyte solutions (i.e., ethylene oxide, propylene oxide, acetone, tetrahydrofuran and *t*-butyl alcohol) and found anomalous water properties occurring at 94–97 mole% water. Figures 4.19 and 4.20 show plots of the water activity coefficient and water excess molar volume, respectively, in dilute aqueous solutions of the nonelectrolytes mentioned above. The minima observed in the water activity coefficient plots indicate an increase in stabilization of solvent water relative to pure water. The increase in water stabilization was rationalized in terms of enhanced intermolecular interaction between water molecules adjacent to the solute molecules rather than arising from the weak hydrogen bonds formed between the hydrophilic groups of the solute molecules and water.

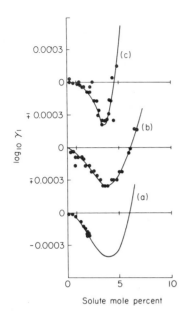

Solute mole percent

Fig. 4.19. Water activity coefficient in dilute aqueous ethylene oxide (a) at 0°C, dioxane (b) at 0°C, and t=butyl alcohol (c) at 10°C.

Fig. 4.20. Water excess molar volume in dilute aqueous ethylene oxide (a) at 10°C, dioxane (b) at 25°C, acetone at 0°C, tetrahydrofuran (d) at 10°C, and t = butyl alcohol (e) at 15°C.

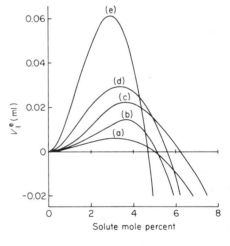

Solute mole percent

Figure 4.20 shows that the excess molar volume of water in dilute aqueous solutions of dioxane passes through a maximum at approximately 4 mole% of solute. This increase in excess molar volume of water produced by dioxane and the other nonelectrolytes also shown in the figure was suggested to be the result of reorientation of water molecules through hydrogen-bonded shell formation adjacent to solute molecules such that within the shell the water–water hydrogen bonds bend less and are stronger than in pure water with the

solute occupying the interstitial volume only partially filled in pure water
[21–23]. Nuclear magnetic resonance evidence was also obtained which
supported the interstitial model. For dilute aqueous solutions of the dioxane
and the other nonelectrolytes mentioned the proton magnetic resonance
chemical shift of water moved toward lower applied magnetic field, with
maxima at 3–6 mole% solute, indicating stronger water–water hydrogen
bonding.

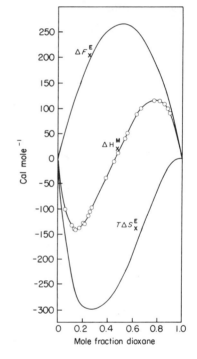

Fig. 4.21. Thermodynamic properties of water–dioxane at 25°C.

The interstitial-solute water-shell model for aqueous solutions was applied
to the excess enthalpy of mixing for aqueous dioxane solutions, Fig. 4.21, as
obtained by Goates and Sullivan [21] and similar data for aqueous acetone
[22], aqueous tetrahydrofuran [23] and *t*-butyl alcohol [24]. The negative
excess enthalpies in the low solute concentration region are consistent with the
interstitial model in which the strengthening of the many intrashell water–
water hydrogen bonds more than compensate for the weak hydrogen bonding
to the organic solute hydrophilic group. The model was also successfully
applied to the observation of anomalous concentration dependence of sound
velocity and absorption in dilute aqueous solutions of *t*-butyl alcohol [25]

and inert-gas solubility in aqueous solutions of ethanol, dioxane and methanol [26–28]. Assuming the interstitial model to be adequate in describing the dilute aqueous dioxane system, the heat of vaporization data of Stallard and Amis [19] are most difficult to explain.

Even though there are general trends in structural effects produced by nonelectrolytes in aqueous nonelectrolyte systems, there are subtle differences depending upon the intrinsic molecular structure of the nonelectrolytes. A good example of this can be seen again from thermodynamic parameters. A comparison of the excess enthalpy of mixing of aqueous–methanol, –ethanol, –propanol, and -*t*-butanol is displayed in Fig. 4.22 [29, 30a]. Franks [30a] discusses the complexity of these aqueous alcohol mixtures in terms of competing interactions, however, a complete understanding of the structural features over the entire mole fraction range is still not at hand. It is clear, however, that dramatic structural changes do occur in these solutions which are functions of the intrinsic molecular structure of the nonelectrolyte.

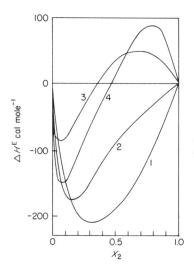

Fig. 4.22. Excess enthalpies of mixing of alcohols and water at 25°C. (1) Methanol; (2) ethanol; (3) propanol; (4) *t*-butanol.

The thermodynamic activation parameters, $\Delta H^{\pm}$ and $\Delta S^{\pm}$, for kinetic processes occurring in aqueous mixed solvent systems can undergo dramatic changes depending upon the mole fraction of the solvent system. Perhaps the classic example is that of the solvolysis of *t*-butyl chloride in aqueous ethanol [31]. The activation parameters for this reaction as a function of the mole fraction of water is shown in Fig. 4.23. The striking dependence of $\Delta H^{\pm}$ and $\Delta S^{\pm}$ on the solvent composition seems extraordinary at first glance, however, such behavior is coming to be considered the rule rather than the exception where aqueous mixed solvent systems are involved. Arnett [30b] summarizes

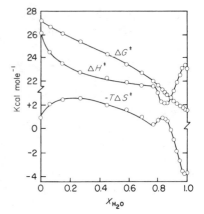

Fig. 4.23. Variation of activation parameters for *t*-butyl chloride solvolysis in aqueous ethanol.

the situation quite well in stating, "There is a remarkable degree of unity in the behavior of a wide diversity of thermodynamic, kinetic, spectral, and other physical properties in highly aqueous organic binary systems. These phenomena are closely related to the degrees of structuredness at various compositions of the binary system relative to water. It is quite possible that much of this behavior will eventually be related to the formation of structures of structures approaching crystalline clathrates hydrates in solution."

The anomolous structural properties of aqueous nonelectrolyte systems may also be observed using a variety of spectroscopic techniques. Of the many techniques available nuclear magnetic resonance spectroscopy has proven to be a very powerful tool for investigating liquid mixtures. Recent nuclear magnetic resonance studies of aqueous amide solutions have shown the effect of the amide on the structure of water to be markedly dependent upon the intrinsic molecular structure of the amide as well as the concentration of the amide in solution [32]. Figure 4.24 shows plots of the water proton chemical shift as a function of the mole fraction of water in twelve aqueous amide systems. An inspection of this figure reveals a similarity in chemical among all of the systems, except formamide, in the water mole fraction range of 1.0 to 0.8. The first addition of the amides to water produces a net low field shift in the water proton resonance position. This low field shift is indicative of structure promotion or a "tightening" of water hydrogen bonds. The interstitial model proposed by Glew *et al.* [20] would seem to adequately explain the low field shift in the dilute amide region.

In these solutions of very low amide concentration the amide molecules may either hydrogen bond with water, producing a high field shift in the water proton resonance position since it is assumed that the water–amide hydrogen bond is not as strong as the water–water hydrogen bond, or become an interstitial species, producing a low field shift. Since the water proton chemical shifts measured are the weighted average of these two effects, the data obtained suggests that the interstitial effect dominates upon the first addition of the amide followed by a net structure breaking effect with increasing amide concentration. Although the water proton chemical shift data obtained for the aqueous-formamide system does not indicate the presence of interstitial formamide molecules, this does not necessarily mean that they do not exist. Because of the relatively small size of formamide it is possible that there is no water expansion produced by interstitial formamide molecules and, therefore, the only effect observed is that of structure breaking due to water–amide bonding.

For all amide systems, except *N*-methylformamide, *N*-methylacetamide, and *N*-ethylformamide, the structure of water is continuously broken as the mole fraction of water decreases from about 0.8 to 0. It is noted that the magnitude of the structure breaking effect is in general related to the size and number of hydrogen bonding sites of the amide, the large disubstituted amides producing the greatest structure breaking effect. The solution composition range over which the interstitial model is used to describe the solutions is also a function of amide size.

With the aqueous *N*-methylacetamide, *N*-methylformamide, and *N*-ethylformamide solutions a very interesting structure promotion effect is observed in the range below 0.3 mole fraction of water.

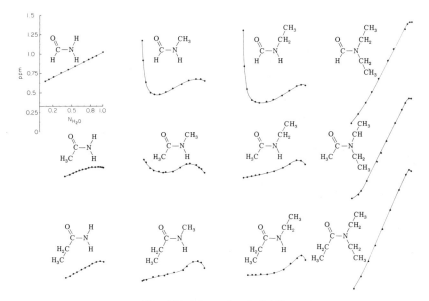

Fig. 4.24. Water chemical shifts.

It is also of interest to note that the line width of water in the *N*-methyl-formamide and *N*-ethylformamide systems begins to broaden significantly in this solution composition range. Figure 4.25 compares the water proton chemical shifts as a function of solution composition for the formamide, *N*-methylformamide, and *N*-ethylformamide systems. Figure 4.26 shows the water proton resonance line width as a function of solution composition for these three systems. It is seen from these two figures that for the formamide system in which only structure breaking is observed, no line broadening occurs. In the other two amide systems line broadening begins to occur in the mole fraction region where the low field shift (structure promotion) is observed. At a water mole fraction of about 0.08, the water proton line width at half height is 8 and 13 Hz in the *N*-methylformamide and *N*-ethylformamide systems, respectively. The line broadening is not produced by an increase in viscosity since a viscosity maximum in these systems occurs at a mole fraction of about 0.55 water where no significant line broadening was observed. The low field shift and line broadening observed in the low water concentration region in these systems would indicate well ordered, tightly bonded water molecules. At present we have no model that would explain this effect.

As stated at the beginning of this chapter, no attempt was made to begin

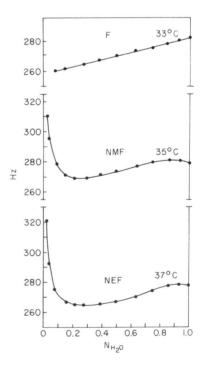

Fig. 4.25. Chemical shifts versus mole fraction of water in water–amide systems.

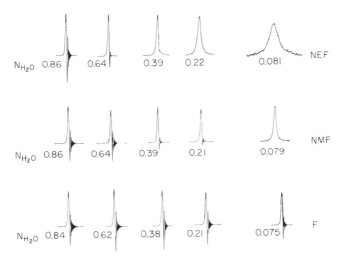

Fig. 4.26. Water proton resonance line width versus mole fraction of water for water–amide systems.

to review and discuss all of the aqueous nonelectrolyte systems for which anomolous water structure effects have been observed. A persual of the data for the systems discussed adequately shows the dependence of the structure of water on the concentration and intrinsic molecular structure of the added nonelectrolyte. In particular it is seen that there are two mole fraction regions in which the structure of water almost invariably undergoes dramatic change; these being at about 0.2 and 0.8 mole fraction water. While such observations are well known to investigators in the field it is hoped that this brief discussion of the unusual structural effects peculiar to aqueous mixed solvent systems will possibly be of some heuristic value and might stimulate those contemplating on taking the "plunge" into solution chemistry to actually do so. At any rate it seems prudent at this point to again listen to the advice of Scatchard [33]: "The best advice which comes from years of study of liquid mixtures is to use any model in so far as it helps, but not to believe that any moderately simple model corresponds very closely to any real mixture."

References

1. D. M. Matthews, J. O. Wear, and E. S. Amis, *J. Inorg. Nucl. Chem.* **13**, 298 (1959).
2. J. O. Wear, C. V. McNully, and E. S. Amis, *J. Inorg. Nucl. Chem.* **18**, 48 (1961).
3. J. O. Wear, C. V. McNully, and E. S. Amis, *J. Inorg. Nucl. Chem.* **19**, 278 (1961).
4. J. A. Bard, J. O. Wear, R. F. Griffin, and E. S. Amis, *Electroanal. Chem.* **8**, 419 (1964).
5. K. P. Mishcheno, *Zh. Fiz. Khim.* **26**, 1736 (1950).

6. D. A. McInnes, "The Principles of Electrochemistry," pp. 85–93. Van Nostrand–Reinhold, Princeton, New Jersey, 1939.

7. O. Ya Somoilov, *Izv. Akad. Nauk. SSSR Ser. Khim.* p. 398 (1952).

8. H. Osaka, *Bull. Chem. Soc. Jap.* **12**, 177 (1937).

9. Y. Kobayasi, K. Toka, and U. Muiri, *J. Sci. Hiroshima Univ. Ser. A* **9**, 33 (1939).

10. E. S. Amis, *J. Phys. Chem.* **60**, 428 (1956).

11a. N. Goldenberg and E. S. Amis, *Z. Phys. Chem. (Frankfurt am Main)* **31**, 10 (1962).

11b. L. G. Longsworth, *J. Amer. Chem. Soc.* **69**, 1288 (1947).

12. J. Casteel, Ph.D. Dissertation, Univ. of Arkansas, Fayetteville, Arkansas, 1972.

13. J. F. Hinton and E. S. Amis, *Z. Phys. Chem. N.F.* **60**, 159 (1968).

14. J. M. McIntyre and E. S. Amis, *J. Chem. Eng. Data* **13**, 371 (1968).

15. S. L. Melton and E. S. Amis, *J. Chem. Eng. Data* **13**. 429 (1968).

16. D. M. Matthews, J. D. Hefley, and E. S. Amis, *J. Phys. Chem.* **63**, 1236 (1959).

17. S. L. Melton, J. O. Wear, and E. S. Amis, *J. Inorg. Nucl. Chem.* **17**, 117 (1961).

18. Z. C. Ho Tan and E. S. Amis, *J. Inorg. Nucl. Chem.* **28**, 2889 (1966).

19. R. D. Stallard and E. S. Amis, *J. Am. Chem. Soc.* **74**, 1781 (1952).

20. D. N. Glew, H. D. Mak, and N. S. Rath, *in* "Hydrogen-Bonded Solvent Systems" (A. K. Covington and P. Jones, eds.), p. 195. Taylor and Francis, London, 1968.

21. J. R. Goates and R. J. Sullivan, *J. Phys. Chem.* **62**, 188 (1958).

22. D. O. Hanson and M. Van Winkle, *J. Chem. Eng. Data* **5**, 30 (1960).

23. J. Erva, *Suom. Kemistilehti B* **28**, 131 (1955); **29**, 183 (1956).

24. J. Kenttamoa, E. Tommila, and M. Mantti, *Ann. Acad. Sci. Fenn. Ser. A2* **93** (1959).

25. M. J. Blandamer, D. E. Clarke, N. J. Hidden, and M. C. R. Symons, *Chem. Commun.* p. 342 (1966).

26. A. Ben-Naim and S. Baer, *Trans. Faraday Soc.* **60**, 1736 (1964).

27. A. Ben-Naim and G. Moran, *Trans. Faraday Soc.* **61**, 21 (1956).

28. A. Ben-Naim, *J. Phys. Chem.* **71**, 4002 (1967).

29. F. Franks and D. J. G. Ives, *Quant. Rev.* **20**, 1 (1966).

30a. F. Franks, *in* "Physico-Chemical Processes in Mixed Aqueous Solvents" (F. Franks, ed.), p. 54. Amer. Elsevier, New York, 1967.

30b. E. M. Arnett, *in* "Physico-Chemical Processes in Mixed Aqueous Solvents" (F. Franks, ed.) p. 105. Amer. Elsevier, New York, 1967.

31. S. Winstein and A. H. Fainberg, *J. Amer. Chem. Soc.* **79**, 5937 (1957).

32. J. F. Hinton and K H. Ladner, *J. Magn. Resonance* **6**, 586 (1972).

33. G. Scatchard, *Chem. Rev.* **44**, 7 (1949).

Chapter 5

SOLVENT INFLUENCE ON RATES AND MECHANISMS

I. Introduction

Solvent effects on rates and mechanisms are many and varied. They encompass electrostatic effects, including: the modification of the forces arising from the dielectric constants of pure and mixed solvents between charged particles [between charged and dipolar (permanent and induced) particles, and between dipolar (permanent and induced in all combinations) particles], the effects of changing from protic to dipolar aprotic solvents, the effects of internal pressure or cohesion of the solvent, the effects of external pressure, the effects of the viscosity of the solvent, the effects of selective solvation by one component in a mixed solvent or of the different extent of solvation by different solvents of the reactants, the transition (activated complex) state, and the products, cage effects, hydrogen bonding effects, ionization effects of the solvent, the effects of solvolysis by the solvent, and the effects of the acidity of the medium. Some of these effects are related. For example, the effects of changing from protic to dipolar aprotic solvents and the different extent of solvation by different solvents of the reactants, the activated complex, and the products are intricately intertwined. Some effects apply particularly to a certain type of reaction. Thus substitution nucleophilic second-order (S_N2) reactions are very sensitive to changes from protic to dipolar aprotic solvents, i.e., from solvents which are hydrogen-bond donors to highly polar solvents which are nonhydrogen-bond donors. Solvent effects on chemical reaction rates may vary in magnitude from a fraction to several powers of ten. Thus for the reaction of diacetone alcohol with hydroxide ion [1, 2] the specific velocity constant is one-tenth greater in pure water than in 80% ethyl alcohol–20% water by weight at 25°C, while the rate constant for the displacement of iodine in methyl iodide by chloride ion [3] at 25°C is over 10^6 times as fast in dimethylacetamide as in water.

The solvent effects on reaction rates and mechanisms will be discussed in the order in which they are listed above. A complete discussion of the effects cannot be given in the alloted space; however, the fundamental principles will be presented in each instance so that the reader will have the necessary knowledge to build upon, and pertinent references will be listed so that the reader may quickly turn to more complete discussions of the topics.

II. Electrostatic Effects on Ion–Ion Reactions

A. THE SCATCHARD EQUATION FOR IONIC REACTANTS

The effect of the dielectric constant of the solvent on the rates of reaction between ionic reactants has been mathematically formulated by Scatchard [4] using the expression for the potential Ψ in the vicinity of an ion from the Debye–Hückel theory. The equation for Ψ is

$$\Psi = \frac{Z_i \varepsilon}{Dr} \frac{\exp[\varkappa(a_i - r)]}{1 + \varkappa a_i} \tag{5.1}$$

where Z_i is the valence of the ith type of ion, ε is the electronic charge, D the dielectric constant, r the distance from the ion at which the potential is Ψ, a_i the distance of closest approach to the ith type of ion, and the Debye $\varkappa$ is given by

$$\varkappa = \left(\frac{4\pi\varepsilon^2}{DkT} \sum n_i Z_i^2 \right)^{1/2} \tag{5.2}$$

where k is the Boltzmann gas constant, T the absolute temperature, n_i is the number of the ith type of ion per cubic centimeter, and the other terms have been defined above or have their usual meanings.

Using this expression the activity coefficient f_i of the ith type of ion can be found as follows:

$$\ln f_i = \frac{1}{kT} \int_0^{Z_i\varepsilon} \Psi \, d(Z_1\varepsilon)$$

$$= \frac{1}{kT} \int_0^{Z_i\varepsilon} \frac{Z_i\varepsilon}{Dr} \frac{\exp[\varkappa(a_i-r)]}{1+\varkappa a_i} \, d(Z_i\varepsilon)$$

$$= \frac{Z_i^2\varepsilon^2 \exp[\varkappa(a_i-r)]}{2DkT(1+\varkappa a_i)r} \tag{5.3}$$

Considering the reaction

$$A + B \rightleftarrows X \rightarrow \text{products} \tag{5.4}$$

we can write from Eq. (5.3),

$$\ln\frac{f_A f_B}{f_X} = \frac{[Z_A^2 + Z_B^2 - (Z_A+Z_B)^2]\varepsilon^2 \exp[\varkappa(a_i-r)]}{2DkT(1+\varkappa a_i)r}$$

$$= -\frac{Z_A Z_B \varepsilon^2 \exp[\varkappa(a_i-r)]}{DkT(1+a_i)r} \tag{5.5}$$

or

$$\frac{f_A f_B}{f_X} = \exp\left(-\frac{Z_A Z_B \varepsilon^2}{DkT}\right) \exp\left[\frac{(-\varkappa r)}{r}\right] \exp\left[\frac{(\varkappa a_i)}{1+\kappa a_i}\right] \tag{5.6}$$

which for $\varkappa = 0$ becomes

$$\frac{f_A f_B}{f_X} = \exp\left(-\frac{Z_A Z_B \varepsilon^2}{DkTr}\right) \tag{5.7}$$

The reaction in Eq. 5.4 at zero ionic strength using the procedure of Brönsted [5] yields

$$\frac{C_X{}^0 f_X}{C_A{}^0 f_A C_B{}^0 f_B} = K = \frac{1}{K'} \tag{5.8}$$

and

$$\ln\frac{f_A f_B}{f_X} = \ln K' \frac{C_X{}^0}{C_A{}^0 C_B{}^0} = -\frac{Z_A Z_B \varepsilon^2}{DkTr} \tag{5.9}$$

The difference between the logarithmic term when the dielectric constant of the solvent is D and when the dielectric constant is some standard reference value D_0 where the activity coefficients of all solutes are unity is

$$\ln\frac{f_A f_B}{f_X} - \ln\left(\frac{f_A f_B}{f_X}\right)_0 = \frac{Z_A Z_B \varepsilon^2}{kTr}\left(\frac{1}{D_0} - \frac{1}{D}\right) \tag{5.10}$$

According to Brönsted [5] the velocity constant $k'_{\varkappa=0}$ for a reaction at absolute temperature T and zero ionic strength is given by

$$k'_{\varkappa=0} = k'_{\varkappa=0;\, D=D_0} \frac{f_A f_B}{f_X} \tag{5.11}$$

where $k'_{\varkappa=0;\, D=D_0}$ is the velocity at the standard reference state of dielectric constant $D = D_0$, zero ionic strength, and absolute temperature T.

The substitution of $f_A f_B / f_X$ from Eq. (5.10) into Eq. (5.11) and use of the logarithmic form yields

$$\ln k'_{\varkappa=0} = \ln k'_{\varkappa=0;\, D=D_0} + \frac{Z_A Z_B \varepsilon^2}{kTr}\left(\frac{1}{D_0} - \frac{1}{D}\right) \tag{5.12}$$

Substituting for the standard state D_0 unity [6] and infinity [7, 8] yields respectively,

$$\ln k'_{\varkappa=0} = \ln k'_{\varkappa=0;\, D=1} + \frac{Z_A Z_B \varepsilon^2}{kTr}\left(1 - \frac{1}{D}\right) \tag{5.13a}$$

and

$$\ln k'_{\varkappa=0} = \ln k'_{\varkappa=0;\, D=\infty} - \frac{Z_A Z_B \varepsilon^2}{kTrD} \tag{5.13b}$$

In the above equations r is the radius $r_A + r_B$ of the complex and is the distance to which two ionic reactants must approach to react.

Christiansen's [9] method of calculating the concentration of the complex, $C_X{}^0$, directly from the equations of Debye and Hückel was used by Scatchard [4].

B. LAIDLER–EYRING APPROACH FOR IONIC REACTANTS

For the reaction

$$A + B + C + \cdots \rightleftarrows X + N + O + \cdots \rightarrow K + L + M + \cdots \tag{5.14}$$

Laidler and Eyring [6] wrote the constant K^* for the equilibrium between the reactants and activated complex

$$K^* = \frac{a_X a_N a_O \cdots}{a_A a_B a_C \cdots} = \frac{c_X c_N c_O \cdots}{c_A c_B c_C \cdots} \frac{f_X' f_N' f_O' \cdots}{f_A' f_B' f_C' \cdots} \tag{5.15}$$

where X represents the intermediate complex, N and O are intermediates formed with the complex and K, L, and M are final products.

The instability of the complex differentiates it from an ordinary molecule. In the complex, the degree of vibrational freedom of the bond which is broken when the complex molecule decomposes is transformed into an extra degree of translational freedom, with respect to which the complex is considered to be a free particle in a one-dimensional space of length S. The extra degree of freedom corresponds to the decomposition of the complex along the abscissae of the energy coordinate of the decomposition plot. Within the space of length S, the partition function F_X of the complex is $(2\pi m_X kT)^{1/2}S/h$.

In terms of the partition functions of the molecules K^* becomes

$$K^* = \frac{F_X F_N F_O \cdots}{F_A F_B F_C \cdots} \tag{5.16}$$

If we separate out from the partition function that part which appertains to the degree of translational freedom corresponding to decomposition we obtain

$$F_X = F_X' \frac{(2\pi MkT)^{1/2}S}{h} \tag{5.17}$$

and hence

$$K^* = \frac{F_X' F_N F_O \cdots}{F_A F_B F_C \cdots} \frac{(2\pi mkT)^{1/2}S}{h} \tag{5.18}$$

and letting

$$K_0^* = \frac{F_X' F_N F_O \cdots}{F_A F_B F_C \cdots} \tag{5.19}$$

K^* can be written

$$K^* = K_0^* \frac{(2\pi mkT)^{1/2}S}{h} \tag{5.20}$$

and from Eqs. (5.15) and (5.20) C_X becomes

$$C_X = K_0^* \frac{(2\pi mkT)^{1/2}}{h} S \frac{C_A C_B C_C \cdots}{C_N C_O \cdots} \frac{f_A' f_B' f_C' \cdots}{f_X' f_N' f_O' \cdots} \tag{5.21}$$

The product of concentration of complex per unit length of one-dimensional space C_X/S, the translational velocity in the direction of decomposition, $(kT/2\pi m_X)^{1/2}$, and the transmission coefficient, $\varkappa$, gives the rate of the reaction. Thus

$$\begin{aligned}
\text{Rate} &= K_0^* \frac{(2\pi m_X kT)^{1/2}}{h} \frac{C_A C_B C_C \cdots}{C_N C_O \cdots} \frac{f_A' f_B' f_C'}{f_X' f_N' f_O'} \left(\frac{kT}{2\pi m_X}\right)^{1/2} \varkappa \\
&= \varkappa \frac{kT}{h} K_0^* \frac{C_A C_B C_C}{C_N C_O} \frac{f_A' f_B' f_C' \cdots}{f_X' f_N' f_O' \cdots} \tag{5.22}
\end{aligned}$$

when the concentrations $C_A, C_B, C_C \cdots C_N, C_O \cdots$ are all unity the rate becomes the specific velocity constant k' the equation for which is

$$k' = \varkappa \frac{kT}{h} K_0 * \frac{f_A' f_B' f_C' \cdots}{f_X' f_N' f_O' \cdots} \qquad (5.23)$$

and for dilute gases, the activity coefficients are unity and

$$k'_{(gas)} = \varkappa \frac{kT}{h} K_0 * \qquad (5.24)$$

It might be pointed out that $\varkappa$ is the total probability for decomposition in the forward direction.

For the reaction

$$A^{Z_A} + B^{Z_B} \leftrightarrows X^{(Z_B + Z_A)} \rightarrow K + L \qquad (5.25)$$

the equation for the specific velocity constant from Eq. (5.25) becomes

$$k' = \varkappa \frac{kT}{h} K_0 * \frac{f_A' f_B'}{f_X'} \qquad (5.26)$$

and in logarithmic form

$$\ln k' = \ln \left(\varkappa \frac{kT}{h} K_0 * \right) + \ln \left(\frac{f_A' f_B'}{f_X'} \right) \qquad (5.27)$$

Splitting the activity coefficient f' into the activity coefficient β of the infinitely dilute solution with reference to the infinitely dilute gas; and the activity coefficient f of the solution being studied with reference to the infinitely dilute solution, gives

$$f' = \beta f \qquad (5.28)$$

and

$$\ln f' = \ln \beta + \ln f \qquad (5.29)$$

The free energy of transfer of an ion from a vacuum to a medium of dielectric constant D is composed of the following four parts: (1) $-Z^2 \varepsilon^2 / 2r$ free energy change due to discharging an ion in a vacuum, (2) ϕ due to transfer of the uncharged particle into infinitely dilute solution where it is subjected to van der Waal's and other forces, (3) $Z^2 \varepsilon^2 / 2rD$ due to charging an ion in a medium of dielectric constant D, and (4) ϕ' due to reorientation of the dipoles around the charged ion. The total free energy change ΔF due to the transfer becomes, therefore,

$$\Delta F = -\frac{Z^2 \varepsilon^2}{2r} + \frac{Z^2 \varepsilon^2}{2rD} + \phi + \phi' \qquad (5.30)$$

The activity coefficient of the ion is

$$\ln \beta = \frac{Z^2 \varepsilon^2}{2kTr}\left(\frac{1}{D}-1\right) + \frac{\phi}{kT} + \frac{\phi'}{kT} \tag{5.31}$$

so that

$$\ln\left(\frac{\beta_A \beta_B}{\beta_X}\right) = \frac{\varepsilon^2}{2kTr}\left(\frac{1}{D}-1\right)\left[\frac{Z_A^2}{r_A} + \frac{Z_B^2}{r_B} - \frac{(Z_A+Z_B)^2}{r_X}\right]$$

$$+ \frac{\phi_A + \phi_A' + \phi_B + \phi_B' - \phi_X - \phi_X'}{kT} \tag{5.32}$$

The activity coefficient f which refers the solution of ionic strength μ to infinitely dilute solution can be neglected in deriving the solvent effect since in the solvent effect it will be assumed that the infinitely dilute solution can be used.

From Eq. (5.23)

$$k' = \varkappa \frac{kT}{h} K_0^* \frac{\beta_A \beta_B}{\beta_X} \tag{5.33}$$

and Eqs. (5.32) and (5.33)

$$\ln k' = \ln\left(\varkappa \frac{kT}{h} K_0^*\right) + \ln\left(\frac{\beta_A \beta_B}{\beta_X}\right)$$

$$= \ln\left(\varkappa \frac{kT}{h} K_0^*\right) + \frac{\varepsilon^2}{2kT}\left(\frac{1}{D}-1\right)\left[\frac{Z_A^2}{r_A} + \frac{Z_B^2}{r_B} - \frac{(Z_A+Z_B)^2}{r_X}\right]$$

$$+ \frac{\phi_A + \phi_A' + \phi_B - \phi_B' - \phi_X - \phi_X'}{kT} \tag{5.34}$$

Applying Eq. (5.13a), Amis and LaMer [7] plotted $\log k_{\varkappa=0}$ versus $1/D$ for the reaction between negative divalent tetrabromophenolsulfonthalein ions and negative univalent hydroxide ions in water–methanol and water–ethanol at 25°C. The plots are given in Fig. 5.1 and are straight lines down to a dielectric constant of about 65 with the requirements of Eq. (5.13a). From the slopes of these straight lines the values of r were calculated to be 1.49 and 1.22 Å for water–methanol and water–ethanol, respectively.

Amis and Price [10] made similar plots for the ammonium ion–cyanate ion reaction in water–methanol and water–glycol at 30°C. The data were those of Svirbely and Shramm [11] in water–methanol and for Lander and Svirbely [12] in water–glycol. The curve for the water–methanol solvent, except for a slight curvature in the region of pure water, was straight down to a dielectric constant of about 45. The data for the water–glycol solvent gave a perfectly straight line down to dielectric constant of about 55. The slopes of both curves were positive as required by Eq. (5.13a). The values of r from the slopes of the lines were 2.2 and 2.5 Å for water–methanol and water–glycol solvents, respectively. The application of this theory to the bromoacetate–thiosulfate reaction has been discussed by Laidler and Eyring [13] and by Laidler [14]. The value for r was 5.1 Å as determined from the slope of the $\log k$ versus $1/D$ plot for this reaction.

Fig. 5.1. Log $k'_{k=0}$ versus $1/D$ at 25°C for the reaction $(B\phi B)^{2-} + OH^-$. $\bullet = H_2O–CH_3OH$; $\circ = H_2O–C_2H_5OH$.

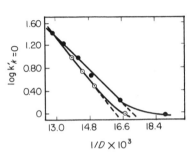

Deviations from linearity of the $\log k$ versus $1/D$ plot can be attributed to failure of the simple approximations used in the derivation of Eq. (5.13a), and to a change in reaction mechanism in some cases as the composition of the solvent is varied. On the whole the relationship holds to a good approximation, although generally there are marked deviations in the region of low dielectric constants.

C. ELECTROSTATICS AND ENERGY OF ACTIVATION FOR ION–ION REACTIONS

Moelwyn-Hughes [15] sets the total energy of activation E equal to the sum of two components: E_n, of which we know nothing, and E_e, an electrostatic contribution, which can be evaluated from the Debye–Hückel expression. The equation for E is

$$E = E_n + E_e = E_n + \frac{Z_A Z_B \varepsilon^2}{Dr}(1 - \varkappa r) \tag{5.35}$$

Using Z_0', the standard collison frequency, and E, the second-order rate constant, we can find k' with the expression

$$k' = Z_0 \exp(-E_n/kT) \exp\left(\frac{Z_A Z_B \varepsilon^2}{DrkT}\right)(1 - \varkappa r) \tag{5.36}$$

With respect to ionic strength, all terms not involving $\varkappa$ in the rate equation can be included in a constant k_0'. This gives

$$k' = k_0' \exp\left(\frac{Z_A Z_B \varepsilon^2 \varkappa}{DkT}\right) \tag{5.37}$$

an equation identical with that of Brönsted for the salt effect.

Differentiating the logarithmic form of Eq. 5.36, assuming Z_0 to be constant, taking note of the fact that $\varkappa$ is a function of temperature and dielectric constant, and multiplying through by kT^2, we obtain the equation for E_A, the apparent energy of activation, at constant temperature. Thus

$$E_A = E_n + \frac{Z_A Z_B \varepsilon^2}{Dr}\left(1 + \frac{\partial \ln D}{\partial \ln T}\right)(1 - \tfrac{3}{2}\varkappa r) \tag{5.38}$$

Letting

$$L = \left(\frac{\partial \ln D}{\partial T}\right)_P \tag{5.39}$$

we have, for Eq. 5.38,

$$E_A = E_n + \frac{Z_A Z_B \varepsilon^2}{Dr}(1 - LT)(1 - \tfrac{3}{2}\varkappa r) \tag{5.40}$$

Solving for E_n from Eq. (5.40) and substituting this value for E_n in Eq. (5.36), we obtain

$$k' = Z_0 \exp(-E_A/kT) \exp\left(-\frac{Z_A Z_B \varepsilon^2 L}{DkT}\right) \exp\left(\frac{Z_A Z_B \varepsilon^2 \varkappa}{2DkT}\right)(1 - 3LT) \tag{5.41}$$

For zero ionic strength Eq. (5.40) becomes

$$E_{A;\varkappa=0} = E_n + \frac{Z_A Z_B \varepsilon^2}{Dr}(1 - LT) \tag{5.42}$$

When the logarithmic form of Eq. (5.36) is differentiated with respect to temperature at constant pressure, ionic strength, and dielectric constant, and the resulting expression multiplied through by kT^2, the equation

$$E_D = E_n + \frac{Z_A Z_B \varepsilon^2}{Dr}(1 - \tfrac{3}{2}\varkappa r) \tag{5.43}$$

results. Here E_D is the energy of activation for the reaction occurring in constant dielectric media. At infinite dilution, $\varkappa = 0$ and Eq. (5.43) becomes

$$E_D = E_n + \frac{Z_A Z_B \varepsilon^2}{Dr} \tag{5.44}$$

True energy of activation, $E_{D;\varkappa=0}$, was obtained by Svirbely and Warner [16] by letting

$$\ln k_{\varkappa=0} = f(D, T) \tag{5.45}$$

and differentiating with respect to T. The resulting expression multiplied by $2.303RT^2$ gave

$$E_{A;\varkappa=0} = E_{D;\varkappa=0} + 2.303RT^2\left(\frac{\partial \log k'_{\varkappa=0}}{\partial D}\right)\frac{dD}{dT} \tag{5.46}$$

Calculating the difference $E_{A;\varkappa=0} - E_{D;\varkappa=0}$ for the ammonium ion–cyanate ion reaction at infinite dilution in methanol–water solvents, Svirbely and Warner found excellent agreement between the calculated and observed values.

The electrostatic contribution to the energy of activation was derived by LaMer [17] as follows:

$$\Delta E_D = \frac{\partial(\Delta F_D, T)}{\partial(1/T)} \tag{5.47}$$

But

$$\Delta F_D = \frac{Z_A Z_B \varepsilon^2}{Dr} \tag{5.48}$$

therefore

$$\Delta E_D = \Delta F_D\left(1 + \frac{\partial \ln D}{\partial \ln T}\right) \tag{5.49}$$

The difference in the energy of activation at constant dielectric constant and the apparent energy of activation at any dielectric constant can be found by subtracting Eq. (5.40) from Eq. (5.43) giving

$$E_D - E_A = \frac{Z_A Z_B \varepsilon^2 LT}{Dr}(1 - \tfrac{3}{2}\varkappa r) \tag{5.50}$$

The difference in the two energies of activation at infinite dilution, $\varkappa = 0$, is given by the equation

$$E_{D;\varkappa=0} - E_{A;\varkappa=0} = \frac{Z_A Z_B \varepsilon^2 LT}{Dr} \tag{5.51}$$

An equation comparable to Eq. (5.50) was obtained [8, 18] by differentiating the Brönsted–Christiansen–Scatchard equation with respect to T (T and D variable), differentiating the equation with respect to T (D constant), taking the difference, and multiplying by RT^2. The equation, where the ionic strength is μ, becomes

$$E_A - E_D = \frac{Z_A Z_B \varepsilon^2 NT}{D^2 J}\left[\frac{1}{r} - \frac{3\varepsilon}{10}\left(\frac{3\pi N\mu}{10DkT}\right)^{1/2}\right]\frac{dD}{dT} \tag{5.52}$$

Here J, the mechanical equivalent of heat in ergs per calorie, was introduced so that Eq. (5.52) yields the difference in energy of activation in calories per mole, and L of Eq. (5.50) was replaced by $(\delta \ln D / \delta T)_p$. Equation (5.52) was modified [19] to contain the coulombic energy of activation and gave

$$E_A - E_D = \frac{Z_A Z_B \varepsilon^2 N}{D_c J} \left\{ \left[\frac{1}{r} - \frac{3\varepsilon}{10} \left(\frac{2\pi N \mu}{10 D_c kT} \right)^{1/2} \right] \frac{T_c}{D} \frac{dD}{dT} - \frac{\Delta D}{D_D r} \right\} \quad (5.53)$$

In this equation $D = D_c - D_D$, and D_c and D_D are the dielectric constants of isocomposition and isodielectric runs, respectively. Both Eqs. (5.52) and (5.53) have been applied to experimental data [18, 19].

The coulombic energy of activation term, Eq. (5.53), was calculated using the expression [20]

$$E_c = -329.7 \frac{Z_A Z_B}{D_1 D_2 r} \Delta D \quad (5.54)$$

Here ΔD is the difference between two dielectric constants D_1 and D_2 of two different isodielectric media in which the temperature coefficient of a reaction is measured so as to obtain the coulombic contribution E_c to the energy of activation of the reaction.

The contribution to the energy of activation E_n made by the ionic atmosphere has been derived [21]

$$E_n = \frac{\partial (F_n / T)}{\partial (1/T)} \quad (5.55)$$

But

$$F_n = \frac{N Z_A Z_B \varepsilon^2 \kappa}{2D} \quad (5.56)$$

therefore

$$\Delta E_n = -\frac{3 N Z_A Z_B \varepsilon^2}{4D} \varkappa \left[1 + \frac{\partial \ln D}{\partial \ln T} + \frac{\partial \ln V}{\partial \ln T} \right] \quad (5.57)$$

Moelwyn-Hughes [22] derived a similar equation in a more explicit fashion employing the collision theory.

Figure 5.2 presents the data [7] for the total energy of activation–square root of the ionic strength plot in the case of the reaction of the negative divalent tetrabromophenalsulfonphthalein ion with negative univalent hydroxide ion. At low ionic strengths the slopes of the curves are those predicted by the Debye–Hückel limiting law.

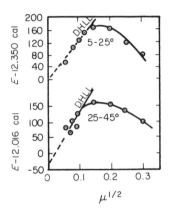

Fig. 5.2. Energy of activation of the reaction $(B\phi B)^{2-} + OH^-$ as a function of the ionic strength. Dashed lines correspond to the Debye–Hückel limiting law for activation energy for reaction between bivalent and univalent ionic reactants as a function of ionic strength.

D. NONELECTROSTATIC EFFECTS AND REACTION RATES

The dielectric constants of solvents enter into the calculation of the influence of nonelectrostatic effects on reaction rates. These effects are represented by the ϕ/kT terms in Eq. (5.34) and can be considered as repulsions that begin suddenly at the distance of closest approach modified to allow for van der Waal's attractions. Assume [23] that a Brönsted complex is formed between reactants undergoing reaction, that spherical symmetry exists around each particle, and that the mutual potential of the two particles at a distance r can be represented by the equation

$$\Psi(r) = \phi_1(r) + \phi_2(r) \tag{5.58}$$

The $\phi_2(r)$ term arises from electrostatic forces in the case the reactants A and B are charged. Thus, if Z_A and Z_B are the valencies of ions A and B, respectively, then

$$\phi_2(r) = \frac{Z_A Z_B \varepsilon^2}{Dr(1 + \varkappa a_i)} \exp[\varkappa(a_i - r)] \tag{5.59}$$

$\phi_1(r)$ was assumed to be positive for distances r at which the formation of the intermediate complex results and to be independent of the dielectric constant of the solvent. This latter assumption was reasonable since for distances necessary for complex formation, the dielectric constant of the solvent would have no significance.

If C_A and C_B represent the concentrations of A and B, respectively, the probability of finding molecule B at a distance between r_0 and $r_0 + dr_0$ from molecule A can be obtained from C_A and C_B and the potential represented in Eq. (5.58). This probability is

$$C_A C_B \exp\left[-\frac{\Psi(r_0)}{kT}\right] r_0^2 \, dr_0 \tag{5.60}$$

If for values of r within these limits, formation of X occurs, then applying the method of Christiansen [24] and of Scatchard [4], the expression for the reaction rate is

$$\text{rate} = KT^{1/2}C_A C_B \exp\left[-\frac{\Psi(r_0)}{kT}\right] \tag{5.61}$$

from which the specific velocity constant can be written

$$k' = KT^{1/2} \exp\left[-\frac{\Psi(r_0)}{kT}\right] \tag{5.62}$$

If for some standard reference temperature T_0 the rate constant is k_0' and can be considered as including the $T^{1/2}$ term, the final expression is

$$\ln k' = \ln k_0' + \frac{\phi_1(r_0)}{kT_0} - \frac{\phi_1(r_0)}{kT} - \frac{Z_A Z_B \varepsilon^2}{DkTr_0(1+\varkappa a_i)} \exp[\varkappa(a_i - r_0)] \tag{5.63}$$

For purposes considered here it is not necessary that k_0' be extrapolated to standard reference state of concentration or of dielectric constant.

Differentiation of Eq. (5.63) separately with respect to $1/D$ and with respect to $1/T$ gives the two equations

$$a_D = \left[\frac{\partial \ln k'}{\partial(1/D)}\right] = -\frac{Z_A Z_B \varepsilon^2}{kTr_0(1+\varkappa a_i)} \exp[\varkappa(a_i - r_0)] \tag{5.64}$$

and

$$a_T = \left[\frac{\partial \ln k'}{\partial(1/T)}\right] = -\frac{1}{k}\left[\phi_1(r_0) + \frac{Z_A Z_B \varepsilon^2 \exp[\varkappa(a_i - r_0)]}{Dr_0(1+\varkappa a_i)}\right] \tag{5.65}$$

A sufficiently large value of the nonelectrostatic compared to the electrostatic potential, will yield a negative value of a_T irrespective of what the charges Z_A and Z_B are.

From Eqs. (5.64) and (5.65) can be found the following inequalities relating measurable quantities:

$$a_T/a_D > T/D \tag{5.66}$$

for A and B of like charge sign, and

$$a_T/a_D < T/D \tag{5.67}$$

for A and B of unlike sign.

Using Eqs. 5.64 and 5.65 and measured values of a_T and a_D both the electrostatic potential $Z_A Z_B \varepsilon^2/Dr_0$ and the nonelectrostatic potential $\phi_1(r_0)$ for the critical distance can be found.

The reaction between persulfate ion and iodide ion in water–ethanol, and the reaction between the tetrabromophenolsulfonphthalein ion and hydroxide ion in water–methanol and in water–ethanol were used to test inequality (5.66). The reaction between ammonium ion and cyanate ion in water–methanol and in water–glycol was used to test inequality (5.67). These results are given in Table 5.1. From the table it is clear that for ions of like sign, a_T/a_D is about 3 to 4 times as large as T/D. For ions of unlike sign it is evident that $a_T/a_D < T/D$ to the extent of the inversion of the sign of a_T/a_D.

Table 5.1

Tests on Inequalities

Reaction	Medium	T	D	a_T	a_D	T/D	a_T/a_D
A. Ions of like sign
$S_2O_8^{-2} + I^-$	EtOH–H$_2$O	313.1	73.12	$-10,140$	-624	4.27	16.24
BϕB^{-2} + OH$^-$	MeOH–H$_2$O	298.1	71.42	$-11,000$	-737	4.18	15.24
BϕB^{-2} + OH$^-$	EtOH–H$_2$O	298.1	71.42	$-11,120$	-921	4.18	12.08
B. Ions of unlike sign
NH$_4^+$ + CNO$^-$	MeOH–H$_2$O	323.1	63.5	$-9,370$	260	5.09	-36.0
NH$_4^+$ + CNO$^-$	Glycol–H$_2$O	323.1	63.5	$-10,140$	184	5.09	-55.2

Table 5.2

Comparison of Nonelectrostatic Potentials at Distance $r = r_0$ with Electrostatic Potentials between Ions

Reaction	Medium	T	D	$\phi_2 \times 10^{13}$	$\phi_1 \times 10^{12}$	$\phi_1(r_0)/\phi_2(r_0)$
NH$_4^+$ + CNO$^-$	MeOH–H$_2$O	323.1	63.4	1.83	1.29	7.0
NH$_4^+$ + CNO$^-$	Glycol–H$_2$O	323.1	63.4	1.28	1.39	10.9
$S_2O_8^{2-}$ + I$^-$	EtOH–H$_2$O	313.1	73.12	3.66	1.39	3.8
BϕB^{2-} + OH$^-$	EtOH–H$_2$O	298.1	73.42	5.27	1.52	2.9
BϕB^{2-} + OH$^-$	MeOH–H$_2$O	298.1	73.42	4.22	1.51	3.6

Using Eqs. (5.64) and (5.65), electrostatic potentials $\phi_2(r_0)$ and nonelectrostatic potentials $\phi_1(r_0)$ were calculated [23] for the same reactants in the same solvents as were used in the calculation of the inequalities. These potentials together with their ratios are listed in Table 5.2. As would be expected, the ratio of nonelectrostatic to electrostatic potential is much greater (about three times as great) for reactions between univalent ions than for reactions between bivalent and univalent ions. It can be noted that the solvent plays a rather important part in determining these ratios. For example, the nonelectrostatic effect is relatively greater in water–glycol solvent than in water–methanol solvent in the case of the ammonium ion–cyanate ion reaction. It has been pointed out [6] that specific solvation of the ions by water in water–organic solvent mixtures reduces the electrostatic effect produced by the change in dielectric constant to a value below that expected from theory. In the case of the ammonium ion–cyanate ion reaction, if a reduced electrostatic effect produced the higher nonelectrostatic effect–electrostatic effect ratio, then there must be a greater specific solvation of the ions by water in the water–glycol than in the water–methanol solvent. However, the larger ratio of $\phi_1(r_0)/\phi_2(r_0)$ in water–glycol compared to that in water–methanol might be due to a relative enhancement of the nonelectrostatic effect in the water–glycol. This would suggest relatively greater repulsive forces as compared to van der Waals' attractions at distance r_0 in water–glycol as compared to water–methanol. Water of solvation occurs in larger amounts in this case in the water–methanol rather than in the water–glycol solution. Due to hydrogen bonding tendencies in their solvent sheaths, the large amount of water on the ions increases the attractive and decreases the repulsive forces among the ions. Hence there is relatively less repulsive than attractive van der Waals' forces in the water–methanol case, and the nonelectrostatic effect is reduced relative to the electrostatic effect. The opposite is true in water–glycol with its lesser solvation of the ions by water.

In the case of reactions between bivalent and univalent ions as compared to reactions between univalent and univalent ions the electrostatic influences are greater and hence the ratio $\phi_1(r_0)/\phi_2(r_0)$ is less. Nevertheless, there is a solvent effect in the bivalent–univalent reactions as shown by the $\phi_1(r_0)/\phi_2(r_0)$ ratios of 2.9 and 3.6 in water–ethanol and water–methanol solvents, respectively, for the bivalent tetrabromophenolsulfonthalein ion–univalent hydroxide ion reaction.

By use of Eq. (5.64), $\phi_1(r_0)$ can be calculated from a_T alone if a_D is not known provided r_0 is known from some other source or provided the second term on the right-hand side of Eq. (5.64) is neglected because of its relatively small magnitude. This latter calculation should yield $\phi_1(r_0)$ values which are the same for all types of reactants, charged or uncharged, in solution or in the gaseous state, since this nonelectrostatic potential arises from repulsive forces at very close distances where electrostatic forces depending on charge and dielectric constant are perhaps no longer appreciable. Table 5.3, column 6, shows that the variation is less than a power of ten among the $\phi_1(r_0)$ values calculated from a_T alone even though the reactions vary from those between uncharged particles in the gaseous state to those between trivalent and bivalent ions in solution.

Table 5.3

Comparison of the Nonelectrostatic Potential between Reactants Calculated from Rate Dependence on Temperature Alone and from Rate Dependence on Both Temperature and Dielectric Constant

Reaction	Range		Value used in calculating column 7		$\phi_1(r_0)$ from rate dependence on $T \times 10^{12}$	$\phi_1(r_0)$ from rate dependence on T and $D \times 10^{12}$	Medium	Reference
	Temperature (°C)	Dielectric constant	Temperature (°C)	Dielectric constant				
$S_2O_8^{2-} + I^-$	20–40	55.0–76.7	40	73.12	1.39	1.04	Ethanol–water	(19)
$B\phi B^{2-} + OH^-$	5–45	64.5–78.5	25	71.42	1.52	0.93	Ethanol–water	(7, p. 901)
$B\phi B^{2-} + OH^-$	5–45	64.5–78.5	25	71.42	1.51	1.13	Methanol–water	(7, p. 901)
$NH_4^+ + CNO^-$	30–60	40.0–63.5	50	63.5	1.39	1.70	Glycol–water	(12)
$NH_4^+ + CNO^-$	30–70	35.0–69.9	50	63.5	1.29	1.32	Methanol–water	(11)
$B\phi BOH^- + H_3O^+$	25–45				2.12		Methanol–water	(10)
Sucrose $+ H_3O^+$	21–41				1.83		Ethanol–water	(18)
Sucrose $+ H_3O^+$	21–41				1.67		Dioxane–water	(18)
$AsO_3^{3-} + FeO_4^{2-\,a}$	89–120				0.74		Water	(24a)
$Fe^{3+} + Sn^{2+\,a}$	0–25				1.50		Water	(24b)
N_2O_5	25–65				1.73		Gaseous	(24c) (see 24d)
$2N_2O$	565–852				4.13		Gaseous	(24e)
C_2H_5OCOCl	150–195				1.99		Gaseous	(24f)
$2HI$	283–508				3.09		Gaseous	(24g) (see 24h)
$H_2 + I_2$	283–508				3.06		Gaseous	(24g) (see 24h)
$H + H_2(P)$	10–100				0.49		Gaseous	(24i) (see 24h)

^{a} k was not extrapolated to $k_{\kappa=0}$ in these cases.

Table 5.3 contains the results of the two methods described above for the calculation of $\phi_1(r_0)$. In column 6 of the table are values of $\phi_1(r_0)$ calculated using Eq. (5.65) alone neglecting the second term on the right-hand side while in column 7, the $\phi_1(r_0)$ values are obtained using both Eqs. 5.64 and 5.65. The variation in $\phi_1(r_0)$ calculated by the two methods is not great and is that predicted by electrostatics, i.e., $\phi_1(r_0)$ calculated from both a_T and a_D is less than that calculated from a_T alone, neglecting the electrostatic term, for ions of like sign and vice versa for ions of unlike sign.

One could speculate from the data in Tables 5.2 and 5.3 that as the valence of ionic reactants increased, the electrostatic effect would approach or even exceed the nonelectrostatic effect. In the latter instance, if the dielectric constant of the solvent changed markedly with temperature for a reaction involving high valence reactants of unlike sign the interesting phenomena of an increase of $\ln k'$ with $1/T$ would be observed since for these charge signs of reactants, the electrostatic effect represented by a $\ln k'$ versus $1/D$ plot has a positive slope. In the case of reactants of like sign an excessively large slope of the $\ln k'$ versus $1/T$ plot might be observed since this plot would have the same sign as the $\ln k'$ versus $1/D$ plot, and the nonelectrostatic and electrostatic effects would seem to enhance one another.

E. COMPONENTS OF ENERGY OF ACTIVATION

From the Brönsted–Christiansen–Scatchard equation it has been shown [17] that the free energy of activation F can be represented as a sum of three parts. Thus

$$F = F_0 + F_D + F_\mu \tag{5.68}$$

where F_0 is the free energy freed from all electric charge effects, F_D is the free energy arising from the electrostatic forces between reacting particles, and F_μ is the free energy resulting from ion atmosphere effects. From the free energy relation, the energy of activation may be written

$$E = E_0 + E_D + E_\mu \tag{5.69}$$

For $\mu = 0$, $E_\mu = 0$, and for isodielectric solvents

$$E_D = F_D = -RT \ln k_D' \tag{5.70}$$

since as will be shown below $S_D = 0$ for these solvents. Thus for isodielectric solvents and $\mu = 0$

$$E_0 = E + RT \ln k_D' \tag{5.71}$$

From experimental values of the total energy of activation and of k_D', E_0 may be calculated; or from experimental values of E, E_D, and E_μ, E_0 may be calculated. For the tetrabromophenolsulfonthalein ion–hydroxide ion reaction, E_D was found to be 6330 cal/mole, E_μ was 120 cal/mole, and E was 21,100 cal/mole. Therefore, E_0 was $21,100 - 6450$ cal/mole or 14,650 cal/mole.

F. Entropy of Activation

The contribution of the solvent to the influence of the ionic strength on the entropy of activation was formulated by LaMer [17]. Since

$$- S = \partial F / \partial T \tag{5.72}$$

and because in dilute solutions the contribution of ionic strength to free energy of activation F_μ is

$$F_\mu = NZ^2 \varepsilon^2 \varkappa / 2D \tag{5.73}$$

from the Debye–Hückel theory, and $\varkappa$ being proportional to $(DTV)^{-1/2}$, the contribution of the ionic strength to the entropy of activation S_μ is

$$S_\mu = \frac{F_\mu}{T} \left[\frac{3}{2} \frac{\partial \ln D}{\partial \ln T} + \frac{1}{2} \frac{\ln V}{\ln T} + \frac{1}{2} \right] \tag{5.74}$$

The Arrhenius equation is

$$k = Z \exp(-E/RT) \tag{5.75}$$

and the absolute rate equation is

$$k = \varkappa (kT/h) \exp(S/R) \exp(-E/RT) \tag{5.76}$$

Thus the Arrhenius frequency factor is related to the total entropy of activation by the expression

$$\ln Z = \ln \varkappa + \ln(kT/h) + (S/R) \tag{5.77}$$

In Eqs. (5.76) and (5.77), $\varkappa$ and kT/h are independent of concentration, and at constant temperature are constant, while S depends on both concentration and temperature and is the sum of three parts. Thus

$$S = S_0 + S_D + S_\mu \tag{5.78}$$

where S_0 appertains to reactions between uncharged reactants, S_D arises from interionic forces being charged reactants, and S_μ is due to the ionic atmosphere of charged reactants. S_0 will be constant for given reactants at a given temperature, S_D will be constant for a given media and therefore a change in S and hence in $\ln Z$, with changing ionic strength will arise from a change in S_μ for given reactants, in a given solvent, at a given temperature. Now from Eqs. 5.73 and 5.74, F_μ and S_μ are linear with $\mu^{1/2}$ at constant T, V, and D, and a plot of $\ln Z$ versus $\mu^{1/2}$ should be a straight line with slope obtainable from Eqs. (5.74) and (5.77). Substituting in Eq. (5.74) the values for water for $\partial \ln D / \partial \ln T$ and $\partial \ln V / \partial \ln T$ gives

$$S_\mu / 2.303 R = Z_A Z_B \varepsilon^2 A \mu^{1/2} \qquad (5.79)$$

where A depends upon D and T and equals 1.44, 1.53, and 1.64 for water at 15, 20, and 30°C, respectively.

An example of a reaction which obeyed the requirements of Eq. (5.79) is the tetrabromophenolsulfonpthalein ion–hydroxide ion reaction for which the $\log S_\mu$ versus $\mu^{1/2}$ curves in the temperature ranges 5–25°C and 25–45°C for aqueous solutions yielded in dilute solutions straight lines with the predicted slopes.

When $\alpha\varkappa$ cannot be neglected in the first approximation of Debye and Hückel but when the volume change with temperature is negligible, the equation for S_D is

$$S_D = \frac{3}{2} \frac{F_\mu}{T} \left[\left(\frac{1}{3} + \frac{\partial \ln D}{\partial \ln T} - \frac{1}{3} \frac{a\varkappa}{1 + a\varkappa} \left(1 - \frac{\partial \ln D}{\partial \ln T} \right) \right] \qquad (5.80)$$

Due to the contribution of the dielectric constant to the interionic forces between charged reactants, the dielectric constant makes a contribution S_D to the entropy of activation, which can be derived from the contribution F_D of the dielectric constant of the media to the free energy of activation. Now

$$F_D = N Z_A Z_B \varepsilon^2 / Dr \qquad (5.81)$$

where $r = r_A + r_B$. Applying Eq. (5.72), S_D is found to be [17]

$$S_D = \frac{F_D}{T} \left[\frac{\partial \ln D}{\partial \ln T} \right] \qquad (5.82)$$

For isodielectric media $\partial \ln D / \partial \ln T$ is zero and hence $S_D = 0$. If for such media the rate constants are extrapolated to $\mu = 0$, S_μ becomes zero and $S = S^0$, the entropy of activation for reactants with the same chemical properties except for a charge such as those of the ions. The frequency factor for these conditions should be, except perhaps for some minor dipolar effects, that for reactants with like chemical characteristics to those of the ions except for charge. This method should permit comparison of data on ionic reactions with those predicted by collision theory between uncharged reactants.

F. FREQUENCY FACTOR AND THE SOLVENT

An equation has been derived [25] from the difference in the logarithm of the frequency factor at fixed composition $\ln Z_C$ and in the logarithm of the frequency factor at fixed dielectric constant $\ln Z_D$. The derivation was made by multiplying the Brönsted–Christiansen–Scatchard equation through by T, differentiating the resultant equation with respect to T (T and D variable), then differenting the equation with respect to T (D constant), and then subtracting the latter differential equation from the former. This gave

$$\ln Z_C - \ln Z_D = \frac{Z_A Z_B N \varepsilon^2}{D^2 R} \left[\frac{1}{r} - \frac{3\varepsilon}{10} \left(\frac{2\pi N^2 \mu}{10 D R T} \right)^{\frac{1}{2}} \right] \frac{dD}{dT} \qquad (5.83)$$

Ordinarily the first term in the brackets of Eq. (5.83) is larger than the second term, and since for ordinary solvents dD/dT is negative, the frequency factor in isocomposition media is less than in isodielectric media. Qualitative and quantitative agreement with the predictions of this equation were found [25].

G. Deviations between Theory and Experiment

Selective solvation and a microscopic dielectric constant around the reactive particles different to the macroscopic dielectric constant of the bulk media will cause a deviation between observation and theory in the theoretical treatments presented previously. In general the higher dielectric constant more polar component of the solvent is selectively bound by the ions, and it would be expected that at lower dielectric constants of mixed solvents, the deviations from theory would be in a direction to favor the observed results in the more polar solvent component when this component alone is the solvent. These expectations are generally realized in practice. Thus in water–alcohol and water–dioxane solvents, the reaction rates in organic-rich-component solvents ordinarily deviate in the direction of results observed in pure water at the same temperature.

It has been suggested [26] that in dealing with change of rate caused by changing dielectric constant it must be considered that theories of solution give the ratio of activities to mole fraction and not the ratio of activities to relative volume concentrations as activity coefficients. Rate constants in terms of concentration can be converted to rate constants in terms of mole fractions by multiplying the former by $\sum (N/V)^{\nu-1}$, where N is the total number of moles, including solvents, in volume V, and ν is the order of the reaction. Such calculations eliminate the sudden change in slope of $\log k'$ versus $1/D$ for the bromoacetate–thiosulfate rate data given by Laidler and Eyring [6]. A more exact extrapolation to zero concentration of the data for some reactions, eliminate or greatly reduce less marked changes in curvature at lower dielectric constants. These corrections for other reactions do not eliminate the change in slope at other lower dielectric constants, and for these Scatchard believes that Laidler and Eyring's assumption of selective solvation is probably correct.

Scatchard's model for the critical complex was two spheres, each with the charge of one of the original ions distributed about its center. Laidler and Eyring assume one sphere with the net charge distributed symmetrically about its center. These are the two extremes with respect to both shape and charge distribution. Scatchard discussed intermediate possibilities [27].

If one component of the solvent is protic and the other is dipolar aprotic, this effect [3], especially with respect to $S_N 2$ reactions, may tremendously exceed the electrostatic effect exerted through the change of dielectric constant with change of composition of the solvent.

III. Salt Effects

A. DIFFERENT TREATMENTS

The primary salt effect is influenced by the solvent in a secondary manner caused by the modification of the activity coefficients of the reactant particles of the dielectric constant of the solvent. The Debye–Hückel [28] theory has been applied to the influence of neutral salts upon the rates of reactions in solution by Brönsted [29], Bjerrum [30], and Christiansen [24]. Brönsted's treatment of the neutral salt effect is the one generally presented and will first be treated here. There are two kinds of salt effects: (1) the primary salt effect due to alteration by the salt of the activities of the reactant particles whether ions or molecules; (2) the secondary salt effect in which the effective concentration of a reactant ion coming from a weak electrolyte is decreased by the reduced ionization of the electrolyte due to added salt. The latter effect is illustrated by the decreased catalytic effect on the inversion of cane sugar by acetic acid in the presence of alkali acetates. In this instance the activity of the hydronium ion reactant is increased by the added salt, but this increased activity is more than offset by the decreased concentration of the hydronium ion due to the repression of the ionization of the acetic acid by the common acetate ion, so that, depending on the relative concentrations of acid and salt, the inversion rate constant may be decreased by as much as 50%.

Brönsted's theory for the primary salt effect assumes that an intermediate complex is formed which may decompose reversibly to reactants, or which may decompose irreversibly to products. Thus

$$A + B \leftrightarrows X \rightarrow C + D \qquad (5.84)$$

The first step is treated as a thermodynamic equilibrium, and the equilibrium constant is

$$K = \frac{a_X}{a_A a_B} = \frac{C_X f_X}{C_A f_A C_B f_B} \qquad (5.85)$$

The reaction rate, $r = dC/dt$, is proportional to the concentration of X, therefore,

$$r = k''C_X = k''KC_A\,C_B(f_A\,f_B/f_X) \tag{5.86}$$

The product, $k''K$, may be combined into one constant k_0', hence

$$r = k_0'C_A\,C_B(f_A\,f_B/f_X) \tag{5.87}$$

Now k_0' is independent of concentration and the factor $f_A\,f_B/f_X = F$ is the correction, due to added salt, to the classical rate equation. Since in concentrated solutions nonthermodynamic factors, with which his theory is not concerned, may influence the rate, Brönsted limits his treatment to dilute solutions.

The bimolecular rate constant k' can be obtained by dividing Eq. (5.87) by $C_A\,C_B$ yielding

$$k' = \frac{dC}{dt}\frac{1}{C_A\,C_B} = k_0'\frac{f_A\,f_B}{f_X} = k_0'F \tag{5.88}$$

It should be pointed out that k_0' though independent of concentration is nevertheless composite and must yet be referred to a standard reference state of dielectric constant and of temperature to free it from those factors which are so influential in reaction rates.

For ionic reactants, the equilibrium equation for the reaction is

$$A^{Z_A} + B^{Z_B} \rightleftarrows X^{(Z_A+Z_B)} \tag{5.89}$$

where the charges on A, B, and X are Z_A, Z_B, and Z_A+Z_B, respectively, and from the Debye–Hückel theory

$$-\ln f_i = \frac{Z_i^2 A\mu^{1/2}}{1 + Ba_i\,\mu^{1/2}} \tag{5.90}$$

and therefore

$$\ln\frac{f_A\,f_B}{f_X} = \frac{-Z_A^2 - Z_B^2 + (Z_A+Z_B)^2 A\mu^{1/2}}{1 + Ba_i\,\mu^{1/2}} = \frac{2Z_A\,Z_B\,A\mu^{1/2}}{1 + Ba_i\,\mu^{1/2}} \tag{5.91}$$

or

$$\frac{f_A\,f_B}{f_X} = \exp\left[\frac{2Z_A\,Z_B\,A\mu^{1/2}}{1 + Ba_i\,\mu^{1/2}}\right] \tag{5.92}$$

Using Eqs. (5.88) and (5.92)

$$k' = k_0'\exp\left[\frac{2Z_A\,Z_B\,A\mu^{1/2}}{1 + Ba_i\,\mu^{1/2}}\right] \tag{5.93}$$

which can be written

$$\ln k' = \ln k_0' + \frac{2Z_A Z_B A\mu^{1/2}}{1 + Ba_i\,\mu^{1/2}} \tag{5.94}$$

using A' as $(2A/2.303)$ we can write

$$\log k' = \log k_0' + \frac{Z_A Z_B A'\mu^{1/2}}{1 + Ba_i\,\mu^{1/2}} \tag{5.95}$$

For dilute solutions limiting forms may be written as

$$\ln k' = \ln k_0' + 2Z_A Z_B A\mu^{1/2} \tag{5.96}$$

and

$$\log k' = \log k_0' + Z_A Z_B A'\mu^{1/2} \tag{5.97}$$

In the previous equations

$$A = \frac{\varepsilon^2}{DkT}\left(\frac{2\pi N}{1000 DkT}\right)^{1/2} \tag{5.98}$$

and

$$B = \left(\frac{8\pi N\varepsilon^2}{1000 DkT}\right)^{1/2} \tag{5.99}$$

where ε is the electronic charge, N Avogadro's number, D the dielectric constant of the media, k the Boltzmann gas constant, and T the absolute temperature.

In dilute solutions plots of $\log k'$ versus $\mu^{1/2}$ should, for ionic reactants, give a straight line with slope $Z_A Z_B A'$. For water at 25°C, A' is about unity and the slopes of the plots mentioned should be the product $Z_A Z_B$. LaMer's [31] classical figure showing the agreement of data with the predictions of Eq. (5.97) is shown in Fig. 5.3. Since ion–dipolar molecule reactions will be discussed under the theory dealing with those reactions, the data for sucrose inversion are omitted from further discussion at this time. Some of these data fail to reach the theoretical slopes, and in these cases the Debye–Hückel theory does not give the correct activity coefficients due to specific solvent effects or other causes.

Weller [32] using Eq. (5.95) in the form

$$\log k' = \log k_0' + Z_A Z_B \frac{1.02\mu^{1/2}}{1 + 2\mu^{1/2}} \tag{5.100}$$

for water at 25°C, plotted $\log k'$ versus $1.02\mu^{1/2}/(1 + 2\mu^{1/2})$ for the reactions indicated, the rates of which were studied by the fluorescence method. The

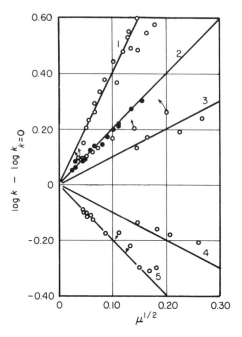

Fig. 5.3. The influence of ionic strength on the velocity of the ionic reactions. Curves: (1) $2[Co(NH_3)_5Br]^{2+} + Hg^{2+} + 2H_2O \rightarrow 2[Co(NH_3)_5H_2O]^{3+} + HgBr_2$ (bimolecular); no foreign salt added. (2) Circles; $CH_2BrCOO^- + S_2O_3^{2-} \rightarrow CH_2S_2O_3COO^- + Br^-$ as the sodium salt; no foreign salt added. Dots, $S_2O_8^{2-} + 2I^- \rightarrow I_2 + 2SO_4^{2-}$ as $Na_2S_2O_8$ and KI. (3) Saponification of nitrourethane ion by hydroxyl ion; $[NO_2{=}N{-}COOC_2H_5]^- + OH \rightarrow N_2O + CO_3^{2-} + C_2H_5OH$. (4) $H_2O_2 + 2H^+ + 2Br^- \rightarrow 2H_2O + Br_2$. (5) $[Co(NH_3)_5Br]^{2+} + OH^- \rightarrow [Co(NH_3)_5OH]^{2+} + Br^-$.

data fit straight lines of the required slopes within experimental accuracy proving the applicability of the first approximation form of the Brönsted theory to this kind of reaction as is shown in Fig. 5.4.

Details of the derivation of the Debye–Hückel theory are extant in the literature [33].

Bjerrum arrived at an expression for the salt effect by assuming that a thermally induced spontaneous monomolecular decomposition of a collision or Stosscomplex determines the rate of reaction. The Stosscomplex is of a physical nature rather than a chemical compound. The reaction is

$$A + B \rightleftarrows X_S \rightarrow C + D \tag{5.101}$$

At any instant the rate of the reaction is proportional to the concentration of the collision complex C_{X_S} at that instant. Therefore,

$$-dC_A/dt = -dC_B/dt = k'_{X_S} C_{X_S} \tag{5.102}$$

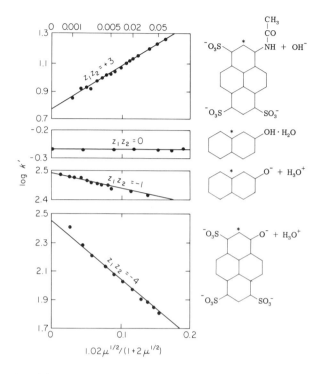

Fig. 5.4. Plot of the first approximation form of the Brönsted theory for various valence types of reactants.

From the mass law the collision equilibrium in terms of activities is

$$a_{X_s}/a_A a_B = C_{X_s} f_{X_s}/C_A f_A C_B f_B = K_a \qquad (5.103)$$

Solving the above equation for C_X and substituting of this in Eq. (5.102) results in an equation identical with the Brönsted formula assuming no distinction between the critical complex of Brönsted and the Stosscomplex X_S of Bjerrum.

Brönsted makes the rate of formation of complex depend on activities, and assumes this rate of formation determines both the reaction velocity and the temperature coefficient of the reaction. Bjerrum assumes that reaction velocities are proportional to concentrations, that the ionic strength influences only the primary equilibrium, and that the reaction of the X_S-complex to yield products is strongly dependent upon temperature.

Bjerrum's concentration hypothesis has been criticized [31] from the standpoint of monomolecular reactions such as

$$A \rightleftharpoons Y \qquad (5.104)$$

For the hypothesis to be valid, the mass law in terms of concentration must hold and

$$C_A/C_Y = \text{const} \tag{5.105}$$

regardless of the change in concentration of the medium [31]. Upon addition of salts, however, displacement does occur except in the insignificant case where the two activity coefficients are affected equally. Bjerrum's theory, according to LaMer is hardly analogous to radioactivity in which transformations are due to instability of the nucleus for which there is a finite probability of decomposition quite independent of external conditions.

Christiansen's theory does not involve the concentration of a critical or of a collision complex but the number of collisions of reacting molecules. Christiansen sets the reaction velocity V equal to

$$V = C_A\, C_B\, Z_{AB}\, T^{\frac{1}{2}} \exp(-E_{AB}/RT)\,\Omega\, \frac{\rho}{\rho + \Sigma NP} \tag{5.106}$$

where $C_A\, C_B\, Z_{AB}\, T^{\frac{1}{2}}$ is the number of collisions per second per liter, E_{AB} is the energy of activation, Ω is the steric factor, ρ is the probability that the complex will decompose to products, and ΣNC is the probability that it will be deactivated again to reactants by molecules of concentration C each with an individual probability P. It should be noted that C is not the bulk concentration but the concentration of the various molecules in the vicinity of the reacting molecules.

At low concentrations the factor $\rho/(\rho + \Sigma NP)$ approaches unity, and at high concentrations it approaches the value $\rho/\Sigma NC$. Therefore, for sufficiently low concentrations Eq. (5.106) becomes much simplified.

C_A and C_B in the vicinity of colliding reactant ions are not the bulk concentrations C_A' and C_B' of the reactants, but in the case of ionic reactants are calculated from these bulk concentrations using the Debye–Hückel theory,

$$C_A = C_A' \exp\left(\frac{\varepsilon Z_A\, \phi_B}{kT}\right) \tag{5.107}$$

where Z_A is the valence of ion A, and ϕ_B is the electrostatic potential at the distance a from ion B and is given by the equation

$$\phi_B = \frac{\varepsilon Z_B}{D}\frac{\exp(-\varkappa a)}{a} \tag{5.108}$$

In the above equation, $\varkappa$ is the Debye kappa and D is the dielectric constant of the medium.

In Christiansen's theory, Eq. (5.107) requires a concentration of ions around ions of opposite sign, but the average kinetic energy of collision between ions of opposite sign is no greater because of the electrostatic attraction between them.

Christiansen sets $\rho/(\rho + \Sigma NP)$ equal to zero and eventually obtains

$$\ln k' = \ln k_0'' - \frac{Z_A Z_B \varepsilon^2}{DkTa} + 2.31 Z_A Z_B \mu^{\frac{1}{2}} \qquad (5.109)$$

where k_0'' is the sum of several factors all of which are constant for a given set of conditions. In the limit of infinite dilution

$$\ln k_{x=0}' = \ln k_0'' - \frac{Z_A Z_B \varepsilon^2}{DkTa} \qquad (5.110)$$

For infinite dielectric constant D or any other condition for which the last term on the right-hand side of Eq. (5.110) to become zero, k_0'' would be the observed rate constant at infinite dilution. LaMer cautioned that even though k_0'' was the value of k' extrapolated to infinite dilution and to a state freed from all net electric charges, yet k_0'' will be affected by influences arising from electric moments.

The velocity of reaction is predominantly influenced by the distribution of concentrations arising from interionic attractions according to Christiansen. The last term in Eq. (5.109) is concentration dependent and corresponds to Brönsted activity coefficient term, i.e., to $f_A f_B / f_X$.

According to Scatchard it is unimportant whether the rate is calculated from the concentration of reacting species or from the number of collisions with the necessary orientation and energy multiplied by a term which takes into account the duration of a collision. Reaction will ensue upon collision if the reactant particles become sufficiently deformed and correctly oriented and obtain the necessary energy before colliding. On the other hand, if collision occurs before any of the other steps has taken place, preliminary complex formation will result.

B. Ion Association and the Salt Effect

Anomalous behavior of many reactions with respect to the primary salt effect are numerous. Included in such reactions are the exchange reactions. Association between oppositely charged ions to form ion pairs or complexes has been used as one explanation of these anomalies. The formation of ion pairs could exert two influences on the reaction rate. First, the actual ionic strength of the medium must be calculated allowing for their presence, and second, the rate determining step might involve one or more such ion-pair or pairs. These effects could alter the charge-type of the activated complex and thus the ionic type of reaction and the $f_A f_B / f_X$ correction to the rate equation. They could also alter the reactivity of the system in a more drastic manner, for example, by changing the potential of the complex to decompose to products by modifying its chemical composition and stability.

Olson and Simonson [34] have discussed in some detail the plot of $\log(k'/k_0')$ versus $\mu^{1/2}$ for various charge type reactants given in Fig. 5.3, and have added further data of their own on the two bromopentamine cobaltic ions. They conclude that for reactions between ions of the same sign the effect of added inert salt is due almost entirely to the concentration and character of salt ions of charge sign opposite to that of reactants and that the rate is not dependent on the ionic strength of the solution. They felt that the salt effects could be interpreted in terms of an ion association constant and specific rate constants for associated and nonassociated reactants, and that the further introduction of activity coefficients is not necessary.

However, even when quantitative allowance is made for the role played by ion pairs, the concept of activity coefficients must without question be retained [35, 36].

Davies [35] treats the specific case of the reaction between bromoacetate-thiosulfate ions as follows. In dilute solution the reaction obeys the Brönsted theory when the sodium or potassium salts of the ions are used. In concentrated solutions of the potassium salts or when bivalent or especially trivalent lanthanum cations are present the deviations from theory become marked. These deviations have been explained on the basis of incomplete dissociation. Necessary dissociation constants for these corrections were listed by Davies.

All possible associations must be allowed for in each reacting system. As equation of the type

$$A^{Z_A} + B^{Z_B} \rightleftarrows AB^{(Z_A + Z_B)} \tag{5.111}$$

will give the concentration of each ion pair. Thus

$$[AB] = \frac{[A][B]f_A f_B}{Kf_C} \tag{5.112}$$

where K is the dissociation constant of the ion pair AB. The activity coefficients can be calculated with fair precision from the equation

$$\log f_i = -0.5Z_i^2 \left(\frac{\mu^{1/2}}{1 + \mu^{1/2}} + 0.2\mu^{1/2} \right) \tag{5.113}$$

for the solvent water at 15–25°C when the solutions are dilute. Using the above equation and the ion-pair dissociation constants, the concentration of each ion-pair can be calculated by successive approximations, and the actual ionic strength of the initial reacting mixture obtained. The equilibrium constants for the products of this reaction have not been reported, but the ionic strength during the reaction is constant from experimental evidence in accordance with the theory of Wyatt and Davies [37] who showed such would be the case for the same dissociation constants for MS_2O_3 and $M(S_2O_3)Ac$ and likewise for the same dissociation constants for $MBrAc^+$ and MBr^+, where M is any metal. For simple cations this would be expected to be the case.

While the effect of the formation of ion pairs relative to change in ionic strength may be minor, a more marked influence of the cation will be exerted due to its inclusion in the activated complex. Thus a bivalent cation's participation in the formation of complexes between thiosulfate and bromoacetate ions will enhance the possibility of complexation from the standpoint of electrostatics by reducing the charge on the complex from -3 to -1, and thus possibly account for the catalytic effect of multiply charged cations. Davies, therefore, wrote a two-term velocity equation, one term of which represented the contribution of the simple reaction, and the other term the contribution of the reacting complexes which contain a cation. Using f_1, f_2, f_3 for the activity coefficients of the uni-, bi-, and trivalent ions, respectively, the Brönsted equation was written

$$-\frac{d[S_2O_3^{2-}]}{dt} = k_0[S_2O_3^{2-}][BrAc^-]\frac{f_2 f_1}{f_3} + k'[M^{2+}][S_2O_3^{2-}][BrAc^-]f_2^2$$

(5.114)

From the calculated values of the true ionic concentrations and the experimental value of k_0, the first term on the right-hand side of Eq. (5.114) can be calculated. The overall rate of reaction $-d[S_2O_3^{2-}]/dt$ can be measured, and then the last term on the right-hand side of Eq. (5.114) found from the difference.

For the complex formation

$$M^{2+} + S_2O_3^{2-} + BrAc^- \rightleftarrows MS_2O_3BrAc^-$$

(5.115)

the activity coefficient term is $f_2 f_2 f_1/f_1 = f_2^2$ as is shown in the second term on the right-hand side of Eq. (5.114). If $\log f_i$ is taken as that given in Eq. (5.113), then $\log f_2^2$ would be given by the equation

$$\log f_2^2 = 2\log f_2 = -2 \times 0.5(2)^2 \left(\frac{\mu^{1/2}}{1+\mu^{1/2}} + 0.2\mu^{1/2}\right)$$

$$= -4\left(\frac{\mu^{1/2}}{1+\mu^{1/2}} + 0.2\mu^{1/2}\right)$$

(5.116)

and $\log f_2^2$ from the second term on the right-hand side of Eq. (5.114) plotted against the function of ionic strength shown in parenthesis in Eq. (5.113) or (5.116) should yield a straight line with a negative slop of 4. A plot of $\log k f_2^2$ versus this function of the ionic strength [35] for the data of LaMer and Fessenden [38] for the magnesium, calcium, and barium salts and for von Kiss and Vass for the reaction of sodium salts in the presence of magnesium nitrate and magnesium sulfate [39] gave satisfactory agreement with theory in spite of the fact that the errors of experiment and calculation are included in the values, and in spite of the fact that only a small part of the reaction in dilute solution results from the metal catalyzed process. Data for the reaction between the potassium salts evidenced less ion pairing, but did show deviation from the simple reaction rate at higher concentrations. These data yielded a satisfactory value of k.

Davies also attacked reactions of this kind by considering the probable mechanism of formation of the cation containing complex to be bimolecular collisions between ion and ion-pair. Thus

$$BrAc^- + MgS_2O_3$$

and $\qquad\qquad\qquad\qquad\qquad\qquad$ Complex $\qquad\qquad$ (5.117)

$$S_2O_3^{2-} + MgBrAc^+$$

Using these species, we find the rate equation, neglecting the termolecular process, to be

$$-\frac{d[S_2O_3^{2-}]}{dt} = k_0[S_2O_3^{2-}][BrAc^-]\frac{f_2 f_1}{f_3} + k_1[MS_2O_3][BrAc^-]$$

$$+ k_2[S_2O_3^{2-}][MBrAc^+]f_2 \qquad\qquad (5.118)$$

where M is any divalent metal. The mass action constants for the dissociation of the complexes MS_2O_3 and $MBrAc^+$ are, respectively,

$$K_1 = \frac{[M^{2+}][S_2O_3^{2-}]f_2^2}{[MS_2O_3]} \quad \text{and} \quad K_2 = \frac{[M^{2+}][BrAc^-]f_2}{[MBrAc^+]}$$

$$(5.119)$$

Solving Eq. (5.119) for $[MS_2O_3]$ and $[MBrAc^-]$ and inserting in Eq. (5.118), gives us

$$-\frac{d[S_2O_3^{2-}]}{dt} = k_0[S_2O_3^{2-}][BrAc^-]\frac{f_2 f_1}{f_3} + \frac{k_1[M^{2+}][S_2O_3^{2-}][BrAc^-]f_2^2}{K_1}$$

$$+ \frac{k_2[M^{2+}][S_2O_3^{2-}][BrAc^-]f_2^2}{K_2}$$

$$= k_2[S_2O_3^{2-}][BrAc^-]\frac{f_2 f_1}{f_3}$$

$$+ [M^{2+}][S_2O_3^{2-}][BrAc^-]f_2^2\left(\frac{k_1}{K_1} + \frac{k_2}{K_2}\right) \qquad (5.120)$$

and comparing with Eq. (5.114), we find

$$k' = \frac{k_1}{K_1} + \frac{k_2}{K_2} \qquad\qquad (5.121)$$

The constants k_1 and k_2 cannot be determined separately from kinetic data, though k_2 would be expected to be greater than k_1 since the encounters with which k_2 is concerned are between ions of opposite sign. K_2 is much greater than K_1. It has been estimated [37] that for the barium reaction k_2/K_2 makes up 3% of the whole, and for the calcium reaction k_2/K_2 makes up 7%. From these data and the known values of the equilibrium constants K_1 and K_2, the mean value of k_1 was calculated to be 1.27 liters/mole/sec for the BaS_2O_3 $BrAc^-$ reaction and 1.34 liters/mole/sec for the $CaS_2O_3 + BrAc^-$ reaction. These values of k_1 are approximately the same. In this collision treatment the greater catalytic effect of barium as compared to calcium ion was attributed to the greater concentration of BaS_2O_3 as compared to other MS_2O_3 complexes, rather than to the exertion of specific effects of the various cations on the reactivity of the activated complex.

Assuming k_1 and k_2 in Eq. (5.120) are not functions of the chemical nature of the cation M^{2+}, the magnitudes of k_1 and k_2 may be calculated by proper manipulation of Eq. (5.121) in which it is applied separately to data for different bivalent cations to obtain ratios of the k' constants, as shown below for calcium and barium.

$$\frac{k'_{Mg}}{k'_{Ca}} = \left(\frac{K_2 k_1 + K_1 k_2}{K_1 K_2}\right)_{Mg} \bigg/ \left(\frac{K_2 k_1 + K_1 k_2}{K_1 K_2}\right)_{Ca}$$

$$= \frac{K_{2Mg} k_1 + K_{1Mg} k_2}{K_{2Ca} k_1 + K_{1Ca} k_2} \cdot \frac{K_{1Ca} K_{2Ca}}{K_{1Mg} K_{2Mg}} \tag{5.122}$$

since the K values are and the k values are not functions of the cations. Such ratio equations employing K and k values recorded yielded [35] $k_2 = 6.0k_1$ for the magnesium–calcium comparison, and $k_2 = 6.4$ from the magnesium–barium comparison. These results are in reasonable agreement, and if the mean value $k_2 = 6.2k_1$ is inserted into Eq. (5.121), yields consistent values of $k_1 = 1.23$ and $k_2 = 6.63$. k_0 was found to be 0.247. Davies has also found in harmony with Eq. (5.97) the gradients of $d \log k/d\mu^{1/2}$ to have the values of $+2, 0, -2$. After taking into account the role of ion pairs other reactions have been found to conform with the Brönsted–Bjerrum theory.

C. RELATIVE SOLVATION, POLARIZATION, AND THE SALT EFFECT

Experimental energies of activation of a reaction can be used to detect nonelectrostatic effects on the reaction produced by adding various salts over ranges of concentrations to the reaction mixture in order to vary the ionic strength [40].

Substituting the values of D and $\partial \ln D / \partial \ln T$ and $\partial \ln V / \partial \ln T$ for a given solvent into Eq. (5.57), the equation can be written in the form [33]

$$\Delta E_{In} = 2.303 Z_A Z_B A_1 \mu^{\frac{1}{2}} \qquad (5.123)$$

where A_1 is a function of temperature and dielectric constant and is 0.45, 0.52, and 0.60 at 15, 25, and 35°C, respectively, for water. From the above equation for reactants of like charge sign, ΔE_{In} and therefore the overall activation energy should increase with increasing ionic strength. Figure 5.5

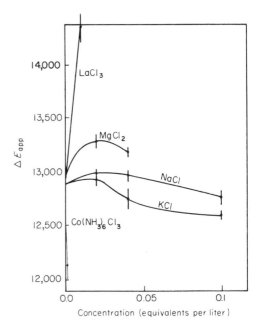

Fig. 5.5. Experimental activation energies versus the equivalent concentration for the reaction between iodide and persulfate ions in the presence of different salts. (Vertical lines indicate error.)

is a plot of the apparent energy of activation ΔE_{app} and their errors indicated by the vertical lines versus concentrations of added salt for the iodide ion–persulfate ion reaction. It is apparent that ΔE_{app} increases with the first small increment of concentration of any of the added salts, but at higher concentrations of salt, ΔE_{app} shows an obvious decrease in the cases of magnesium, sodium, and potassium chlorides; the latter salt exhibits this phenomenon to a more marked degree. Other data in the literature [41, 42] also exhibit a decrease in ΔE_{app} with added potassium chloride. This decrease in energy of activation has been explained [1, 40] as due to the polarization which the cation exerts on the electrons of the anion when ion pairs are formed or when the cation comes closely enough to the activated complex. If an ion of opposite sign approaches closely enough to a reacting ion or a complex both polarization and electrostatic effects are present. It has been shown [43, 44] that for thermodynamic properties of bi-bivalent solutions, the magnitude of this critical distance is rather arbitrary.

There must be the transfer of an electron from an iodide ion to the persulfate ion in the case of the reaction between the ions. The free persulfate ion should be a poorer electron acceptor than an ion pair such as

$$
+\delta \quad O=\overset{\overset{O^{+\delta}}{\|}}{\underset{\underset{O_{-2\delta}}{|}}{S}}-O-O-\overset{\overset{O^{+\delta}}{\|}}{\underset{\underset{2\delta-O}{|}}{S}}=O \quad +\delta \qquad \qquad M^{+}
$$

ST. 1-5, sf 1 (fe)

formed between the metal and persulfate ions. The shape of the persulfate ion $S_2O_8^{2-}$ is such that for equal charges on the cations, those of small radius can induce a greater negative charge toward the center of persulfate ion than cations of larger radius. In solution it is difficult to unambiguously determine ionic radii [45]; and it is here that the solvent must be considered; even though in many solvation determinations only the relative and not the absolute solvation numbers of ions are found. From transference numbers, using the Hittorf procedure involving an inert reference substance, it has been found [46] that sodium ion has a greater solvation number than potassium ion in aqueous solutions at like concentration and temperature. Therefore, the sodium ion would be more bulky than the potassium ion when both are in solution, even though the unsolvated sodium ion has a radius of 0.9 Å and the unsolvated potassium ion a radius of 1.3 Å. Hence in the above depicted model the hydrated potassium ion would cause more polarization of the persulfate

Table 5.4

*Comparison of the Salt Effect on the Rate of
Reaction between Persulfate and Iodide Ions and the
Limiting Equivalent Conductance at Infinite
Dilution of the Cations*

Salt	$10^3 k'$	λ°
KCl	2.70	73.5
NaCl	2.12	50.1
LiCl	1.91	38.6
$N(CH_3)_4Br$	1.89	44.9
$N(C_2H_5)_4Br$	1.48	32.6
$MgCl_2$	2.89	53.0
$Co(NH_3)_6Cl_3$	38.7	101.9
$LaCl_3$	$\sim 6.5^a$	69.7

[a]Extrapolated.

ion than would the hydrated sodium ion. On the assumption that ion conductivity might be used as a reasonable indication of the ionic radius in a given
solvent, the salt effects on the reaction rate between persulfate and iodide ions
were compared with the ionic conductances of cations at infinite dilution of
several salts. The results for the kinetic runs are presented in Table 5.4 at
25°C and for different salts at 0.04 equivalents per liter. There is a marked
parallelism between decrease in rate constant and decrease in limiting equivalent conductance for cations of like valence at infinite dilution. There is only
one exception which is the inversion between Li^+ and $N(CH_3)_4^+$.

IV. Electrostatic Effects on Ion–Molecule Reactions

A. INTRODUCTION

Since there are ion-dipolar and ion-induced dipolar forces exerted between
ions and molecules, an electrostatic term allowing for the effect of these forces
on the rates of chemical reactions between ions and molecules should appear
in rate equations involving these species of reactants. Ion–molecule electrostatic interactions are smaller than ion–ion ones, and the effects arising from
these interactions in the ion–molecule case will be more readily obscured by
other solvent and structural effects. Nevertheless these ion–molecule electrostatic–interaction effects on ion–molecule reaction rates are appreciable and
should be calculable. Some of the theoretical expressions for the influence of
the solvent on electrostatic forces which affect rate processes between ions
and dipolar molecules are presented below.

B. LAIDLER–EYRING EQUATION

Using the type reaction

$$A^{Z_A} + B^0 \rightarrow M^{*Z_A} \rightarrow X + Y \tag{5.124}$$

Laidler and Eyring [47] wrote for the activity coefficient of A and M*,
including the semiempirical $b\mu$ term of Hückel [48], the respective equations,

$$\ln f_A = \frac{Z_A^2 \varepsilon^2}{2r_A kT}\left(\frac{1}{D}-1\right) - \frac{Z_A^2 \varepsilon^2}{2DkT}\frac{\varkappa}{1+a_A \varkappa} + b_A \mu + \frac{\phi_A}{kT} \tag{5.125}$$

$$\log f_{M^*} = \frac{Z_A^2 \varepsilon^2}{2r_{M^*} kT}\left(\frac{1}{D}-1\right) - \frac{Z_A^2 \varepsilon^2}{2DkT}\frac{\varkappa}{1+a_{M^*}\varkappa} + b_{M^*} \mu + \frac{\phi_{M^*}}{kT} \tag{5.126}$$

where Z_A is the charge on A and M, μ is the ionic strength and the bs are constants. The activity coefficient f_B of the dipolar molecule in a medium of dielectric D, containing n_i ions of type i per cubic centimeter, charge Z_i, and radius r_i, can be calculated using Laidler and Eyring's modification of Kirkwood's [49] expression for the free energy of transfer F_1 of a strong dipole of moment m_B from a vacuum to a medium having dielectric constant D_0 after the molecules, but not the ions, have been added, and by adding to this the approximate expression of Debye and McAulay [50] for the increase of free energy F_2 engendered by addition of the ions. In the case of the activity coefficient of an ion, the Hückel $b\mu$ term was introduced to take into consideration the effect of dielectric saturation.

The free energy of transfer is given by

$$F_1 = kT \ln \beta = -\frac{1}{kT} \frac{m^2_B}{a^3_B}\left[\frac{D_0-1}{2D_0+1}\right] + \frac{\phi_B}{kT} \tag{5.127}$$

and the free energy F_2 caused by addition of the ions is represented by

$$F_2 = kT \log f = \frac{\bar{\alpha}\varepsilon^2}{2D} \sum_i \frac{Z_i^2 n_i}{r_i} \tag{5.128}$$

where $\bar{\alpha}$ is defined by

$$D' = D_0'(1 - \bar{\alpha}n_B) \tag{5.129}$$

in which D_0' is the dielectric constant of the pure solvent, and D' that of a solution containing n_B molecules of the dipolar substance per cubic centimeter. The dipolar molecule has an activity coefficient given by the equation

$$\ln f_B = -\frac{1}{kT}\frac{m^2_B}{a^3_B}\left[\frac{D_0-1}{2D_0+2}\right] + \frac{\bar{\alpha}\varepsilon^2}{2DkT}\sum \frac{Z_i^2 n_i}{r_i} + \frac{\phi_B}{kT} \tag{5.130}$$

and the rate expression becomes

$$\ln k' = \ln\left(\varkappa \frac{kT}{h} K_0^*\right) + \frac{Z_A^2\varepsilon^2}{2kT}\left(\frac{1}{D}-1\right)\left(\frac{1}{r_a}-\frac{1}{r_{M^*}}\right)$$

$$-\frac{Z_A^2\varepsilon^2}{2DkT}\left(\frac{\varkappa}{1+a_A\varkappa}-\frac{1}{1+a_{M^*}\varkappa}\right)$$

$$-\frac{1}{kT}\left[\frac{D_0-1}{2D_0+1}\right]\frac{m_B^2}{a_B^3} + \left(b_B - b_{M^*} + \frac{\bar{\alpha}\varepsilon^2}{DrkT}\right)\mu$$

$$+\frac{\phi_A+\phi_B-\phi_{M^*}}{kT} \tag{5.131}$$

where r is the average of the r_i values. Assuming $a_A = a_{M^*}$ since their effects at $\mu \to 0$ is vanishingly small, the rate equation becomes

$$\ln k' = \ln\left(\varkappa\frac{kT}{h}K_0^*\right) + \frac{Z_A^2\varepsilon^2}{2kT}\left(\frac{1}{D}-1\right)\left(\frac{1}{r_A}-\frac{1}{r_{M*}}\right)$$
$$-\frac{1}{kT}\frac{m_A^2}{a_B^3}\left[\frac{D_0-1}{2D_0+1}\right] + \left(b_A - b_{M*} + \frac{\bar{\alpha}\varepsilon^2}{DkT}\right)\mu + \frac{\phi_A+\phi_B-\phi_{M*}}{kT}$$

$$(5.132)$$

and for ionic strength $\mu = 0$, the equation can be written

$$\ln k' = \left(\varkappa\frac{kT}{h}K_0^*\right) + \frac{Z_A^2\varepsilon^2}{2kT}\left(\frac{1}{D}-1\right)\left(\frac{1}{r_A}-\frac{1}{r_{M*}}\right)$$
$$-\frac{1}{kT}\frac{m_B^2}{a_B^3}\left[\frac{D-1}{2D+1}\right] + \frac{\phi_A+\phi_B-\phi_{M*}}{kT}$$

$$(5.133)$$

This equation predicts that $\ln k'$ extrapolated to $\mu = 0$ plotted against $1/D$ for a molecule of zero moment reacting with an ion of charge Z_A should yield a straight line, the slope of which should be

$$\frac{Z_A^2\varepsilon^2}{2kT}\left(\frac{1}{r_A}-\frac{1}{r_{M*}}\right)$$

$$(5.134)$$

Laidler and Eyring [47] point out that data involving varying dielectric constant that can be extrapolated to ionic strength zero are scarce, and reasoned that since the Hückel and Debye–McAulay terms in Eq. (3.132) tend to cancel, the contribution of ionic strength to these types of reactions would ordinarily be very small. Hence a plot of $\ln k'$ versus $1/D$ at finite ionic strength could be tested for obedience to theory. To test the theory, these authors plotted $\ln k'$ versus $1/D$ for the N-chloroacetanilide and N-chloro-propionanilide conversions into the p-chloro compounds in the presence of hydrochloric acid [51]. These reactions may involve an initial attack by chloride ion. Straight lines with positive slopes were obtained as required by Eq. (5.132) which predicts the lines should always have positive slopes since the charge on the ion is squared. In these instances the moments of the reactant molecules N-chloroacetanilide and N-chloropropionanilide are not zero, however, the Kirkwood, as compared to the ionic term, is only slightly influenced by the dielectric constant, and hence there is linear dependence of $\ln k'$ on $1/D$. A quantitative test of obedience of data to theory would require values of r_A and r_{M*}.

C. LAIDLER–LANDSKROENER EQUATION

The Laidler–Landskroener equation [52, 53] was derived to relate the rate constant of the reaction to the dielectric constant for reactions in which the electrostatic interactions are more important than nonelectrostatic ones, such as ion–ion, ion–dipole and for certain dipole–dipole reactions. The equation was derived using Kirkwood's [54] and Kirkwood and Westheimer's [55] equation for the activity coefficient of a molecule possessing a distribution of charges embedded in a sphere of radius b and of dielectric constant D_i submerged in a medium of dielectric constant D, and referred to a standard solvent of dielectric constant D_0. For zero ionic strength Kirkwood's equation for the activity coefficient is

$$\ln f_i = \frac{1}{2kT} \sum_{n=0}^{\infty} \frac{(n+1)Q_n}{D_i b_i^{(2n+1)}} \left[\frac{D_i - D}{(n+1)D + nD_i} - \frac{D_i - D_0}{(n+1)D_i + nD_0} \right]$$

(5.135)

where the Q terms are given by

$$Q_n = \sum_{k=1}^{M} \sum_{l=1}^{M} e_k e_l r_k^{n} r_l^{n} P_n(\cos \theta_{kl})$$

(5.136)

where the e's are charges on the sphere, the r's their distance from the center, and $P_n(\cos \theta_{kl})$ is the Legendre polynomial. Barring the case where the charges are concentrated near the surface of the sphere the summation in Eq. (5.135) converges rapidly, and only infrequently is it necessary to go beyond $n = 1$. Therefore, only the polynomials

$$P_0(x) = 1$$

(5.137)

and

$$P_1(z) = \cos x$$

(5.138)

were used by Laidler in his treatment. The principal requirement is that $D_i \ll D$, the actual dielectric constant of the medium D_i was given the value of 2, and for the standard state of a gas with $D_0 = 1$, $Q_0 = Z_i^2 \varepsilon^2$ and $Q_1 = G_1 \varepsilon^2$, Eq. (5.135) can be written

$$\ln f_i = \frac{1}{2kT} \left[\frac{Q_0}{2b_i} \left(\frac{2-D}{D} - 1 \right) \right]$$

$$+ \frac{Q_1}{b_i^3} \left(\frac{2-D}{2D+2} - \frac{1}{4} \right) = \frac{Z_i^2 \varepsilon^2}{2kTb_i} \left(\frac{1-D}{D} \right) + \frac{3G_1 \varepsilon^2}{8kTb_i^3} \left(\frac{1-D}{D+1} \right)$$

(5.139)

where G_1 is a number which depends on the charge distribution and is proportional to the dipole moment. In the final form of Eq. (5.139) the first term on the right-hand side is the Born term, and the second term, which for a single charge is zero, corresponds to the effect of the charge distribution within the sphere.

Expanding the term $(1-D)/(D+1)$ we obtain p'.

$$\frac{1-D}{1+D} = \frac{1}{1+D} - \frac{D}{1+D} = \frac{1}{D} - \frac{1}{D^2} + \frac{1}{D^3} - \frac{1}{D^4} + \cdots$$

$$-\left(1 - \frac{1}{D} + \frac{1}{D^2} - \frac{1}{D^3} + \cdots\right)$$

$$= \frac{1}{D}\left(1 - \frac{1}{D} + \frac{1}{D^2} - \frac{1}{D^3} + \cdots\right) - \left(1 - \frac{1}{D} + \frac{1}{D^2} - \frac{1}{D^3} + \cdots\right)$$

$$= \left(\frac{1}{D} - 1\right)\left(1 - \frac{1}{D} + \frac{1}{D^2} - \frac{1}{D^3} + \cdots\right) \tag{5.140}$$

and for D large

$$\frac{1-D}{1+D} = \left(\frac{1}{D} - 1\right)\left(1 - \frac{1}{D}\right) = -\frac{1}{D^2} + \frac{2}{D} - 1 \cong \frac{2}{D} - 1 \tag{5.141}$$

Substitution into Eq. (5.139) yields

$$\ln f_i = \frac{Z_i^2 \varepsilon^2}{2kTb_i}\left(\frac{1}{D} - 1\right) + \frac{3G_i \varepsilon^2}{8kTb_i^3}\left(\frac{2}{D} - 1\right) \tag{5.142}$$

considering the equation

$$A + B \rightleftarrows X \rightarrow \text{products} \tag{5.143}$$

$$\ln k' = \ln k_0' + \ln\left(\frac{f_A f_B}{f_X}\right) \tag{5.144}$$

where k' is the specific rate constant for the reaction in a medium of dielectric constant D, k_0' is the specific rate constant for a medium of reference dielectric constant D_0 taken as unity and the f's are the activity coefficients of the respective species. Suitable values of f from Eq. (5.142) substituted into Eq. (5.144) gives an equation for the effect of dielectric constant on reaction rates. The equation is

$$\ln k' = \ln k_0' + \frac{\varepsilon^2}{2kT}\left(\frac{1}{D} - 1\right)\left(\frac{Z_A^2}{b_A} + \frac{Z_B^2}{b_B} - \frac{(Z_A + Z_B)^2}{b_X}\right)$$

$$+ \frac{3\varepsilon^2}{8kT}\left(\frac{2}{D} - 1\right)\left(\frac{G_A}{b_A^3} + \frac{G_B}{b_B^3} - \frac{G_X}{b_X^3}\right) \tag{5.145}$$

Using the relationship between the G factors and the moments μ of the respective species Laidler [53] wrote Eq. (5.145) in the form

$$\ln k' = \ln k_0' + \frac{\varepsilon^2}{2kT}\left(\frac{1}{D} - 1\right)\left[\frac{Z_A^2}{b_A} + \frac{Z_B^2}{b_B} - \frac{(Z_A + Z_B)^2}{b_X}\right]$$

$$+ \frac{3\varepsilon^2}{8kT}\left(\frac{2}{D} - 1\right)\left[\frac{\mu_A^2}{b_A^3} + \frac{\mu_B^2}{b_B^3} - \frac{\mu_X^2}{b_X^3}\right] \tag{5.146}$$

Equation (5.146) predicts that a plot of $\ln k'$ versus $1/D$ should yield a straight line the slope S of which can be expressed in terms of radii, charges, and dipole moments thus

$$S = \frac{\varepsilon^2}{2kT}\left[\left(\frac{Z_A^2}{b_A} + \frac{Z_B^2}{b_B} - \frac{(Z_A + Z_B)^2}{b_X}\right) + \frac{3}{4}\left(\frac{\mu_A^2}{b_A^3} + \frac{\mu_B^2}{b_B^3} - \frac{\mu_X^2}{b_X^2}\right)\right]$$
$$\tag{5.147}$$

Laidler and Landskroener [52] point out that in order to evaluate b_X and μ_X a model for the activated complex has to be constructed. They have constructed such models, for example, for the acidic and basic hydrolyses of esters. Laidler [53] believes it best to consider Eq. (5.146) a semiquantitative and as giving only a rough indication of the effect of changing dielectric constant on the rate of reaction. To illustrate, if the moment of the complex μ_X is greater than that of either reactant, that is if μ_X is greater than either μ_A or μ_B, an increase in the dielectric constant will cause an increase in the reaction rate since a higher dielectric constant favors the formation of any highly polar species, in this case, the activated complex.

D. The Frequency Factor

Electrostatic interactions between neutral molecules usually have a very small effect on entropy of activation, however, an appreciable effect may be observed if either a reactant or the activated complex is highly polar. Thus, when a quaternary ammonium salt which is highly polar is formed from a tertiary amine and an alkyl iodide, the complex binds solvent molecules more strongly than do the reactants, thus causing an ordering effect and hence a decrease in entropy of activation and a correspondingly small frequency factor. Laidler [53] summarizes the effect of electrostatic interactions on frequency factors by saying that for a reaction in which there is a separation of opposite charges or the approach of like charges there will be an unusually low frequency factor; and that for a reaction in which there is a separation of like charges or an approach of opposite charges in the formation of the activated complex the frequency factor will be unusually high. For the first type of reaction an increase in the dielectric constant accelerates the rate of reaction, while for the second type of reaction, an increase in dielectric constant decelerates the rate of reaction.

E. IONIC STRENGTH INFLUENCE ON RATES OF ION–MOLECULE REACTIONS

Equation (5.146) indicates no ionic strength effect on ion–molecule reactions. However, Laidler [53] extends the treatment of ion–molecule theory to include an ionic strength dependence, by adding to the Debye–Hückel expression for the activity coefficient of a species the $b\mu$ term introduced by Hückel. Thus,

$$\log f_A = Z_A{}^2 A' \mu^{\frac{1}{2}} + b_A \mu \tag{5.148}$$

The Debye–McAulay [50] equation

$$\log f_B = b_B \mu \tag{5.149}$$

was used for the activity coefficient for reactant B which has no net charge. An equation entirely equivalent to Eq. (5.148) applied to the activity coefficient of the activated complex which has the same net charge Z_A as reactant A. Substituting these activity coefficients into the base ten logarithmic form of Eq. (5.88) yields

$$\log k' = \log k_0' + (b_A + b_B - b_X)\mu \tag{5.150}$$

The Debye–Hückel term involving $\mu^{1/2}$ canceled out since it occurred in the activity coefficient terms of A and X which have the same charge; hence, the importance of including the terms including the first power of μ in this approach. According to Eq. (5.150) in a plot of $\log k'$ for an ion–molecule reaction versus μ, a straight line should result with a slope given by $(b_A + b_B - b_X)$, however, the ionic strength effect on the rate should be much less in this case than for the case of ion–ion reactions. Hydrolysis of acetals by hydroxide ions [56] lend credence to these conclusions.

Equation (5.150) was written [53] in the exponential form

$$k' = k_0' e^{b'\mu} \tag{5.151}$$

where

$$b' = 2.303(b_A + b_B - b_X) \tag{5.152}$$

using a Maclaurin expansion and discarding all powers in $b'\mu$ but the first, $e^{b'\mu}$ can be set equal to $1 + b'\mu$, and Eq. (5.151) becomes

$$k' = k_0'(1 + b'\mu) \tag{5.153}$$

The above equation would imply a linear dependence of k' on μ, which relationship has been observed for several kinetic processes [48]. While a quantitative treatment for the $b\mu$ term in case of an ion has been formulated [57], no satisfactory method has been devised for the reliable determination of its magnitude, either experimentally or theoretically. Discussions including a review [58] have been made of this term.

F. Ion–Dipolar Molecule Reaction Rates from the Combined Onsager Dipole and Debye–Hückel Ionic Atmosphere Theories

In deriving the rate equation [59], the potential Ψ_0 in the neighborhood of any one dipolar molecule was first obtained. Except for the boundary conditions at the surface of the molecule, the differential equation for Ψ_0 is the same as the potential around an ion when interaction between dipoles are neglected. Let n_i be the number of ions of species i per cubic centimeter and Z_i be the valence of the species. As in the potential around an ion, the differential equation for the potential is

$$\Delta^2 \Psi_0 = \varkappa^2 \Psi \tag{5.154}$$

where $\varkappa$ is the Debye kappa.

Assuming a point singularity for the dipole, a particular solution of Eq. (5.154) is

$$\Psi_0 = C_1 e^{(-\varkappa r)}/r \tag{5.155}$$

For $r > a$, a solution having the same dependence as that of a dipole potential can be found by differentiating Eq. (5.155) with respect to z, provided that the idealized dipole is represented as a sphere of radius a, that the position of the dipole coincides with the origin, and provided its direction is that of the positive z axis. Thus

$$\frac{\partial \Psi_0}{\partial z} = -\frac{C_1 e^{-\varkappa r}}{r^2}(1+\varkappa r)\frac{dr}{dz} \tag{5.156}$$

but

$$\frac{\partial \Psi_0}{\partial z} = \frac{\partial \Psi_0}{\partial r}\frac{\partial r}{\partial z} \tag{5.157}$$

and

$$r^2 = x^2 + y^2 + z^2 \tag{5.158}$$

Thus

$$dr/dz = z/r = \cos\theta \tag{5.159}$$

and Eq. (5.156) becomes

$$\Psi_0 = \frac{C_1 e^{-\varkappa r}}{r^2}(1+\varkappa r)\cos\theta, \qquad r \geqslant a \qquad (5.160)$$

The solution [Eq. (5.160)] for the special case considered above was given by Bateman *et al.* [60]. Kirkwood [49] gave the solution for the potential of a particle with the most general distribution of charge.

For the interior of the molecule following the Debye–Hückel procedure and holding D constant, the potential for the interior of the molecule becomes

$$\Delta^2\psi_i = 0, \qquad r \leqslant a \qquad (5.161)$$

In this domain ψ_i must be continuous except for a singularity at the origin corresponding to a dipole of given strength. From this procedure a reasonable expression for the ion atmosphere is obtained, however, to fit the formula to experimental kinetic data, dipole values about five times those of ordinary dipoles must be chosen. Some other influential electrostatic forces such as the interaction of the dipoles with the dielectric solvent, are not being correctly accounted for.

Onsager [61] has shown that the external moment with which an immersed dipole acts upon distant charges is different from its moment *in vacuo*. The medium modifies both the permanent and induced moments. Therefore, the Onsager model of a dipole immersed in a dielectric liquid must be combined with the Debye–Hückel theory of ionic atmospheres.

The model like that of Onsager is a spherical molecule of permanent moment in vacuo μ_0 and polarizability α. The polarizability is related to the refractive index n by the equation

$$\alpha = \frac{n^2-1}{n^2+2}a^3 \qquad (5.162)$$

The solution for μ_0 in Eq. (5.160) must be joined properly at $r = a$ to a solution of Eq. (5.161) having a dipose singularity at $r = 0$. To do this it is necessary that we find the external characteristic moment μ^*.

The exterior potential of a rigid dipole placed in a cavity of radius a is given by Eq. (5.160), and the exterior potential is represented by

$$\psi_i = \left(\frac{m}{r^2} + Br\right)\cos\theta, \qquad r \leqslant a \qquad (5.163)$$

where B is a constant.

Employing the usual boundary conditions, namely $\psi_i = \psi_0$ and $\partial\psi_i/\partial r = \partial\psi_0/\partial r$, constants C_1 and B can be determined as follows:

$$\frac{C_1 e^{\varkappa a}}{a^2}(1+\varkappa a) = \frac{m}{a^2} + Ba \tag{5.164}$$

and

$$\frac{DC_1 e^{-\varkappa a}}{a^3}(2+2\varkappa a+\varkappa^2 a^2) = \frac{2m}{a^3} - B \tag{5.165}$$

and hence

$$C_1 = -\frac{3me^{\varkappa a}}{D(2+2\varkappa a+\varkappa^2 a^2) + (1+\varkappa a)} \tag{5.166}$$

and

$$B = -\frac{m[D(2+2\varkappa a+\varkappa^2 a^2) - 2(1+\varkappa a)]}{a^3[D(2+2\varkappa a+\varkappa^2 a^2) + (1+\varkappa a)]} \tag{5.167}$$

The Onsager condition for internal equilibrium is

$$m = \mu_0 + \alpha F_z \tag{5.168}$$

where F_z is the force on the dipole from self-created polarization. In an electric field F_z, the total electric moment of a molecule is therefore the sum of the permanent moment μ_0 and the induced moment αF_z.

The local field F_z is

$$F_z = -\frac{\partial}{\partial z}\left(\psi_i - \frac{m\cos\theta}{r^2}\right) = -B \tag{5.169}$$

Hence from Eqs. (5.162), (5.168), and (5.169)

$$m = \mu_0 - \frac{n^2-1}{n^2+2}a^3 B \tag{5.170}$$

The value of B substituted from Eq. (5.167) into (5.170) gives

$$m = \mu_0 \frac{[n^2+2][D(2+2\varkappa a+\varkappa^2 a^2) + (1+\varkappa a)]}{3[D(2+2\varkappa a+\varkappa^2 a^2) + n^2(1+\varkappa a)]} \tag{5.171}$$

m is then found as a function of μ_0 and n^2, and its value from Eq. (5.171) substituted into Eq. (5.166) yields

$$C_1 = \mu^*/D \tag{5.172}$$

where

$$\mu^* = \frac{\mu_0(n^2+2)De^{\varkappa a}}{D(2+2\varkappa a+\varkappa^2 a^2) + n^2(1+\varkappa a)} \tag{5.173}$$

and substituting C_1 from Eq. (5.172) into Eq. (5.160) gives

$$\psi_0 = \frac{\mu^* e^{-\varkappa r}(1 + \varkappa r)}{Dr^2} \cos \theta \qquad (5.174)$$

where ψ_0 is the potential of the ionic atmosphere around a dipolar molecule of permanent moment μ_0; μ^* is the external moment of the molecule in the dielectric solvent.

When there is no ionic atmosphere, $\varkappa = 0$ and Eqs. (5.163) and (5.174) become, as required, identical with Onsager's solutions. Equations (5.173) and (5.174) correspond to dipolar molecules with no polarizability.

Let the concentrations of the intermediate complex and the reactants A and B be C_X, C_A, and C_B, respectively. By the Boltzmann principle, and using the procedure of Christiansen and Scatchard, the probability of finding ion B in a specified element of volume, defined by the limits r and $r + dr$, θ and $\theta + d\theta$, and ψ and $\psi + d\psi$, is proportional to

$$C_A C_B \exp\left(-\frac{\psi_0 \varepsilon Z_B}{kT}\right) r^2 \sin \theta \, dr \, d\theta \, d\Phi \qquad (5.175)$$

where ψ_0 is defined by Eq. (5.174) and r, s, and Φ are the polar coordinates about the center of the dipole.

The rate of the formation of X must be determined in order to obtain the rate of the reaction. Two different approaches may be used in doing this. It may be assumed (1) that there are sensitive zones on the surface of the molecule A which, if touched by B, the complex X is formed or (2) for the formation of X to occur, there is a critical distance of approach from each direction of ion B to molecule A. In the first case it is necessary that r, θ, and Φ have specified values between rather narrow limits r_0 and $r_0 + dr_0$, θ_0 and $\theta_0 + d\theta_0$, and Φ_0 and $\Phi_0 + d\Phi_0$. This alternative will be followed first.

Assuming rotational symmetry about the z axis, Φ need not be specified, and the integration of Eq. (5.175) with respect to Φ gives 2π. Thus C_X is

$$C_X = 2\pi k' C_A C_B r_0{}^2 \sin \theta_0 \, \Delta r_0 \, \Delta \theta_0 \exp\left(-\frac{\psi_0 \varepsilon Z_B}{kT}\right) \qquad (5.176)$$

where k' is a proportionality constant. Letting

$$K = 2\pi k' r_0{}^2 \sin \theta_0 \, \Delta r_0 \, \Delta \theta_0 \qquad (5.177)$$

there results

$$C_X = K C_A C_B \exp\left(-\frac{\psi_0(r_0, \theta_0)\varepsilon Z_B}{kT}\right) \qquad (5.178)$$

or in logarithmic form

$$\ln\frac{C_X}{C_A C_B} = \ln K - \frac{\psi_0(r_0,\theta_0)Z_B\varepsilon}{kT} \qquad (5.179)$$

From Eq. (5.174) when κ becomes zero

$$\psi_0 = \frac{\mu_0^*}{Dr_0^2}\cos\theta \qquad (5.180)$$

where μ_0^* is Onager's value, and from Eq. (5.173) is

$$\mu_0^* = \frac{\mu_0(n^2+2)D}{2D+n^2} \qquad (5.181)$$

Substituting the value of ψ_0 from Eq. (5.180) into Eq. (5.179) yields

$$\ln\frac{C_X}{C_A C_B} = \ln K - \frac{Z_B\varepsilon\mu_0^*\cos\theta}{DkTr_0^2} \qquad (5.182)$$

Therefore

$$\ln\frac{f_A f_B}{f_X} = \ln\frac{C_X}{C_A C_B} - \ln\frac{C_X^0}{C_A^0 C_B^0}$$

$$= \frac{\varepsilon Z_B\cos\theta}{DkTr_0^2}[\mu_0^* - \mu^*\exp(-\kappa a)(1+\kappa a)]$$

$$= \mu_0\frac{\varepsilon Z_B\cos\theta}{DkTr_0^2}\left[\frac{(n^2+2)D}{2D+n^2} - \frac{(1+\kappa a)(n^2+2)D}{D(2+2\kappa a+\kappa^2 a^2)+n^2(1+\kappa a)}\right]$$

$$\qquad (5.183)$$

Substituting for μ_0 in terms of μ_0^* in Eq. (5.183) gives us

$$\ln\frac{f_A f_B}{f_X} = \frac{\varepsilon Z_B\mu_0^*\cos\theta}{DkTr_0^2}\left[\frac{D\kappa^2 a^2}{2D(1+\kappa a+\kappa^2 a^2/2)+n^2(1+\kappa a)}\right] \qquad (5.184)$$

and since

$$\ln k' = \ln k_0' + \ln\frac{f_A f_B}{f_X} \qquad (5.185)$$

Eqs. (5.184) and (5.185) yield

$$\ln k' = \ln k_0' + \frac{\varepsilon Z_B\mu_0^*\cos\theta}{DkTr_0^2}\left[\frac{D\kappa^2 a^2}{2D(1+\kappa a+\kappa^2 a^2/2)+n^2(1+\kappa a)}\right]$$

$$\qquad (5.186)$$

which can be put in the form

$$\frac{\ln k' - \ln k_0'}{\varepsilon Z_B \mu_0^* \cos \theta / 2DkTr_0^2} = W = \frac{x^2 a^2}{1 + xa + (x^2 a^2 / 2D) + (n^2 / 2D)(1 + xa)}$$

(5.187)

A reference point defined by the double transition $x = 0$, $D = \infty$ has proven satisfactory in correlating the dielectric dependence of k' for ion–ion reactions. Such a double transition in the case of ion–dipole reactions indicates a change in k' with $1/D$ opposite in sense to the change in k' with x. Data indicates that the changes in k' should be in both cases in the same direction. Electrostatic forces, either repulsive or attractive, tend toward zero as D increases indefinitely for some ion–ion reaction. In the case of dipoles, the external moments increase with increasing D, and though $\lim_{D=\infty} \mu^*$ exists, the point $D = \infty$ is not an adequate reference point since it accentuates rather than eliminates the effect of D. One approach assumed that $x = 0$ was a satisfactory reference point, and that Eq. (5.186) states the relationship to both D and x [59].

Dimensionless variables have been used to transform Eq. (5.186) to test the dependence of k' upon D. Putting

$$x^2 = \lambda^2 / D$$

(5.188)

where λ has the dimensions cm^{-1} though it is free of D, then if

$$S = \lambda a = \lambda r_0$$

(5.189)

and

$$W' = \frac{2kT(\ln k' - \ln k'_{x=0})}{Z_B \varepsilon \mu_0^* \lambda^2 \cos \theta}$$

(5.190)

the general expression can be written

$$W' = \frac{1}{D^2} \frac{1}{1 + S/D^{1/2} + S^2 / 2D}$$

(5.191)

Equation (5.191) implies that in the limit $\lambda = 0$ and hence $x = 0$, W' will vary as $1/D^2$ at higher concentrations, the increase of W' with decreasing D will become less; and the increase will depend on the parameter S which, according to Eqs. (5.188) and (5.189) is proportional to the square root of the ionic strength.

Taking the dimensionless quantity

$$z = xa$$

(5.192)

and the dimensionless quantity

$$W = \frac{(\ln k' - \ln k'_{\varkappa=0})2DkTr_0{}^2}{Z_B \varepsilon \mu_0{}^* \cos \theta} \tag{5.193}$$

Eq. (5.187) can be written

$$W = \frac{z^2}{1 + z + z^2/2 + (n^2/2D)(1+z)} \tag{5.194}$$

In applications of the theory to data, the dielectric constant and ionic strength dependencies of the rates of ion–dipolar molecule reactions have been made employing Eq. (5.187) in the respective forms of Eqs. (5.191) and (5.194).

The equation has been applied [8, 59] to both the dielectric constant dependence of the rates of ion–dipolar molecule reactions for both positive and negative ionic reactants.

There was some uncertainty about the extrapolation to a reference dielectric constant in the above theory of ion–dipolar molecule reactions that did not apply in the ionic strength extrapolation to $\varkappa = 0$. A derivation of the dielectric constant dependence of k' has been made [62] using Coulombic energy considerations. The approach is simple and can be applied to various electrically charged and electrically unsymmetrical combinations. The complex formation–activity coefficient approach is perhaps to be preferred in formulating the salt effect.

The mutual potential (potential energy) E_c between an ion and a dipole can be written

$$E_c = -\frac{Z\mu_0 \cos \theta}{Dr^2} \tag{5.195}$$

when higher than first powers of the distance of separation of charge centers in the dipole are neglected. At two different dielectric constants D_1 and D_2, the difference in Coulombic energy ΔE_c of the ion dipole is

$$\Delta E_c = -\frac{Z\varepsilon\mu_0}{r^2}\left(\frac{1}{D_2} - \frac{1}{D_1}\right) \tag{5.196}$$

The quantitative effect of Coulombic energy on specific reaction rates can be formulated by considering a reaction at dielectric constant D_2 having energy requirements E_{D_2} and at dielectric constant D_1 having energy requirement E_{D_1}. Then the energy of activation E_{D_2} and E_{D_1} are related by the equation

$$E_{D_2} = E_{D_1} + \Delta E_c \tag{5.197}$$

At dielectric constant D_2, the specific velocity constant is given by

$$k'_{D_2} = Z' \exp\left(-\frac{E_{D_2}}{RT}\right) \tag{5.198}$$

and

$$\ln k'_{D_2} = \ln Z' - \frac{E_{D_2}}{RT} = \ln Z - \frac{E_{D_1}}{RT} - \frac{\Delta E_c}{RT} \tag{5.199}$$

also

$$\ln k'_{D_1} = \ln Z' - \frac{E_{D_1}}{RT} \tag{5.200}$$

In Eqs. (5.198), (5.199), and (5.200) the factor Z' represents the Arrhenius frequency factor. From the above equations we can write,

$$\ln k'_{D_2} = \ln k'_{D_1} + \frac{Z\varepsilon\mu_0}{kTr^2}\left(\frac{1}{D_2} - \frac{1}{D_1}\right) \tag{5.201}$$

If D_2 be taken as a reference dielectric constant with infinite magnitude and if D_1 be taken as any dielectric constant, D, then from Eq. (5.201)

$$\ln k_\infty' = \ln k_D' - \frac{Z\varepsilon\mu_0}{kTr^2 D} \tag{5.202}$$

or

$$\ln k_D' = \ln k_\infty' + \frac{Z\varepsilon\mu_0}{kTr^2 D} \tag{5.203}$$

Taking the charge of the ion into account, Eq. (5.203) predicts that a plot of $\ln k_D'$ versus $1/D$ should be a straight line of positive slope if $Z\varepsilon$ is positive and of negative slope if $Z\varepsilon$ is negative. These predictions are in agreement with the predictions of Eq. (5.186). The specific velocity constants used should preferably be those which have been extrapolated to $\varkappa = 0$. The slope of the plot should yield a reasonable value of r. The preferable moment at zero ionic strength used in Eq. (5.203) should probably be μ_0^* which is related to μ_0 by the Onsager theory Eq. (5.181), when $\varkappa = 0$, thusly

$$\mu_0^* = \frac{\mu_0(1+n^2)D}{2D+n^2} \tag{5.204}$$

and Eq. (5.203) would become

$$\ln k_{D=D; \varkappa=0} = \ln k_{D=\infty} + \frac{Z\varepsilon\mu_0^*(2D+n^2)}{D^2 kTr^2(n^2+2)} \tag{5.205}$$

The parameter $\mu_0{}^*$, r, D, and n occur in this equation, and assumptions would have to be made concerning them since their values are ordinarily not recorded in the literature. Selected values of r and n might be inserted in Eq. (5.205) and

$$\ln k_{D=D;\varkappa=0} \quad \text{versus} \quad \frac{Z\varepsilon(2D+n^2)}{D^2kTr^2(n^2+2)}$$

plotted, and if a straight line with a slope that gives reasonable values of $\mu_0{}^*$ is obtained, theory is in agreement with experiment.

The simple expression Eq. (5.203) has been incorporated [63] into Eq. (5.186) giving

$$\ln k' = \ln k'_{\varkappa=0;\,D=\infty} + \frac{Z\varepsilon\mu_0}{DkTr_0{}^2} + \frac{Z\varepsilon\cos\theta}{DkTr_0{}^2}\left(\mu_0{}^* - \frac{\mu^*(1+\varkappa\mu_0)}{e^{\varkappa r_0}}\right)$$

(5.206)

where the second term on the right-hand side gives the dielectric constant dependence, and the last term on the right-hand side the ionic strength dependence of the rate.

A word might be said about the difference in the assumptions made in the derivations of Eqs. (5.186) and (5.203). The former was derived assuming a critical complex was formed and following the procedures of Christiansen and of Scatchard in calculating the concentration of the complex; while the latter equation was derived from purely electrostatic considerations without particular assumptions as to the formation of an intermediate complex. The combined Eq. (5.206) is similar in appearance and function to the Brönsted–Christiansen–Scatchard equation [equivalent to Eq. (5.109)].

The equation has been applied widely to both positive and negative ionic reactants reacting with dipolar molecules [1, 8] to calculate both salt effects

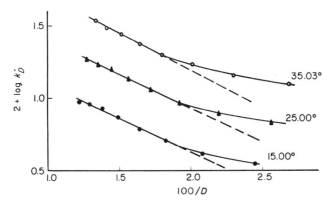

Fig. 5.6. Variation of the logarithm of specific rate constant with reciprocal of dielectric constant.

and dielectric constant effects on rates. Illustrative of the applicability of the equation to these effects is the data for the alkaline hydrolysis of methyl propionate [63] in acetone water at three temperatures: 15, 25, and 35°C. The plot of $\log k_D'$ versus $100/D$ is shown in Fig. 5.6. From the slope, S, of the straight line portion of the curve r was calculated using the relationship

$$r = \left(\frac{Z\mu\varepsilon}{2.303kTS} \right)^{1/2} \tag{5.207}$$

For methyl propionate the dipole moment μ was taken as 1.8 Debye units in harmony with the recorded values [64] of 1.7, 1.8, and 1.79 Debye units for methyl acetate, ethyl acetate, and ethyl propionate, respectively. The values of r calculated from these and other ester hydrolyses data are given in Table 5.5.

Table 5.5

Values of the Parameter r for Both the Acid and Alkaline Hydrolysis of Esters

Ester	Solvent	Temperature (°C)	$r \times 10^8$ cm
A. Acid hydrolysis			
Ethyl acetate	Water and water–dioxane	35.0	9.08
		45.0	9.18
		55.0	9.50
Ethyl acetate	Water and water–acetone	35 0	5 21
		45.0	6.19
		55.0	Unreasonably small
Methyl propionate	Water and water–acetone	25.13	3.43
		35.21	3.42
		45.48	3.01
B. Alkaline hydrolysis			
Ethyl acetate	Water and water–ethanol	0.00	0.94
		9.80	0.93
		19.10	0.98
Ethyl acetate	Water and water–acetone	0.00	2.0
		15.87	1.8
		25.10	1.5
Methyl propionate	Water and water–acetone	15.00	1.4
		25.00	1.4
		35.00	1.3
Ethyl fluoroacetate	Water and water–ethanol	15.00–30.00	1.29
Ethyl chloroacetate	Water and water–ethanol	15.00–30.00	2.64
Ethyl bromoacetate	Water and water–ethanol	15.00–30.00	2.64

Only one datum shows real conflict between theory and experiment. The data for the alkaline α-haloacetates [65] were calculated using the average slopes of the slightly curved $\log k$ versus $1/D$ lines at 15.00, 25.00, and 30.00°C for each of the α-haloacetates, and the dipole moments used in the calculations were 2.09, 2.64, and 2.64 Debye units for the respective esters, ethyl fluoroacetate, ethyl chloroacetate, and ethyl bromoacetate. These moments were estimated from the ratios of the dipole moments of *p*-fluoroanisole, *p*-chloroanisole, and *p*-bromoanisole which were available in the literature [66] and the dipole moment listed in the literature for ethyl chloroacetate [67].

From the dielectric constant effect on the acid hydrolysis of esters the method proposed by Bender [68] is satisfactory in correlating experiment with theory [69]. The mechanism is

$$\tag{5.208}$$

The first step would be rate controlling. Any similar mechanism in which the rate controlling step is between a positive ion and a dipolar molecule would be satisfactory with respect to the above theory.

In the case of alkaline hydrolysis in these dielectric constant dependencies of the rate constant, the rate controlling step would be that between a negatively charged ion and a dipolar molecule, though the overall mechanism might involve several steps. These steps could be [70]

$$\tag{5.209}$$

The first step or some other step involving the reaction of a negative ion with a dipolar molecule would be rate controlling from the standpoint of the effect of the dielectric constant on the reaction rate.

Other reactions which were in agreement with the predictions of this theory concerning the effect of dielectric constants on the rate of reaction are [1] the rate of sucrose inversion by hydronium ion, the diacetone alcohol–hydroxide ion reaction, and the reaction of ethyl bromomalonate–thiosulfate ion reaction.

In Eq. (5.191) the sign of the ion has to be taken into account for checking data against theory. This was not foreseen from electrostatics, and probably is a consequence of the chemical rather than the electrical properties of the system. As illustrated by the examples cited above, much chemical rate data conforms to the requirements of Eq. (5.191) with respect to the effect of the dielectric constant of the solvent on ion–dipolar molecule reaction rates.

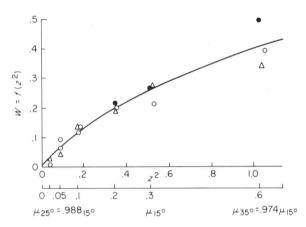

Fig. 5.7. Agreement of ionic strength data with theoretical curve: $W = f(z^2)$ versus z^2. ● = 15.00°; △ = 25.00°; ○ = 35.03°.

Figure 5.7 illustrates the application of Eq. (5.187) applied to the salt effect on ion–dipolar molecule reaction rates. The plot is for W versus z^2, and the data was for the alkaline hydrolysis of methyl propionate in water at 15.00, 25.00, and 35.03°C. The data were made to fit the curve by adjusting the values of the parameters $k'_{\varkappa = 0}$, μ_0^*, and r_0. These values together with the square of the refractive indexes calculated from the enhanced moments μ_0^* [using Eq. (5.204)] at zero ionic strength and at the temperatures indicated are given in Table 5.6. The theory has been applied to various reactions involving both positive and negative ions reacting with dipolar molecules. These include: the positive hydrogen ion reacting with sucrose, ethylene acetal and glucose [59]; the negative hydroxide ion reacting with diacetone alcohol [71]; and the negative bivalent thiosulfate ion reacting with ethyl bromoalonate [72].

Table 5.6

Parameters Used to Fit Ionic Strength Data to Theoretical Curve

Temperature (°C)	$r \times 10^8$ cm	$\mu_0{}^*$ (Debye units)	n^2	$k'_{\kappa=0}$
15.00	4	9.8	9.5	0.0947
25.00	4	10.6	11.6	0.187
35.03	4	11.6	11.6	0.344

G. The Solvent and Energy of Activation

The effect of the dielectric constant on the energy of activation for the reaction between an ion and a dipolar molecule can be calculated as a coulombic energy of activation as presented in Eq. (5.196). The approach permits the derivation of Eq. (5.203) which allows the calculation of the effect of the dielectric constant of the solvent upon ion–dipolar molecule reaction rates.

The energy of activation for a series of reactions between a common ion and aromatic molecules separately substituted by groups of dipole moment μ has been represented by the equation [73]

$$E_s = E_u + K_1 \mu + K_2 \mu^2 \tag{5.210}$$

where E_s is the energy of activation of the substituted molecule reaction and E_u the energy of activation of the unsubstituted molecule reaction with the common ion.

For meta- and para-substituted molecules the energy of activation has been written in terms of two separate linear equations [74]

$$E_s = E_u + K_m \mu \quad \text{and} \tag{5.211}$$

$$E_s = E_u + K_p \mu \tag{5.212}$$

where the gradient K_m refers to the meta- and the gradient K_p to the para-substituted molecules. Ions of opposite sign reacting with somewhat similar molecules exhibits a similar linear relationship but with slopes of opposite sign [75, 76].

These facts were explained in terms of electrostatic principles by resolving the total energy of activation E into the components [74]

$$E = E_0 + E_e(r_1, \theta_1, D_1) + E_e(r_2, \theta_2, D_2) \tag{5.213}$$

where E_0 is all the energy of activation except the two electrostatic contributions, $E_e(r_1, \theta_1, D_1)$ is the electrostatic energy of activation resulting from the interaction between the reactant ion and the dipole of the group which it is attacking, and $E_e(r_2, \theta_2, D_2)$ is the electrostatic energy of activation originating from the ion at the distance of reaction and the second dipole situated at some distance from the seat of action in the polar molecule. $E_e(r_1, \theta_1, D_1)$ may be assumed constant for all members in the series, and may be combined with E_0 to give E_u. Dropping the subscript 2 the equation

$$E_s = E_u + E_e(r, \theta, D) \tag{5.214}$$

may be written, where r is the distance from the center of the attacking ion to the center of the nonreacting dipole when the activated complex is formed between the ion and the reacting dipole, θ is the angle between the line of centers and the polar axis of the substituted dipolar molecule, and D is the dielectric constant of the intervening medium, which, for the case under consideration, is primarily the benzene ring. Now

$$E(r, \theta, D) = -\frac{NZ\mu \cos \theta}{Dr^2} \tag{5.215}$$

and from Eqs. (5.214) and (5.215)

$$E_s = E_u - \frac{NZ\mu \cos \theta}{Dr^2} \tag{5.216}$$

and for head-on alignment of ion and dipole becomes

$$E_s = E_u - \frac{NZ\mu}{Dr^2} \tag{5.217}$$

Thus the magnitudes and signs of the constants in Eqs. (5.211) and (5.212) are explained. The energy of activation plotted against the dipole moments of dipolar molecules reacting with ions should yield straight lines with negative slope if the ion is positive in charge, and with positive slope if the ion is negative in charge. These are in harmony with the predictions of the electrostatic influences on ion–dipolar molecule reactions derived by Amis and Jaffe [59] and by Amis [62].

The theory was applied to the alkaline hydrolyses of the para-substituted ethyl benzoate [74, 77] and to acid catalyzed enolization of para-substituted acetophenones. Both plots of E_s versus μ were linear, but the alkaline hydrolysis gave the steeper slope. Apparently the hydroxide ion which attacks the carbon atom of the carboxyl atom in the saponification reaction approaches more nearly the center of the substituted dipole than does hydronium ion that attacks the oxygen atom of the carboxyl group in the enolization reaction. Moelwyn-Hughes shows how substitution in the benzene ring affects the velocities of chemical reactions by indicating how the fields of group dipoles aid or hinder the approaching reactant ion.

With respect to the influence of the solvent on reaction rates, our chief concern is the evaluation of the energy of activation E_u for the reaction between an ion and an unsubstituted polar molecule. The problem has been approached [74] from the viewpoint of Ogg and Polanyi's [78] adaptation of London's treatment to the problem of substitution, with optical inversion at a saturated carbon atom. A mechanistic representation can be written:

$$Y^- \; + \; \Big(\overset{H}{\underset{H-H}{\diagdown}}\Big)\!\!\!>\!C-X \longrightarrow \left[\, Y\cdots \Big(\overset{H}{\underset{H_-H}{}} C\Big) \cdots X \,\right]^-$$

$$(5.218)$$

$$Y-C\!\!<\!\!\Big(\overset{H}{\underset{H_-H}{}}\Big) \; + \; X^-$$

The solvated ion Y^- at infinite distance from dipolar substituted methane molecule in the ground electronic and vibrational state approaches in the direction indicated because of the nature of the ion–dipole interaction. The hydrogen and carbon atoms are coplanar in the critical complex. The Morse function represents the energy of the covalent bonds and the Born–Heisenberg function represents the energy of the ion–dipolar interactions that occur between the ion and the solvent-molecules as well as those between the ion and reactant molecule. The energies of activation for the iodide ion–methyl fluoride and iodide ion–methyl bromide reactions were calculated to be 31.6 and 26.5 kcal/mole, respectively. The energies of activation for the latter reaction have been found to be approximately 18.3, 18.3, and 14.3 kcal/mole in water, methanol, and acetone, respectively. For the complete methyl halide series reacting with hydroxide ion, the observed energies of activation are not indicative of the strengths of the covalent bonds which are being broken. It has been observed [79] that the energy of activation was less for the methyl fluoride than for the methyl iodide reaction. Calculations of complete energy surfaces show that in the gas phase, the energy of activation is zero for the hydroxide ion–alkyl halide reactions [74]. These extremely rapid reactions in the gas phase are slowed down by solvent to the point where they are measurable in solution. Moelwyn-Hughes explores the hypothesis that, in the case of anions reacting with dipolar molecules, the energy of activation applies to the escape of the ion from its solvent sheath. The reorientation and rearrangement of the solvent molecules solvating the ion accounts for most of the energy of activation, and its magnitude depends on the intrinsic properties of the ion and solvent, though it is modified by the polar molecule with which the ion reacts during or following its escape from the solvation shell.

If c is the number of molecules of solvent normally coordinated with each ion or molecule of solute, and if ϕ be the energy of one solute–solvent interaction, then the concentration n^* of solute molecules which are deficient by one solvent molecule if given approximately by the Boltzmann equation:

$$\frac{n^*}{n} = \frac{\exp[-(c-1)\phi/kT]}{\exp(-c\phi/kT)} = \exp(\phi/kT) = \exp(-E'/kT) \quad (5.219)$$

where n is the total concentration of solute and E' is the energy required to remove one solvent molecule from the solvation shell of one solute molecule. According to Moelwyn-Hughes the reaction rate, when reaction takes place upon collision between an ion and a solute molecule each of which is deficient by one molecule of solvent solvation, is given by

$$-dn/dt = zn_1^* n_2^* = zn_1 n_2 \exp[-(E_1' + E_2')/kT] \quad (5.220)$$

where z is the collision frequency, and the energy of activation E' is

$$E' = E_1' + E_2' \quad (5.221)$$

Properties of the reactants and presumably of the solvent used would have to be known in order to calculate E_1' and E_2'. Moelwyn-Hughes has found such calculations encouraging, and believes that the difference between the energies of activation of hydrogen ion and hydroxide ion catalysis of esterification is mainly due to the difference in the heats of hydration of these ions. The difference in these energies of activation is roughly equal to the difference in the heats of hydration divided by the coordination number.

It is evident that the energies of reorientation and rearrangement of solvent molecules around reactant particles and the heat of solvation of reactant species are important in explanations of reaction rates. Such a process would no doubt demonstrate large specific effects arising from selective solvation of either or both reactant species by one component of a mixed solvent.

It has been pointed out [74] that ion–dipolar molecule reactions show greater regularity than other types of reactions in solution. Depending on the ion and dipole reactants selected and on the solvent used, the rates in solution of certain second-order ion–dipolar molecule reactions may differ enormously. In such cases the difference in rates may arise principally from variations in the energy of activation E, while the Arrhenius frequency factor A in the equation for the second-order velocity constant, namely $k_2' = A \exp(-E_A/RT)$, may remain almost constant. The case of the catalytic chlorination of certain ethers of the type ROC_6H_4X by Jones [80] has been cited [74]. In these the rates, depending on the nature of R and X, varied by 3300. The frequency factor A remained constant.

Since E is so strongly and in so many aspects solvent dependent, it is obvious that rates would depend markedly upon the solvent.

The real difficulty of applying electrostatic theory to ion–dipolar molecules, Benson [81] indicates, is that the Coulombic term is of the same order of magnitude as the difference in free energy of hydration of the ion A^{Z_A} and the transition complex X^{Z_A}. He suggests that a more realistic model of the transition complex might result from considering it as the product of the displacement in the solvent sheath of the ion of a solvent molecule of dipole moment μ_s by a reactant molecule of dipole moment μ_B.

This procedure, neglecting dipole-dipole interactions, would give the free energy of formation of the transition state as

$$F_\mu = \frac{Z_A \, \varepsilon \mu_B \, \cos \theta}{D r_B{}^2} \left(1 - \frac{\mu_s}{\mu_B} \frac{r_B{}^2}{r_s{}^2} \right) \qquad (5.222)$$

This equation indicates that the dipole moments of the dipolar reactant and of the solvent molecules and the distance of closest approach of the dipolar reactant and of the solvating solvent molecules of the ion are related to the effect of the dielectric constant on the reaction rate. For $\mu_s/r_s{}^2 = \mu_B/r_B{}^2$ the dielectric constant would not affect the rate since F_μ would be zero. If $\mu_s/r_s{}^2$ and $\mu_B/r_B{}^2$ are different, opposite effects would be expected depending on which is greater.

Since solvation or coordination is not the whole story of the effects of solvents on the rates of reaction in solution, the last equation cannot be comprehensive or all inclusive. To the extent that such a theory does govern rates, it would seem that the proposed model would give rise to pronounced specific effects in mixed solvents if there were selective solvation by one of the solvent components.

V. Electrostatic Effects on Dipolar Molecule–Dipolar Molecule Reactions

A. INTRODUCTION

The electrostatic effects exerted through the dielectric constant of the solvent on dipolar molecule–dipolar molecule reactions will be much less than on ion–ion reactions and considerably less than on ion–dipolar molecule reactions. This is true because Coulombic forces between dipolar molecules are very much smaller than those between ions and markedly smaller than those between ions and dipolar molecules. Nevertheless, the coulombic forces between dipolar molecules are significant and must be treated in any comprehensive presentation of electrostatic effects on reaction rates. Because of their small magnitude electrostatic effects, arising from changes in the dielectric constant of the solvent, on rates of reaction between dipolar molecules, compared to other charge type reactants already treated, will tend to be more readily obscured by specific solvent, solvation, and structural effects, which in themselves may be difficult or impossible to elucidate.

B. LAIDLER AND EYRING EQUATION

Although, except as a rough approximation, van der Waals forces in the case of dipolar substances cannot be neglected, Laidler and Eyring [47] believe that electrostatic forces are somewhat stronger than nonelectrostatic ones when strongly polar molecules are involved, and treated certain reactions involving certain strong dipoles in terms of electrostatic dipolar forces. Kirkwood's [49] expression for the transfer of a strong dipole of moment μ from a vacuum to a medium of dielectric constant D was employed. This expression for a molecule of radius a and a symmetrical charge distribution, when a nonelectrostatic term ϕ is included, can be written.

$$\Delta F = kT \ln \beta = -\frac{\mu^2}{a^3}\left[\frac{D-1}{2D+1}\right] + \phi \tag{5.223}$$

For low dielectric constant media it has been shown [82] that Eq. (5.223) cannot be applied. If electrostatic forces are predominant and the solutions of polar molecules are fairly concentrated, the equation is very useful [47]. Satisfactory straight lines with reasonable slopes have been obtained [82–87] by plotting the logarithms of the activity coefficients of various polar substances against $(D-1)/(2D+1)$. Deviations from linearity occurred only at low dielectric constants.

Using the bimolecular reaction

$$A + B \rightarrow M^* \rightarrow X + Y \tag{5.224}$$

and applying Eq. (5.223) the equation for the specific velocity becomes

$$\ln k' = \ln\left(\varkappa\frac{kT}{h}K_0{}^*\right) - \frac{1}{kT}\frac{D-1}{2D+1}\left[\frac{\mu_A^2}{a_B^3} + \frac{\mu_B^2}{a_B^3} - \frac{\mu_{M^*}^2}{a_{M^*}^3}\right] + \frac{\phi_A + \phi_B - \phi_{M^*}}{kT} \tag{5.225}$$

here D is the dielectric constant of the final solution formed and is effectively the dielectric constant of the pure solvent for dilute solutions.

This equation would predict that, if nonelectrostatic terms are neglibly small, a plot of $\ln k'$ against $(D-1)/(2D+1)$ should give a straight line. Such plots were made [47] for the quaternary ammonium salt formation at 29°C in alcoholbenzene mixtures, for the acid hydrolysis of ethyl orthoformate [$H_2O + HC(OEt)_3$], the alkaline hydrolysis of ethyl benzoate

$[H_2O + PhCOOEt]$ and the water hydrolysis of tertiary butyl chloride $[H_2O + (CH)_3CCl]$, the last three in ethanol–water mixtures. It is agreed that, from the standpoint of solvent influence, ester hydrolysis should be considered as a reaction between the two dipoles, ester and water. While the plots for the data in ethanol–benzene mixtures at 29°C for the reactions of trimethylamine and benzyl bromide and of pyridine and benzyl bromide to give the corresponding quaternary ammonium salts were linear, the data for the same reactions in benzene–nitrobenzene solvent were curved considerably. In nitrobenzene rich solvents the reactions were much faster than predicted by theory. In quaternary ammonium salt formation specific solvent effects are common. Thus the triethylamine–ethyl iodide reaction in a variety of solvents showed rates and dielectric constants that were not in the same order [88]. However, there is some correlation between the two as shown by Menschutkin [89] who pointed out that the reaction mentioned above does tend to be accelerated by solvents of high dielectric constants. Laidler and Eyring attributed the deviations to the nonelectrostatic contributions ϕ to which the solvating power of the solvent makes the most important contribution. In nitrobenzene the quaternary salts are soluble and ionized [90, 91] but are only slightly soluble in benzene. Solutions of amines and iodides are probably nearly ideal. The activated complex in quaternary salt formations, while not ionic, has a large dipole moment, and hence approximates the salt in general properties [92]. The complex is perhaps selectively solvated by the highly polar nitrobenzene in solutions containing this compound as the solvent or as a solvent component in conjunction with less polar substances. In such solvents, the complex has a low activity coefficient, and the velocity of the reaction is therefore high. In terms of the ϕ function, ϕ_A and ϕ_B may be almost normal in nitrobenzene-containing solvents, while ϕ_{M*} may be small and hence the rate of reaction fast. The product of the reaction and hence probably the activated complex have low solubility in benzene. Thus β_{M*} and ϕ_{M*} would be large and, hence the reaction rate in this solvent would be slow. Since the Menschutking reaction will not take place in the gas phase, even the solvent benzene must effect some sort of a stabilization of the highly polar or partially ionized complex.

For the above theory to be applied, activity coefficients of reactants and complexes are necessary but meager, especially for the complex.

C. COULOMBIC ENERGY APPROACH

In the Coulombic energy approach [93] it is assumed that the effect of the dielectric constant on the rate is given by the equations which follow, where all powers higher than the second of distances of separation of charge centers in dipolar molecules have been neglected. For the dipolar reactants the coulombic energy E_c is

$$E_c = \frac{2\mu_1\mu_2 \cos\theta_1 \cos\theta_2}{Dr^3} + \frac{\mu_1\mu_2 \sin\theta_1 \sin\theta_2}{Dr^3} \qquad (5.226)$$

where μ_1 and μ_2 are dipole moments of the respective dipolar molecules, and for head-on alignment

$$E_c = 2\mu_1\mu_2/Dr^3 \qquad (5.227)$$

Applying this restricted equation, the difference in Coulombic energy at two different dielectric constants D_1 and D_2 is

$$\Delta E_c = \frac{2\mu_1\mu_2}{r^3}\left(\frac{1}{D_2} - \frac{1}{D_1}\right) \qquad (5.228)$$

The energy of activation for the dipole–dipole reaction at dielectric constant D_2 can be related to that at D_1 by the expression

$$E_{D_1} = E_{D_2} + \Delta E_c \qquad (5.229)$$

The rate constant is related to the energy of activation per mole by the Arrhenius equation, which for the medium of dielectric constant D_2 can be expressed

$$k'_{D_2} = Z' \exp\left(-\frac{E_{D_2}}{RT}\right) \qquad (5.230)$$

and

$$\ln k'_{D_2} = \ln Z' - \frac{E_{D_2}}{RT} = \ln Z' - \frac{E_{D_1}}{RT} - \frac{\Delta E_c}{RT} \qquad (5.231)$$

where ΔE_c is the coulombic energy per mole engendered by changing the dielectric constant from D_1 to D_2. In Eqs. (5.230) and (5.231), Z' is the Arrehius frequency factor. But

$$\ln k'_{D_1} = \ln Z' - \frac{E_{D_1}}{RT} \qquad (5.232)$$

Therefore

$$\ln k'_{D_2} = \ln k'_{D_1} - \frac{\Delta E_c}{RT} \qquad (5.233)$$

and substituting into Eq. (5.233) the value of ΔE_c from Eq. (5.228), we have

$$\ln k'_{D_2} = \ln k_1' - \frac{2\mu_1\mu_2 N}{RTr^3}\left(\frac{1}{D_2} - \frac{1}{D_1}\right) \qquad (5.234)$$

Selecting D_2 as a reference dielectric constant with a value of infinity and letting D_1 become any other dielectric constant D, we have for Eq. (5.234)

$$\ln k_\infty' = \ln k_D' + \frac{2\mu_1\mu_2}{kTDr^3} \tag{5.235}$$

since $N/R = 1/k$. Therefore, the velocity constant k_D' at any dielectric constant by the equation

$$\ln k_D' = \ln k_\infty' - \frac{2\mu_1\mu_2}{kTDr^3} \tag{5.236}$$

where k in Eqs. (5.235) and (5.236) is the Boltzmann gas constant.

To the extent that the assumptions made in deriving Eq. (5.236) are valid, and to the extent that the moments μ_1 and μ_2 of the dipolar reactants in vacuum can be used to represent their moments in solution, Eq. (5.236) should for a plot of $\ln k_D'$ versus $1/D$ give a straight line, and the slope $(2\mu_1\mu_2)/(kTr^3)$ should yield a reasonable value of $r = r_A + r_B$, the distance of approach for the two dipoles, A and B to react. The distance r should have the magnitude of a molecular dimension.

In Fig. 5.8 a plot is presented of $\log k_D'$ versus $1/D$ for the reaction between the dipolar reactants water and tertiary butylchloride studied by Hughes [94]. The scatter of points is perhaps due to lack of precision. Equating the slope of the line to $-(2\mu_1\mu_2)/(kTr^3)$ as required by Eq. (5.236) yields a value of r of 0.75 Å when the dipole moments of water and of tertiary butyl chloride are taken as 1.84 and 2.10 D.U. [95], respectively.

The value of r is low by a factor of perhaps 3 or 4, but is nevertheless gratifying because of the assumptions made in deriving Eq. (5.236), its simplicity, and the readiness with which its predictions are visualized.

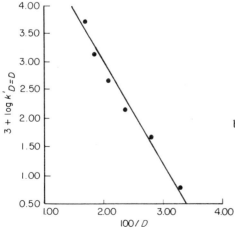

Fig. 5.8. The reaction of tertiary butyl chloride with water.

The mechanism must, from these electrostatic considerations, show the rate-controlling step to be a dipolar molecule reacting with a dipolar molecule and could be step 1 presented in the following mechanism:

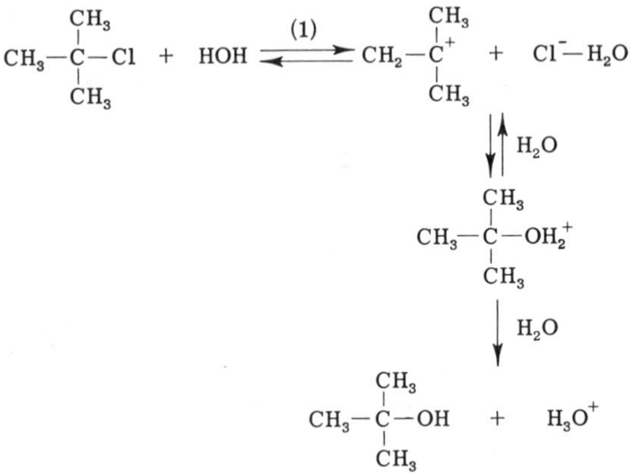

$$
\begin{array}{c}
\underset{\underset{CH_3}{|}}{\overset{\overset{CH_3}{|}}{CH_3-C-Cl}} + HOH \underset{(1)}{\rightleftharpoons} \underset{\underset{CH_3}{|}}{\overset{\overset{CH_3}{|}}{CH_2-C^+}} + Cl^- - H_2O
\end{array}
$$

$$ \updownarrow H_2O $$

$$
\underset{\underset{CH_3}{|}}{\overset{\overset{CH_3}{|}}{CH_3-C-OH_2^+}} \qquad (5.237)
$$

$$ \downarrow H_2O $$

$$
\underset{\underset{CH_3}{|}}{\overset{\overset{CH_3}{|}}{CH_3-C-OH}} + H_3O^+
$$

The plot of data according to Eq. (5.225) by Laidler and Eyring [47] showed somewhat better precision than that in Fig. 5.8. In a quantitative test of Eq. (5.225), however, the moments and radii of both reactants and complex would have to be known.

The hydrolysis of the diacid chlorides of phthalic and of terephthalic acids in water dioxane mixtures have been studied [96]. In both reactions the rate was first order with respect to each reactant and second order overall. The Kirkwood relationship for the increase of the rate for reaction between dipolar molecules with increased polarity of the medium was in accord with the data, and resulted in dipole moments of 6.85 and 6.95 D.U. for the activated complexes of the diacid chlorides of phthalic and terephthalic acids respectively.

It has been pointed out [74] that if the energy of activation between two polar molecules includes a coulombic term $E_c = \mu_1\mu_2/Dr^3$, then the energy of activation should vary linearly with $\mu_1\mu_2$ for a series of chemically similar reactions, when the solvent and temperature remain the same. The apparent energy of activation should vary linearly with $(1 - LT)/D$ for a given reaction measured in a variety of solvents of different dielectric constants at a fixed temperature. Moelwyn-Hughes points out that both these and other consequences of electrostatic theory have been confirmed for reactions between molecules in solution. The extent to which coulombic energy is involved in the formation of the critical complex, and the interpretations of the other components of energy of activation are furthered by the magnitudes of electrostatic effects. It is the coulombic contribution to the energy of activation that we have quantitatively related to the rates of dipolar molecule–dipolar molecule reactions, the electrostatic interactions among reactant molecules are of the same order of magnitude as are van der Waals interactions [97]. It is thought to be grossly oversimplifying to neglect the van der Waals interactions in the case of such reactants and to throw the behavior of solutions wholly on the influence of dielectric constant. Benson questions whether the correlation between the rate of reaction of dipolar molecule–dipolar molecule reactants and the dielectric constant of the solvent can contribute markedly to either the elucidation of the nature of the transition state complex or the theory of solutions. He feels that more interesting information would probably result without introducing more complexity using a detailed molecular model for the solution involving only near-neighbor interactions. The dipole moments and the radii of the solute and solvent species would be the only parameters for such a model.

A perusal of *Chemical Abstracts* reveals that a vast number of applications of electrostatics to reaction rates are constantly being made. Thus, while there is a large body of investigators of phenomena, who believe the electrostatic approach to the theoretical elucidation of chemical phenomena in solution is inadequate due to the complexity of these phenomena, yet much can be done with electrostatic forces among different charge types of chemical particles to clarify their behavior in solution. This behavior appertains not only to chemical reaction rates but to other phenomena as well, such as conductance and electrode potentials.

In the present treatment of the effect of electrostatics on reaction rates, an exhaustive listing of applications of the theories presented would be superfluous. The purpose here was to make the reader aware of these theories, give examples where the theories have been successfully applied, and prepare the reader to recognize these solvent influences and to be familiar with the approaches used in the explanation of such phenomena. It is also the intention of the authors to make the reader aware of the causes of deviation from theory. These causes include ion-pairing, solvation of reactants, selective solvation of reactants, solvolysis, lack of knowledge of the microscopic dielectric constant in the intermediate vicinities of reactant particles, and many other phenomena. Some of these effects have already been discussed or will be presented in this treatment of solvent effects on chemical phenomena. It might be well in this closing section on electrostatics and reaction rates to discuss briefly the microscopic as contrasted to the macroscopic dielectric constant.

VI. Dielectric Constant and Dielectric Saturation

The macroscopic dielectric constant is roughly an average value arrived at from measurements on a large group of molecules. The microscopic dielectric constant is the dielectric constant in the vicinity of a particular solute particle due to the partial or complete orientation of the solvent molecules by the field of the solute particle. If the solute particle is an ion, the electric field in its vicinity is high and solvent molecules approach complete orientation, termed dielectric saturation, in which state all orientation polarization vanishes and the dielectric constant has a low value comparable to that of nonpolar molecules such as hydrocarbons and other molecules of perfect charge symmetry.

Frank and co-workers [98] have treated thermodynamic and other properties of solutions by assuming that the solvent has molecular structure and considering in detail the interactions between the ions and the individual solvent molecules. They have shown that next to an ion in aqueous solution there is a fixed shell of molecules having low entropies and volumes, and constituting a hydration shell. Between this shell and bulk solvent, there is a disordered zone of solvent molecules, resulting from the competitive forces on these molecules of the attracting ion and the hydrogen bonding pull of the bulk water molecules. The water molecules in these abnormally disordered zones, have greater entropy than do those in the remainder of the solution. These ideas do not seem to have been applied to rates of reaction in solution; though they have been used to calculate the energies between two ions in aqueous solution by taking into account explicitly the hydration shell and treating the remainder of the solvent as continuous.

A second approach used in accounting for the microscopic dielectric constant in the vicinity of an ion has been to treat the dielectric constant as a function of the distance from an ion quantatively in terms of the variation of the dielectric constant with field strength [53, 99]. These calculations were made for simple ions ranging in valence from 1 to 4. The calculations showed near the ions, irrespective of valence, the dielectric constant had a value of the square of the refractive index, n^2, namely, 1.78 and rose steeply with distance from the ion to the value of the pure solvent, namely 78.5 for water at 25°C. The lower the valence of the ion, the shorter the distance from the ion at which the minimum value of the dielectric constant was reached. For all valences of the ions, the curves were similar in shape and were not unlike the shape of potentiometric titration curves when strong acids are titrated with strong bases. Especially at short distances, due to this decrease in dielectric constant, the electrostatic forces between ions in water solutions are markedly different from what they would be if the dielectric constant were throughout the solution uniformly that of the pure solvent. This effect also magnifies the free energy necessary for bringing ions together particularly to within a few angstroms of each other. Sacher and Laidler [100] have applied this method of calculating free energy increase to electron exchange reactions, but application to bond-making and bond-breaking reactions apparently have not been made. The distances from ions for which dielectric saturation prevails ranges from about

2 Å for univalent ion to about 4 Å for a tetravalent ion. The dielectric constant reaches its maximum values at about 71 and 14 Å, respectively, for these two-charge types of ions.

Laidler thinks that the two approaches will lead to quite similar interpretations since the low dielectric constant region and layer of frozen water molecules will coincide approximately. The low dielectric constant approach may be easier to apply and may prove more fruitful in ion–ion kinetics.

Approximate theories of solutions may yield satisfactory results so far as rates of reactions, free energies of activation, and free energies of reaction are concerned, but may give less satisfactory results when applied to energies of activation and frequencies, in which cases more accurate and improved methods may be required. The reason for this conclusion is that enthalpy and entropy changes tend to compensate each other. Thus a change in a parameter which may cause a small change in free energy of activation, may produce a large change in both enthalpy and entropy.

Thus Indelli and Amis [40] resorted to energy of activation to detect nonelectrostatic effects produced by adding various salts over wide concentration ranges to ionic reactants in solution.

VII. Protic and Dipolar Aprotic Solvent Effects

A. INTRODUCTION

Parker and co-workers [101, 102] have found bimolecular reactions between anions and polar molecules, especially bimolecular nucleophilic substitution reactions ($S_N 2$ reactions) at saturated carbon atoms, to be much faster in dipolar aprotic solvents than in polar solvents. The bimolecular reaction can be visualized as in Eq. (5.128), generalized by using R- groups for the hydrogen. Protic solvents which are hydrogen bond donors include such solvents as water, alcohols, ammonia, amides, N-alkly amides, and carboxylic acid. Dipolar aprotic solvents are not hydrogen-bond donors and are typified by such substances as dimethylsulfide, acetone, 2,3-dioxane, dialkyl amides (as dimethylacetamide and dimethylformamide), nitrobenzene, nitromethane, sulfur dioxide, and acetonitrile. The one-step general $S_N 2$ reaction can take place through transition states of various charge types, e.g., $+1, 0, -1, -2, -3$, which are equal to, more than, or less than the charge on the reactants. In Eq. (5.218), Y^- can be a halogen, SCN^-, N_3^-, $S_2O_3^{2-}$ or other anions or neutral molecules as NR_3, R_2S, C_5H_5N, $C_5H_{10}NH$, or other neutral molecules; and the substrate can be an alkyl halide, 4-nitrofluoro-benzene, trimethylsulphonium cation ($CH_3^+SMe_2$), *trans*-dichlorobis(piperi-dene)platinum(II) {*trans*-[Pt(pip)$_2$Cl$_2$]}, and other substances.

B. FUNDAMENTAL CONSIDERATIONS

It is accepted [101] that the influence of the solvent on reaction rates is determined by the difference in the free energies, enthalpies and entropies of solvation of the reactants and of the transition states. This approach to solvent effects on rates is based on the fact that the specific rate of a chemical reaction depends on the difference of standard free energy of the reactants and of the standard states and has had wide application by many workers in the field of kinetics [95, 103–107].

From the absolute rate theory the equation for the reaction is

$$A + B \rightleftharpoons X^* \rightarrow \text{products} \tag{5.238}$$

with the velocity v of the reaction being proportional to the concentration of the complex $[X^*]$

$$v \propto [X^*] \tag{5.239}$$

where $[X^*]$ is given by

$$[X^*] = K^*[A][B]\frac{\gamma_A \gamma_B}{\gamma^*} \tag{5.240}$$

Therefore

$$v \propto K^*[A][B]\frac{\gamma_A \gamma_B}{\gamma^*} \tag{5.241}$$

and the specific velocity constant k' is given by

$$k' \propto \frac{v}{[A][B]} \propto K^*\frac{\gamma_A \gamma_B}{\gamma^*} \tag{5.242}$$

For ideal systems

$$k_0' \propto K^* \tag{5.243}$$

and the rate constant in solution in general can be related to that in the ideal solutions by the equation

$$k' = k_0'\frac{\gamma_A \gamma_B}{\gamma^*} \tag{5.244}$$

The activity coefficient γ relates the behavior of a solute species to its behavior in ideal solution, and k_0' is the specific velocity constant for the reaction in ideal solution.

Laidler [104] states that the rate of reaction is pronouncedly influenced by the degree of solvation of the reactants and activated complex. If the activated complex is relatively much more solvated than the reactants, the activity coefficient of the complex is much less than in solvents in which it is not solvated and by Eq. (5.244) the rate of reaction is much greater in the former solvents. Thus the intermediate complex in quaternary ammonium salt formation is highly solvated by nitrobenzene but not by benzene, and the specific velocity constant for reaction between triethylamine and ethyl iodide at 100°C is 70.1 in nitrobenzene and only 0.0058 in benzene. In this reaction the activated complex probably is polar like the product and is therefore considerably solvated by nitrobenzene but not by benzene. For a solvent which solvates reactants to a higher degree than it solvates activated complexes, the reaction will take place more slowly than in a solvent that does not solvate the reactants. Laidler [104] cautions that the preceding analysis of the relationship between solvation and reaction rates holds strictly only for reactions in an inert solvent which has little effect upon the kinetic behavior. He points out that ionizing power of the solvent, electrostatic forces, and other solvent influences may predominate in solvent influence on reaction rates.

Parker [101, 108] in considering solvent effects on reaction rates starts with solvent activity coefficients, $^{o}\gamma_i^{s}$, for reactants, transition states, and products. Parker defined solvent activity coefficients such that $^{o}\gamma_i^{s}$ represents the change in the standard chemical potential $\bar{\mu}_i$ of the solute i upon transfer from an arbitrarily chosen reference solvent O to other solvents. The solute i is hypothetically ideal with respect to Henry's law unimolar solution. Thus

$$\bar{\mu}_i^{s} = \bar{\mu}_i^{o} + RT \ln {^{o}\gamma_i^{s}} \tag{5.245}$$

When solvent activity coefficients were applied to reaction rates in terms of the absolute rate theory, an equation entirely similar to Eq. (5.244), but in terms of the solvent activity coefficients $^{o}\gamma_i^{s}$ was obtained:

$$k^{s} = k^{o} \frac{^{o}\gamma_{A^+}^{s} \; ^{o}\gamma_{B^-}^{s}}{^{o}\gamma_{X^*}^{s}} \tag{5.246}$$

In the quantitative application of Eq. (5.246) to solvent effects on rate of reactions, it is necessary that the values of $^{o}\gamma_i^{s}$ for anions, cations, molecules, and transition states be known or calculable. At present very few measurements of such quantities, especially for organic substances, are available. Parker shows how estimates of $^{o}\gamma_i^{s}$ may be made and how they may be applied by physical organic chemists.

For S_N2 reactions both reactants and transition states are considered as anions, cations, or polar molecules. In considering the effect of the transfer from protic to dipolar aprotic solvent of the reaction represented in Eq. (5.218) the changes in the solvation of all species must be taken into account. While the solvation energies of ions is a power of 10 or more larger than those of molecules, yet the difference in solvation energies upon transfer ($RT \ln {^{o}\gamma_i^{s}}$)

of the solutes from one solvent to another may be greater in the case of a polar molecule than in the case of an ion [107, 109]. Parker cites examples of $^{o}\gamma_i{}^{s}$ of 10^2 or greater for dipolar molecules.

For obtaining difference in rates of S_N2 reactions in protic and dipolar aprotic solvents, the reference solvent is chosen by Parker as the dipolar aprotic dimethylformamide (DMF) and the standard state is taken as the hypothetically ideal unimolar solution in DMF at 25°C. Unless otherwise designated, this is the standard state to which all solvent coefficients $^{o}\gamma_i{}^{s}$ are referred.

Polar molecules, especially those which are not hydrogen-bond acceptors, have been found [101, 110, 111] to be more soluble in dipolar aprotic solvents such as DMF and dimethylsulfoxide (DMSO) than in protic hydrogen bonding solvents such as water, methanol, and formamide.

The effect of nonhydrogen-bond donor or acceptor polar solvents or the structure of hydrogen-bond acceptor and donor protic solvents produces a somewhat different standard chemical potential from what might have been expected from electrostatics alone [101, 107]. Highly polar aprotic solvents do exert dipole–dipole interactions among individual molecules, but the molecules are not as strongly interacting as are solvent molecules of hydrogen bond donor and acceptor protic solvents. Polar molecules are more soluble and more solvated in dipolar aprotic than in protic solvents since dipolar aprotic solvents permit the incorporation of polar solutes in their structure more readily than do protic solvents.

Dividing both sides of Eq. (5.246) by k^0 and multiplying through by $C_{A^+}^s C_{B^-}^s / C_{A^+}^o C_{B^-}^o$ gives us

$$\frac{k^s C_{A^+}^s C_{B^-}^s}{k^o C_{A^+}^o C_{B^-}^o} = \frac{\text{rate}^s}{\text{rate}^o} = \frac{C_{A^+}^s \, {}^{o}\gamma_{A^+}^s + C_{B^-}^s \, {}^{o}\gamma_{B^-}^s}{C_{A^+}^o C_{B^-}^o \, \gamma_{X*}^s} \tag{5.247}$$

or

$$\frac{\text{rate}^s}{\text{rate}^o} = \frac{a_{A^+}^s + a_{B^-}^s}{C_{A^+}^o C_{B^-}^o} \cdot \frac{1}{{}^{o}\gamma_{AB*}^s} \tag{5.248}$$

For all saturated solutions of salt AB in equilibrium with the same crystalline form of solid AB, the chemical potential of salt AB is the same, but in the standard solution, $C_i{}^o = a_i{}^o$, and since both solutions are saturated, $C_i{}^o = a_i{}^o = a_i{}^s$. Therefore,

$$\frac{\text{rate}^s}{\text{rate}^o} = \frac{1}{{}^{o}\gamma_{AB*}^s} \tag{5.249}$$

By measuring the rate of the reaction in solvent S and in standard solvent O where all solutions are saturated and at the same temperature, the solvent activity coefficient for the transition state of an S_N2 reaction can be found.

Evans and Parker [101, 112] have applied this to the determination of

$$D_\gamma E_{(BrCH_3SMe_2)^*} = 35.7, \qquad D_\gamma DMAC_{(BrCH_3SMe_2)^*} = 0.73, \qquad and$$

$$D_\gamma N_{(BrCH_3SMe_2)^*} = 1.0$$

for the $S_N 2$ decomposition of trimethylsulphonium bromide at 25°C, where D, E, DMAC, and N represent the respective solvents dimethylformamide, ethanol, dimethylacetamide, and nitromethane, dimethylformamide being the reference solvent. The respective γ values indicate that the transition state is about 2 kcal/mole more solvated in the dipolar aprotic solvents D, DMAC, and N than in the protic solvent E. For the back reaction which is one between two polar molecules with the same transition state as the forward reaction, the product of the activity coefficients $D_\gamma E_{CH_3Br} \cdot D_\gamma E_{CH} = 36$ which is in agreement with the finding that the back reaction has about the same rate in ethanol as in dimethylformamide [99]. The $S_N 2$ mechanism for the forward and back reactions were written [99]:

$$CH_3 \overset{+}{-} \overset{}{S}(CH_3)_2 + Br^- \; \rightleftharpoons \; \overset{\delta -}{Br} \cdots \overset{H}{\underset{H}{C}} \cdots \overset{\delta +}{S}(CH_3)_2 \; \rightleftharpoons \; CH_3Br + (CH_3)_2 S \qquad (5.250)$$

Reactions of the $S_N 1$ type are less suitable than the $S_N 2$ type for interpretation in terms of solvent activity coefficients and to correlate using free energy relationships since the $S_N 1$ type reaction shows greater flexibility in the nature of the transition state due to the fact that the process involves more than one step.

The nature of the $S_N 1$ transition state in solvolysis reactions for example, can be somewhat elucidated [101] by measuring the solvent activity coefficients of reactants RX, and then from the rate data estimating the solvent activity coefficient for the polar transition state $\overset{\delta +}{R} \cdots \overset{\delta -}{X}$.

Much solvolysis data has been correlated by free energy relationships such as the Grunwald-Winstein [113] Y-relationship, which Parker [101] writes in terms of solvent activity coefficients as

$$\log \frac{^0 \gamma^S_{RX}}{^0 \gamma^S_{RX^*}} = m \log \frac{^0 \gamma^S_{t-BuCl}}{^0 \gamma^S_{t-BuCl^*}} \qquad (5.251)$$

for the process

$$RX \rightleftharpoons [^{\delta +}R \cdots ^{\delta -}X] \rightleftharpoons \underset{\text{ion pair}}{R^+ \cdot X^-} \rightleftharpoons R^+ + X^- \xrightarrow[\text{fast}]{SH} RS + HX$$

$$(5.252)$$

The reference solvent was 80% ethanol by volume and 20% water by volume.

The rate constants in solvents S_1 and S_2 are related [101] by the equation

$$\frac{k_{S_1}}{k_{S_2}} = \frac{{}^{S_2}\gamma_{RX}^{S_1}}{{}^{S_2}\gamma_{RX*}^{S_1}} \tag{5.253}$$

which would not hold if the process were not purely $S_N 1$ but depended on solvent nucleophilicity [101, 114]. Parker emphasizes the shortcomings of this approach to $S_N 1$ type reactions. These include the fact that although ${}^{S_2}\gamma_{RX}^{S_1}$ can be found from Henry's law constants and Eq. (5.253) yields an activity coefficient for some transition state, this transition state is not necessarily the simple one corresponding to the ionization of RX. It is emphasized that regardless of the nature of k_{S_1}, there is no surety that it is related to the one-step ionization of RX. Then ion–pair return and solvent nucleophilicity causes the solvolysis rate to be less than the ionization rate. Also, the greater flexibility in the nature of the transition state in the $S_N 1$ as compared to the $S_N 2$ type of reaction makes the $S_N 1$ type of reaction to be less suited to free energy relationships between different reactions and to interpretation in terms of solvent activity coefficients.

C. ANIONS—THEIR SOLVATION AND REACTION WITH POLAR MOLECULES

Parker [101, 110, 111, 115, 116] discusses solvation of anions in hard protic solvents that hydrogen-bond with small anions and soft, dipolar aprotic, highly polarizable solvents that demonstrate mutual polarizability with large polarizable anions. Hard [117, 118] solvents show strong interactions with hard anions and soft [117, 118] solvents show strong interactions with soft anions.

A comparison is made of anion solvation in the protic solvent methanol and the dipolar aprotic solvent dimethylformamide, the dielectric constants of which are approximately equal, being 32.6 and 36.7, respectively, at 25°C. From the Born equation which gives the electrostatic contribution $\Delta \bar{\mu}^{el}$ to the free energy of solvation per mole of ion, it would be concluded that the solvation of an anion would be the same in MeOH and DMF. The model for the derivation is for a sphere of charge $Z\varepsilon$ and radius r to be transferred from a vacuum ($D = 1$) to a medium of uniform dielectric constant D. Thus

$$-\Delta \bar{\mu}^{el} = \frac{N Z^2 \varepsilon^2}{2r}\left(1 - \frac{1}{D}\right) \tag{5.254}$$

According to this equation for solvents of the same D and ions of the same $Z e$, the electrostatic contribution to the free energy of solvation per mole of ion should be the same for all ions and all solvents. However, chemical interactions, i.e., specific solvent effect such as mutual polarizability or hydrogen bonding may produce solvation energies greater than those calculated wherever and whenever the specific effects occur. Hard protic solvents will show greatest hydrogen bonding interactions with small hard anions F^-, OH^-, and Cl^- with high charge density, and least hydrogen bonding with large soft anions I^-, I_3^-, SCN^-, with low charge density. Mutual polarizability is greatest for these latter ions with either dipolar aprotic solvents or with polarizable protic solvents. For an anion A^- the change in chemical potential on transfer from DMF to MeOH can be expressed in terms of the solvent activity coefficient $^D\gamma_{A^-}^M$. Thus

$$\bar{\mu}_{A^-}^M - \bar{\mu}_{A^-}^D = RT \ln {}^D\gamma_{A^-}^M \qquad (5.255)$$

While it may never be possible to evaluate without some misgivings [119], the individual solvent activity coefficients of anions and cations, the problem has been approached by making certain extrathermodynamic assumptions which have some reasonable foundations [101, 120–126]. Three of the assumptions are:

1. Ferrocene and ferricinium cations are both complexed by the same ligand. The complexes are spherical, symmetrical, and of comparable size. Both solutes are assumed to be affected in the same manner by transfer from one solvent to another. Thus

$$\frac{{}^o\gamma_{\text{ferricinium}}^s}{{}^o\gamma_{\text{ferrocene}}^s} = \frac{{}^o\gamma_{F^+}^s}{{}^o\gamma_F^s} = 1 \qquad (5.256)$$

2. It has been suggested [125, 126] that large, unit-charged, symmetrical, nonpolarizable cations, e.g., ceasium cation, are solvated to the same extent in all solvents, especially if the solvents have equal dielectric constants. It has been proposed [121, 125, 126] that $^o\gamma_{Cs^+}^s \approx {}^o\gamma_{Rb^+}^s \approx 1$. In water, methanol, formamide, and acetonitride this last assumption gives results in harmony with assumption 1.

3. The symmetrical ions, cation and anion, of tetraphenylarsonium tetraphenylboride, $Ph_4As^+ \cdot B^-Ph_4$, are of comparable structure and size. The phenyl groups form an enclosing shell around the central charged atoms. It is proposed [101, 122] that the anion and cation are influenced quite similarly by transfer from one solvent to another, that is,

$$^o\gamma_{Ph_4As^+}^s \approx {}^o\gamma_{B-Ph_4}^s = ({}^o\gamma_{Ph_4As^+}^s \cdot {}^o\gamma_{B-Ph_4}^s)^{1/2}$$

Except that different charge types are involved, this assumption is similar to the ferrocene–ferrocenium in concept. Individual enthalpies of anions and cations have been calculated [109, 125] using assumption 3.

Electrochemical methods are directed at measuring the junction potentials between two half-cells containing different solvents and depend largely on assumptions 1 and 2 [124]. Electromotive forces of pertinent cells and polarography are the most acceptable methods of obtaining solvent activity coefficients [120, 123, 124, 127–129]. Some such measurements have been made.

Parker [101] lists several assumptions, which, if valid, permit the calculation of solvent activity coefficients based on assumptions 1–3, and which use the solubilities of the solute in two solvents [130], one of which is taken as the standard reference solvent. The necessary assumptions are: (1) the saturated solutions in which salt AB has the same activity are not too concentrated, (2) there is no ion association, (3) the solutes do not react with the solvent, and (4) the solid phase in equilibrium with the saturated solution is the same in each solvent. The solubility concentration C^o of a solute in the standard reference solvent O is also the activity a^o of the solute in that solution. The concentration C^s of the solute in solvents must be multiplied by the solvent activity coefficient γ^s, or what is its equivalent $^o\gamma^s$ to give its activity in solvent S. Then

$$a_{A^+}^s = C_{A^+}^s \gamma_{A^+}^s = C_{A^+}^s {}^o\gamma_{A^+}^s \tag{5.257}$$

and

$$a_{B^-}^s = C_{B^-}^s \gamma_{B^-}^s = C_{B^-}^s {}^o\gamma_{B^-}^s \tag{5.258}$$

since the activity coefficients in the standard reference solvent are unity. Hence

$$a_{A^+}^o a_{B^-}^o = C_{A^+}^o C_{B^-}^o = a_{A^+}^s a_{B^-}^s = C_{A^+}^s {}^o\gamma_{A^+}^s \cdot C_{B^-}^s {}^o\gamma_{B^-}^s \tag{5.259}$$

and

$$\frac{C_{A^+}^o C_{B^-}^o}{C_{A^+}^s C_{B^-}^s} = \frac{K_s^o}{K_s^s} = {}^o\gamma_{A^+}^s \cdot {}^o\gamma_{B^-}^s \tag{5.260}$$

Applying Eqs. (5.256) and (5.260) to a salt F^+B^-, we have

$$\frac{C_{F^+}^o C_{B^-}^o}{C_{F^+}^s C_{B^-}^s} \cdot \frac{a_F^s}{a_F^o} = \frac{C_{F^+}^o C_{B^-}^o}{C_{F^+}^s C_{B^-}^s} \cdot \frac{C_{F^+}^s {}^o\gamma_{F^+}^s}{C_{F^+}^o} = {}^o\gamma_{F^+}^s {}^o\gamma_{B^-}^s \tag{5.261}$$

or

$$\frac{C_{F^+}^o C_{B^-}^o}{C_{F^+}^s C_{B^-}^s} \cdot \frac{C_{F^+}^s}{C_{F^+}^o} = \frac{{}^o\gamma_{F^+}^s {}^o\gamma_{B^-}^s}{{}^o\gamma_{F^+}^s} = {}^o\gamma_{B^-}^s \tag{5.262}$$

Using assumption 2, $^o\gamma_{Cs^+}^s = 1$, we see from Eq. (5.260) that

$$\frac{C_{Cs^+}^o C_{B^-}^o}{C_{Cs^+}^s C_{B^-}^s} = {}^o\gamma_{Cs^+}^s {}^o\gamma_{B^-}^s = {}^o\gamma_{B^-}^s \tag{5.263}$$

From assumption 3, $^o\gamma_{Ph_4As^+}^s = ({}^o\gamma_{Ph4As^+}^s {}^o\gamma_{Ph_4B^-}^s)^{1/2}$ we have

$$({}^o\gamma_{Ph_4As^+}^s)^2 = {}^o\gamma_{Ph_4As^+}^s {}^o\gamma_{Ph_4B^-}^s \tag{5.264}$$

and

$$^{o}\gamma^{s}_{Ph_4As^+} = {}^{o}\gamma^{s}_{Ph_4B^-} \tag{5.265}$$

From Eqs. (5.263) and (5.265)

$$\frac{C^{o}_{Ph_4As^+} C^{o}_{Ph_4B^-}}{C^{s}_{Ph_4As^+} C^{s}_{Ph_4B^-}} = {}^{o}\gamma^{s}_{Ph_4As^+} {}^{o}\gamma^{s}_{Ph^4B^-} = ({}^{o}\gamma^{s}_{Ph_4As^+})^2$$

$$= ({}^{o}\gamma^{s}_{Ph_4B^-})^2 \tag{5.266}$$

and for a salt $Ph_4As^+B^-$ from Eq. (5.266)

$$\frac{C^{o}_{Ph_4As^+} C^{o}_{B^-}}{C^{s}_{Ph_4As^+} C^{s}_{B^-}} = {}^{o}\gamma^{s}_{Ph_4As^+} {}^{o}\gamma^{s}_{B^-} \tag{5.267}$$

Thus any of the three assumptions will for any anion give $^{o}\gamma^{s}_{B^-}$.

Parker [101] tabulated solvent activity coefficients $^{D}\gamma_i{}^{s}$ at 25°C, calculated from solubilities for different ions in different solvents, S, relative to a standard state of 1 mole solute in dimethylformamide, D.

An example of how the solubility data was used to calculate the solvent activity coefficient at 25°C for bromide ion $^{D}\gamma^{M}_{Br^-}$ in methanol, M, relative to dimethylformamide can be illustrated as follows. The logarithm of the solubility product constant for CsBr in M at 25°C was found to be $\log K_M \doteq -2.19$. The value in D was $\log K_D = -3.25$. From Eq. (5.260)

$$\log K_D/K_M = \log K_D - \log K_M = -3.25 - (-2.19) = -1.06$$

$$= \log{}^{D}\gamma^{M}_{Cs^+} \cdot {}^{D}\gamma^{s}_{Br^-} \tag{5.268}$$

But $\log{}^{D}\gamma^{M}_{Cs^+} = 0.0$ and hence $\log{}^{D}\gamma^{M}_{Br^-} = -1.06$ or -1.1. Knowing $^{D}\gamma^{M}_{Br^-}$ and the solubilities of slightly soluble silver salts from potentiometric or other measurements, the $^{o}\gamma^{s}_{Ag^+}$ can be calculated and $^{o}\gamma^{s}_{B^-}$ found for all anions B^- that form slightly soluble silver salts. The method can be extended stepwise to include other cations and anions which form slightly soluble salts, the solubilities of which have been measured.

As for reactions of small anions with dipolar molecules by the S_N2 mechanism, the rates are much faster in dipolar aprotic than in protic solvents. For any reaction there is a distinct difference in the effect on rate between classes of solvents; however, within a class of solvents, irrespective of its physical properties—polarity, dipole moment, dielectric constant, molecular weight, structure, density, viscosity, refractive index, etc., there is very little effect on any given reaction caused by change of solvent within the class. Except for some correlation of rates with the physical parameter of polarity, there is no correlation of rates with any ordinary physical property of the solvent [101].

Thus for the protic, hydrogen-bond donor solvents, the S_N2 anion–polar molecule reactions are slow, while for nonhydrogen-bond donor solvents, such reactions are fast.

For example, the reaction

$$N_3^- + \; \underset{H}{\overset{H}{H-C-I}} \; \longrightarrow \; \left[N_3 \overset{H \quad H}{-C-I} \right]^- \longrightarrow \; N_3-C\overset{H}{\underset{H}{-H}} + I^- \qquad (5.269)$$

in methanol and dimethylformamide shows a difference in the logarithm of the specific velocity constant, $\log k^M - \log k^D$, at 25°C of -4.6. In other words, the transfer from the protic methanol solvent to the dipolar aprotic dimethylformamide solvent increased the rate constant for the reaction by a factor of almost 10^5. For the same solvent transfer the reaction

$$N_3^- + \; \underset{H}{\overset{H}{H-C-OTs}} \; \longrightarrow \; \left[N_3 \overset{H \quad H}{-C-OTs} \right]^- \longrightarrow \; N_3-C\overset{H}{\underset{H}{-H}} + OTs^-$$
$$(5.270)$$

shows $\log k^M - \log k^D = 2.0$, or a 10^2 increase in the rate constant in changing from methanol solvent to dimethylformamide solvent. In the equation CH_3OTs is methyl tosylate. For S_N2 reactions the agreement of solvent effects was found at both alkyl and aryl carbons even though the former are one-step and the latter two-step processes. With respect to charge distribution and bond breaking, the transition states in the two cases are greatly different [101, 131]. Writing Eq. (5.246) in terms of an anion Y^- and a neutral molecule R_3CX, and in terms of the solvents methanol M and dimethylformamide D gives

$$\log k^M - \log k^D = \log{}^D\gamma_{Y^-}^M + \log{}^D\gamma_{R_3CX}^M - \log{}^D\gamma_{(YR_3CX^-)*}^M \quad (5.271)$$

and

$$\log{}^D\gamma_{R_3CX}^M - \log{}^D\gamma_{(YR_3CX^-)*}^M = \log({}^D\gamma_{R_3CX}^M / {}^D\gamma_{(YR_3CX^-)*}^M)$$
$$= \log k^M - \log k^D - \log{}^D\gamma_{Y^-}^M \quad (5.272)$$

or for X = Cl, Br, or I,

$$\log({}^D\gamma_{R_3CX}^M / {}^D\gamma_{(YR_3CX^-)*}^M) = \text{const} \pm 0.5 \quad (5.273)$$

Parker [101] indicates that the difference in solvation by methanol relative to DMF of aryl and alkyl halides with different substrates is compensated for by the difference in solvation of the pertinent transition state. Equation 5.273 was used in deriving several free energy relationships between reaction rates and equilibria.

$S_N 2$ reactions of methyl tosylates with azide ion or thiocyanate ion show much less effect on transfer from solvent methanol to solvent DMF than do reactions of alkyl or aryl halides, due no doubt to the difference in the solvation of the respective transition states since $^D\gamma^M_{CH_3OTs}$ and $^D\gamma^M_{CH_3I}$ are similar in value. Methanol relative to DMF solvates to a greater extent the transition state for the reaction of Y^- with methyltosylate than the transition state for the reaction of Y^- with methyl iodide. Parker [101] proposes the two following transition states, either of which would be in harmony with the above conclusion

$$\underset{I}{\overset{\text{H}}{\underset{\text{H}}{\overset{}{}}} \quad ^-Y\cdots\overset{\text{H } \delta\delta-}{\underset{|}{C}}\cdots OTs \qquad\qquad \underset{II}{\overset{\delta\delta- \text{H}}{Y\cdots\overset{|}{C}}\cdots OTs^-} \tag{5.274}$$

Transition state (I) strongly resembles reactants and

$$\frac{^D\gamma^M_{Y^-} \cdot {}^D\gamma^M_{CH_3OTs}}{^D\gamma^M_{(YCH_3OTs^-)*}} \approx 1 \tag{5.275}$$

and transition state (II) conforms closely to products so that

$$\frac{^D\gamma^M_{OTs^-} \cdot {}^D\gamma^M_{CH_3Y}}{^D\gamma^M_{(YCH_3OTs^-)*}} \approx 1 \tag{5.276}$$

From Eqs. (5.275) and (5.276), $^D\gamma^M_{Y^-} \cdot {}^D\gamma^M_{CH_3OTs} \approx {}^D\gamma^M_{OTs^-} \cdot {}^D\gamma^M_{CH_3Y}$, which must be true if $k^M/k^D \approx 1$ as can be seen from Eq. (5.246). While the ratios k^M/k^D are not unity for the methyl tosylate reactions, they are much closer to unity than for the methyl halide reactions, from which Parker [101] concludes that $S_N 2$ reactions of CH_3OTs with azide or thiocyanate ions compared to like reactions of methyl halides have transition states which resemble much more markedly reactants or products.

Not all displacement reactions of the anion–molecule charge type, e.g., chloride ion displacement from platinum, are sensitive to transfer from dipolar aprotic to protic solvent [101, 132, 133]. Parker [101] attributes this insensitivity to solvent type to either a large and positive value of $^D\gamma^M_{R_3PtX}$, a transition state considerably less than unity, or the choice of a wrong model for the transition state, i.e., some other than a simple $S_N 2$ exchange unassisted by electrophilic catalysis, ion pairing, or some specific solvent effect is determining the rate. The reactions unlike $S_N 2$ reactions at saturated carbon do show unusual salt effects in acetone [101, 132]. They have activation entropies and enthalpies which are unlike $S_N 2$ reactions when both are of the anion–cation charge type [101, 134], and show nucleophilic tendencies of halide ions toward platinum [101, 132] which are at variance with those for $S_N 2$ at carbon [135].

In amide solvents involving amides, methyl amides, and dimethyl amides, it was found that the only property, other than the empirical polarity factor, which could be consistently correlated with the rate of $S_N 2$ reaction between methyl iodide and chloride ion was the ability of the solvent molecules to donate hydrogen bonds to the solute particles and/or to other solvent particles. For the second-order rate constant k_2 l./mole/sec^{-1}, $\log k_2$ for the above reaction was -4.28, -3.84, $+0.38$, $+0.91$, and $+1.3$ in formamide, methylformamide, dimethylformamide, dimethylacetamide, and N-methylpyrrolidone, respectively [101]. The first two mentioned solvents are protic and the last three solvents, in which the rate of reaction is much faster, are dipolar aprotic. The maximum difference in the rates in the two types of solvents is of the order of $10^{5.6}$.

These large differences in the rates of this reaction in the two kinds of solvent were attributed [101, 110, 136] to a strong hydrogen bonding interaction of a protic solvent with chloride ion which interaction is not present with the large polarizable transition state. Values of about 1 and about 10^{-6} were assigned to $^D\gamma^F_{(ClCH_3I^-)}$ and $^D\gamma^F_{Cl^-}$, respectively, with the solvation of CH_3I being ignored. To be taken in account also are the polarized, highly ordered structures of the hydrogen bonded solvents such as formamide, and their consequent lesser solvation of large polarizable solutes and large polarizable transition states, as compared to the solvation of these by the weaker structured, more polarizable, nonhydrogen-bond donor solvents like DMF [101, 107, 137].

For the reaction

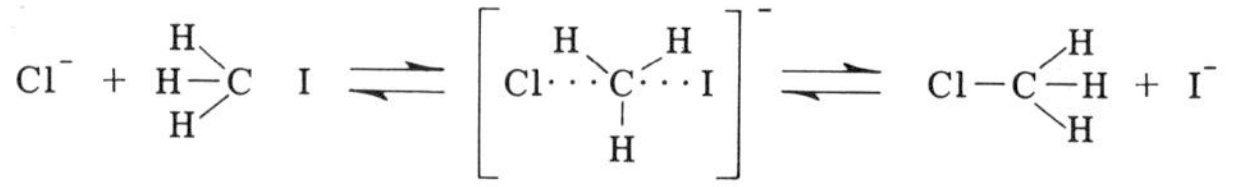

$$Cl^- + H-\underset{\underset{H}{|}}{\overset{\overset{H}{|}}{C}}\ I \ \rightleftharpoons \ \left[Cl\cdots\underset{\underset{H}{|}}{\overset{\overset{H}{}}{C}}\cdots I \right]^- \ \rightleftharpoons \ Cl-\underset{\underset{H}{}}{\overset{\overset{H}{}}{C}}-H + I^- \qquad (5.277)$$

at 25°C the difference in the logarithms of the specific velocity constants in methanol and in DMF, i.e., $\log k^{MeOH} - \log k^D = -5.5 - (+0.4) = -5.9$. The assumption (1) that the Cs^+ ion is equally solvated in MeOH and DMF, that is that $\log {}^D\gamma^M_{Cs^+} = 0$, gives $\log {}^D\gamma^M_{Cl^-} = -3.3$ and $\log {}^D\gamma^M_{(ClCH_3I^-)} = +3.9$. The chloride ion is more solvated by MeOH than by DMF and the transition state is more solvated by the DMF than by MeOH as is indicated by the signs of the logarithms of the solvent activity coefficients of the two solute species. That the rate of the reaction [Eq. (5.277)] is so much faster in DMF than in MeOH can be attributed equally to the difference in solvation of the chloride ion and the difference in solvation of the transition state [101].

The assumption (3) that the tetraphenylarsonium cation and the tetraphenylboride anion are equally solvated in any particular solvent, i.e., that $\log {}^D\gamma^M_{Ph_4As^+} = \log {}^D\gamma^M_{Ph_4B^-}$, yields $\log {}^D\gamma^M_{Cl^-} = -7.1$ and $\log {}^D\gamma^M_{(ClCH_3I^-)*} = +0.1$. The signs and magnitudes of the logarithms of the solvent activity coefficients in this case indicate that the Cl^- ion is much more solvated by the MeOH than by the DMF, while the transition state is only slightly, if at all, more solvated by DMF than by MeOH. The rate difference of the reaction [Eq. (5.277)] in the two solvents would then be due almost exclusively to the greater solvation of the chloride ion in the MeOH as compared to DMF.

Based on the assumption selected one arrives at different conclusions. It is thus difficult to decide whether the difference in rate of a given reaction in dipolar aprotic versus protic solvents arises from differences in solvation in the two solvents of a reactant, of the transition state, or to a combination of these. Parker [101] prefers the tetraphenylarsonium–tetraphenylboride assumption, though he cautions that the results of either assumption cannot at present be taken too seriously, and that further experimental work needs to be done before correct conclusions can be reached. The tetraphenyl-arsonium–tetraphenylboride assumption does take into account size effects such as making and breaking solvent structure, and at the same time allows for electrostatic solvation and other features possessed by the ferrocene-ferricinium couple and the cesium ion assumptions.

For reactions of the type

$$Y^- \ + \ R_3CX \ \longrightarrow \ \left[\begin{array}{c} R \quad R \\ Y \cdots C \cdots X \\ | \\ R \end{array} \right]^- \ \longrightarrow \ YCR_3 \ + \ X^- \qquad (5.278)$$

Parker [101] suggests that Y^- will be less solvated in methanol relative to DMF, as Y^- becomes a weaker hydrogen-bond acceptor and more polariz-able, and thus $\log {}^D\gamma_{Y^-}^M$ will be less negative. From Eqs. (5.271) and (5.273)

$$\log \frac{k^M}{k^D}(Y^-) = \log {}^D\gamma_{Y^-}^M + \log \frac{{}^D\gamma_{R_3CX}^M}{{}^D\gamma_{(YR_3CX^-)*}^M}$$

$$= \log {}^D\gamma_{Y^-}^M + \text{const} \pm 0.5 \qquad (5.279)$$

Thus as $\log {}^D\gamma_{Y^-}^M$ becomes less negative and Y^- larger and more polarizable, the rate of reaction (5.278) in methanol should approach more nearly its rate in DMF. Parker [101] found this expectation to be realized. The determination of nucleophilic properties and base strengths of anions toward Lewis acids are related in a marked way to the large differences in solvent response of reactions in which large, polarizable anions are involved.

It is demonstrable that the ability of the anion to accept hydrogen bonds has a great influence in determining the effects on anion–dipolar molecule reaction rates on transfer from protic to dipolar aprotic solvents [101].

D. REACTIONS OF THE S_N2 TYPE BETWEEN POLAR MOLECULES

Discussions of solvent effects on this type of reaction are numerous [101, 138–140]. The Hughes–Ingold [141] theory for solvent effects on reactions of the type Eq. (5.278) in which anions are reacting with dipolar molecules with only a dispersal of negative charge in the transition state predicts an insensitivity to changes in the polarity of the solvent. For reactions of the type

$$Y: + RX \rightleftharpoons [\overset{\delta+}{Y} \cdots R \cdots \overset{\delta-}{X}]^* \rightleftharpoons \overset{+}{Y} - R + X^- \qquad (5.280)$$

in which uncharged reactants yield charged products by going through a transition state in which charge is being formed, the Hughes–Ingold theory predicts that rate will be greater in more polar solvents. Parker [101] points out that the Hughes–Ingold theory does a remarkable job of correlating the rates of these types of reactions with solvent polarity in either protic or dipolar aprotic solvents, but fail completely when rates in protic solvents and dipolar aprotic solvents are compared. Thus the dipolar molecule–dipole molecule reaction between n-butyl bromide and pyridine has the values at 50°C of the log k' (k' in sec^{-1} $mole^{-1}$) of -4.8, -4.7, and -4.5 in the polar solvents MeOH and 88% MeOH–12% H_2O, and the dipolar aprotic solvent dimethyl formamide, respectively, while the methyl bromide–azide ion reaction at 0°C has the values of $\log k'$ (k' in sec^{-1} mole) of -5.6, -5.6, -1.0, and -1.3 in the polar solvents MeOH and 88% MeOH–12% H_2O, and in the dipolar aprotic solvents dimethylacetamide and dimethylformamide, respectively. In general the rates of the reaction between nonhydrogen-bond donor polar molecules are little influenced by a change from protic to dipolar aprotic solvents of the same dielectric constant but less polarity, for example 88% MeOH–12% H_2O to DMF, while in contrast the azide ion–alkyl halide reaction is sped up by a factor of over 10^4 in k' when such a solvent transfer takes place.

Parker [101] criticizes the theories [142, 143] that increased rates of reaction in dipolar aprotic solvents arise from some ground state interaction between polar substrate and solvent followed by nucleophilic attack on the resulting high energy species since from the thermodynamic rate theory, the existence of such a species in dipolar aprotic solvents in contrast to protic solvents would not be in harmony with the greater solubility of polar organic molecules in dipolar aprotic solvents, and the possibility of recovery unchanged of organic substrates from solution in dipolar aprotic solvents would be less. Rather, Parker suggests that there is a sort of selective solvation by which the dipolar aprotic solvents in or around the transition state for a reaction in dipolar aprotic–protic solvent reduces the free energy of the transition state relative to that for reaction in pure protic solvent. There is not much influence on the representative S_N2 dipolar molecule–dipolar molecule reactions of the tertiary anime and of dimethyl sulfide with alkyl halides when transferred to protic from dipolar aprotic solvents of the same dielectric constant. It is believed [101] that the rates of only anionic nucleophile bimolecular reactions are markedly influenced by protic–dipolar aprotic solvent effects. The lack of effect on the rate of dipolar molecule–dipolar molecule reactions when transferred from protic to dipolar aprotic solvents is the approach to a small number or to unity of the solvent activity term in equations corresponding to Eq. (5.244) applied to reactions between polar molecules.

E. ANION–CATION S_N2 REACTIONS

An anion–cation reaction and a polar molecule–polar molecule reaction which go through the same transition state is represented by the reversible reaction [101]

$$CH_3 - \overset{+}{S}(CH_3)_2 + Br^- \rightleftharpoons [\overset{-\delta}{Br} \cdots CH_3 \cdots \overset{+\delta}{S}(CH_3)_2]^* \rightleftharpoons CH_3Br + (CH_3)_2S$$

$$(5.281)$$

This reaction has been studied in 88% CH_3OH–12% H_2O [134], in dimethylacetamide [134], in methanol [144], in ethanol [112, 144], in dimethylformamide [112], dimethylacetamide (DMAC) [134], and in nitromethane [101].

At 100°C 88% MeOH–12% H_2O and DMAC have the same dielectric constant, but the forward reaction rate diminished by $10^{4.68}$ (log k changes from $+1.24$ to -3.44) on transfer from the dipolar aprotic solvent DMAC to the protic solvent 88% MeOH–12% H_2O. The rate of the back reaction does not change markedly (log k_b changes from -2.39 to -2.17) for the same solvent transfer [101]. The equilibrium constant for the process, Eq. (5.281), shows about the same order of magnitude change, $\sim 10^5$ on transfer from DMAC to 88% MeOH–12% H_2O as does the forward reaction, which was accounted for [101] on the assumption that $^D\gamma_{Br^-}^M$ is about 10^{-5}, and that the solvent activity coefficients of the other species taking part in the reaction either cancel each other or are about unity.

The charge distribution in the transition state for the process, Eq. (5.281), is probably similar to that of an ion-pair—approximately that between free ions of opposite charge and two polar molecules, as is indicated by the entropy of activation for the forward process, $+31$ e.u., being somewhat less than one half the entropy for the overall forward process, 71 e.u. [101]. The transition state is probably more solvated than the polar products but less solvated than the ionic reactants, and these differences in solvation are reflected in the forward and backward enthalpies of activation, $\Delta H_f^* = 32$ kcal/mole and $\Delta H_b^* = 11.6$ kcal/mole. The latter enthalpy is low, perhaps because the energy of solvation favors the formation of the transition state which is more polar than reactants in the back reaction. The large value of $\Delta H_f{}^*$ is partly due to the desolvation necessary to produce a polar transition state from ionic reactants.

As to reactions between cations and polar molecules, these have been found to be rather insensitive to polar and to dipolar aprotic–protic solvent effects [101].

VIII. Mixtures of Dipolar Aprotic Solvents and S_N2 Reactions

Cavell and Speed [145] attributed the effect of increasing the mole fraction of protic component to a specific interaction between n moles of the protic component of the solvent and one mole of Y^-. However, Parker [101] disputes this claim and says that there is a general interaction between Y^- and ROH in ROH–dipolar aprotic solvent mixtures as is verified by the electronic absorption spectra of the systems [146]

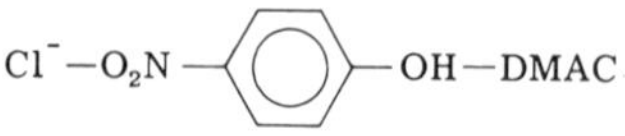

$$Cl^- - O_2N - \langle\bigcirc\rangle - OH - DMAC \quad \text{and} \quad As^- - CH_3OH - DMAC.$$

Representing the solution of the chloride ion by MeOH using the equation

$$Cl^- + n\text{-MeOH} \rightleftharpoons [Cl\cdot(MeOH)_n{}^-] \qquad (5.282)$$

and assuming both Cl^- and $Cl(MeOH)_n{}^-$ to be relative species, the rate expression for the reaction, Eq. (5.277) in DMAC–MeOH mixtures is [101, 115]

$$k_2'[Cl^- + Cl\cdot(MeOH)_n{}^-][CH_3I] = k_0'[Cl][CH_3I]$$
$$+ k_\infty'[Cl(MeOH)_n{}^-][CH_3I]$$

$$(5.283)$$

but from Eq. (5.282)

$$[Cl(MeOH)_n{}^-] = K[Cl^-][MeOH]^n \qquad (5.284)$$

Substituting $[Cl(MeOH)_n{}^-]$ from Eq. (5.284) into Eq. (5.283) and solving for $(k_0' - k_2')/(k_2' - k_\infty')$ gives

$$\frac{k_0' - k_2'}{k_2' - k_\infty'} = K[MeOH]^n \qquad (5.285)$$

or

$$\log \frac{k_0' - k_2'}{k_2' - k_\infty'} = \log K + n \log[MeOH] \qquad (5.286)$$

A plot [101, 115] of $\log[(k_0' - k_2')/(k_2' - k_\infty')]$ versus $\log[MeOH]$ gave a curve the slope *n* of which was found to increase continually with increasing [MeOH]. Thus the interaction of MeOH with chloride ion increases and involves more MeOH molecules as the concentration of MeOH increases. Parker generalizes this to include mole fractions of all alcohols and all anions.

In these equations k_0', k_∞', and k_2' are the specific velocity constants for reaction Eq. (5.277) in pure DMAC, pure MeOH, and mixtures of DMAC and MeOH, respectively.

IX. Factors Affecting the Reactivity of Anion–Dipole S_N2 Reactions

A. SOLVATION OF TRANSITION STATES

Parker [101] lists transition state solvation, mobility of leaving groups and nucleophilic tendencies as important factors in determing the reactivities of anion–dipole S_N2 reactions. Some of these, for example, nucleophilicity, will be discussed later, but here these effects will be applied to the effect on the rate of S_N2 reactions when the solvent is changed from protic to dipolar aprotic. Reactivities in nucleophilic displacements at carbon has been discussed [117, 131, 147–149] extensively. Limited application has been assigned [131] to changing the replaced group or the nucleophile, essentially because of the drastic effects of the solvent, which tend to mask empirical correlations, especially when changing from water, alcohols, and their mixtures to dipolar aprotic solvents.

If in Eq. (5.246) the product of the solvent activity coefficients for reactants in the numerator of the term on the right-hand side of the equality sign of the equation equals the solvent activity coefficient for the transition state in the denominator of the term there will be no solvent effect on change of solvent. If the transition state closely resembles in charge distribution and structure either reactants or products both of which respond similarly to solvent change, there will be no effect of solvent change on the reaction rate.

Let $k^M_{(CH_3I)}$, $k^D_{(CH_3I)}$, $k^M_{(CH_3Cl)}$, and $k^D_{(CH_3Cl)}$ be the rate at 0°C for the reactions in DMF (D) and in MeOH (M) of methyl chloride and of methyl iodide with the azide ion N_3^-, respectively. Then

$$\frac{k^M_{(CH_3I)}}{k^D_{(CH_3I)}} = \frac{{}^D\gamma^M_{(N_3^-)} \; {}^D\gamma^M_{(CH_3I)}}{{}^D\gamma^M_{(N_3 \cdots CH_3 \cdots I^-)^*}} \tag{5.287}$$

$$\frac{k^M_{(CH_3Cl)}}{k^D_{(CH_3Cl)}} = \frac{{}^D\gamma^M_{(N_3^-)} \; {}^D\gamma^M_{(CH_3Cl)}}{{}^D\gamma^M_{(N_3 \cdots CH_3 \cdots Cl^-)^*}} \tag{5.288}$$

Then

$$\frac{k^M_{(CH_3I)}}{k^D_{(CH_3I)}} \cdot \frac{k^D_{(CH_3Cl)}}{k^M_{(CH_3Cl)}} = \frac{{}^D\gamma^M_{(CH_3I)}}{{}^D\gamma^M_{(CH_3Cl)}} \cdot \frac{{}^D\gamma^M_{(N_3 \cdots CH_3 \cdots Cl)^*}}{{}^D\gamma^M_{(N_3 \cdots CH_3 \cdots I^-)^*}} \tag{5.289}$$

But $^{D}\gamma^{M}_{(CH_3I)}/^{D}\gamma^{M}_{(CH_3Cl)}$ has been found [101] from solubility data on CH_3I and CH_3Cl in methanol and DMF to be ~ 5. Hence

$$\frac{k^{M}_{(CH_3I)}}{k^{D}_{(CH_3I)}} \cdot \frac{k^{D}_{(CH_3Cl)}}{k^{M}_{(CH_3Cl)}} \approx 5 \frac{^{D}\gamma^{M}_{(N_3\cdots CH_3\cdots Cl^{-})^{*}}}{^{D}\gamma^{M}_{(N_3\cdots CH_3I\cdots I^{-})^{*}}} \qquad (5.290)$$

Now [99]

$$\log\frac{k^{M}_{(CH_3I)}}{k^{D}_{(CH_3I)}} = -5.52 - (-0.51) = -5.01 \qquad (5.291)$$

and [99]

$$\log\frac{k^{M}_{(CH_3Cl)}}{k^{D}_{(CH_3Cl)}} = -4.00 - (-7.70) = 3.70 \qquad (5.292)$$

Thus

$$\log\frac{k^{M}_{(CH_3I)}}{k^{D}_{(CH_3I)}} \cdot \frac{k^{D}_{(CH_3Cl)}}{k^{M}_{(CH_3Cl)}} = -5.01 + 3.70 = -1.31 \qquad (5.293)$$

From Eqs. (5.290) and (5.293),

$$-1.31 = \log 5 + \log {}^{D}\gamma^{M}_{(N_3\cdots CH_3\cdots Cl^{-})^{*}} - \log {}^{D}\gamma^{M}_{(N_3\cdots CH_3\cdots I^{-})^{*}} \quad (5.294)$$

Hence

$$\log {}^{D}\gamma^{M}_{(N_3\cdots CH_3\cdots Cl^{-})^{*}} - \log {}^{D}\gamma^{M}_{(N_3\cdots CH_3\cdots I^{-})^{*}} = -1.31 - 0.70 = -2$$
$$(5.295)$$

Similar measurements on the relative solubilities of methyl bromide and methyl tosylate in MeOH and DMF and their reaction rates with the azide ion led to [99]

$$\log {}^{D}\gamma^{M}_{(N_3\cdots CH_3\cdots Br^{-})^{*}} - \log {}^{D}\gamma^{M}_{(N_3\cdots CH_3\cdots I^{-})^{*}} = -1 \qquad (5.296)$$

and

$$\log {}^{D}\gamma^{M}_{(N_3\cdots CH_3\cdots OTs^{-})^{*}} - \log {}^{D}\gamma^{M}_{(N_3\cdots CH_3\cdots I^{-})^{*}} = -3 \qquad (5.297)$$

Parker [101] gives data for SCN^{-} anions reacting with the same molecules. The conclusions drawn from the above calculations and from similar ones involving the SCN^{-} anion are that transitions states involving chloride and tosylate ions as leaving groups are to a much greater extent solvated by methanol relative to DMF than are transition states having iodide ions as leaving groups, and that this is not correlated with the ability of the leaving group to accept hydrogen bonds from methanol since the latter tendencies are in the order $Cl^{-} \gg OTs^{-} \approx I^{-}$. The difference in relative solvation in MeOH and DMF is not so much greater in the case of the transition state involving bromide ions as the leaving group compared to the transition state possessing iodide ion as the leaving group [see the relatively small negative value of the difference in solvent activity coefficients for the pertinent transition states in Eq. (5.296)]. These data indicate that the order of susceptibility to the change of solvent from DMF to MeOH of the rates of the reaction with anions would be $CH_3I > CH_3Br > CH_3Cl > CH_3OTs$.

Parker [101] emphasizes that the two ends of the transition state may be solvated differently by different solvents. There appears to be an inverse relationship in the solvation by given solvent of the entering and departing groups of anion–dipolar molecule transition states when the entering group is more anionic in one than in the other transition state. Thus, in the two transition states $(N_3 \cdots CH_3 \cdots Cl^-)^*$ and $(N_3 \cdots CH_3 \cdots I^-)^*$, the azide group is more anionic in the latter than in the former transition state, and is perhaps more solvated by MeOH in the latter transition state even though the iodide end of the $(NH_3 \cdots CH_3 \cdots I^-)^*$ transition state is less solvated by MeOH than is the chloride end of the $(N_3 \cdots CH_3 \cdots Cl^-)^*$ transition state.

B. LEAVING GROUP MOBILITIES

The mobilities of leaving groups are solvent dependent such that they are not easily related by linear free energy relationship for leaving group tendencies [99, 129]. The mobility in DMF of bromide relative to iodide for the entering groups N_3^-, SCN^-, and C_5H_5N does not vary markedly as measured by the difference in the logarithms of the rate constants for the pairs of reactions

$$CH_3Br + N_3^- \rightarrow CH_3N_3 + Br^-$$
$$CH_3I + N_3^- \rightarrow CH_3N_3 + I^-$$
(5.298)

$$CH_3Br + SCN^- \rightarrow CH_3SCN + Br^-$$
$$CH_3I + SCN^- \rightarrow CH_3SCN + I^-$$
(5.299)

and

$$n\text{-}BuBr + C_5H_5N \rightarrow n\text{-}BuC_5H_5N^+ + Br^-$$
$$n\text{-}BuI + C_5H_5N \rightarrow n\text{-}BuC_5H_5N^+ + I^-$$
(5.300)

The difference in the logarithms of the rate constants for the above three pairs of reactions at 0°C in DMF being -0.83, -0.85, and -0.66, respectively. However, this near-constancy did not prevail for the pair of reactions

$$2,4(NO_2)_2C_6H_3Br + SCN^- \rightarrow 2,4(NO_2)_2C_6H_3SCN + Br^-$$
$$2,4(NO_2)_2C_6H_3I + SCN^- \rightarrow 2,4(NO_2)_2C_6H_3SCN + I^-$$
(5.301)

for which the difference in logarithms of the rate constants was $+0.11$. When Cl^- or OTs^- replaced Br^- in the above first three pairs of reactions, the difference in rate constants at 0° in DMF were again nearly constant for either Cl^- or OTs^- for the first three pairs of reactions but each ion had its own characteristic difference which were in the order $Cl^- > OTs^- > I^-$ in absolute value though all were negative. Thus in DMF the natures of the leaving groups seemed to be more influential in determining their mobilities than did the nature of the entering groups at least for the first three pairs of reactions. In MeOH the entering groups seemed influential also in determining the mobilities of the leaving groups. The orders of mobilities of leaving groups were not the same in MeOH as in DMF and are not so great in absolute values of the difference in the logarithms of the rate constants in MeOH as in DMF. The difference in the logarithms of the rate constants for the first three pairs of reactions at 0°C in MeOH are -0.05, -0.48, and -0.04, and for the fourth pair 0.00. It is pointed out [99, 129] that MeOH is a leveling solvent for halogen mobilities since it promotes the departure of the smaller, more strongly bound halide ions from a transition state compared to the departure of the larger less tightly bound halide ions. The leaving group tendencies in DMF and MeOH seem unsuited to correlate using a linear free energy relationship such as that of Davis [150].

C. NUCLEOPHILICITY

There has been an extensive, thorough review of the influence of the nucleophilic reagent on reactivity in nucleophylic reactions [133]. It is believed [99, 129] that nucleophilicity is as much a function of external factors such as solvent, the leaving group, and the type of carbon undergoing attack as it is a function of the properties of the nucleophiles such as hardness and softness, polarizability, redox potential, and hydrogen basicity [151, 152]. For different reactions of the same nucleophile these external factors necessitate a variety of nucleophilic "constants" since they cannot always be accommodated by a reaction constant. Linear free energy relations have, in protic solvents, been successful in predicting relative rates of reactions of nucleophiles in a variety of reactions. However, nucleophilic characteristics change appreciably in changing from protic to dipolar aprotic solvents, and hence, quite different parameters are needed in the latter type solvents. The nature of the substrate R_3CX also help determine nucleophilic tendencies.

The general equation [Eq. (5.246)]

$$\log k^S = \log k^O + \log {}^O\gamma^S_{Y^-} + {}^O\gamma^S_{R_3CX} - \log {}^O\gamma^S_{(YR_3CX-)^*} \qquad (5.302)$$

where S represents any solvent, O a standard solvent, Y^- any anionic reactant, and R_3CX any substrate, can be particularized to any solvent, standard solvent, anionic reactant, and substrate. Using dimethylamine (D) as the standard solvent, methanol as the general solvent, SCN^- as the reference anion and Cl^- as the general ion, an equation for the difference of $\log k$ for the reaction of SCN^- and Cl^- with methyl iodide can be written as follows:

$$\log \frac{k_{Cl^-}^M}{k_{SCN^-}^M} = \log \frac{k_{Cl^-}^D}{k_{SCN^-}^D} + \log \frac{^D\gamma_{Cl^-}^M}{^D\gamma_{SCN^-}^M} + \log \frac{^D\gamma_{(SCNCH_3I^-)^*}^M}{\gamma_{(ClCH_3I^-)^*}^M} \qquad (5.303)$$

Now [99]

$$\log k_{Cl}^M - \log k_{SCN^-}^M = 2.25, \qquad \log k_{SCN^-}^D = +1.51,$$

and

$$\log \frac{^D\gamma_{Cl^-}^M}{^D\gamma_{SCN^-}^M} = -4.1$$

hence

$$-2.25 = 1.51 - 4.1 - \log \frac{^D\gamma_{(ClCH_3I^-)^*}^M}{^D\gamma_{(SCNCH_3I^-)^*}^M} \qquad (5.304)$$

which yields $\log[^D\gamma_{(ClCH_3I^-)^*}^M / ^D\gamma_{(SCNCH_3I^-)^*}^M] = -0.3$. Several such calculations have been made and the values tabulated [101, 129] with similar results for many Y^- nucleophiles such as Oac^-, Br^-, and N_3^-. Therefore, the transition states for these nucleophiles are influenced similarly by transfer from DMF to MeOH. Thus, the last term on the right-hand side of Eq. (5.303) is small enough to be uninfluential in determining the changes in nucleophilicity of Y^- with respect to SCN^- with this solvent transfer. These changes are, therefore, dictated by the term $(^M\gamma_{Y^-}^D / ^M\gamma_{SCN^-}^D)$ which can be as large as $10^{5.3}$ as is the case of the acetate ion relative to the thiocyanate ion. Within the same solvent nucleophilicities toward carbon may vary by $\sim 10^7$. Thus, for $2,4(NO_3)_2C_6H_3O^-$, $\log[k_{(2,4(NO_3)_2C_6H_3O^-)}^M / k_{SCN^-}^M] = -4.18$, showing relatively small reactivity compared to SCN^-, while for $C_6H_5S^-$, $\log[k_{C_6H_5S}^M / k_{SCN^-} = 3.2$, showing high reactivity with respect to SCN^-. The difference of neuchleo-philic tendencies of these two nucleophiles is then $10^{7.4}$.

What Parker [101] emphasizes is that since the solvent can affect nucleo-philic tendencies by 10^5, it is of no profit to discuss chemical reaction rates in terms of inherent properties of the nucleophile such as bond strength, ease of adjustment of valence shells to transition state configurations, hardness and softness, size, α effects, polarizability, charge type, structure, etc., unless the properties of the specific solvent is included in the analysis. This is obvious when changes in solvent can produce an effect of 10^5 in nucleophilic tendencies.

X. Correlations Involving Linear Free Energy Relationships

Parker [101] has correlated through the solvent activity coefficients and by use of linear free energy relationship changes in properties of energetic processes relative to other processes, when these changes arise from the effect of solvation of anion Y^- with transfer of solvent in which the processes occur. Straight line plots were obtained between the logarithms of the relative solubility product, constants of silver salts AgY ($\log K_s^D/K_s^M$), and the logarithms of the relative specific velocity constants for the reactions of Y^- with methyl iodide ($\log k_M'/k_D'$) in DMF and in MeOH at 25°C, and also between the logarithms of relative ionization constants of HY ($\log K_i^D/K_i^M$) at 20–25°C [153] and the logarithms of the relative reaction rates of Y^- with methyl iodide ($\log k_M/k_D$) at 0°C [154]. The slopes of the lines in each case were unity. The anions involved in the solubility-rate correlation were SCN^-, Br^-, N_3^-, Cl^-, and Oac^-; the anions involved in the ionization-rate correlation were Cl^-, $CH_3CO_2^-$, $ClCH_2CO_2^-$, $PhCO_2^-$, Br^-, PhS^-, N_3^-, $Ar'O^-$, ArO^-, and $Ar'S^-$.

Let us consider the solubility-rate relationship to show that a linear relationship should exist. From Eq. (5.260),

$$\log[K_s^D/K_s^M](\text{AgY}) = \log {}^D\gamma_{Y^-}^M + \log {}^D\gamma_{Ag^+}^M \qquad (5.305)$$

and

$$\log[K_s^D/K_s^M](\text{AgSCN}) = \log {}^D\gamma_{SCN^-}^M + \log {}^D\gamma_{Ag^+}^M \qquad (5.306)$$

Thus,

$$\log[K_s^D/K_s^M](\text{AgY}) - \log[K_s^D/K_s^M](\text{AgSCN}) = \log {}^D\gamma_{Y^-}^M - \log {}^D\gamma_{SCN^-}^M \qquad (5.307)$$

$$\log[K_s^D/K_s^M](\text{AgY}) = \log {}^D\gamma_{Y^-}^M - \log {}^D\gamma_{SCN^-}^M + \log[K_s^D/K_s^M](\text{AgSCN}) \qquad (5.308)$$

The $\log[K_s^D/K_s^M](\text{AgSCN})$ has been found [101] to be 2.46, and letting

$$^D P^M = \log {}^D\gamma_{Y^-}^M - \log {}^D\gamma_{SCN^-}^M \qquad (5.309)$$

we find

$$\log[K_s^D/K_s^M](\text{AgY}) = {}^D P^M + 2.46 \qquad (5.310)$$

From Eq. (5.246)

$$\log[k^M/k^D](Y^-) = \log {}^D\gamma_{Y^-}^M + \log {}^D\gamma_{CH_3I}^M - \log {}^D\gamma_{(YCH_3I^-)}^M \qquad (5.311)$$

for the rates of reaction of, and

$$\log[k^M/k^D](SCN^-) = \log {}^D\gamma_{SCN^-}^M + \log {}^D\gamma_{CH_3I}^M - \log {}^D\gamma_{(SCNCH_3I^-)}^M \qquad (5.312)$$

$$\log[k^M/k^D](Y^-) - \log[k^M/k^D](SCN^-) = \log {}^D\gamma_{Y^-}^M - \log {}^D\gamma_{SCN^-}^M$$
$$+ \log {}^D\gamma_{(YCH_3I^-)}^M - \log {}^D\gamma_{(SCNCH_3I^-)}^M \qquad (5.313)$$

But, since $\log {}^{D}\gamma^M_{(ACH_3I^-)}$ is a pproximately constant for the reaction of all anions and since [101] $\log[k^M/k^D](SCN^-) = -2.15$, then

$$\log[k^M/k^D](Y^-) = {}^{D}P^M - 2.15 \tag{5.314}$$

Therefore, from Eqs. (5.310) and (5.314) we can solve for ${}^{D}P^M$ and obtain

$$\log[K_S^D/K_S^M](AgY) = \log[k^M/k^D](Y) + 4.61 \tag{5.315}$$

and a plot of $\log[K_S^D/K_S^M]$ versus $\log[k^M/k^D]$ should yield a straight line with a slope of unity. The intercept cannot be taken as exact since an approximation was made concerning the equality of the logarithms of the solvent activity coefficients of all anion transition states. However, the intercept of 4.61 is not too different from the actually observed intercept of about 3.5. In like manner, the ionization-rate linear relationship can be derived. The P^- values range from ${}^{D}P^M_{(CH_3CO_2^-)} = -5.5$ at 0°C to ${}^{D}P^M_{(I_3^-)} = +4.9$ at 25°C. The difference in these extremes extend over $10^{10.4}$ range.

XI. Enthalpy and Entropy of Activation and the Solvent

Enthalpies of transfer from water to DMSO for some single ions and molecules have been reported [105], but sufficient data, including the above and available energies and entropies of activation, are not extant to permit calculations of entropies and enthalpies of transfer for S_N2 transition states. Such calculations would extend our knowledge concerning solvent effects upon reaction rates beyond that available from knowledge of changes in free energies of solvation of reactant, product, and transition states. In calculating single ion enthalpies of transfer Arnett and McKelvey [107, p. 222] accepted the assumption [122] that the enthalpies of transfer of from water to DMSO were equal for the tetraphenylarsonium cation and the tetraphenylboride anion. These investigators found the cations K^+, Cs^+, Na^+, and $(C_6H_5)_4As^+$, and the anions I^- and $(C_6H_4)_4B^-$ had greater solvation enthalpies in DMSO than in water, while the cation Et_4N^+ and the anions Cl^- and Br^- had greater solvation enthalpies in water than in DMSO. Parker [101] assumes that since the enthalpy of solvation of I^- ion is 2.52 kcal/mole greater in DMSO than in water, that I^- is more solvated, enthalpy-wise in DMSO than in water, in agreement with his qualitative concepts about anion solvation in these solvents [108, 109]. Parker [101] gives a table of data illustrating that the increase of rates upon transfer from water or methanol to DMF of S_N2 reactions of anions at saturated carbon atoms is attributable to decreased enthalpy of activation, with only a partially compensatory effect of a small decrease in entropy of activation. Change in both enthalpy and entropy of activation promote greater rates in DMF than in MeOH for reaction of anions at aromatic carbon. For SCN$^-$ reacting at aromatic carbon, the change in entropy is the dominant factor favoring a faster reaction rate on tranfer from MeOH to DMF. Since for $S_N 2$ reactions of anions at saturated carbon atoms the change in entropy is relatively small compared to change in enthalpy upon solvent transfer, the change in free energy upon transfer will parallel change in enthalpy. Parker [101] makes a comparison of the enthalpies [101] of transfer of tetraethyl-ammonium bromide and free energies [155] of transfer of the silver halides at 25°C from water to DMSO. He concludes that while the enthalpy of transfer depends on the size and polarizability of the ion, the entropy of transfer is the same for all halide ions, supporting his contention that the enthalpy and not the entropy determines in a large measure the solvent effect on anion–dipolar molecule $S_N 2$ reactions, and that the effect depends largely on the solvation of the reactant anion rather than on the solvation of the molecule or transition state.

Solvent effects on rates of 78 $S_N 2$ and E2 reactions in the solvents water, methanol, formamide, dimethylformamide (DMF), dimethylacetamide (DMAC), sufalane (TMS), acetonitrile, dimethylsulfoxide (DMSO), N-methyl-2-pyrralidone (NMePy), acetone, hexamethylphosphoramide (HMPT), nitramethane, and 80% v:v DMSO–methanol were interpreted in terms of solvent activity coefficients, taking methanol as the reference solvent for reactant and for transition states [156].

Using certain extrathermodynamic assumptions [157], it was found that the solvation of small "hard," reactant anions, for example, Cl$^-$ ion decreased strongly in the order of solvents: $H_2O > MeOH > HCONH_2 \gg DMSO$, $CH_3NO_2 > CH_3CN$, DMF > TMS > DMAC > HMPT, NMePy. For polar reactants and uncharged transition states the order of the solvation decreases slightly and roughly in the order of solvents HMPT, NMePy > DMSO, DMAC, DMF > CH_3CN > CH_3NO_2 > MeOH > HCONH$_2$ $\gg$ H$_2$O. The large polarizable S_NAr transition state anions decrease in solvation in the order of solvent HMPT, DMSO > DMF > CH_3OH, NMePy > CH_3NO_2, CH_3CN, just as does a model transition state anion. Observations independent of extrathermodynamic assumptions are listed as follows: Entering and displaced groups influence the solvation of $S_N 2$ transition state anions. The type of carbon atom being attacked influence the susceptibility of $S_N 2$ and E2 transition states to transfer from protic to dipolar aprotic solvents. Transition states for reaction tend to be more solvated by methanol than by DMF in the order benzene Hal > secondary R Hal > primary R Hal > CH_3Hal > 4 nitrobenzyl Hal. Similar substrate transition states for $S_N 2$ and E2C reactions respond in much the same way to solvent transfer. Depending on the nature of the atom being attacked, the solvation of $S_N 2$ transition states varies by more than 4 kcal/mole. The rates of $S_N Ar$ reactions are influenced by solvation of substituents both in reactant and transition states. Changes in enthalpy rather than changes in entropy of activation usually determine protic-dipolar aprotic effects on rates. The "fastest" solvent studied was HMPT.

XII. Finkelstein Reactions

Finkelstein reactions have been used to help in the understanding of nucleophilic reactions at saturated carbon atoms [101, 158–165]. It is [101, 165] believed that the criticism of the interpretation of the halide exchange reactions at a saturated carbon atom was partly justified [166] but partly unjustified [167, 168]. The strong influence, now well known, of ion pairing effects on Finkelstein reactions of lithium halides was one basis for criticism. Some chemists view with some skepticism the successful calculations [164] of enthalpy and entropy differences between Finkelstein reactions of α- and β-branched alkyl halides, and especially when such calculations give insufficient attention to the discussion [164] of steric effects because of reservations about ion pairing effects. A second basis for criticism was that the reaction of *t*-butyl halides with halide ions were not $S_N 2$ but E2 reactions followed by an additional reaction:

$$t\text{-BuBr} + Cl^- \;\rightleftharpoons\; \begin{matrix} H_3C \\ {}^{}C=CH_2 \\ H_3C \end{matrix} + HCl + Br^- \;\rightleftharpoons\; t\text{-BuCl} + Br^-$$

$$(5.316)$$

Calculations have been made [169] for differences in a characteristic entropy and enthalpy difference between a series of alkyl halides and their transition states, on the basis that differences in activation energy and in activation entropy be independent of the solvent and of salt effects as the alkyl group is changed for any one set of Finkelstein reactions.

Parker [101] correlates the logarithms of the specific velocity constants of Finkelstein reactions of alkyl bromides with chloride ion at 25°C in acetone and in DMF. The plot of $\log k$ in acetone versus $\log k$ in DMF was a straight line with a slope of one and an intercept of about -1.9 for methyl, ethyl, *n*-propyl, isobutyl, isopropyl, and neopentyl bromides reacting with chloride ions. The corresponding plots of the energy of activation in acetone versus the energy of activation in DMF and of the Arrhenius $\log B$ in acetone versus the Arrhenius $\log B$ in DMF are also straight lines with slopes of 1.00, but the scatter of points are much more marked in the two latter graphs. The intercepts for the two latter lines are about 15.7 kcal/mole for the energy plot and about 6.6 for the $\log B$ plot.

Parker [101] found also that the change in the Arrhenius parameters and rates with change in alkyl group was roughly the same in each solvent when the reactions of azide ion with alkyl bromide is transferred from DMF to MeOH, which is a more substantial solvent change than from acetone to DMF. Thus, calculated rate data for Finkelstein reactions of chloride ion with alkyl bromides do approximately parallel changes in energy differences between each alkyl bromide ion and its transition state. One exception found was *t*-butyl bromide.

The rate data also indicates [101] that acetone is a typical dipolar aprotic solvent which influences reactions in much the same way as DMF, DMAC, and DMSO provided ion-pairing effects are accounted for. Thus, were ion pairing taken into account, reaction of alkyl bromides with chloride ion in infinitely dilute solutions would be at 25°C about six times as fast in acetone as in DMF.

XIII. Summary

Bimolecular reactions between anions and polar molecules are very much faster in dipolar aprotic than in protic solvents. This effect is especially prominent in substitution nucleophilic reactions of the second order, $S_N 2$ reactions, between anions and dipolar molecules. Anion–cation $S_N 2$ reactions are strongly influenced by solvent transfer, while cation–polar molecule $S_N 2$ reactions are not markedly sensitive to dipolar aprotic–protic solvent transfers. Rates of reaction between nonhydrogen-bond donor polar molecules are not much influenced by transfer from dipolar aprotic to protic solvents of the same dielectric constants, and only bimolecular reactions of anionic nucleophiles have rates which are markedly sensitive to dipolar aprotic–protic solvent effects.

The problem of the solvent influence on reaction rates is reduced to determination of the free energies, entropies, and enthalpies of solvation of the reactants and of the transition states according to Parker, because the specific reaction rate depends on the standard free energy difference between reactants and the transition state. Therefore, any consideration of solvent effects on rates or equilibria must start from solvent activity coefficients of reactants, transition states, and products. Few values of solvent activity coefficients are known for anions, cations, and transition states. By measuring the rate constant for the formation of the transition state for saturated solution of a reactant cation and a reactant anion in a solvent S relative to a reference solvent and remembering that reactants in a saturated solution all start at the same free energy level in all solvents, the solvent activity coefficients of transition states can be measured. This method is not in general a practical approach for the determination of solvent activity coefficients of transition states, and has not been widely applied. One instance of its application is the work of Evans and Parker [112] on the $S_N 2$ trimethylsulphonium bromide decomposition. With regard to the statement that the influence of solvents on reaction rates reduces to the determination of the free energies, entropies and enthalpies of solvation of the reactants and of the transition state, the authors would like to make the following observations. The extent and nature of solvation of a solute species depends on the method used to measure it [170]. One method may show no solvation of a solute species but rather a solvent structure breaking effect, while another method may show extensive solvation of a solvent species [171, 172]. In calculating the extent of solvation of a solvent species one has to assume that the extent of solvation of some reference species is known. This becomes questionable with change of solvent or of method of measurement [170, 173]. Some solvent effects, such as solvolysis and purely electrostatic influence do not arise from solvation.

From the above statements, it is apparent that if difference of solvation of reactants and transition states is to be used to explain solvent effects on chemical reaction rates, then other solvent effects must be proven to be absent, and the extent of solvation of the species concerned must be measured in a manner compatible with the rate process being studied, in fact, by a method involving the rate process itself, and that the standard reference of solvation must be unquestionably known with reference to its extent and nature of solvation in all solvents used in the study.

In spite of the above stringent requirements for absolute knowledge of solvation of reactants and transition state species to explain solvent effects on chemical reaction rates, reasonable assumptions concerning the extent and nature of solvation of certain reactant transition state, and product species, have lead to remarkable insight into the effect of transition from protic to dipolar aprotic solvents on $S_N 2$ anionic–dipolar molecule reaction rates [101].

XIV. Differential Attraction of Solutes for Solvents and Rates

Differential attraction of solutes for solvents may take the form of certain solutes in a reaction process being relatively more solvated than others, that is, the degree of solvation may vary among the pertinent solutes.

It has been pointed out that the solvation of the activated complex to a higher degree than that of the reactants will cause an increased rate of reaction [33, 36, 47, 101, 102]. It has been indicated [47, 102] that the greater rate of reaction of triethylamine and ethyl iodide at 100°C to give tetraethylammonium iodide in polar solvent nitrobenzene (specific rate constant 70.1) compared to the rate in nonpolar solvents as, for example, benzene (specific rate constant

0.0058) arises from a differential solvation in the polar nitrobenzene of the activated as compared to the reactant species. This differential, or specific solvation, does not occur in the nonpolar solvents. Thus, the activated complex, like the product, is probably a highly polar substance and is, therefore, solvated extensively by polar nitrobenzene, but not by benzene. The rate of the reaction is, thus, greater in nitrobenzene according to Eq. (5.244) since α^* will be decreased due to solvation.

In contrast, a solvent which solvates the reactant species to a greater degree than the activated species will cause a reaction to occur less rapidly than will a solvent which does not show a differential attraction toward reactant as compared to activated species.

Sometimes in changing solvents the difference in relative solvation by the two solvents of the reactant states is found to be compensated for by an equivalent difference in the solvation by the two solvents of the appropriate transition states. Thus, though aryl and alkyl halides must vary in solvation by methanol relative to dimethylformamide this variation is balanced by an equivalent difference in solvation of the appropriate transition state [101].

Much less susceptible to solvent transfer from methanol to dimethyl-formamide are the S_N2 reactions of methyl tosylate than are the reactions of alkyl or aryl halides. The difference in behavior of the methyl tosylate and methyl iodide with azide or thiocyanate ions arises from the fact that the transition state of tosylate with one of the anions is better solvated with methanol relative to dimethylformamide than is the transition state of one of the anions with methyl iodide [101]. The reactions are illustrated in Eqs. (5.269) and (5.270).

Thus, if the activated complex is polar and highly solvated its activity coefficient is small; and if at the same time the less polar reactants are not solvated their activity coefficients are not reduced; the ratio of the rate in polar solvating solvent to that in the gas phase or less polar solvent should be large [36, 79].

The mechanistic picture for the polar activated complex formation and for the overall reaction in the kinetics of the triethylamine–ethyl iodide reaction would perhaps be as follows [36]

$$
\begin{array}{c}
\begin{array}{c} CH_3CH_2 \\ CH_3CH_2-N \\ CH_3CH_2 \end{array}
+ CH_3CH_2I \longrightarrow
\begin{array}{c} CH_3CH_2 \quad\quad I^{-\delta} \\ {}^{+\delta}CH_3CH_2-N^{-\delta}\cdots CH_2CH_3 \\ CH_3CH_2 \quad\quad {}^{+\delta} \\ \downarrow \\ (CH_3CH_2)_4 \; N^{+\delta} \; I^{-\delta} \end{array}
\end{array}
\qquad (5.317)
$$

In the double-sphere model for the activated complex, the reactant ions of like sign give more electrostriction in the activated complex than do reactant ions of unlike sign. These strong electrostatic forces in the neighborhood of the highly charged complex formed from like charged ions restrict the freedom of motion of the solvent molecules and cause a consequent loss of entropy which is greater the higher the charge [102]. The ions forming the complex having smaller charges are not so marked in their electrostrictive properties, and there is, therefore, a loss of entropy when the complex is formed.

For ions of unlike charge there is less charge on the complex than on the reactant ions. Thus electrostriction is decreased when the complex is formed and hence there is an increase in entropy [102].

The model in which the two spherical reactant ions form a one sphere activated complex would presumably show this increase in electrostriction and consequent decrease in entropy for ions of like charge and a decrease in electrostriction and an increase in entropy for ions of unlike charge.

The surface charge density would be about 1.26 greater for the sphere formed from ions of unit charge if the charge were located in the center of the sphere than for the two single ion spheres of like sign and size, though the areas of the two single ions would be correspondingly greater. These statements assume no total volume constriction or expansion on ion fusion to form the complex. Charge density rather than total area would probably be more influential in determining the total solvation.

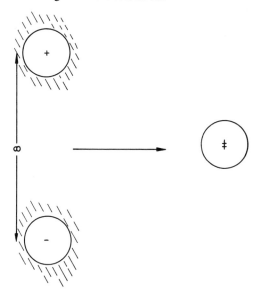

Fig. 5.9. Entropies of activation in terms of electrostriction of the solvent molecules. In the single ions there is electrostriction and in the activated complex there is not. There is, therefore, an increase in entropy upon the formation of the complex. For ions of like sign the hatching would be more pronounced around the complex than around the single ions.

For fusion of two unlike charged ions of like size and magnitude of charge to form a single sphere complex, with centrally located charges in all spheres and no change in total ionic volume, the activated sphere complex would be neutral, and hence solvation would decrease markedly and entropy increase greatly on the formation of the complex if charge density only is considered. The situation is illustrated in Fig. 5.9. The hatching around the spheres represent solvation considering only the influence of surface charge density.

XV. Polarity of Solvent and Reaction Rates

A. SOLVENT POLARITY

Solvent polarity is a term which may refer to several empirical parameters which characterize solvents. Dielectric constant is one of these parameters and its influence on rates when controlled by electrostatic influence has been discussed in detail in the sections on electrostatic effects on rates and mechanisms of ion–ion, ion–dipolar molecule, and dipolar molecule–dipolar molecule reactions.

Dipole moment is a second parameter that is used to designate polarity. Moments include induced and permanent moments, the former arising from temporarily induced electrical charge assymetry in a chemical entity brought about by the close proximity of an electrical charge or dipole. This induced moment is a function of the distortion polarization of the molecule, and for other than monatomic molecule involves both atomic and electronic polarization. The permanent moment arises from permanent asymmetric distribution of charges within the molecule and is measured by orientation polarization of the molecule in an applied field. Substances having appreciable dielectric constants usually have appreciable moments, but there is not always correspondence between the two. Thus, the respective dielectric constants for formamide, N-methylformamide water, N-dimethylformamide, and N-dimethylacetamide are 111.3, 185.5, 78.5, 37.6, and 37.8, and the respective dipole moments of the compounds are 3.68, 4.2, 1.8, 3.82, and 3.79 D.U. The data are for 25°C.

The refractive index is a third empirical parameter of solvent polarity. The square of this parameter is equal to the dielectric constant ($N^2 = D$) for those nonpolar substances whose molecules are not permanent dipoles and when the refractive index is measured at sufficiently long wavelengths.

The molar transition energy E_T, from the visible spectrum of pyridinium N-phenol betaines in solvents [174], is a parameter related to the polarity of those solvents [101].

Listed [101] among the empirical parameters of solvent polarity are molar transition energies from the spectrum of 1-ethyl-4-methoxycarbonyl-pyridinium iodide [175], a Y factor obtained from rates of $S_N 1$ solvolysis of alkyl halides and tosylates [176], a Ω function from the ratio of endo to exo products in the Diels–Alder addition of cyclopentadiene to methyl acrylate in different solvents [177], an S factor from Brownstein's [178] empirical generalization of all solvent parameters, and $\log k$ ion ($\sec^{-1}$) for the rate of ionization of p-methoxyneophyl-p-toluenesulfonate [179].

Of these empirical polar aparameters of the solvent, only E_T was found [101] to consistently correlate with the rate of the reaction

$$CH_3I + Cl^- \rightarrow CH_3Cl + I^- \qquad (5.318)$$

In amide solvents at 25°C over a fairly narrow range of values, E_T is found to decrease with increasing $\log k'$ for the reaction. Thus, in going from formamide, to N-methylformamide, to N-dimethylformamide as solvents; $\log k'$ for the above reaction changes from -4.28 to -3.84, to -0.38, and E_T changes from 56.6, to 54.1, to 43.8 kcal/mole. The dielectric constants and the dipole moments of the solvents change, respectively, in this series of solvents from 111.3 to 185.5 to 37.6, and moments from 3.68 to 4.2 to 3.82. The ability of these and other amide solvents to hydrogen-bonds correlates better with $\log k'$ for the above reaction in the respective solvents. As the ability to hydrogen-bonds goes down, the value of $\log k'$ goes up.

It has been proposed [180] that both $S_N 2$ and $S_N 1$ reactions can be divided into subclasses according to charge types as a helpful method of rationalizing certain kinetic variations. This classification involved substituting reagents which may be either negatively charged, Y^-, or neutral, Y, and eliminated atoms or groups before reaction which may be formally neutral, RX, or positive, RX^+. See Table 5.7. When X is neutral but R is negatively charged, as in the α-bromoacetate ion, then depending on whether Y is neutral or

Table 5.7

Charge Types of Reactants, Complex, and Products in S_N2 Reactions[a]

Type	Reactants	Complex	Products
1	$Y^- + RX$	$\overset{-\delta}{Y} \cdots R \cdots \overset{-\delta}{X}$	$YR + X^-$
2	$Y + RX$	$\overset{+\delta}{Y} \cdots R \cdots \overset{-\delta}{X}$	$^+YR + X^-$
3	$Y^- + RX^+$	$\overset{-\delta}{Y} \cdots R \cdots \overset{+\delta}{X}$	$YR + X$
4	$Y + RX^+$	$\overset{+\delta}{Y} \cdots R \cdots \overset{+\delta}{X}$	$^+YR + X$

[a] Pocker [181].

negative, special treatment is required [33]. The assumptions which have been made concerning the reactants and respective intermediate complexes are (1) increase in magnitude of charge will increase solvation, (2) increase in dispersal of charge will decrease solvation, and (3) destruction of charge will decrease solvation more than dispersal of charge. With respect to the solvent, the term polarity was synonymous with power to solvate solute charges and was assumed to increase with the dipole moment of the solvent molecule, and to decrease with increased depth of shielding of the charges of the dipole.

Applying these assumptions, the gross effect of the solvent on reactions of different charge types have been summarized [181] as follows: For ion–molecule reactions the total magnitude of charge is present in the initial and transition states, but is dispersed over more atoms in the transition state, and the rate should show a small decrease with an increase in the polarity of the solvent. For $S_N 2$ reactions between two neutral molecules polar solvents would greatly increase the rates, since charges are generated in the transition states.

As examples of type 1 reactions listed in Table 5.7, one might list the iodine ion–methyl iodide reaction

$$I^- + CH_3I \rightarrow \overset{-\delta}{I} \cdots CH_3 \cdots \overset{-\delta}{I} \rightarrow ICH_3 + I^- \qquad (5.319)$$

the rate constant for which increases by about 10^4 in changing from water to acetone as the solvent [182], and the azide ion–ethyl bromide reaction

$$N_3^- + C_2H_5Br \rightarrow \overset{-\delta}{N_3} \cdots C_2H_5 \cdots \overset{-\delta}{Br} \rightarrow N_3C_2H_5 + Br^- \qquad (5.320)$$

the second-order rate constant for which increases by about 5×10^2 when the solvent is changed from ethanol to nitroethane.

At 25°C the dielectric constant of water is 78.5 and that of acetone is 30.3. Since the reaction represented in Eq. (5.319) is one between an ion and a dipolar molecule the rate should increase slightly in going from much more polar water to much less polar acetone as measured by difference in dielectric constant. However, the large effect on the rate constant in these two solvents is not governed by the difference in polarity, but by the fact that water is a protic and acetone a dipolar aprotic solvent.

The ion–dipolar molecule reaction represented in Eq. (5.320) increases in rate constant by about 5×10^2 in changing from ethanol solvent with a dielectric constant of 25 to nitroethane solvent with a dielectric constant of 37. This increase of rate constant with increase of dielectric constant is not in harmony with the theory of the influence of the polarity of solvents on rates of chemical reactions. The large increase of the rate constant in going from ethanol to nitroethane is mainly dependent upon the fact that ethanol is a protic solvent and nitroethane is a dipolar aprotic solvent. The relative solvating power of the two media also are of importance. For ion–dipolar molecule reactions several cases are known in which the rate of reaction increases with decrease in the solvating power of the medium. Relatively, other factors being equal, nucleophilicity should increase with increase in the energy of solvation of an anion. The ability to solvate is low in most nonhydroxylic solvents, and this promotes ion pairing; which in turn would lead to a reduction of nucleophilicity of the anionic reagent; and would become evident by a decrease in the second-order rate constant with increase in total concentration of the saline species [181–185].

It has been found that the activity of the free ion is greater than that of the ion-pair, though the contribution of the latter is difficult to evaluate [181, 184–186]. In the case of the lithium radiobromide-n-butyl bromide reaction in acetone at 25°C, the second-order rate constant decreased 360% for a 400 fold increase in salt concentration. Evans and Sugden [184] demonstrated that a constant value of rate constant was obtained, assuming that only the dissociated bromide ion is reactive. Thus, in the case of the reaction represented by Eq. (5.140), if the nonhydroxylic solvent nitroethane enhanced ion-pairing, the rate of the reaction could be expected to be slower in nitroethane than in ethanol. The fact that the reverse is true is as said above, attributable to the protic nature of ethanol and the dipolar aprotic nature of nitroethane.

We present another example of a class 1 reaction. For the n-propyl tosylate–bromide ion reaction at 50°C in DMSO and in v:v 70% DMSO:30% H_2O, it was found [187] that the rates in DMSO were in the order $Cl^- > Br^- > I^-$ and in the aqueous DMSO the rates were in the opposite order. Thus, in DMSO the rate constants 10^3k in l./mole/sec are 7.93, 4.88, and 1.53 for Cl^-, Br^-, and I^-, respectively; while in v:v 70% DMSO:30% H_2O, 10^3k had the values 0.250, 0.318, and 1.663 for Cl^-, Br^-, and I^-, respectively. The reversed nucleophilic order was attributed to small difference in halide solvation in DMSO and in water. The rates were than rationalized by assuming the order of nucleophilicities of the unsolvated ions toward n-propyl tosylate to be $Cl^- > Br^- > I^-$.

Menshutkin and solvolytic reactions are of type 2 as listed in Table 5.7. The former involves the reaction between an alkyl halide and a tertiary amine to yield a quaternary ammonium salt:

$$RX + R_3N \rightleftarrows R_4N^+X^- \rightleftarrows R_4N^+ + X^- \qquad (5.321)$$

The reaction between ethyl iodide and triethylamine has been studied [188] in a variety of solvents. Reaction rates were obtained which are not even in the same order as the dielectric constant. These results are, therefore, opposite to the predictions of electrostatics. Laidler and Eyring [47] mention Menshutkin's [189] conclusion that there is a rough relation between the two. Nonelectrostatic influences, especially solvation, are given by Laidler as the probable explanation of this nonconformity with electrostatic theory.

Kirkwood [190] suggested that the free energy of activation was proportional to the term $(D-1/2D+1)$ where D is the macroscopic dielectric constant. As we have seen earlier in this chapter Laidler and Eyring [47] included this term in a rate equation to account for electrostatic effects on dipolar molecule–dipolar molecule reaction rates. These authors applied the theory to Menshutkin reactions and found a linear dependence between $\log k'$ and $(D-1)/(2D-1)$ for benzene–alcohol mixed solvents, but in benzene–nitrobenzene the plot was definitely curved. Water and alcohols are both electrophilic and nucleophilic in character and will solvate both cations and anions as well as both developing cations and anions. Acids are electrophilic solvents and will dissolve cations which they readily solvate. Nucleophilic solvents such as amines and ethers will solvate anions, but will also solvate neutral nucleophilic substances and will consequently reduce both the nucleophilicity and the base strength of the substance. This is true, for example, of amines. Depending on the solvent, either primary, secondary, or tertiary amine can become the strongest base [191]. Parallel effects should be evident with respect to the nucleophilicity of the three types of amines toward saturated carbon atoms.

It is evident that solvents cannot be treated as continuous dielectric media, and specific solvent effects arise in the close proximity of the reacting species. Pocker [181] points out that the use as solvents of aromatic and, in general, multiple ring substances, rather than corresponding aliphatic derivatives of similar dielectric constants, leads to higher rates for most functional groups, presumably because aromatic solvent molecules can further stabilize the charge separation in the transition state of a Menshutkin type reaction owing to the high capacity for polarizability of the π-electrons in the aromatic ring.

Hydrogen-bonding compounds such as methanol, phenol, and p-nitrophenol enhance increasingly so with increasing acidity of the hydroxylic addend the rate of reaction of methyl bromide with pyridine in benzene [192]. Nitrobenze likewise catalyzes this reaction to about the same extent as does methanol [193]. The catalytic curves are not linear over a sufficient range. It is believed [181] that these catalytic effects are due to the stabilization of the incipient charges in activated complex by a polar covalent effect rather than by the formation of any stochiometric complex among halide, amine, and one molecule of the second solvent.

A reaction illustrative of type 3 reactions is the reaction of trimethylamine and the trimethylsuphonium cation [194]. Basically this type of reaction should be similar to type 1 substitutions. The reaction is

$$\overset{+}{Me_3N} + MeS(Me)_2 \rightarrow [\overset{+\delta}{Me_3N} \cdots Me \cdots \overset{+\delta}{SMe_2}] \rightarrow \overset{+}{Me_3NMe} + SMe_2$$

$$(5.322)$$

Reactions of this charge type prove to be rather insensitive to changes in the polarity of the solvent and to polar–dipolar aprotic solvent transitions. The rate of the reaction does increase by over a hundredfold in going from water, to methanol, to ethanol, to nitromethane, the rate constants at 44.6°C in the respective solvents being 6.53×10^{-6}, 4.13×10^{-5}, 6.67×10^{-5}, and 7.75×10^{-4} liters/mole/sec. The dielectric constants of these solvents at 25°C are 78.5, 32.6, 24.3, and 46.3, respectively, and hence, this measure of polarity does not completely correlate with the trend of the rate constants. In general, however, the more polar the solvent the lower the rate. This decrease in rate with increased polarity of solvent has been attributed [181] to two effects based on energy and entropy of activation. First, the solvation energy of the smaller reactant ion is larger than that of the larger transition state ion causing an increase in energy of activation with the stronger solvating media and hence the slower reaction rate. The energies of activation for the reaction, Eq. (5.322), in the above solvents have the respective values, 23.07, 21.6, 20.55, and 18.0 kcal/mole. Second, the larger intermediate state ion will have a less ordered solvation shell causing a gain in its statistical probability which would produce an increase in the entropy and of the Arrhenius frequency factor in the stronger solvating media. This is illustrated by the values of the Arrhenius A of 4.9×10^{10}, 3.05×10^{10}, 9.44×10^{9}, and 1.90×10^{9} in the respective solvents given above for the reaction represented in Eq. (5.322). Pocker [181] suggests that a more quantitative treatment would include a consideration of the solvation of the amine whose nucleophilicity is strongly reduced by hydrogen-bonding media.

A reaction which corresponds to a type 4 reaction is the decomposition of a tri-*n*-sulphonium halide. For example, Pocker and Parker [195] and Mac *et al.* [196] have studied the reaction

$$CH_3 - S^+(CH_3)_2 + Br^- \rightleftarrows [\overset{-\delta}{Br} \cdots CH_3 \cdots \overset{+\delta}{S(CH_3)_2}] \rightleftarrows CH_3BR(CH_3)_2S$$

$$(5.323)$$

and found the rate to decrease with solvent polarity in agreement with the Hughes–Ingold theory of solvent effect on reactions of this charge type. It was found in ethanol [195] and in dimethylacetamide [196] that the second-order rate constant markedly decreased from k^0 that at zero ionic strength to k^μ that at ionic strength μ, produced by increasing the concentration of Me_3BrS. This behavior was more satisfactorily explained by taking into account the degree of dissociation, α, of the weak electrolyte Me_3SBr using the equation

$$k^\mu = k^0\alpha^2 \qquad (5.324)$$

than by applying the conventional Bronsted–Bjerrum theory represented by the equation

$$\log k^\mu = \log k^0 + Z_A Z_B A\mu^{1/2} \qquad (5.325)$$

As mentioned before, the forward reaction was about 10^3 times slower in protic 88% MeOH–12% H_2O solvent than in dipolar aprotic DMF or DMAC solvents of like dielectric constants.

B. SUMMARY

Polarity of solvent as represented by dielectric constant, dipole moment, refractive index, and various transition energies do not lend themselves to a quantative correlation of solvent effects upon reaction rates, except when the predominant influence of the solvent is electrostatic as exerted by its dielectric constant on the electrical forces between charged particles, or on forces between charged and electrically unsymmetrical particles, or on forces between electrically unsymmetrical and electrically unsymmetrical particles. These effects have been discussed in previous sections dealing with electrostatic influences on chemical reaction rates. There are some other qualitative correlations of solvent polarity and reaction rates, but generally, there are other than solvent polarity effects simultaneously influencing rates and many times these other solvent effects are the predominant ones. Thus, in the case of $S_N 2$ reactions the protic–dipolar aprotic solvent influence far overshadows any polar solvent effect arising from the dielectric constants of the solvents.

XVI. Pressure and Reaction Rates

A. INTERNAL PRESSURE OR COHESION

There are two classes of pressure effects on reaction rates. There is the effect of internal pressure or cohesion of the solvent which will be discussed in this section, and the effect of external pressure.

The effect of internal pressure or cohesion of the solvent on reaction rates has been discussed by different authors [102, 197–199]. The internal pressure of solvent has been calculated [200] using the expression $E_\sigma/v^{1/3}$, where E_σ is the total surface energy in dynes per centimeter and v is the molar volume in cubic centimeters. Another factor used to calculate [201] cohesion is L/v, where L is the latent heat in joules per mole.

An empirical rule [198] states that solvents of high cohesion cause acceleration of the rates of reactions having products of higher cohesion than reactants, and cause a deceleration of the rates of reactions having products of lower cohesion than the reactants. The opposite would be true for solvents of low cohesion. Solvents would have little influence on reactions rates when reactants and products have like cohesions. These rules have been theoretically justified [202, 203]. The term "internal volume factor" has been used [202] to describe this phenomena.

From the first part of the rule given above, reactions obeying the rule should be accelerated by solvents of high cohesion. In general, this part of the rule is obeyed as is shown by the data in Table 5.8. That the second part of the rule is fairly accurately obeyed is illustrated by the data of Richardson and Soper [198] given in Table 5.9.

Table 5.8

Influence of the Cohesion of the Solvent on the Specific Rates of Quaternary Salt Formation

Solvent	$E_\sigma/v^{1/3}$	L/v	Tetraethylamine and ethyl iodide (100°C)	Tetraethylamine and ethyl bromide
Hexane	9.45	243.3	0.000180	—
p-Xylene	12.12	292.2	0.00287	0.000103
Benzene	15.29	275.0	0.00584	0.000228
Chlorobenzene	14.68	360.8	0.00231	0.000843
Acetone	14.46	408.3	0.0608	0.0024

Table 5.9

Influence of the Cohesion of the Solvent on the Specific Rates of the Esterification of Isopropyl and of Isobutyl Alcohols by Acetic Anhydride

Solvent	$E_\sigma/v^{1/3}$	L/v	Acetic anhydride with isopropyl alcohol	Acetic anhydride with isobutyl alcohol
Hexane	9.45	243.2	0.0855	0.0307
Xylene	12.12	292.5	0.0510	0.0196
Benzene	15.29	275.0	0.0401	0.0148

As to the third part of the rule, there is little solvent effect on the conversion in the presence of ethyl tartrate of anissynaldoxime into anialdoxime for which the cohesion of products and reactants are probably nearly the same. Laidler [104] believes that deviations from these rules often arise from solvation effects in solutions which are not regular, i.e., ones in which the molecular distributions are not entirely random [204]. He writes for the activity coefficient γ of a solute the expression

$$RT \ln \gamma_1 = v_1 \left(\frac{x_2 v_2}{x_1 v_1 + x_2 v_2} \right)^2 \left\{ \left(\frac{E_1}{v_1} \right)^{1/2} - \left(\frac{E_2}{v_2} \right)^{1/2} \right\} \qquad (5.326)$$

where x_1 and x_2 are the respective mole fractions of solute and solvent, v_1 and v_2 their respective molar volumes, and E_1 and E_2 the respective heats of vaporization in the pure state of the solute and solvent. For dilute solutions x_1 is small and $x_1 v_1 \ll x_2 v_2$, hence

$$RT \ln \gamma_1 = v_1 \left\{ \left(\frac{E_1}{v_1}\right)^{1/2} - \left(\frac{E_2}{v_2}\right)^{1/2} \right\} \tag{5.327}$$

If a is the van der Waals attraction constant for a substance and P its cohesion or internal pressure, $E/v \sim a/v^2 \sim P$, and Eq. (5.327) becomes

$$RT \ln \gamma_1 = v_1 [P_1^{1/2} - P_2^{1/2}]^2 \tag{5.328}$$

but

$$\ln k' = \ln k_0' + \ln(\gamma_1 \gamma_2/\gamma_x) \tag{5.329}$$

where k_0' is the specific velocity for the ideal solution. Substituting from Eq. (5.328) for the pertinent species into Eq. (5.329) gives

$$\ln k' = \ln k_0' + (1/RT)\{v_A [P_A^{1/2} - P_S^{1/2}]^2 + v_B [P_B^{1/2} - P_S^{1/2}]^2$$
$$- v_x [P_x^{1/2} - P_S^{1/2}]^2\}$$
$$= \ln k_0' + (1/RT)[v_A \Delta_A + v_B \Delta_B - v_x \Delta_x] \tag{5.330}$$

In these equations the subscript S refers to the solvent and the Δ's refer to the respective $[P_i^{1/2} - P_S^{1/2}]^2$ terms, which are always positive.

Let symbols marked with subscripts c and d represent quantities pertaining to the combination and the disproportionation states, and k_s' and k_g' the ratios k_d'/k_c' in solution and the gas phase, respectively, for the cage reaction between two geminate ethyl radicals as shown in the following equation:

$$(2C_2H_5) \begin{array}{c} \nearrow C_4H_{10} \\ \searrow C_2H_4 + C_2H_6 \end{array} \tag{5.331}$$

The source of the two caged ethyl radicals is the photolysis of azeoethane [205]. The equation for the ratio k_d'/k_c' is written

$$RT \ln(k_s'/k_g') = (V_d - V_c) P_S - 2(V_d P_d^{1/2} - V_c P_c^{1/2}) P_S^{1/2} + (V_d P_d - V_c P_c) \tag{5.332}$$

The quadratic Eq. (5.332) may be converted into linear form using the reasonable assumption that $V_d = V_c = V$ giving

$$RT \ln(k_s'/k_g') = 2V(P_c^{1/2} - P_d^{1/2}) P_S^{1/2} - V(P_c - P_d) \tag{5.333}$$

This equation was tested by plotting $\ln(k_s'/k_g')$ versus $P_S^{1/2}$ for the combination and disproportionation reaction represented in Eq. (5.331). The ratio (k_d'/k_c') was taken as the product ratio C_2H_4/C_4H_{10}, which was determined with a gas chromatograph packed either with silica gel, firebrick, or alumina. This ratio for a given solvent divided by the ratio for the gas phase gave (k_s'/k_g'). The data for the photolysis represented in Fig. 5.10 was for

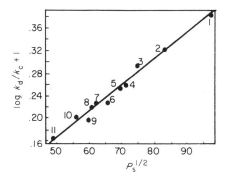

Fig. 5.10. Combination and dispropor-tionation of ethyl radicals in solution. $\log(k_d/k_c)$ as a function of the square root of the internal pressure of the solvent. (1) Ethylene glycol, (2) acetonitrile–styrene scavenger, (3) acetonitrile, (4) aniline–styrene scavenger, (5) 1-propanol–styrene scavenger, (6) 2-butanol, (7) toluene, (8) m-xylene, (9) ethylbenzene, (10) limonene, and (11) isooctane.

65°C. The values of internal pressure used were taken from Hildebrand and Scott [204] or were calculated using Hildebrand's method. The empirical, linear equation for Fig. 5.10 was found to be

$$\log(k_d'/k_c') = -1.08 + 4.76 \times 10^{-3} P_s^{1/2} \qquad (5.334)$$

and it is observable that within the validity of the approximations made, the internal pressure of the combination transition state is greater than the dis-proportionation transition state. $V(P_c - P_d)$ and $V(P_c^{1/2} - P_d^{1/2})$, respectively, were found to be 12.1 and 0.152, taking the value of k_d'/k_c' to be 0.13; at 65°C the molar volume V lies between that of butane (ca. 100 cc) and twice that of ethane (ca. 140 cc), but perhaps closer to the latter value [206, 207]. It was, therefore, estimated [196] that the difference $(P_c - P_d)$ is less than 100 atm and that the difference $(P_c^{1/2} - P_d^{1/2})$ is approximately 1. Thus the transition states are very similar, and their potential energies differ at most by only a few kT, as is supported by the fact that the rate ratio k_d'/k_c' in the gas phase is independent of the deactivating collisions and independent of the efficiency of the inert molecules acting as energy transfer agents.

For this combination-disproportionation reaction couple, the nature of the transition states involved has been discussed [208, 209] for the case of the gaseous system. It was suggested that the two competing processes pass through a common set of intermediates of the nature suggested by the equili-brium of the two partially bonded species depicted in the following equation

$$CH_3CH_2 + CH_3CH_2 \longrightarrow \underset{\underset{CH_3}{|}}{H_2C \cdots \overset{\overset{CH_3}{|}}{CH_2}} \rightleftharpoons \underset{\underset{H_2C \cdots H}{|}}{H_2C \cdots \overset{\overset{CH_3}{|}}{CH_2}} \qquad (5.335)$$

$$\downarrow \qquad\qquad\qquad \downarrow$$

$$C_4H_{10} \qquad\qquad C_2H_4 + C_2H_6$$

Stefani [199] assumes that the mechanism implied in Eq. (5.335) is correct, although there is a controversy with respect to the question of a head-to-tail versus a four center disproportionation state [210].

The cage diffusion reaction system for trifluoromethyl radicals was investigated at 65°C in 27 solvents and at 30° in 6 solvents [211]. The trifluoromethyl radicals were prepared by the photolysis of hexafluoroazomethane. Quantitative measurements were made on the fraction of the radicals undergoing the cage reaction and diffusion at different temperatures and in different solvents. An energy of activation defined as $E_a = E_{cage} - E_{diff}$ was calculated to be constant in nearly all the solvents examined (27 solvents). There were discrepancies observed in hydroxylic solvents. The logarithms of the probability ratios F_{cage}/F_{diff} were found to be linearly related to the reciprocal of the rate of change of boiling point of the solvents with pressure, which was interpreted to mean that solvation interactions were similar in all solvents. Constancy of E_a for both ethyl and trifluoromethyl radicals in the different solvents was thought to be accidental and resulted from a compensating effect between the transfer and immediated local consequences of an excess energy in the radicals which controls the extent of the cage reaction, and the self cohesion of the solvent which controls the size and relaxation of the cage.

B. External Pressure

From Le Chatelier's principle, the rate of a reaction will be increased by an increase of external pressure if the volume of the critical complex molecules is less than the sum of the volumes of the reactant molecules, while the rate of reaction will be decreased by increase of external pressure if the volume of the critical complex molecules is greater than the sum of the volumes of the reactant molecules. For decrease in external pressure, the opposite would be true. Evans and Polanyi [212] first studied the effect of external variables on reaction rates. Their observations were that reaction rates increased by increase of pressure and hence, that the volume of the critical complex molecules was less than the sum of the volumes of the reactant molecules. The statement that reactions accompanied by contraction should be accelerated by increase of pressure was limited to those reactions in which the density of the transition state is intermediate between that of the initial and final states. For example there is a contraction between the initial and final states in the cis–trans isomerization of fumaric acid. However, the reaction is not accelerated by pressure, perhaps due to the extension of the C–C linkage in the transition state, which causes this state to be less dense than either the initial or final states. On the other hand increase of pressure will accelerate the combination of hydrogen with carbon monozide to give formaldehyde and polymerization reactions, where it is reasonable to suppose that the transition state will have a density intermediate between that of the initial and final states of the system.

The equation for the dependence of the rate constant on pressure was written [212]

$$\frac{\partial \ln k'}{\partial P} = \frac{\Delta V}{RT} \qquad (5.336)$$

where $\Delta V = V_1 - V_2$, the difference in the volumes in the initial and transition states. In the case of changes in the nature and magnitude of forces between solvent and solute molecules as well as changes in the solute molecules themselves as they go from initial to transition state, ΔV must consist of two terms accounting for both resultant volume changes. For neutral particles reacting to give neutral particles, the solute–solvent term will presumably not be very great, but it could be large for charged particles giving a neutral intermediate, or, in general, when reactants and intermediate complex are markedly different in charge or polarity. The necessary equation is

$$\frac{\partial \ln k'}{\partial P} = \frac{\Delta_1 V}{RT} + \frac{\Delta_2 V}{RT} \qquad (5.337)$$

where $\Delta_1 V$ is the difference in the volumes of the initial and final states independent of the solvent, and $\Delta_2 V$ is determined by the difference in the electrostatic forces between the reactant species and the solvent and the electrostatic forces between the transition state and the solvent. In the case of an organic substitution reaction of the type $X^- + RY \rightarrow X^- RY$ (or XRY^-) $\rightarrow$ $XR + Y^-$, the second term would be important since the electrostatic forces between the transition state and the solvent will be less than those between reactants and solvent and since in the transition state the molecule RY would screen off the solvent on one side of X^-. The theory of hydrostatic pressure effects on rates of reaction was first formulated by van't Hoff [213] who suggested an equation similar to Eq. (5.336). Other internal variables can be treated in a manner analogous to pressure.

Equation (5.336) was derived by Benson [81] who divided the transition state rate equation into kinetic and thermodynamic contributions and considered the effect of change of external thermodynamic variables of state, here confined to pressure, in the rate. The equation for the rate constant k' for the rate of appearance of products from the transition state is

$$k' = \varkappa_X K_X \nu_X \frac{f_A f_B}{f_X \cdots} \qquad (5.338)$$

where $\varkappa_X$ is the transmission coefficient, K_X is the equilibrium constant between the transition state X and the reactants A, B, ..., ν_X is the mean frequency with which X passes through the critical configuration on the potential energy diagram, and $f_A, f_B, f_X \cdots$ are the activity coefficients of the respective species.

K_X and f's are thermodynamic and $\varkappa_X$ and v_X are kinetic in character. Setting $\varkappa_X$ equal to one and letting

$$k_X' = k_X{}^* \frac{kT}{v_X} \tag{5.339}$$

since v_X can be taken as a normal vibrational frequency whose partition function can be factored out of K as kT/h, gives

$$k' = \frac{kT}{h} k_X{}^* \frac{f_A f_B}{f_X} \tag{5.340}$$

The method of partition functions for the resolution of K_X would necessitate the analysis and factoring of v_X, however, as an approximation Eq. (5.340) contains only thermodynamic factors, and only these need be analyzed in the theoretical analysis of specific velocity constants. With respect to their influence on reaction rate, the only external variables that need be treated are the thermodynamic variables of state of which pressure is the only one to be considered here. Taking the partial derivative with respect to pressure of the logarithmic form of Eq. (5.340) yields

$$\frac{\partial \ln k'}{\partial P} = \frac{\partial \ln T}{\partial P} + \frac{\partial \ln k_X{}^*}{\partial P} + \left(\frac{\partial \ln f_A}{\partial P} + \frac{\partial \ln f_B}{\partial P} - \frac{\partial \ln f_X}{\partial P} \cdots \right) \tag{5.341}$$

If we limit ourselves to dilute solutions at constant temperature

$$\left(\frac{\partial \ln k'}{\partial P} \right) = \left(\frac{\partial \ln k_X{}^*}{\partial P} \right)_T \tag{5.342}$$

But, from thermodynamics

$$RT \ln K_X{}^* = -\Delta \bar{F}_X \tag{5.343}$$

and

$$(\partial \bar{F}^*/\partial P)_T = \Delta \bar{V}_X{}^* \tag{5.344}$$

Hence,

$$RT \left(\frac{\partial \ln k'}{\partial P} \right)_T = RT \left(\frac{\partial \ln k_X{}^*}{\partial P} \right)_T = -\left(\frac{\partial \bar{F}_X}{\partial P} \right)_T = -\Delta \bar{V}_X{}^* \tag{5.345}$$

where $\Delta \bar{V}_X{}^* = \bar{V}_X - \bar{V}_A - \bar{V}_B - \ldots$, and $\bar{V}_i$ is the partial molal volume of the ith component of the solution. For other than strong interactions between $A, B, \ldots, \bar{V}_X, \bar{V}_A, \bar{V}_B$ approximately equal $V_X, V_A, V_B, \ldots$, respectively, then

$\Delta V_s{}^* = \Delta V_X$, and Eq. (5.345) becomes identical with Eq. (5.336) when solute–solvent interactions are neglected. For small enough interactions between A, B, ..., the $V_X - V_A - V_B = \Delta V_X{}^* = 0$, and pressure does not affect the rate of reaction.

For dilute, nonideal solutions the equation

$$RT\left(\frac{\partial \ln k'}{\partial P}\right)_{T, x_i} = -\Delta V_X{}^* = RT\left(\frac{\partial \ln k'}{\partial P}\right)_{T, c_i} + (n-1)\,RT\beta_s \quad (5.346)$$

can be written [81], where n is the order of the reaction and β_s is the coefficient of compressibility of the solution. Thus from Eq. (5.346) the change of the logarithm of the rate constant with pressure at constant temperature will depend on the change of the partial molal volume for the transition-state reaction. Since $\Delta V_x/RT$ is of the order of magnitude of 0.00/atm, change of pressure will be influential in changing the rate constant only at pressures of 100 atm or greater.

By measuring rates over a sufficiently wide range of pressures, V_X and β_s may be found since V_A and V_B can be determined independently.

Moelwyn-Hughes [214] using the van Laar [215] relations writes for the chemical potential μ_2 (partial molal free energy) of component 2 in a binary liquid mixture the equation

$$\mu_2 = kT \ln f_2 + kT \ln x_2 + u_2{}^0 + Pv_2 + x_1{}^2 \Delta u^0 \quad (5.347)$$

where f_2 is the activity coefficient, x_2 the mole fraction, $u_2{}^0$ the average potential energy of one molecule in the pure state, and $\bar{v}_2$ is the partial molecular volume of component 2 in the solution; x is the mole fraction of component 1 in the solution, P is the external pressure, and Δu^0 is the interchange energy or the average increase in the energy per molecule of either kind when it exchanges all its neighbors for neighbors of another kind. At the same temperature and pressure, the chemical potential of pure component 2 may be written

$$\mu_2{}^0 = -kT \ln f_2 + u_2{}^0 + Pv_2{}^0 \quad (5.348)$$

where $v_2{}^0$ represents the molecular volume of pure component 2. Using Eq. (5.347) and Eq. (5.348) the equation

$$\mu_2 = \mu_2{}^0 + kT \ln x_2 + P(v_2 - v_2{}^0) + x_1{}^2 \Delta u^0 \quad (5.349)$$

can be obtained. For an activated molecule of the same species under identical conditions, a similar equation can be written:

$$\mu_2{}^* = \mu_2{}^0 + kT \ln x_2{}^* + P(\bar{v}_2{}^* - v_2{}^0) + x_2{}^2 \Delta u^0 + e' \quad (5.350)$$

where e' is the additional internal energy of an active molecule and $\bar{v}_2{}^*$ is its partial molecular volume at the temperature and pressure of the system.

For dilute solutions $x_1 = 1$ and at equilibrium $\mu_2 = \mu_2^*$, and hence from Eqs. (5.350) there is obtained

$$\mu_2^0 + kT \ln x_2^* + P(\bar{v}_2^* - v_2^0) + x_2^2 \Delta u^0 + e' = \mu_2^0 + kT \ln x_2$$
$$+ P(v_2 - v_2^0) + x_2^2 \Delta u^0 \tag{5.351}$$

which can be solved to give

$$\frac{n_2^*}{n_2} = \frac{x_2^*}{x_2} = \exp\left[-\frac{P(\bar{v}_2^* - v_2)}{kT}\right]\exp\left[-\frac{e'}{kT}\right] \tag{5.352}$$

since the ratios of mole fractions of activated and normal molecules and of the concentrations of activated and normal molecules are equal. The number of molecules decomposing per cubic centimeter per second is vn^*, where v is the probability per second that an activated molecule will react. The specific first-order velocity constant k' is the fraction of the molecules decomposing per cubic centimeter per second and is given by vn^*/n. Thus multiplying both sides of Eq. (5.352) by v gives

$$k' = v \exp\left[-\frac{P(\bar{v}_2^* - \bar{v}_2)}{kT}\right]\exp\left[-\frac{e'}{kT}\right] \tag{5.353}$$

In units of moles per liter, Eq. (5.353) becomes

$$\ln k' = \ln v - \frac{E}{RT} - \frac{P(\bar{V}_2^* - \bar{V}_2)}{RT} \tag{5.354}$$

where the V's refer to partial molal volumes. If v' is independent of pressure and if subscripts are dropped Eq. (5.354) reduces to

$$\ln k' = \ln k_0' - \frac{P(\bar{V}^* - \bar{V})}{RT} \tag{5.355}$$

This is Moesveld's [216] law. A plot of $\ln k'$ versus P should be a straight line with a negative slope if $\bar{V}^* > V$, and with a positive slope if $\bar{V}^* < V$. From the slope $\bar{V}^* - \bar{V}$, and if $\bar{V}$ for the normal molecules is known, $\bar{V}^*$ for the activated molecules should be obtainable. Moelwyn-Hughes [214] lists $(k'_{2000\ atm}/k'_{1\ atm})$ and $\bar{V}^* - \bar{V}$ at 25°C for various reactions. These range from 5.98 to 0.76 for $(k'_{2000\ atm}/k'_{1\ atm})$ and from -21.8 to $+3.3$ cm^3/mole for dimerization of cyclopentadiene in the liquid state and for the formation of the quatenary ammonium salt $(CH_3)(C_2H_5)(C_6H_5)(C_6H_4CH_2)NBr$ in chloroform, respectively.

Reactions have been classed [217] into three groups with respect to the effect of pressure on their rates. One group composed of "normal" bimolecular reactions take place at a rate which can be calculated from the rate of collision between molecules having the required energy of activation. The Arrhenius frequency factors are of the order of 10^{10} liters/mole-sec. The hydrolysis of sodium monochloroacetate by sodium ethoxide and the reaction of sodium ethoxide and ethyl iodide are examples of this class of reaction. "Slow" reactions which occur at a rate several powers of ten less than that calculated from the rate of collision between molecules having the necessary energy of activation constitutes a second group of reactions. Examples of this group are the esterification of acetic anhydride with ethanol and the formation of quaternary ammonium salts. The Arrhenius frequency factor for this group of reactions is often much smaller than 10^{10} liters/mole-sec being 1.33×10^{7} for the reaction between pyridine and *n*-butyl bromide in acetone at 60°C and 3.85×10^{9} for the esterification of acetic anhydride with ethyl alcohol in excess ethanol solvent at 20°C. Decomposition such as that of phenylbenzyl-methylallylammonium bromide, which are slow and unimolecular are in the third group.

For the "normal" or first group reactions the reaction rates increase to a relatively small extent and in a roughly linear manner with increase of pressure, but tend to fall off at higher pressures. The increase in rate arises from a decrease of energy of activation with pressure, since for these reactions, the Arrhenius frequency factor shows a tendency to decrease only slightly with increase of pressure.

The "slow" or second group reactions show a much larger increase of rate with increase of pressure than do the "normal" reactions. For "slow" reactions the increase of the Arrhenius frequency factor more than compensates for the increase of energy of activation with increase of pressure permitting the rate to increase. For the Menshutkin reaction in acetone, both the frequency factor and the energy of activation increase with increasing pressure, but in methanol the frequency factor is increased but the energy of activation is little affected by pressure [218]. Both the frequency factor and energy of activation decrease with increasing pressure for the esterification of acetic anhydride by ethanol in excess ethanol as solvent. In the case of the Menschutkin reaction, structural variations were attributed [218] not to the commonly accepted theory of steric interference between nonbonded atoms and groups in the transition state, but to differences in degree and type of solvation of transition states imposed by the configurations of the reacting molecules. For more complex molecules, the acceleration of rate by increase of pressure is greater. Thus acceleration of reaction rate with pressure increases in the order Me $<$ Et $\approx$ Bu $<$ Pr for the reaction of the four iodides with dimethylaniline. It has been shown [218] that for these four alkyl iodides reacting with dimethylaniline in methanol solvents the volume of activation $\Delta \bar{V}^{*}$ varies inversely as the entropy of activation ΔS^{*}, so that greatest volume decrease is accompanied by the smallest entropy change. For $NPhMe_{2}$ reacting with

MeI, EtI, BuI, and i-PrI, $-\Delta V^*$ equals 26, 34, 34, and 47 cm³/mole respectively, and $-\Delta S^*$ equals 30.4, 29.0, 28.5, and 20.4 e.u., respectively, at 52.5°C in methanol.

According to Laidler weaker binding will be associated with greater values of the temperature coefficient of ΔV^* which can be obtained from the relationship $\partial \Delta V^*/\partial T = -\partial \Delta S^*/\partial P$. For the reaction of dimethylaniline with methyl iodide between 1 and 2875 atm $-\Delta S^*$ has been found to be 6 e.u., while for the reaction with ethyl and butyl iodide ΔS^* decreases by 11 e.u. [218]. These data are in harmony with Laidler's suggestion. However, $\Delta S^*_{2875} - \Delta S_1 =$ 4.8 e.u. for the reaction with isopropyl iodide, and it has been suggested [218] that the correlation does not apply when there are large variations between the binding of different solvent molecules.

The third group which has negative pressure coefficients are not well represented. One such reaction [219] is the first-order decomposition of benzoyl peroxide in carbon tetrachloride at 70°C. In the case of the decomposition of phenylbenzylmethylallylammonium bromide, the decrease in rate with increase of pressure is not marked. In this unimolecular decomposition both the frequency factor and the energy of activation decrease with increasing pressure, but the decrease in the frequency factor dominates and causes the decrease in the rate.

Perrin's observations can be summarized in the statement that the volume of activation and entropy of activation parallel each other. Laidler [104] illustrates this statement by a plot of volume of activation in cubic centimeters per mole versus entropy of activation in calories per degree per mole. The points lie on a straight line in spite of scatter.

Evans and Polanyi [220, 221] indicate two phenomena that must be considered in the interpretation of volume of activation. In the first place structural factors may cause a change in the volumes of the reactant molecules themselves as they become activated. This effect would cause a volume increase for unimolecular processes, and a volume decrease for bimolecular processes. In the second place rearrangement of the solvent molecules may cause a volume change. For reactions involving ions or fairly strong dipoles, the solvent effects are generally more pronounced than structural ones.

The parallelism between volumes and antropies of activation arise from the similarity of the solvent effects on the two [102]. If the charge is enhanced by reaction, as in the fusion of ions of like sign or the separation of ions of unlike sign, there is an increase in electrostriction due to an intensification of the electric field, and there is a decrease in both volume and entropy due to loss in freedom of solvent molecules. These reactions are generally of the "slow" type and are illustrated by reactions between ions of like sign, ester hydrolyses, esterifications, Menschutkin reactions, and unimolecular solvolyses. When two ions of opposite sign come together or any other process takes place that weakens the electric field when the activated complex is formed, bound solvent molecules will be released and both volume and entropy of activation will be increased. Such reaction types are represented by reactions between ions of opposite sign or by reverse Menschulkin reactions, and are included in the "fast" group. Those reactions in which there is little change in electric field and which undergo little change when the activated complex is formed, show small negative changes in both volume and entropy of activation. These are "normal" reactions and are represented by negative ion replacements.

Volumes of activation are much more constant than are entropies of activation for a given type of reaction since volumes depend primarily on electrostriction effects while entropies of activation are quite sensitive to the weakening or strengthening of chemical bonds and other similar phenomena [102]. Reaction mechanisms have been elucidated using pressure effects on reaction rates [222]. There are two groups of solvolysis reactions which are specifically catalyzed by oxonium ion, i.e., reactions in which there is a pre-equilibrium proton transfer to the substrate forming its conjugate acid. In group A-1, the conjugate acid decomposes bimolecularly and has a solvent molecule covalently bonded in the transition state. The volume of activation for the A-1 group will be close to zero and perhaps positive while the volume of activation for the A-2 group will be negative by at least several cubic centimeters per mole. Whalley [222] pointed out that for some reactions at least these differences in volume of activation could be used to distinguish between the A-1 and A-2 mechanisms. The activation quantities [223] for the hydrolysis of the substances shown in Table 5.10 are illustrative of this use of volumes of activation to determine mechanism. These volumes of activation are all close to zero or positive confirming that the hydrolyses take place by the unimolecular A-1 mechanism. The entropies of activation are all positive which is consistent with but not a confirmation of the A-1 mechanism.

The volumes of activation in Table 5.11 for the acid-catalyzed hydrolysis of epoxides are negative and so nearly equal that the mechanism was assumed

Table 5.10
Activation Volumes

Substance	Temperature (°C) for ΔE^* and ΔS^*	ΔE^* (kcal/mole)	ΔS^* (cal/deg/mole)	Temperature (°C) for ΔV^*	ΔV^* (cm³/mole)
Dimethoxymethane	10–60	26.39	+ 6.8	25	−0.5 ± ~0.5
Diethyoxymethane	10–50	25.39	+ 6.9	25	−0.0 ± ~0.5
Dimethoxyethane	0–25	22.70	+ 13.1	0	+1.5 ± ~0.5
				15	+1.8 ± ~0.5
Triethoxymethane	0–25	15.71	+ 6.1	0	+2.4 ± ~1.7

Table 5.11

Volume of Activation ΔV^* and Volume Charge ΔV° for the Complete Reaction for Acid-Catalyzed Hydrolysis of Epoxides

Epoxide	Temperature (°C)	ΔV^*	ΔV° (cm³/mole at 1 atm)
Ethylene	0	$-5.9 \pm \sim 1$	$-9.4 \pm \sim 0.5$
	15	$-7.4 \pm \sim 0.9$	—
	25	$-7.9 \pm \sim 0.7$	$-10.3 \pm \sim 0.5$
	40	$-8.9 \pm \sim 1.3$	$-11.2 \pm \sim 0.7$
Propylene	0	$-8.4 \pm \sim 1.3$	$-11.2 \pm \sim 0.5$
Isobutylene	0	$-9.2 \pm \sim 1.7$	—

to be the same for all the epoxides [224], although it had been suggested that a change of mechanism did occur in the series ethylene, propylene, isobutylene epoxides [225]. Since the volumes of activation are all around -6 to -9 cm³/mole, the mechanism was accepted to be the A-2 type for all three epoxide reactions. The A-2 mechanisms using ethylene oxide as an example involves the following steps

$$\text{H}_2\text{C}\text{—O}\text{—CH}_2 + \text{H}_3\text{O}^+ \rightleftharpoons \text{H}_2\text{C}\text{—}\overset{+}{\text{O}}\text{H}\text{—CH}_2 + \text{H}_2\text{O} \quad \text{equilibrium} \tag{5.356}$$

$$\text{H}_2\text{O} + \text{H}_2\text{C}\text{—}\overset{+}{\text{O}}\text{H}\text{—CH}_2 \xrightarrow{\text{slow}} \begin{array}{c}\overset{+}{\text{C}}\text{H}_2\text{OH}_2\\ |\\ \text{CH}_2\text{OH}\end{array} \tag{5.357}$$

$$\begin{array}{c}\overset{+}{\text{C}}\text{H}_2\text{OH}_2\\ |\\ \text{CH}_2\text{OH}\end{array} + \text{H}_2\text{O} \xrightarrow{\text{fast}} \begin{array}{c}\text{CH}_2\text{OH}\\ |\\ \text{CH}_2\text{OH}\end{array} + \text{H}_3\overset{+}{\text{O}} \tag{5.358}$$

It might be of interest to compare the above A-2 mechanism with the mechanism had it been A-1. The first step Eq. (4.356) for the A-1 mechanism would have been the same. Further steps in the A-1 mechanism would have been

$$\text{H}_2\text{C}\text{—}\overset{+}{\text{O}}\text{H}\text{—CH}_2 \xrightarrow{\text{slow}} \begin{array}{c}\overset{+}{\text{C}}\text{H}_2\\ |\\ \text{CH}_2\text{OH}\end{array} \tag{5.359}$$

$$\begin{array}{c}\overset{+}{\text{C}}\text{H}_2\\ |\\ \text{CH}_2\text{OH}\end{array} + \text{H}_2\text{O} \xrightarrow{\text{fast}} \begin{array}{c}\overset{+}{\text{C}}\text{H}_2\text{OH}_2\\ |\\ \text{CH}_2\text{OH}\end{array} \tag{5.360}$$

$$\begin{array}{c}\overset{+}{\text{C}}\text{H}_2\text{OH}_2\\ |\\ \text{CH}_2\text{OH}\end{array} + \text{H}_2\text{O} \xrightarrow{\text{fast}} \begin{array}{c}\text{CH}_2\text{OH}\\ |\\ \text{CH}_2\text{OH}\end{array} + \text{H}_3\text{O}^+ \tag{5.361}$$

The effect of pressure on the rate of solvolysis of benzyl chloride has been studied in the mixed solvents, acetone–water, dimethylsulfoxide–water, ethanol–water [226], methanol–water, isopropanol–water, t-butanol–water [227], and glycerol–water [228], and the rate of the solvalyzes of t-butyl chloride, benzyl chloride, and p-chlorobenzyl chloride in ethanol–water mixtures [226].

For the last three reactions mentioned, the transition state partial molal volume behavior, $\overline{V}_t$ as a function of the ethanol–water solvent composition was obtained by a dissection of the activation volume ΔV^* into initial $\overline{V}_g$ and transition $\overline{V}_t$ state components using the relation

$$\Delta V^* = \overline{V}_t - \overline{V}_g \tag{5.362}$$

The volume parameters for the three solvolyzes are shown [223] in Fig. 5.11. In all cases ΔV^* passes through a minimum as a function of the solvent composition of the solvent, and the initial state partial molal volume passes through a maximum. The transition state partial molal volume behaves differently for the three reactions. For the more $S_N 2$ type p-chlorobenzyl chloride $\overline{V}_t$ shows a maximum, while for the $S_N 1$ type butyl chloride solvolysis $\overline{V}_t$ demonstrates a distance minimum. For benzyl chloride $\overline{V}_t$ exhibits a behavior which is intermediate between these two extremes. The behavior of $\overline{V}_t$ as a function of solvent composition is apparently related to the position occupied by the reaction in question on the mechanistic scale from $S_N 2$ to $S_N 1$. It was therefore suggested [228] that the solvent dependence of the partial molal volume of the transition state $\overline{V}_t$ can serve as a useful yardstick in the characterization of reaction mechanism.

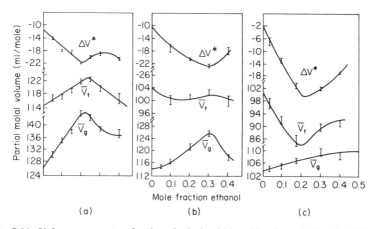

Fig. 5.11. Volume parameters for the solvolysis of (a) p-chlorobenzyl chloride, (b) benzyl chloride, and (c) t-butyl chloride in aqueous ethanol.

Extrema in the variation ΔV^* with solvent composition seems to be the rule for solvolyzes reactions in aqueous–organic binary solvent systems, and appear often in the activation parameter dependence on solvent composition [226].

In dilute solutions, rates of chemical reactions are markedly influenced by hydrostatic pressure, the influence being greater for reactions involving large and complex organic molecules and which at atmospheric pressure occur at much slower rates than would be expected from elementary considerations of collision frequencies and energy relationships.

In abnormal reactions the nascent particles must be stabilized by solvent, and for reactions of the type

$$A + B \underset{k_2'}{\overset{k_1'}{\rightleftharpoons}} AB \overset{k_3'}{\longrightarrow} C + D \tag{5.363}$$

The rate is subject to a probability factor $T_1/(T_1+T_2)$, which will increase steadily with increase of hydrostatic pressure [229]. Here T_1 is the average life of the activated complex AB and increases with hydrostatic pressure, and T_2 is the time required for the solvation of C and D and for a given solvent is practically constant. Such a probability factor is not involved in "normal" reactions, which are, therefore, independent of pressure. The requirement of the stabilization of the nascent particles by solvent would result in a marked solvent dependence of the effect of pressure on "abnormal" reactions.

It has been suggested that the role of the influence of solvent with respect to the pressure effects on reaction rates might be strongly dependent on the reversibility of reactions [230]. For example, the association of dimethyl-aniline and methyl iodide, which has been studied in polar and nonpolar solvents [231], shows appreciable reversibility in nonpolar solvents.

Summarizing, it can be stated that if there is an overall contraction in volume when a reaction occurs, the rate or equilibrium process will be promoted by increase in pressure; and if there is an overall increase of volume when a reaction takes place, the rate or equilibrium process will be inhibited by increase in pressure.

Differentiating Eq. (5.354) with respect to pressure at constant temperature gives

$$\left(\frac{\partial \ln k'}{\partial P}\right)_T = \left(\frac{\partial \ln V}{\partial P}\right)_T - \frac{1}{RT}\left(\frac{\partial E}{\partial P}\right)_T$$

$$- \frac{(\bar{V}^* - \bar{V})}{RT} - \frac{P}{RT}\left[\frac{\partial(V^* - V)}{\partial P}\right]_T \tag{5.364}$$

If the energy of activation is independent of pressure or nearly so, then the second term on the right-hand side of Eq. (5.364) is zero or negligible and Eq. (5.364) becomes

$$\left(\frac{\partial \ln k'}{\partial P}\right)_T = \left(\frac{\partial \ln V}{\partial P}\right)_T - \left(\frac{\overline{V}^* - \overline{V}}{RT}\right) - \frac{P}{RT}\left[\frac{\partial(\overline{V}^* - \overline{V})}{\partial P}\right] \qquad (5.365)$$

which is the equation given by Moelwyn-Hughes [214] who points out that the last term on the right is negligibly small relative to the second term, but that the first term on the right may be significant.

Defining E_V and E_P, the apparent energies at constant volume and constant pressure, respectively, as

$$E_V = RT^2\left(\frac{\partial \ln k'}{\partial T}\right)_V \qquad (5.366)$$

and

$$E_P = RT^2\left(\frac{\partial \ln k'}{\partial T}\right)_P \qquad (5.367)$$

from thermodynamic procedures the relation between them can be proven to be [214]

$$E_V = E_P + RT^2 \frac{\alpha}{\beta}\left(\frac{\partial \ln k'}{\partial P}\right)_T \qquad (5.368)$$

where α is the coefficient of isobaric expansion and β is the coefficient of isothermal compression. This equation has been applied to the dimerization of cyclopentadiene in the liquid phase [232]. It was found that both E_V and A_V, the parameters of the Arrhenius equation, obtained at constant volume, were unaffected by pressure within experimental error, while E_P and A_P, the parameters determined at constant pressure, increased markedly with increase of pressure. Thus E_P increased from 16.6 ± 0.4 to 18.8 ± 0.3 kcal/mole when the pressure was changed from 1 to 3000 atm. The change in log A_P (A_P in l./mole/sec) was from 6.1 ± 0.3 to 8.8 ± 0.3 for the same pressure change. E_V and A_V were considerably larger than the corresponding E_P and A_P. Thus E_V was 20.7 ± 0.7 kcal/mole and log A_V (A_V in l./mole/sec) was 9.2 to ± 0.5.

If in Eq. (5.364) $(\partial \ln V/\partial P)$ and $-(P/RT)[\partial(V^* - V)/\partial P]_T$ are negligible compared to

$$-[(\overline{V}^* - \overline{V})/RT]$$

then

$$-(V^* - V) = -\Delta \overline{V} = RT\frac{\partial \ln k'}{\partial P} \qquad (5.369)$$

and from Eqs. (5.368) and (5.369) there results [212]

$$E_V = E_P - T(\alpha/\beta)\,\Delta\overline{V} \qquad (5.370)$$

It was pointed out [214] that since $\Delta \overline{V}$ is usually negative, E_V is generally greater than E_P as was noted [232] for the dimerization of cyclopentadiene in the liquid state.

It has been observed [233] that the change of energy of activation with pressure is to a first approximation equal to the change in volume of activation at $0°K$.

Numerous studies have been made [234–236] on many-step enzymatic reactions as to the effect of temperature and pressure on parameters, which are in no sense to be regarded as rate constants or equilibrium constants. Before the results can be accepted as chemically meaningful more thorough experimental work will have to be carried out [237].

XVII. Viscosity and Reaction Rates: Cage Effects

Reactions in solution which are diffusion controlled have received considerable attention in the literature, and more recently reactions in which the primary event was the recombination of a pair of genimate free radicals, that is reactions resulting from a primary cage effect, have attracted much interest. Some of these studies had as their objective either to identify the primary cage reaction or to determine its importance relative to cognate nondiffusion controlled reactions. Later investigations have emphasized solvent effects [238–245]. Noyes [244] in considering the solvent as a continuum, and thus avoiding solvation phenomena and other specific solute–solvent interactions, formulated a theoretical analysis of diffusion controlled reactions which showed that under certain conditions a linear relationship should exist between the reciprocal of the probability of occurrence of a reaction in a primary cage and the reciprocal of the viscosity coefficient of the solvent. Discrepancies which have been observed in obedience of data to theory have been attributed to solute–solvent interactions that the theory neglects [239, 240, 246]. Thus if the perturbations due to the solvent does not change with changing viscosity, presumably the expected linear relationship will be observed. Such linear or nearly linear correlations have been employed [241, 243, 247]. In the preceding cases the obedience to theory resulted from a rather limited number of observations involving a particular precursor for the cage reaction in mixed solvents or in which the range of pure solvents and/or viscosities was very narrow. The correlation between data and theory apparently often depends on the relative position of the viscosity scale. In other cases the linear relationship involves the reciprocal of the square root of the viscosity coefficient [242, 248].

The quenching of fluorescence and coagulation have rates which depend on viscosity. The viscosity of the medium will presumably influence any rate process which is governed by the rate of diffusion of interacting particles to within reacting distance. Such processes are characterized by relatively small energies of activation. The Smoluchowski equation [249] for the frequency of collision z of particles A and B

$$z = \frac{kT}{3\eta} \frac{(r_A + r_B)^2}{r_A r_B} \eta_A \eta_B \tag{5.371}$$

where η_A and η_B are the numbers of the A and B molecules, respectively, per cubic centimeter, r_A and r_B are their respective radii, and η is the viscosity of the solvent at temperature T was modified by Moelwyn-Hughes [214] to give an equation for the specific velocity constant k' for a second-order reaction by multiplying both sides by $(N/1000) \exp(-\Delta E/RT)$, which yielded

$$k' = \frac{N}{1000} \frac{kT}{3\eta} \frac{(r_A + r_B)^2}{r_A r_B} \exp\left[-\frac{\Delta E}{RT}\right] \tag{3.372}$$

This equation does not reproduce rate data for chemical reactions involving appreciable energies of activation as well as does the collision bimolecular rate expression from kinetic theory, to wit,

$$k' = \frac{N}{1000}(r_A + r_B)^2 \left[8\pi kT\left(\frac{1}{r_A} + \frac{1}{r_B}\right)\right]^{\frac{1}{2}} \exp\left[-\frac{\Delta E}{RT}\right] \tag{5.373}$$

According to Eq. (5.372) a plot of $\ln(k'\eta/T)$ versus $1/T$ should yield a straight line from the negative slope of which the energy of activation ΔE can be calculated provided r_A and r_B are known. Equation (5.373) would require a plot of $\ln(k'/T^{1/2})$ to produce a straight line from the negative slope of which ΔE can be calculated provided r_A and r_B are known. It has been found [214] that the former plot of the data for the reaction of methyl iodide and sodium ethoxide in ethanolic solution [249] gave a continuous curve, the changing slope of which indicated an increase in ΔE with increasing T. However, a plot of $\ln(k'/T^{1/2})$ versus $(1/T)$ yielded a straight line which gave a value of ΔE of 18,970 cal/mole, confirmed by later data to within 100 cal.

Moelwyn-Hughes [214] points out that the kinetic theory of gases does not relate collision frequency and viscosity, but will lead to equations in which the collision frequency for spherical molecules of the same kind is related both directly and inversely to the viscosity. The more familiar binary collison frequency equation is obtained by elimination of the viscosity term.

An approximate equation for the specific velocity constant of quenching k' as a function of the viscosity of the solution has been derived [250] namely,

$$k' = T/200\eta \tag{5.374}$$

Assuming spherical particles of equal size with encounter radii the same as the diffusional radii resulted in an equation for the number of collisions n cm^3/mole/sec [251]. Thus

$$n = 8RT/3\eta \qquad (5.375)$$

where R is the gas constant in ergs/mole/deg, η the viscosity of the medium in poise, and T the absolute temperature. Equation (5.375) would give the maximum velocity constant for the process if each encounter were affective.

A general equation

$$k' = \frac{P(1-\alpha n_b{}^0)}{f} \frac{8NkT}{3000\eta}\left(\frac{1+r_a/r_b+r_b/r_a}{4}\right) \qquad (5.376)$$

has been proposed [252] for the molar rate constant k' in l./sec/gram-mole to apply to diffusion-controlled reaction rates in solution. In Eq. (5.376) P is the probability factor, α is the fraction of the bulk concentration $n_b{}^0$ of reactant B at the surface of reactant A, f is the factor expressing the effect of interionic forces on rate of diffusion, r_a is the radius of species A, r_b is the radius of species B, and the other terms have their usual meanings. If both P and f are taken as unity, and if the concentration $\alpha n_b{}^0$ of species B at reaction distance a from species A is kept zero by the reaction, then assuming $r_a = r_b$ and remembering that $Nk = R$, Eq. (5.376) becomes identical to Eq. (5.375) where the volume is in cm^3 and it is assumed that every collision produces reaction.

The photoreduction of eosin Y with thiourea to give leuco eosin has been studied [251], and it has been postulated that a long-lived excited state of the dye is formed by the transition of the first electronically excited state of the dye. This metastable state of eosin could react with thiourea to give leuco eosin, or the metastable state could revert to the ground state by radiationless transition or by a delayed fluorescence or phosphorescence. In addition, the metastable state could be quenched by the eosin Y in the ground state. If D′ stands for the dye in the metastable state and A represents the alkyl thiourea, the three steps can be written

$$D' + A \xrightarrow{k_1'} \text{products} \qquad (5.377)$$

$$D' \xrightarrow{k_2'} D \qquad (5.378)$$

$$D' + D \xrightarrow{k_3'} 2D \qquad (5.379)$$

Now if Eq. (5.375) is used to calculate the number of collisions/cm^3/mole/sec the result is 6.6×10^9 collisions/sec in one liter of one molar acqueous solution at room temperature. If in Eq. (5.379) each collision is effective in quenching the metastable state, the rate constant is given by the collision frequency. The ratio of k_3'/k_2' was found to be 4.88×10^4 and using k_3' as 6.6×10^9, k_2' is

1.35×10^5. The reciprocal of k_2', namely 7.04×10^{-6} sec is the mean lifetime for the metastable state. This mean lifetime was found to be about fifty times that of the first electronically excited singlet state. The solvent has been treated as a conventional structureless continuum in explaining the effects of the solvent on the quantum yields for dissociation [239, 244, 253]. For a molecule consisting of two identical atoms each with radius b and with a minimum energy E to cause dissociation, then during photochemical dissociation, the energy between the two atoms separating in opposite directions equals E. Due to viscous drag the solvent will decrease the velocity u of the atom as given by the equation

$$\frac{du}{dt} = \frac{6\pi\eta bu}{m} \qquad (5.380)$$

where η is the viscosity of the medium and m is the mass of the atom. The separation μ_0 of the atoms before the beginning of random diffusion is given by the equation

$$\mu_0 = 2b + \frac{m(hv - E)}{3\pi\eta b} \qquad (5.381)$$

The probability of subsequent recombination varies inversely as this separation, and therefore the quantum yield ϕ can be expressed as

$$\phi = 1 - \frac{2b\beta'}{\mu_0} \qquad (5.382)$$

where β' is the probability that a pair of atoms separating with normal thermal energies will recombine. If, as would be the case for macroscopic bodies moving through the medium, it is assumed that two atoms are influenced by a frictional drag proportional to the viscosity, and that for a particular initial velocity the body will be brought to rest at a distance inversely proportional to this viscosity, then each atom will lose its kinetic energy within the distance $(\mu_0 - 2b)/2$ and this distance will be inversely proportional to η for a constant energy of the absorbed quantum. Equation 5.201 becomes

$$\phi = 1 - \frac{\beta'}{1 + A/\eta} \qquad (5.383)$$

where A is independent of the solvent and is a constant for a particular wavelength of light. If the frictional force per unit area on the atoms of microscopic diameters is the same as that on a macroscopic sphere, then

$$A = \frac{2m(hv - E)}{3\pi P^2} \qquad (5.384)$$

For thermal velocities of separation of the two atoms, the center of mass of the system as well as the separation of the particles has an average kinetic energy of $\frac{3}{2}kT$. Hence,

$$hv - E = \tfrac{3}{2}kT \tag{5.385}$$

and

$$A = \frac{GkTm}{3\pi P^2} \tag{5.386}$$

Therefore

$$\phi = 1 - \frac{\beta'}{1 + VGRTm/3\pi P^2 \eta} \tag{5.387}$$

The quantity β' was shown to be given by the equation

$$\beta' = \frac{1}{1 + A/\eta} \tag{5.388}$$

by reasoning based on the assumption that two atoms produced by thermal dissociation separate with an average kinetic energy component of $\frac{3}{2}kT$ along the line of centers.

In applying this theory to the dissociation of iodine in inert solvents varying by a factor of 10^3 in viscosity, it was found, that although there were quantitative discrepancies, the theory predicted the quantum yield well [254, 255].

The magnitudes of proximity effects were suggested by the quantitative discrepancies. The quantum yields were less than predicted at long wavelengths, and were greater than predicted at short wavelengths in hexane. The long wavelength deviations arise because iodine atoms whose perimeters have not been separated by half a molecular diameter tend to be forced back together again by surrounding solvent molecules. The short wavelength discrepancies originate since the most probable distribution of solvent molecules will tend to force the atoms apart rather than together.

A further weakness of the theory is that it predicts a very small quantum yield for viscosities greater than 0.1 poise, while experiments up to viscosities of 3.8 poise indicate that the quantum yield falls with increasing viscosity down to a limiting value that is not affected by further viscosity increase. It appears that the behavior of solute particles are little affected by the increase of molecular complexity beyond a certain value, although there may be considerable influence on macroscopic viscosity.

Calculated and observed values of quantum yields for the dissociation of iodine at different wavelengths in hexane and hexachlorobutadiene was presented in Table 5.12. The first solvent has a viscosity of 0.0029 poise and the

Table 5.12

Comparison of Theory and Experiment on Quantum Yields

λ(Å)	Hexane		Hexachlorobutadiene	
	Observed	Calculated	Observed	Calculated
4047	0.83	0.54	—	—
4358	0.66	0.52	0.075	0.087
5461	0.46	0.46	0.036	0.070
5790	0.36	0.44	0.018	0.065
6430	0.14	0.40	0.023	0.055
7350	0.11	0.31	0.020	0.040

second solvent has a viscosity of 0.030 poise. The correlation between experiment and theory is good in hexane at the wavelengths 4358, 5461, and 5790 Å. The agreement in hexachlorobutadiene is good only at 4358 Å. At the short wavelength 4047 Å, the observed yield is less than the calculated value.

If the energy absorbed is high enough, and if the atoms separate energetically enough and far enough apart, the separation of the atoms is apparently promoted by the solvent; however, if the energy and perhaps the distance of separation are less than critical values, the solvent deters the separation. These results are in agreement [254, 255] with the original cage picture of Rabinowich and Wood [256].

The rate of the reaction of iodine with silver cyanide in mixtures of isobutyl alcohol and carbon tetrachloride at various temperatures was dependent on the viscosity of the medium. That diffusion limited the reaction rate was indicated by the similarity between the heat of diffusion and the activation energy of the reaction as is demonstrated in Table 5.13.

It has been found [247] that the logarithm of the ratio of the probabilities of the cage reaction and of the diffusive process $F = F_{cage}/F_{diff}$ is a linear

Table 5.13

Heat of Diffusion H and Energy of Activation E for the
I_2–AgCn reaction in iso-BuOH–CCl_4 Solutions

Volume present CCl_4 in liquid	H (kcal/mole)	E (kcal/mole)
0	—	3.10
67	3.00	2.80
91	2.91	2.67
100	2.50	2.65

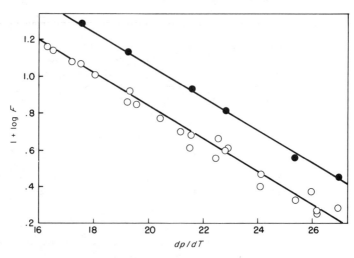

Fig. 5.12. Linear dependence of log F versus the reciprocal of the rate of change of the boiling point with pressure of the solvents. Open circles indicate reaction at 65°C; solid circles indicate reaction at 30°C.

function of the cohesion of the solvent as measured by the reciprocal of the rate of change of the boiling point of the solvent with pressure, $\partial P/\partial T$. See Fig. 5.12. The reactions involved the primary cage and diffusion-controlled reactions of the trifluoromethyl radical in six solvents at 30°C (solid circles) and in over twenty solvents at 65°C (open circles). The equation for this correlation was obtained from the combination of the Arrhenius expression for the temperature dependence of F with the vapor pressure equation and the empirical relationship between the barrier to viscuous flow and the heat of vaporization [257].

$$\frac{\partial P}{\partial T} = \frac{\Delta H}{RT^2} p \tag{5.389}$$

$$\Delta H = nE_\eta \tag{5.390}$$

Combining these equations with the Arrhenius expression for the ratio F yields

$$\ln F = \ln Z - B\frac{\partial P}{\partial T} = C - B\frac{\partial P}{\partial T} \tag{5.391}$$

where C is the logarithm of the Arrhenius per-exponential factor; this is a function of temperature and pressure. In spite of the fact that the solvents have quite different chemical properties, the effects of the solute–solvent interactions on the ratio of the rates of the two competing processes are similar in all the solvents the data for which are plotted, as is supported by the fit of the experimental points to the straight line. Descrepancies were observed in hydroxylic solvents. Except for hydroxylic solvents, the activation energy $\Delta E_a = (E_{cage} - E_{diff})$ is the same for all solvents in the cases of both ethyl and trifluoromethyl radical reactions, but for methyl radicals ΔE_a is apparently not constant in different solvents [243, 245].

That the data for the trifluoromethyl radicals conformed to the requirements of Eq. (5.391) suggested that the magnitude of ΔE_a for these radicals might be calculated from the data using Eqs. (5.392), (5.393), and (5.396).

$$\Delta E_a = RT^2 \frac{d \ln F}{dT} \tag{5.392}$$

$$\Delta E_\eta = -RT^2 \frac{d \ln \eta}{dT} \tag{5.393}$$

where ΔE_a is the activation energy which represents the difference betweeu the barrier to the dimerization reaction and the barrier to cage escape and may be represented by

$$\Delta E_a = E_{cage} - E_\eta \tag{5.394}$$

where E_η, the energy of viscous flow of the pure solvent, has been substituted for E_{diff}, the energy of diffusion of CF_3 radicals [245, 258–260]. The assumption was made [245] that the mechanisms of diffusion of the solute and solvent molecules are inseparable, i.e., the solvated radicals in a correlated process are dragged along in the random walk executed by the solvent sheath around them. Combination of Eqs. (5.392) and (5.393) results in

$$F = \alpha \eta^\beta \tag{5.395}$$

where

$$\alpha = \exp\left[-\frac{(\Delta E_a \ln A_\eta - \Delta E_\eta \ln A)}{RT \ln A_\eta} \right] \tag{5.396}$$

and

$$\beta = \ln A / \ln A_\eta \tag{5.397}$$

where $\ln A$ and $\ln A_\eta$ are the logarithms of the constants of integration of Eqs. (5.392) and (5.393). Since ΔE_a is independent of solvent, the variation of α at constant temperature will arise from the quantity $\Delta E_\eta (\partial \ln A / \partial \ln A_\eta)$. The ratio $\partial \ln A / \partial \ln A_\eta$ which occurs as an exponent in Eq. (5.395) is a positive quantity and has the values 2.4, 3.7, and 3.5 for ethyl radicals in isooctane, toluene, and decaline, respectively.

From Eq. (5.395), $\ln F$ plotted against $\log \eta$ should yield a straight line. This expectation is verified in Fig. 5.13 though there is considerable scatter of data due largely to the solvent dependence of β. The relative positions in Fig.

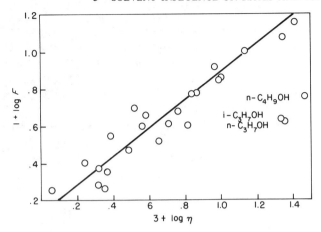

Fig. 5.13. Plot of log F versus log η for the CF_3 radicals at 65°C in various solvents.

5.13 for the propanols and butanol and those of *n*-tetradecane and 1-hexa-decene which lie directly above them in the figure might be noted. Similar characteristics might be expected for their cage effects since, for these five substances, the viscosity coefficients and the viscous energies ($\Delta E_\eta = 3.7$ to 4.2 kcal) are not very different [245]. In fact for these liquid any anomalies would be expected to be the reverse of those observed; since the alcohols are "structured" liquids and the average lifetime of a cavity or a cage is presumably longer than in more normal liquids, the fraction of the radicals undergoing the cage reaction should be larger in the alcohols. Hence for alcohols log $F = \log(F_{cage}/F_{diff})$ should be larger not smaller than for normal liquids of the same viscosities. This anomalous behavior was explained by the statement that in the alcohols the details of the photolytic mechanism of the azo compound to yield geminate radicals are not the same as in hydrocarbon solvents, and most probably this is reflected in the energy of the geminate radicals. In fact it was observed that the rate of photolysis of hexafluoroazomethane in hydroxylic solvents depends on the hydrocarbon content of the solvent molecules, being greater in propanol than in methanol and exhibiting no decomposition in water [260].

XVIII. Hydrogen-Bonding Influence on Reactions

A comprehensive treatment of hydrogen bonding, including its effect on rates and mechanisms of reactions has been written by Pimental and McClellan [261].

In hydrogen bonding, the hydroxylic hydrogen of one molecule of a hydroxy compound will form a bond with an oxygen atom of another molecule. In general, hydrogen bonds can be formed by hydrogen covalently bonded to one highly electronegative atom being strongly attracted to another highly electronegative atom. Atoms which show marked tendencies toward hydrogen bonding are those of fluorine, oxygen, and nitrogen. The boiling points of water and other hydroxy compounds are explained on the basis of hydrogen bonding among the molecules, repeated to the extent that the hydroxy compound forms long chains or agglomerations of molecules. The Grotthus theory of conductance of hydrogen ions postulates that a proton attaches itself to one end of a chain of water molecules and another proton almost simultaneously detaches itself from the other end of the chain, so that effectively the hydrogen ion jumps the length of the molecular water chain in the time required for a bond to form or break. This is a much shorter time than that required for a solvated hydrogen ion to diffuse an equivalent distance under the influence of the field created by an applied potential.

For hydrogen bonding one group must serve as an acidic or proton doning group and include carboxyl, hydroxyl, amine, or amide groups. The other group must function as a basic or proton accepting group and include oxygen in hydroxyls, carbonyls, and ethers, halogens in certain molecular environments, and nitrogen in N-heterocyclic compounds.

In intermolecular hydrogen bonding, hydrogen bonds two molecules of the same or different substances, while in intramolecular hydrogen bonding there is bonding by hydrogen of two groups within the same molecule. Intermolecular bonding may produce chains, cyclic configurations, or three-dimensional structures; and occurs in water, alcohols, carboxylic acids, amides, proteins, polypeptides, and polyhydroxyorganic, and inorganic substances, In intramolecular hydrogen bonding the closing of the gap between two rings resembles a clawlike action and such hydrogen bonding is termed chelation (after the Greek *chele*, claw). The hydrogen bond is much stronger than the van der Waals attractive forces between molecules but is much weaker than common covalent bonds. The enthalpy per mole of hydrogen bond usually ranges from 2 to 7 kcal but can lie on either side of this range.

Both resonance [262–266] and dipole interactions [267–270] have been used to explain the hydrogen bond. The hydrogen is not usually halfway between the bonded atoms in the case of formic acid dimer, and according to some theorists cannot be due to resonance; however, the resonance theory proponents argue that there is an unequal contribution of the two resonance states. Electrostatic interactions contribute to, but cannot be the only explanation of, hydrogen bonding according to the reasoning of Hine [271] since if they were, there should be a correlation between electronegativity and hydrogen bonding tendencies. If hydrogen bonding is determined predominately by electrostatic interaction, the ability to accept hydrogen bonds should be in the order alkyl fluorides > alcohols > amines. Apparently the reverse of this is the case, with stronger hydrogen bonds being formed by the more basic atoms. Thus HF and the strongly basic fluoride ion form a strong hydrogen bond.

The inhibition of phenol alkylation by ethers is an example of the effect on the rate of reaction of hydrogen bonding between reactant and solvent [272]. Dioxane in proportion to its concentration was found to decrease the rate of alkylation of phenol by *t*-butyl chloride. A quantitative account of the decrease in rate was given by assuming the formation of 2:1 phenol:dioxane hydrogen-bonded complex, and that phenol so bound could not participate in the alkylation reaction. To explain the data in the case of tetrahydropyran, it was assumed that a 1:1 complex was formed. For dioxane and tetrahydropyran, respectively, the complexes were written as given:

$$
\begin{array}{c}
\text{H}_2\text{C}-\text{CH}_2 \qquad \text{C}_6\text{H}_5 \\
\text{O}-\text{H}\cdots\text{O} \qquad \text{O}\cdots\text{H}-\text{O} \\
\text{H}_2\text{C} \qquad \text{H}_2\text{C}-\text{CH}_2
\end{array}
\tag{5.398}
$$

$$
\begin{array}{c}
\text{H}_2\text{C}-\text{CH}_2 \qquad \text{C}_6\text{H}_5 \\
\text{H}_2\text{C} \qquad \text{O}\cdots\text{H} \\
\text{H}_2\text{C}-\text{CH}_2
\end{array}
\tag{5.399}
$$

Halflives for the alkylation of phenol were calculated using the equation

$$
\frac{t_{1/2}}{t^o_{1/2}} = \left(\frac{M^o}{M}\right)^n
\tag{5.400}
$$

where $t_{1/2}$ refers to the halflives of reaction, M is the molarity of the phenol, n (found to be 6) is the order of the reaction with respect to phenol, the superscripts refer to the reaction in pure phenol, and the terms without superscripts refer to the reaction with dioxane present. Calculations were made assuming (1) dioxane served only as a diluent for the reactants, (2) dioxane forms a 1:1 complex with phenol, and (3) dioxane forms a 1-dioxane:2-phenol complex with phenol. Calculations were also made using tetrahydropyran as a component of the solvent assuming (1) tetrahydropyran served only as a diluent for reactants, and (2) tetrahydropyran formed a 1-tetrahydropyran:1-phenol complex. The results of the calculations are shown in Table 5.14 [272].

Comparing columns 5 and 6 for experiments 3–6 it is evident that theory and experiment are in agreement for a 2-phenol:1-dioxane complex, and comparing columns 4 and 6 for experiments 7–9 it can be seen that theory and experiment agree most favorably for a 1-phenol:1-tetrahydropyran complex.

From these data one would conclude that phenol bound in a complex with ether does not undergo alkylation. In alkylation the hydroxyl group evidently

Table 5.14

Experimental and Calculated Halftimes for Different Assumptions

Experi-ment no.	Solvent	$t_{1/2}$, assumption 1 (min)	$t_{1/2}$, assumption 2 (min)	$t_{1/2}$, assumption 3 (min)	$t_{1/2}$, observed (min)	Ether/phenol (moles)
1	Phenol	—	—	—	55.2	—
2	Phenol	—	—	—	53.1	—
3	Phenol–dioxane	65.5	78.7	95.4	96.7	0.0287
4	Phenol–dioxane	84.7	118.2	174.3	178.6	0.0621
5	Phenol–dioxane	103.1	217.6	497.7	457.8	0.113
6	Phenol–dioxane	134.0	461.3	1654.7	1497.7	0.168
7	Phenol–tetrahydropyran	85.9	118.4	—	119.0	0.0563
8	Phenol–tetrahydropyran	111.3	247.6	—	268.8	0.116
9	Phenol–tetrahydropyran	137.2	400.2	—	423.7	0.162
10	Phenol–tetrahydropyran	144.5	—	—	146.1	0.152

plays an important role. The solvation of the halogen atom of the alkyl halide is perhaps the function played by the hydroxyl group.

In the kinetics of cyanohydrin formation in aqueous solution acetaldehyde and propionaldehyde react quantitatively with hydrocyanic acid but acetone does not [273]. The second-order reactions were not generally acid catalyzed, but a small general acid catalysis was detected for propionaldehyde. Hydrogen bonding, specifically of water with the carbonyl group, but generally of any molecular acid with the carbonyl group was involved in the mechanism thus:

$$\diagdown C = O + H_2O \underset{\longleftarrow}{\overset{fast}{\longrightarrow}} \diagdown C = O \cdots HOH \text{ (hydrogen bond)} \quad (5.401)$$

$$\diagdown C = O \cdots HON + CN^- \xrightarrow{slow} \diagup{C}\diagdown \begin{array}{c} CN \\ OH \end{array} + OH^- \quad (5.402)$$

or

$$\diagdown C = O + HA \underset{\longleftarrow}{\overset{fast}{\longrightarrow}} \diagdown C = O \cdots HA \text{ (hydrogen bonding)} \quad (5.403)$$

$$\diagdown C = O \cdots HA + B^- \xrightarrow{slow} \diagdown C = O \begin{array}{c} B \\ OH \end{array} + A^- \quad (5.404)$$

In the benzoyl peroxide–trifluoroacetic acid reaction, it is believed [274] that the radical PhCOO from benzoyl peroxide forms hydrogen bonds with carboxylic acids which facilitate the hydrogen transfer between the carboxylic acid and the polar radical. Trifluoroacetic acid, which is the strongest acid investigated, and, therefore, has the greatest tendency to form hydrogen bonds produces the most pronounced hydrogen transfer reaction. CF_3 is formed from the decarboxylation of the resulting CF_3COO radical. The CF_3 attacks the peroxide causing further decomposition or dimerizes. However, fluoroform is formed from the reaction

$$CF_3 + HR \rightarrow CF_3H + R \qquad (5.405)$$

if isooctane is a component of the solvent. On the other hand, acetyl and propionyl peroxides form CH_3 and C_2H_5 radicals since the unstable CH_3COO decarboxylate before any other reaction occurs.

What has been termed a basic principle on the effect of the solvent through hydrogen bonding has been formulated [275]. A particular solvent will diminish the speed of a reaction if the active centers that take part in the reaction are blocked by hydrogen bonding or other interaction with that solvent. However, the speed of a reaction will be increased by a solvent, which by hydrogen bonding or otherwise, promotes electron shift necessary for the reaction. The idea was accepted that hydrogen bonding can take place between any positive hydrogen and any negative element present in the system.

That the self-condensation of glycylglycylglycine methyl ester in methyl alcohol required a low energy of activation was attributed [276] to a hydrogen bonded head-to-tail position complex of the reacting molecules in which the ester and amino groups of the adjoining molecules are held in close proximity thus providing conditions favorable for reaction to occur by the elimination of methanol. The configuration of the complex is represented as

$$(5.406)$$

The carbonyl groups were apparently not blocked by hydrogen bonding with methanol. The depicted reaction would apparently have an energy of activation less than that for a similar reaction not involving such an intermediate by an amount equal to the heat of formation of the hydrogen bonds. It would be of interest to investigate the effect on this reaction of dioxane and tetra-hydropyran with their strong tendencies to form hydrogen bonds.

Formation of hydrogen ion has been postulated [277, 278] in the termo-lecular mechanism of triphenylmethyl halide displacements, and in the termolecular displacement reactions of methyl halides.

Third-order kinetics occur in the case of triphenylmethyl halide displacements. In benzene solution containing excess pyridine at 25°C methanol reacts with triphenylmethyl chloride or bromide to give methyl ether. Phenyl ether is formed at a slower rate when phenol is substituted for methanol. When both methanol and phenol are present, the rate of formation of methyl ether is seven times that of the combined rates with methanol or phenol separately. In this case the rate is proportional to the concentrations of methanol, phenol, and halide, although phenol is not consumed in the rapid phase of the reaction.

A concerted termolecular attack, in which a molecule of phenol hydrogen bonds with the halogen atom weaking its bond with carbon, and a molecule of methanol simultaneously solvates the carbon that is becoming a carbonium ion, is used to explain the high rate of reaction in the presence of both methanol and phenol.

The presence of methanol, phenol, or mecuric bromide promotes the termolecular displacement reactions of methyl halides. Thus a concerted push-pull third-order mechanism involving the formation of a hydronium ion [278] for the displacement of bromide in methyl bromide is shown:

$$(5.407)$$

This reaction is superimposed on the ordinary slower, second-order mechanism that proceeds in the absence of no more effective solvating agents than the solvent, benzene. When no pyridine is present, the reaction of 2.6 M methyl bromide with 24.7 M methanol has a rate constant which is 10^5 less than that with pyridine present.

The reaction shceme for the acid-catalyzed reactions of trialkyl phosphites, diisopropyl fluorophosphate, and acetamide was written [279] to involve hydrogen bond equilibrium as follows:

$$S + H_3O^+ \cdots nH_2O \underset{}{\overset{K_{assoc}}{\rightleftharpoons}} S \cdots H_3O^+ + nH_2O \qquad (5.408)$$

$$S \cdots H_3O^+ + H_2O \xrightarrow{k'} \text{products (slow)} \qquad (5.409)$$

The kinetics of the reaction will conform to the predictions of Euler's theory

[280], however, K_{assoc} in Eq. (5.408) is not the base constant of the substrate, but is the hydrogen-bond association constant between the substrate and the hydronium ion relative to that of water.

In the α-amylase-catalyzed hydrolysis of amylose, a complex involving hydrogen bonding was assumed [281]. The hydrogens of both the imidazolium groups and of solvent water hydrogen bond. The predominent effect on the rate of added methanol was found to be that arising from the dielectric constant. Hiromi's [282, 283] theory for the effect of the dielectric constant of the medium on the rates of reaction correlated the data for the α-amylase-catalyzed hydrolysis of anylose using reasonable values of the parameters involved. Straight lines with negative slopes were obtained when $\log k_3'$ and $\log K_m'$ for the mechanistic equations

$$E + S \underset{k_2'}{\overset{k_1'}{\rightleftarrows}} ES \overset{k_3'}{\longrightarrow} E + P \qquad (5.410)$$

$$E \underset{}{\overset{K}{\rightleftarrows}} E' \qquad (5.411)$$

$$E' + S \overset{K_m'}{\rightleftarrows} E'S \qquad (5.412)$$

were plotted versus the reciprocals of the dielectric constants of the media. In the last equations E represent the enzyme, E' the reversibly denatured form of the enzyme, S the substrate, and P the final product.

For the decomposition of thiourea there has been postulated [284] a hydrogen bonded, ionized intermediate as shown in the reaction scheme given:

$$(CH_3)_2N-CS-NH-CH_3 \underset{k_2'}{\overset{k_1'}{\rightleftarrows}} \left[\begin{array}{c} S \\ \| \\ H_3C \diagdown C \diagup \\ N \diagup \diagdown N-CH_3 \\ H_3C \diagup \diagdown H \end{array} \right]^{2+} \overset{k_3'}{\longrightarrow} CH_3NCS + (CH_3)_2NH$$

$$(5.413)$$

$$CH_3NCS + H_2O \overset{k_4'}{\longrightarrow} CH_3NH_2 + COS$$

$$(5.414)$$

Reaction 5.414 in acid solution is rapid, and there should be equal amounts of methylamine and dimethylamine produced. Assuming the stoichiometry, k_3' and k_4' were obtained. The amines are no longer in the predominately protonated form as the solutions become more basic, and the reverse reaction rate constant k_2' will at the same time be magnified since the forward rate depends on the unionized thiourea. Thus the rate of this reaction involving hydrogen bonding is influenced markedly by the acidity of the media. Tetramethylthiourea in various media at elevated temperatures did not show extensive reaction in neutral solution. Hence it was reasoned that a hydrogen atom bonded to nitrogen was required for the decomposition of thioureas.

A comparison of 4(7) nitrobenzimidazol and 5(6) nitrobenzimidazol showed the former to be more rapidly reduced polarographically, more easily reduced catalytically, and more acidic [285]. These phenomena implied an intramolecular influence that made the nitro group more susceptible to reduction and rendered the removal of the essential proton more difficult. The 4(7) isomer was found to be less associated and hence more volatile in solution. The hydrogens required for meschydric linkages were otherwise involved. As a result of these observations, an intramolecular hydrogen bonding in the 4(7) isomer was postulated. Below structure (I) represents the 4(7) isomer in its unchelated and chelated forms, and (II) depicts the 5(6) isomer which does not chelate but can associate by intermolecular hydrogen bonding.

$$(5.415)$$

I   II

The importance of intramolecular hydrogen bonding in other isomeric compounds has been further developed [286–288].

In the rates of phenylhydrazoneandoxime of aldehydes, the rates of reaction were more rapid in those aldehydes where intramolecular hydrogenbonding could occur then in the case of their associated isomers [289]. The rates of reaction of unsubstituted aldehydes fell between those of the other two classes. In general the rates were greater in alcohol than in chloroform, though no relation was developed between the physical or chemical properties of the solvent and the rate of reaction.

The polarography of nitrophenols of nitrocresols and of nitrodihydroxy-benzenes [290–293] have been studied. It was found that when intramolecular hydrogen bonds are sterically possible, the reduction of nitro groups in carefully buffered solutions at a dropping mercury cathode are more easily reduced than the nitro groups when no such bonds are possible. Thus the nitro groups in 3-nitrocatechol, 2-nitrohydroquinone, and 2-nitroresorcinol are more easily reduced than the nitro groups in 4-nitrocatechol. Apparently there is only one intramolecular hydrogen bond in 2-nitroresorcinol even though the capability exists of having both nitro oxygens tied up in hydrogen bonding with the ortho OH groups. The intramolecular hydrogen bonding, and hence the rate of reduction, is influenced to some extent by the medium, as shown by the observations that within the pH range from 5.5 to 8.5 the same kind of nitro group appears to be present in both 4-nitroresorcinol and 4-nitro-catechol. Hence there is probably no intramolecular hydrogen bonding in 4-nitroresorcinol within this pH range. There is probably a weak intramolecular hydrogen bond in 4-nitroresorcinol at lower pH values.

The kinetics of the hydrolysis of esters in various solvents have been studied [294] at 0–80°C. In the intramolecular hydrogen bonded monoethyl malate ion the reaction rate was slower, and a higher energy of activation was necessary to overcome this hindrance to hydrolysis. The salt effect for this ion decreased in the order $Li^+ > Na^+ > K^+$ in harmony with chelate formation. Preferential absorption of one of the solvent components was proposed to explain the deviations from the absolute rate theory with respect to the Arrhenius frequency factor and the energy of activation. In water the hydrogen ion concentration in general had no effect on the hydrolysis rates of monoethyl malonate and citraconate. However at high acetone concentrations in water–acetone solvents, these rates were increased by increased hydrogen ion concentration. The rate of the monoethylmalonate ion hydrolysis was at varience with the theory relating the change of rates with variation of dielectric constant Intramolecular hydrogen bonding was used to explain the data on the rates of hydrolysis of monoethylmalonate and citraconate.

The contribution of the hydrogen bond, chelation, and solvent effects in chemical kinetics has been reviewed, and the influence of intramolecular hydrogen bond on the rates of chemical reactions studied [295]. The influence of hydrogen bonding on reaction rates has been discussed in other papers by the same authors [296–298]. The systemization of the information on the chemical manifestations of the hydrogen bond was attempted. The purpose was to show its presence in many different reactions, and how it can be discerned in the different reaction types. It was thought that intermolecular hydrogen bonding of the reactants with the solvent might exert important effects on reactions in solution and might even determine the effect of the solvent on reaction velocities.

It has been found [299] in the biochemical field that heat-denatured and ultraviolet-irradiated deoxyribonucleic acid (DNA) was consistent with the hypothesis that the denatured DNA, due to intramolecular hydrogen bonds involving the amino group of the purine and pyrimidine bases, exists as randomly coiled structures. In elucidating the structure of DNA, various denaturation, reactivation, and ultraviolet irradiation reactions were investigated. Intramolecular hydrogen bonding was promoted by rapid cooling after denaturation; these bonds were again broken by heating to 45°C. Formaldehyde which reacts with the amino groups of the bases may be used to prevent the formation of intramolecular hydrogen bonds during rapid cooling.

The inactivation of deuterated and protonated trypsin by ultraviolet light has been studied, and the involvement of the intramolecular hydrogen bond in the disruption has been postulated [300].

The kinetics of the base pairing involving two competing reactions has been studied in the system 2-caprolactam (A) and 2-aminopyrimidine (B) [301]. The reaction sheme,

$$
\begin{array}{c}
& A_2 + B \\
& \nearrow \\
k'_{12} \diagup \diagup k'_{21} \\
A + A + B \\
k'_{31} \diagdown \diagdown k'_{13} \\
& \searrow \\
& AB + A
\end{array}
\tag{5.416}
$$

is characterized by two relaxation times. They belong to "normal coordinates" of reaction and are usually not simply related to the individual reaction pathways. In general each of them depends upon all the individual rate constants of the coupled reaction sheme. A simplification of the expression for the relaxation times is possible since the two times differ by about an order of magnitude. A further simplification of the expression for the relaxation times and an identification of the relaxation process was made possible by a comparison of the measured concentration dependence of the reaction partners as a function of frequency upon application of a high dc field with results obtained with pure E-caprolactam solutions. Expressions for the reciprocal relaxation times were written,

$$
\frac{1}{\tau_1} = k'_{21} + 4k'_{12}\,\bar{c}_A
\tag{5.417}
$$

$$
\frac{1}{\tau_2} = k'_{31} + k'_{13}\left(\bar{c}_A + \frac{1}{1+4K'\bar{c}_A}\,\bar{c}_B\right)
\tag{5.418}
$$

τ_1 is the shorter and τ_2 the longer relaxation time; and $\bar{c}_A$ and $\bar{c}_B$ are the equilibrium concentrations of A and B. K' and K'' the equilibrium constants of the dimerization and of the 1:1 association, respectively, were known, so that the equilibrium concentrations could be calculated from the initial concentrations of A and B. According to Eq. (5.417) a plot of $1/\tau_1$ versus $\bar{c}_A$ should give a straight line of slope $4k'_{12}$ and an intercept of k'_{21}, and according to Eq. (5.418) a plot of $1/\tau_1$ versus the term in parenthesis should yield a straight line of slope k'_{13} and an intercept of k'_{31}. Such plots were made using measured values of τ_1 and τ_2 as functions of $\bar{c}_A$ and $\bar{c}_B$ and the plots yielded $k'_{12} = (4.8\pm0.2)10^9$/mole/sec, $k'_{21} = (3.0\pm0.2)10^7$/sec, $k'_{13} = (7.9\pm0.2)10^8$/mole/sec, and $k'_{31} = (1.1\pm0.2)10^7$/sec. The values of K' and K'' were

162 ± 10/mole and 70 ± 5/mole, respectively. It was concluded from these results that the formation of hydrogen-bonded dimers and molecule pairs is very rapid and diffusion controlled at least in the case of dimerization of E-caprolactam. The second hydrogen bond is formed as fast as or faster than the first hydrogen bond is again broken as is witnessed to by the fact that the pairing takes place in a single step.

For the rapid identification of complementary bases in genetic code reading, the fast pairing is essential. In contrast the lifetime of the paired complex is very short, on the order of 10^{-8}–10^{-7} sec, and involves the breakage of both hydrogen bonds. For fast code reading based upon a "trial and error" mechanism, where wrong combinations must vanish speedily, it is again imperative that there be high dynamic lability expressed in the short lifetime. A high accuracy of the molecular code reading is not favored because of the following conditions. The number of wrong hydrogen-bond pairs formed and disrupted is large compared to the number of right combinations since all combinations of two bases will lead to more or less hydrogen bond pairing. Hence the stability constants must remain comparatively low. The differential stability of the complementary base pairing alone form the basis for the accuracy of such processes as they are observed in living organisms. To establish the small error rate observed in the biological process of genetic information transfer and readout, a code-checking mechanism presumably involving enzymic recognition sites is necessary.

Only nonpolar media was used in this study of base-pairing model systems; however, many properties of the actual base-pairing systems in their biochemical environment were conserved. Also kinetic studies with oligonucleotides carried out in aqueous media [302] support these findings. Individual hydrogen-bond pairing reactions are in the same range of the orders of magnitude. Because of the competitive hydrogen-bond formation with the solvent, the lifetime may be smaller in aqueous media.

In the decarboxylation of β-keto acids

$$RCOCH_2CO_2H \rightarrow RCOCH_3 + CO_2 \qquad (5.419)$$

the rates are enhanced in mixtures of an organic solvent in water generally showing a maximum in 50/50 v/v organic component–water mixtures [303]. It is believed that the organic component of the mixture favors the chelated (reactive) form of the β-keto acid while the water is expected to solvate partial charges which may develop in the transition state. The intramolecular hydrogen-bonded structures can be represented [304] thus:

$$(5.420)$$

Table 5.15

*Comparison of Entropies of Activation for Decarboxylation of
Unsubstituted Phenylmalonic and Unsubstituted Malonic Acids and
Their Monoanion Species*

	Water		80% Dioxane	
Acid	H_2A	HA^-	H_2A	HA^-
Malonic	-2	$+1$	-9	-2
Dibromonalonic	—	$+15$	—	$+4$
Phenylmalonic	-8	$+5$	-14	-2

As far as substituted benzoylacetic acids in benzene solution where the acids
are monomeric and presumably occur as intramolecularly hydrogen bonded
structures as shown above are concerned, only small substituent effects were
noted in the decarboxylation of the acids.

The enhancement effect of organic components of the solvent on the rates
dicarboxylic acid in isopropanol–water mixtures at 40°C, and by Hall and
Hanrahan [306] in the case of phenylnialanic acid in dioxane–water mixtures
at 55 and 65°C. In the latter reaction activation parameters were found to be
consistent with a proposed mechanism which postulated an intramolecular
hydrogen bond in the transition state [307]. The pertinent parameters were
the more negative intropies of activation in 100% water and in 80% dioxane–
20% water for both malonic and phenylmalonic acids than for the monoanion
species of the respective acids. A comparison of the entropies of activation for
the decarboxylation of the unsubstituted phenylmalonic and unsubstituted
malonic acids and for their monoanion species is given [306] in Table 5.15.
A negative entropy of activation indicates a greater degree of ordering in the
transition than in the initial state.

XIX. Nucleophilicity and Reaction Rates

A. INTRODUCTION

Some ions, elements, and compounds having certain properties in common
have been termed "cationiod," while other ions and compounds have been
termed "anioniod" [308]. In the first class were included hydrogen ions,
diazonium ions, ozone, the halogens, hypochlorous acid, carbonyl compounds, and α,β-unsaturated ketones. In the second class were listed CN^-,
$R—C{\equiv}C^-$, EtO^-, OH^-, the negative sodiomalonic ester ions, ammonia,
ethylenic and actylenic hydrocarbons, phenol ethers, and vinyl ethers. Later

it was suggested that the two classes be designated electrophilic (electron seeking) and nucleophilic (nucleus seeking) [309]. These latter two designations are in general use today.

An excellent review of nucleophilic reactivity has been written [310] in which the various factors influencing nucleophilic reactivity have been classified. Nucleophilic tendencies have been discussed, and the solvents in influencing these tendencies have been presented [101]. Linear free energy equations designed to quantitatively correlate nucleophilic tendencies and reaction rates have been discussed in rather complete detail [36].

Electrophilic reagents or electrophiles are electron seeking substances since they are deficient in electrons. Thus hydrogen ions, ozone, and sulfur trioxide are electrophiles:

$$H^+, \quad :\overset{..}{\underset{}{O}}: \overset{\overset{..}{O}..}{} :\overset{..}{O}^-, \quad \text{and} \quad :\overset{..}{O}: \overset{:\overset{..}{O}:}{\underset{:\overset{..}{O}:}{S}} \tag{5.421}$$

Electrophilic reagents include positive metal and nonmetallic ions such as Ag^+, carbonium (R_3C^+), bromonium (Br^+), and nitronium (NO_2^+); double bond substances such as $C=O$, $C=N$, and $N=O$; and triple bond substances such as $C\equiv N$. Electrophilic properties and acidic nature may be evident in halides such as $SnCl_4$ and $TiCl_4$ in which the central atom may hold more than an octet of electrons and in which there are vacant d orbitals of relatively low energy. If an electron attracting group, such as $-CH_3$, $-CH_2R$, $-CHR_2$, or $-C-O^-$, has reduced the electron density in a $C=C$ double bond, the bond may act as an electrophile. Lewis acids are electron deficient and are electrophiles.

Nucleophilic reagents or nucleophiles are nucleus-seeking or electron-donating. This class of substance include negative ions (hydroxide, halide, carbanions, and others); molecules of fifth and sixth group elements having unshared electron pairs (ammonia, phosphine, amines, alcohols, ethers, and mercaptans); and olefins and aromatic hydrocarbons. Generally these reagents attack another molecule at a location where the atomic nucleus is poorly shielded by outer electrons. Nucleophiles are Lewis bases and also react in most cases similar to Brönsted bases. As Lewis bases they form compounds with Lewis acids by donating pairs of electrons, and as Brönsted bases they accept protons from Brönsted acids.

Acid–base, substitution, and addition reactions represent electrophilic and nucleophilic reactions. Thus the Lewis base NH_3 and Lewis acid BF_3 react

$$\begin{matrix} H & :\overset{..}{F}: \\ H:\overset{..}{N}: & + & \overset{..}{B}:\overset{..}{F}: \\ H & :\overset{..}{F}: \end{matrix} \longrightarrow \begin{matrix} H & :\overset{..}{F}: \\ H:\overset{..}{N}: & \overset{..}{B}:\overset{..}{F}: \\ H & :\overset{..}{F}: \end{matrix} \tag{5.422}$$

as do the Brönsted base NH_3 and Brönsted acid HBr

$$H : \overset{\overset{H}{..}}{\underset{\underset{H}{..}}{N}} : \quad + \quad H : \overset{..}{\underset{..}{Br}} : \quad \longrightarrow \quad \left[H : \overset{\overset{H}{..}}{\underset{\underset{H}{}}{N}} : H \right]^+ \quad + \quad : \overset{..}{\underset{..}{Br}} : \overset{..}{} \qquad (5.423)$$

In the last reaction NH_4^+ is the conjugate Brönsted acid of the base NH_3 and Br^- is the conjugate Brönsted base of the acid HBr. In the nucleophilic substitution reaction

$$[Co(NH_3)_5Br]^{2+} + OH^- \rightarrow [Co(NH_3)_5OH]^{2+} + Br^- \qquad (5.424)$$

both the attacking agent and the leaving ion are Lewis bases; while in the electrophilic substitution reaction

$$Br^+ + C_6H_6 \rightarrow C_6H_5Br + H^+ \qquad (5.425)$$

both the attacking bromonium ion and departing hydrogen ion are electron deficient Lewis Acids.

An extension of the terms nucleophile and nucleophilic activity is given in this paragraph. Any reagent which donates its electrons to or shares them with a different atomic nucleus has been defined as a nucleophilic reagent or nucleophile [311, 312]. This has been critized as including reducing agents such as alkali metals and tin(II) chloride [310], and a nucleophile has been defined [310, 313, 314] as a reagent which supplies a pair of electrons to form a new bond between itself and another atom. This nearly corresponds to the Brönsted [315] definition of a base and is almost identical to the Lewis [316] definition. It has been proposed [314] that the term "nucleophilic reactivity" be used when referring to rate phenomena, and "basicity" be used with regard to equilibria. Also it has been suggested [317] that thermodynamic affinity for elements other than hydrogen be designated as carbon-basicity, sulfur-basicity, etc. Reagents having two or more alternative nucleophilic centers have been termed "ambident anions." These are illustrated by nitrite ion which may be alkylated either on oxygen or nitrogen to form a nitrite ester or a nitro compound, respectively [310].

Nucleophilicity and basicity are not directly related. Highly polarizable nucleophiles are often more reactive than their basicities would indicate [319]. A number of nucleophiles in certain reactions have been found to be excessively reactive with respect to their basicities or polarizabilities. These have been found to have an unshared electron pair on the second atom of the nucleophile, and therefore their high reactivities has been termed the alpha effect [320].

B. FACTORS AFFECTING NUCLEOPHILIC REACTIVITY

Factors affecting nucleophilicity have been discussed [97, 310, 320, 321]. The alpha effect has already been defined as arising from an unshared pair of electrons on the second or alpha carbon of the nucleophile. This electron pair promoted reaction by supplying electrons to the first atom, thus allievating the deficiency produced by the first atom's donating electrons to the electrophilic center. Hypochlorite ion and *N,N*-dimethylhydroxylamine are examples of nucleophiles which show the alpha effect of excessive reactivity with respect to either basicity or polarizeability. An analogy was the stabilization of a carbonium ion center by the presence of an alkoxy group $R \cdot O \cdot CH_2^+ \rightarrow RO^+ = CH_2$.

Nucleophilicity correlates with basicity, but basicity per se does not cause nucleophilic reactivity [310]. If N represents the nucleophilic site of the nucleophile and E the electrophilic site of the electrophile, basicity is principally dependent upon the negative charge potential possessed by the nucleophile, and acidity with the positive charge potential of the electrophile [320]. Strong interaction in the transition state occurs between a nucleophile of high negative charge potential, i.e., between a strong base and a strong acid. The factor of basicity increases as the positive charge on E increases. Some strong electrophiles which are positively charged in the transition state are hydrogen, carbonyl carbon, tetrahedral sulfur, tetrahedral phosphorous, and trivalent boron.

Available, low-lying, unfilled d orbitals are accompanied by polarizability, and promote nucleophilic attack [320]. Pauli's exclusion principle of repulsion operates between N and E when a nucleophile having outer shell p electrons approaches an electrophilic center also possessing nonbonding p electrons. With low-lying d orbitals available, the p electrons of N can be promoted with little energy requirement to d orbitals at the rear of N and thus obviate Pauli's repulsion. A second manner in which low-lying d orbitals can assist nucleophilic attack is by overlap of an empty d orbital of N with a filled p orbital of E forming a sort of π bond, in which N shares a pair of electrons from E while concomitantly supplying electrons to E to form a partial σ bond. Substances that can react in this manner have been termed bihilic reagents. Fluorine, oxygen, and bivalent sulfur are electrophilic centers which respond strongly to the polarizability of nucleophiles and at the same time for which the Pauli exclusion would be expected to be marked.

The responsibility for the exceptional reactivity of the catechol monoanion with Sarin has been attributed to acid catalysis by the undissociated hydroxy group [322]. Provided the mechanism is a concerted displacement, it was thought that the —OH group might electrophically help in the elimination of F, or if the mechanism involves the formation of a complex intermediate, the —OH group might hydrogen-bond with the phosphoryl oxygen promoting the addition of the anion. Evidence for and against hydrogen bonding or acid catalysis involving protic nucleophiles, specifically referring to reaction of *p*-nitrophenyl acetate, has been discussed [323].

The energy for the formation of the transition state was considered by Hudson [324] to be dependent on a four-step mechanism: (1) desolvation of N^- requiring the fraction α of the solvation energy E_S or αE_S; (2) removal of the electron from N^- requiring the fraction β of electron removal of the electron affinity E_e or βE_e; (3) donating a fraction of the electron to the electrophile E; and (4) binding together of the new radicals like $N\cdot$ and $E\cdot$ to form a partial N—E bond giving of the fraction at the dissociation state of the dissociation energy E_D of the N—E bond or γE_D. The reception of the fraction of an electron by E which in general has no vacant orbital must be accompanied by the recession from E of equivalent electronic charge either to the departing group X in the case of displacement reaction or into the conjugated system in the case of an addition reaction. This recession requires expenditure of energy to partially break a bond. This energy requirement is to some extent compensated for by the electron affinity of the atom or group receiving the negative charge from E and by the energy of their solvation increment due to increased charge. Factors (a), (b), and (d) are considered the principle ones affecting nucleophilic reactivity since (c) and the other factors mentioned are considered constant for the reaction of one substrate with a series of nucleophiles. For a small extent of bond formation between N and E only nucleophilic desolvation and the ion-induced dipole interaction between N^- and E are considered. E is conceived of as a positive ion and Y^- as having an induced electron displacement. If α_N is the polarizability of N^-, q the charge at E, and r the N—E bond length, the ion induced dipole energy is given by $-\alpha_N q^2/2r^4$ [214, p. 297]. Change in relative nucleophilicity with variation of relative nucleophilicity at the electrophilic center, the character of the transition state, and solvating power of the solvent are implicit in Hudson's approach. The gain of energy from bond formation will depend on the nature and extent of bond formation, and in general will increase with the basicity of the reagent for a given substrate. To desolvate and remove an electron from N will decrease in energy requirements with decrease in extent of bond formation, as Y increases in polarizability, and as the solvent affinity for Y diminishes. "Lowcost" nucleophiles of high polarizability and of low solvent affinity will have advantage in certain reactions, and "high return" reactions of strong basicity in others. Bunnett [310] feels that Hudson's semiquantitative calculations are unconvincing because of the bold assumptions made, though cited data are qualitatively in agreement with his ideas. Further he attributes little if any part to electrostatic attraction, only formally grants the significance of the E basicity of N in certain reactions by the intermediate complex mechanism, and does not recognize the existance of a specific alpha effect.

Highly polarizable nucleophiles such as I^- and $C_6H_5S^-$ are especially reactive compared to reagents of lower polarizability in several reaction series, involving substrates and mechanisms of different types [325, 327], when the substrate possesses a substituent of high polarizability in the vicinity of the reaction site. London forces or dispersion have been proposed as the beneficial interaction between highly polarizable nucleophiles and substrate substituents, which in a concerted displacement may be the leaving group; it may be bonded to the electrophilic center but undisturbed in the rate determining step, or in aromatic substitution it may be an ortho substituent at aromatic or benzylic carbon. The London attraction energy is inversely proportional to the sixth power of the distance separating the atoms or groups concerned and is directly proportional to their polarizabilities [327]. The rate effects expected from theory were confirmed semiquantitatively by calculations [328]; however [310], although fluorine and hydrogen have approximately the same polarizability, and hence no enhancement due to London forces was expected in the case of fluorine, yet o-fluoro and 2,6-difluoro substituents in the benzyl chloride series increased the reactivity of I^- and $C_6H_5S^-$ relative to CH_3O^-. Dipole-induced dipole interaction or some other factor proportional to the polarizability of the reagent could have caused the observed effect [310, 214]. Bunnett [310] believes that the E—N bond distance is fractionally greater than one bond length, and that at this distance London and dipole-induced dipole interactions with E may affect nucleophilicity. These factors would be greater for shorter distances and higher polarizabilities and smaller at longer distances and lower polarizability. Since atoms with lower polarizability generally have shorter bond distances, it does not necessary follow that London and dipole-induced dipole forces will always favor reagents of higher polarizability at distances approaching one bond-length.

It has been pointed out [101] that nucleophilic tendencies change appreciably on transfer from polar to dipolar aprotic solvents, and that quite different parameters are required in the two types of solvents when linear free energy relations are used to predict the rates of reactions of nucleophiles under various conditions. This solvent-type effect on the intrinsic properties of nucleophiles has been discussed in a preceding section.

C. CORRELATION EQUATIONS

1. Swain Equation

Swain's equation is a linear free energy relationship. One equation should permit the correlation of all solvolysis rates including those in which the solvent participates as a nucleophile. The polar displacement reaction of an uncharged organic molecule S with a nucleophilic reagent N and an electrophilic reagent E has been represented [329, 330] by the equation

$$N + S + E \rightleftarrows [\text{Transition state}] \rightarrow \text{products} \tag{5.426}$$

The overall rate constant k^1 of this reaction is related to the rate constant $k_0{}^1$, where water acts as both N and E in the same medium at the same temperature thus:

$$H_2O + S + H_2O \rightleftarrows \text{different transition state} \rightarrow \text{products} \tag{5.427}$$

These rates were correlated using the linear free energy equation

$$\log(k'/k_0') = sn + s'e \tag{5.428}$$

where n, the nucleophilic constant, is a quantitative measure of the nucleophilic reactivity of N ($n = 0$ for water); e, the electrophilic constant, is a measure of the electrophilic reactivity of E ($e = 0$ for water); and s and s', substrate constants, are measures of the discrimination of S among different N and E reagents.

Four corollaries which are useful concepts between structure and reactivity can be obtained from Eq. (5.428).

For fixed N and S when e is proportional to the acid ionization constant of E, that is $e = (-\alpha/s^1)(pKa)$, the Brönsted catalytic law for acids can be derived:

$$\log(k_a'/k_a'') = \alpha \log(K_a/K_a') \tag{5.429}$$

hence

$$\log k_a' = \alpha \log K_a + C \tag{5.430}$$

and

$$k_a' = G_a K_a{}^\alpha \tag{5.431}$$

For fixed S and E and n proportional to the basic ionization constant of N, i.e., $n = (-\beta/S)(pK_b)$, the Brönsted catalytic law for bases can be found:

$$\log(k_b'/k_b'') = \beta \log(K_b/K_b') \tag{5.432}$$

therefore,

$$\log k_b' = \beta \log K_b + C \tag{5.433}$$

and

$$k_b' = G_b K_b{}^\beta \tag{5.434}$$

For $sn \ll s'e$, Eq. (5.428) becomes

$$\log(k'/k_0') = s'e \tag{5.435}$$

If E is the solvent and $s' = m$ ($m = 1$ for tertiary butyl chloride) and $e = Y$ (Y is the solvent "ionizing" power), the Grunwald–Winstein correlation of solvolysis rates results. Thus

$$\log(k'/k_0') = mY \qquad (5.436)$$

For $s'e \ll sn$, then

$$\log(k'/k_0') = sn \qquad (5.437)$$

which would be the case when the electrophilic reagent is held constant. Thus when water is the solvent and acts as the only important E, then $e = 0.00$ for $E = E^0$.

For Eq. (5.437) the standard substance was chosen as methyl bromide with $s = 1.00$. For azide, hydroxide, iodide, and thiosulfate ions and for aniline, values of n were then found from the equation

$$\log(k'/k_0') = n \qquad (5.438)$$

where k_0' was the rate of hydrolysis of methyl bromide and k' was the rate of reaction of methyl bromide with a given one of the substances listed. Thus for each of the substances n could be calculated.

Then the values of the $\log(k'/k_0')$ for the reaction of epichlorohydrin (2,3-epoxypropyl chloride) with water and each of the substances mentioned above were determined and plotted versus the values of n found for the substances (azide, hydroxide, iodide, and thiosulfate ions and aniline) using methyl bromide. According to Eq. (6.437) the slope of this line gave the substrate constant s for epichlorohydrin. The nucleophilic constant n for each nucleophilic reagent; and the substrate constant s for each substrate were determined in similar manner. The method of least squares was applied to the straight lines used in determining the s constant from data. This equation correlated variations of more than 10^5 for the average substrate in the case of 47 reactions studied with a probable error factor of 1.5 in the calculation of k'/k_0' and hence in the calculation of k'.

In the more general equation

$$\log(k'/k_0') = c_1 d_1 + c_2 d_2 \qquad (5.439)$$

146 values of $\log(k'/k_0')$ involving 25 values of c_1, 25 values of c_2, 17 values of d_1, and 17 values of d_2 have been determined [328]. In Eq. (5.439), k_1 is the first-order rate constant for solvolysis of any compound in any solvent, k_0' is the corresponding rate constant in a standard solvent (here chosen as 80% ethanol–20% water by volume) at the same temperature, c_1 and c_2 are constants depending only on the compound undergoing solvolysis, and d_1 and d_2, though not exactly equivalent to n and s, are intended to measure nucleophilicity and electrophilicity.

A wide range of structural variation was represented by the data. The compounds varied from paronitrobenzoyl and methyl to triphenyl and from fluorides to arylsulfonates. Some had strong neighboring group participation, and the pinacolyl compound even rearranged. The solvents ranged from water and anhydrous alcohols to glacial acetic acid and anhydrous formic acid. Since more data existed for 80% by volume ethanol in water than for any other, this solvent was chosen as the standard solvent.

To weight the 146 $\log(k'/k_0')$ values alike the condition

$$\sum \left[\log(k'/k_0')_{\text{obs}} - (c_1 d_1 + c_2 d_2)\right]^2 \qquad (5.440)$$

was taken to define the best fit. Setting the partial derivatives with respect to each of the $25c_1$, $25c_2$, $17d_1$, and $17d_2$ parameters equal to zero resulted in 84 simultaneous equations which were solved by a iterative procedure using a digital computer. The following conditions were assumed to make the solution unique:

$$d_1 = d_2 = 0.00 \qquad \text{for } 80\% \text{ ethanol} \qquad (5.441)$$

$$c_1 = 300c_2 \qquad \text{for MeBr} \qquad (5.442)$$

$$c_1 = c_2 = 1.00 \qquad \text{for } t\text{-BuCl} \qquad (5.443)$$

$$300c_1 = c_2 \qquad \text{for } (\text{Ph})_3\text{CF} \qquad (5.444)$$

Qualitatively, Eqs. (5.442)–(5.444) are in the correct order since sensitivity toward nucleophilic reagents is in the order methyl bromide, t-butyl chloride, trityyl fluoride. The ratio c_1/c_2 is a convenient single number which characterizes the reactivity of the compounds; a compound which discriminates relatively more highly among nucleophiles than electrophiles has a high c_1/c_2 ratio. This ratio diminishes from paranitro to paramethyl in the order methyl, ethyl, isopropyl, tertiary butyl, benzhydryl, trityl, which is the anticipated order.

The difference $(d_1 - d_2)$ is a convenient single number characterizing the solvent. For the most nucleophilic solvents this difference is greatest, decreasing in the order anhydrous alcohols, acetone–water and alcohol–water mixtures, water, glacial acetic acid, anhydrous formic acid.

Table 5.16 lists some of the constants obtained. The values arbitrarily assigned are in parenthesis. The logarithms of the rate constants for the various reactions in the various solvents were calculated using these constants. The mean error was 0.124 in $\log k'_{\text{calc}}$ for 25 compounds in 17 solvents. Thus the error in k' was a factor of 1.33, the antilog 0.124.

It has been pointed out [321] that to the extent that Eq. (5.428) holds, the solvating power for cations (carbonium ions, at least is proportional to the nucleophilicity, and that there is a smooth and continuous transition from

Table 5.16

Values of Compound and Solvent Constants

Compound	c_1	c_2	c_1/c_2
$NO_2PhCOCl$	1.09	0.21	5.2
NO_2PhCOF	1.67	0.49	3.4
$PhCOCl$	0.81	0.52	1.6
$PhCOF$	1.36	0.66	2.1
$MePhCoCl$	0.82	0.65	1.3
$MePhCOF$	1.29	0.80	1.6
$MeBr$	0.82	0.27	3.0
$EtBr$	0.80	0.36	2.2
$EtOTs$	0.65	0.24	2.7
n-$BuBr$	0.77	0.34	2.2
$PhCH_2Cl$	0.74	0.44	1.7
$PhCH_2OTs$	0.69	0.39	1.8
i-$PrBr$	0.90	0.58	1.5
i-$PrOBs$	0.63	0.48	1.33
$MeOCxOBs$	0.57	0.57	1.00
$BrCxOBs$	0.80	0.87	0.92
$PinOBs$	0.76	0.87	0.86
$PhCHClMe$	1.47	1.75	0.84
$(Ph)_2CHCl$	1.24	1.25	0.99
$(Ph)_2CHF$	0.32	1.17	0.27
t-$BuCl$	(1.00)	(1.00)	(1.00)
$(Ph)_3CSCN$	0.19	0.28	0.69
$(PH)_3COPhNO_2$	0.18	0.59	0.31
$(PH)_3CF$	0.37	1.02	(0.33)

Solvent	n	e	$n-e$
MeOH 100	−0.05	−0.73	+0.7
MeOH 96.7	−0.11	−0.05	−0.1
MeOh 69.5	−0.06	+1.32	−1.4
EtOH 100	−0.53	−1.03	+0.5
EtOH 90	−0.01	−0.54	+0.5
EtOH 80	(0.00)	(0.00)	(0.00)
EtOH 60	−0.22	+1.34	−1.6
EtOH 50	+0.12	+1.33	−1.2
EtOH 40	−0.26	+2.13	−2.4
Me_2CO 90	−0.53	−1.52	+1.0
Me_2CO 80	−0.45	−0.68	+0.2
Me_2CO 70	−0.09	−0.75	+0.7
Me_2CO 50	−0.25	+0.97	−1.2
H_2O 100	−0.44	+4.01	−4.5
AcOH 100	−4.82	+3.12	−7.9
Ac_2O 97.5	−8.77	+5.34	−14.1
HCOOH 83.3	−4.44	+6.26	−10.7
HCOOH 100	−4.40	+6.53	−10.9

$S_N 1$ to $S_N 2$ mechanism. There is some correlation of n with basicity and e with acidity, and the values of s and s' show compounds reacting by the $S_N 2$ mechanism are more susceptible to nucleophilic attack [331].

Solvolysis rates have been related [329, 332] using a special two-parameter equation:

$$\log(k'/k_0')_A - \log(k'/k_0')_{A_0} = ab \tag{5.445}$$

The standard methyl bromide is represented by the subscript A_0, and other chlorides and bromides are represented by the subscript A. The first-order rate constant for the solvolysis of any organic chloride or bromide (A) or of the standard compound (A_0) in any solvent is k'; k_0' is the corresponding rate constant in a standard solvent, here chosen as 80% ethanol–20% water by volume at the same temperature; a is a constant depending on only the chloride or bromide and b is a constant depending on only the solvent.

In obtaining the best values of a and b by the method of least squares involving an iterative procedure, the following conditions were imposed to obtain a unique solution:

$$b = 0.00 \text{ for } 80\% \text{ EtOH} \tag{5.446}$$

$$a = 0.00 \text{ for MeBr} \tag{5.447}$$

$$a = 1.00 \text{ for } t\text{-BuCl} \tag{5.448}$$

according to Eqs. (5.445) and (5.448), choosing tertiary butyl chloride for comparison of its rate of solvolysis with that of the standard compound methyl bromide makes $a = 1$ and

$$\log(k'/k_0')_t - \log(k'/k_0')_{\text{MeBr}} = b \tag{5.449}$$

Applying this relation to solvents in which the rates for both tertiary butyl chloride and methyl bromide had been measured yielded crude values of b. Using any solvent for which b values had been found and Eq. (5.445), rough a values were determined.

Using the equation

$$\log(k'/k_0') - \log(k'/k_0')_{t\text{-BuCl}} = (a - 1.00)b \tag{5.450}$$

rough values of b for other solvents in which methyl bromide had not been studied were calculated.

Experimental errors were minimized and better values of b were obtained from the crude values of a for all solvents by applying the method of least squares with equal weighing of all the usable values of $\log(k'/k_0')$. From the better values of b, better values of a were obtained in the same manner.

Table 5.17 compares values of a for different substituents on the carbon at which reaction occurs, and Table 5.18 compares the value of b with the dielectric constants of the solvents.

Table 5.17

Values of a for Compounds $R_1 R_2 R_3 CX$

Compound	Number of reactions	a	R_1—, R_2—, R_3—, X
PicCl	8	−0.42	—$C(NO_2)$=CH—$C(NO_2)$=CH—(NO_2)=, Cl
$NO_2PhCOCl$	7	−0.37	4-$NO_2C_6H_4$—, O=, Cl
$PhCOCH_2Br$	8	−0.04	C_6H_5CO—, H—, H—, Br
MeBr	10	(0.00)	H—, H—, H—, Br
PhCOCl	12	+0.06	C_6H_5—, O=, Cl
EtBr	5	+0.15	CH_3—, H—, H—, Br
i-BuBr	4	+0.16	$(CH_3)_2CH$—, H—, H—, Br
n-BuBr	12	+0.18	$CH_3CH_2CH_2CH_2$—, H—, H—, Br
$PhCH_2Cl$	8	+0.19	C_6H_5—, H—, H—, Cl
MePhCOCl	5	+0.41	4-$CH_3C_6H_4$—, O=, Cl
i-PrBr	5	+0.42	CH_3—, CH_3—, H—, Br
PhCHClMe	5	+0.64	C_6H_5—, CH_3—, H—, Cl
$(PH)_2CHCl$	13	+0.78	C_6H_5—, C_6H_5—, H—, Cl
t-BuBr	7	+0.93	CH_3—, CH_3—, CH_3—, Br
t-BuCl	15	(+1.00)	CH_3—, CH_3—, CH_3—, Cl

Table 5.18

Values of b for Solvents

Solvent	Number of reactions	b	Dielectric constant at or near 20°C
Et_3N	3	−17.27	3.2
n-$BuNH_2$	5	−10.15	5.3
C_5H_5N	5	−9.66	12.4
$PhNH_2$	5	−4.78	7.3
MeOH	6	−0.94	33.7
EtOH	14	−0.74	23.2
Me_2CO 90	7	−0.72	24.6
EtOH 90	4	−0.52	28.0
MeOH 96.7	6	−0.51	34.2
EtOH 80	15	(0.00)	33.9
Me_2CO 80	7	+0.04	30.9
Me_2CO 70	7	+0.42	36.5
AcOH	5	+0.57	9.7
MeOH 69.5	5	+0.61	47.3
EtOH 60	4	+0.88	44.7
Me_2CO 50	8	+1.02	49.5
EtOH 50	8	+1.14	51.3
H_2O	4	+2.95	79.2
HCOOH	6	+4.00	58.5

In general a increases with the increase of the electron-supplying ability of the substituent. The smallest a values occur with substituents possessing nitro groups which are electron attracting. Also, a increases in the stepwise replacement of hydrogen atoms with methyl groups. When a phenyl group replaces a methyl group a resonance rather than an inductive effect could cause the increase in a. A shift of the electron distribution from the phenyl group toward the reaction site could constitute the resonance effect.

The temperature, the nature of the leaving group, and the polar effects of the substituents on the reaction all influence the value of a.

Picryl chloride is less well correlated than the other compounds, perhaps because of excess solvent interaction with polar nitro groups or because of excess resonance in the transition state for reaction with more nucleophilic solvents.

The correlation of the data with the equation was measured by ϕ, where ϕ is given by the equation

$$\phi = (1 - \varepsilon/\theta)100\% \tag{5.451}$$

where

$$\varepsilon = (1/n) \sum_n (|\log q_{obs} - \log q_{calc}|) \tag{5.452}$$

and

$$\theta = (1/n) \sum_n (|\log q_{obs} - (1/n) \sum \log q_{obs}|) \tag{5.453}$$

In these equations n is the number of q values observed and q is the rate or equilibrium constant or a ratio of rate or equilibrium constants. Absolute values, regardless of sign, were used, as is indicated by the vertical bars. For perfect fit of the data to the equation ε is zero and ϕ 100%. Small or even negative values of ϕ may result when the correlation is poor. Typical calculated fits varied from excellent to poor.

2. Edwards Equation

The equation of Edwards [333, 334] can be written

$$\log(k'/k_0') = \alpha E_n + \beta H \tag{5.454}$$

This equation combines the nucleophilic and basic scales to correlate the reaction of electron donors. The ratio (k'/k_0') depicts equilibrium or rate constants for reactions with the given nucleophilic reagent and with water; α and β are constants; and E_n is a characteristic constant of electron donors, related to both their polarizabilities (taken as molar refraction R_∞) and to their basicities ($H = pK_a + 1.74$).

$$E_n = aP + bH \tag{5.455}$$

where P measures the polarizability of the nucleophilic reagent and is given by

$$P = \log(R_\infty/R_{H_2O}) \tag{5.456}$$

From Eqs. (5.454) and (5.455)

$$\log(k'/k_0') = \alpha aP + \alpha bH + \beta H$$
$$= \alpha aP + (\alpha b + \beta) H$$
$$= AP + BH \tag{5.457}$$

where $A = \alpha a$, the product of the constants; and $B = (\alpha b + \beta)$, the combination of the constants. Hence the sensitivity of the nucleophilic reagent to a change of polarizability and basicity, respectively, are measured by A and B.

Rates of displacement reactions of carbon, oxygen, hydrogen, and sulfur; and equilibrium constants for complex ion associations, solubility products, and iodine and sulfur displacements are correlated rather well by the equation.

The αE_n term in all cases except for the displacement of hydrogen dominates in Eq. (5.454), and the βH term makes only a slight contribution to $\log(k'/k_0')$.

It should be mentioned that E_n for a reagent (N^-) was taken as the electrode potential E_0 for the oxidative dimerization (to $N:N$) corrected for the potential, -2.60, of the couple

$$2H_2O \rightleftarrows H_4O_2^{2+} + 2e \tag{5.458}$$

i.e.,

$$E_n = E_0 + 2.60 \tag{5.459}$$

The acidities in water of the conjugate acids of the reagents relative to water determined the H values ($H = pK_a + 1.74$).

The correlations are solvent dependent since E_n and H are determined relative to water.

Equations (5.454) and (5.457) apparently correlate a wider variety of reaction types than, for example, the linear free energy relations of Swain and co-workers. These equations, compared to other free energy relations also provides independent means of parameter determination. The correlation of the hydroxyl ion is a significant achievement of the Edwards equation for displacements on carbon. Neither the E_n or the P values of the cyanide ion are correlated with its nucleophilicity by the Edwards equation. Either resonance interaction with the cyanogen molecule [335] or the necessity of consideration of the polarizability of more than one atom, a general characteristic of polyatomic molecules, could be the source of this failure. There is a small stearic requirement in the case of displacement of hydrogen. No reagent correlation has so far assessed this characteristic.

The Swain and Scott and Edwards equations have been assessed as being valuable to organic chemists since they are very successful for some reaction in protic solvents [101]. However, it is pointed out [101] that different parameters are needed in switching from protic to dipolar aprotic solvents since nucleophilic characteristics are quire different in the two types of solvents. The nature of the substrate also influence nucleophilic tendencies. These last two items have been discussed at some length in the section on protic–dipolar aprotic solvents.

Bunnett [310] states that the Swain and Scott equation (Eq. (5.437) where k_0' is k'_{H_2O}) has given serious deviations including inversions of relative nucleophilicity, and that it is now recognized to have limited applicability. The Edwards equation give better fits than the Swain and Scott equation [310].

It has been pointed out that the equations, as are the other free energy equations, are basically empirical [36]. However Wells [336] indicates that when the variables can be identified, the derived parameters are useful theoretical quantities.

3. Hansson Equation

Another linear free energy equation correlates variation in two substituents [337]. Structural variation in both reactants in the reactions of amines with epoxides is accounted for by the equation which is written

$$\log(k'/k_0') = \rho(\sigma_a + \sigma_0) + \tau \qquad (5.460)$$

where k' is the specific velocity constant for any amine with any epoxide; k_0' is the specific velocity constant for the reaction of ammonia with ethylene oxide; ρ depends only on the reaction and is unity for the reactions of amines with epoxides at 20°C in water; σ_a depends only on the structure of the amine and is zero for ammonia; σ_0 depends only on the structure of the epoxide and is zero for ethylene oxide; and τ zero for ethylene oxide, is a correction factor, apparently necessitated because of steric factors, and required by failure generally in the correlation of trimethylamine and triethylamine.

Since ρ is taken as unity at 20°C in solvent water, it is presumably both temperature and solvent dependent.

Both σ_0 and τ are zero for ethylene oxide, hence value of τ_a were found by averaging the values of $\log(k'/k_0')$ for each of the amines reacting with ethylene oxide, propylene oxide, glycidol, and epichlorohydrin. Also σ_a was zero for ammonia, and hence values of σ_0 were found by averaging the values of $\log(k'/k_0')$ for each epoxide reacting with eight amines including ammonia. The structural parameters for the Hansson equation [336] are presented in Table 5.19.

The attack occurs at the same site in all four epoxides according to the

Table 5.19

Structural Parameters for the Hansson Equation

Amine	a	Amine	σ_2	R	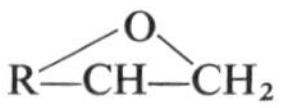 $R{-}\underset{\displaystyle\quad}{\overset{\displaystyle O}{CH}}{-}CH_2$ σ_0
NH_3	0.00	$C_2H_5NH_2$	0.74	H	0.00
CH_3NH_2	0.94	$(C_2H_5)_2NH$	0.63	CH_3	-0.02
$(CH_3)_2NH$	1.47	$(C_2H_5)_3N$	0.36	CH_2OH	-0.04
$(CH_3)_3N$	2.23	Pyridine	0.14	CH_2Cl	0.51

Hansson equation. However it has been pointed out [338] that the structural effects on epoxide reactions must be considered in elucidating the direction ring opening.

Steric factors would presumably be involved in all epoxide-amine reactions. Thus the τ-term was found not to correct for steric factors [339], and calculated σ-terms showed no additive relationships in reactions of a series of methylpyridines with propylene oxide.

4. Grunwald–Winstein Equation

For reactions occurring by the S_N1 mechanism it has been found [340] that when values of $\log k'$ for the solvolysis of any one compound in a number of solvents are plotted against values of $\log k'$ for the solvolysis of teriary butyl chloride in the same solvents, a linear plot resulted. As a useful quantitative measure of the ionizing power of a given solvent $\log k'^{BuCl}$ (BuCl = tertiary butyl chloride) was selected. A function Y was determined by the equation

$$\log k'^{BuCl} = Y + \log k_0'^{BuCl} \tag{5.461}$$

where k'^{BuCl} is the specific velocity constant for the solvolysis of tertiary butyl chloride at 25.0°C in a given solvent and $k_0'^{BuCl}$ is the specific velocity constant for the solvolysis reaction at 25° in an aqueous ethanol solvent which is 80% by volume ethanol. The S_N1 mechanism may be represented by the equations

$$RX \xrightarrow{\text{slow}} R^+ + X^- \tag{5.462}$$

and

$$R^+ + Y^- \xrightarrow{\text{fast}} RY \tag{5.463}$$

in which the slow rate controlling step is the formation of the carbonium ion R^+; and the reaction rate is, under certain conditions, independent of the concentration of the nucleophilic substance Y^-. The mechanism is thus a nucleophilic reaction of the first order and is the type of reaction involved in the Grünwald Winstein work.

Later [341] Eq. (5.461) was written

$$\log k' = \log k_0' + mY \qquad (5.464)$$

where k' and k_0' are the respective specific velocity constants for $S_N 1$, solvolyses of a given compound in a given solvent and in aqueous solvent which is 80% by volume ethanol, Y is the ratio for the solvolyses of t-butyl chloride in the same solvents and m is a constant independent of the solvent but characteristic of the compound. The authors feel that data indicate that Y truly measures ionizing power and does not include terms due to covalent solvent–carbon interaction in the transition state. This equation has been applied successfully to a number of solvents [329, 330, 342].

Later it was found that the original treatment of the mY relationship was limited [343, 344]. The structural limitations of the original mY treatment was ultimately placed on m which proved to be characteristic of RX and of the solvent pair with the retention of one set of Y values, found as originally intended to represent the effect of solvent change, i.e., be characteristic of the solvent. The set of Y values based on t-butyl chloride was selected. The modified mY treatment, employing separate lines for each solvent pair, was found for solvolyses reactions to fit 760 values of $\log k'$ with 372 constants with a probable error of about 0.025 in $\log k'$. The correlation of this data using the four parameter equation Eq. (5.440) would have required 434 constants. The usefulness of four parameter equations from the standpoint of bookkeeping has been questioned.

5. Taft Equation

Steric and resonance effects were assumed to be the same for both acid and basic hydrolyses of esters [345]. If k_B is the specific velocity constant for the basic hydrolyzes of the ester XCO_2R in which R is constant and X is varied, and $k_{0,B}'$ is the specific velocity constant for the basic hydrolysis of ester selected as standard, the constants are related by the equation

$$\log(k'/k_{0,B}') = \rho_\beta \sigma_X + S + R \qquad (5.465)$$

where the S factor measures the steric and the R factor the resonance contribution to the rate. Similarly for the acid hydrolysis the equation

$$\log(k_A'/k_{0,A}') = \rho_A \sigma_X + S + R \qquad (5.466)$$

holds, where the steric [345] and resonance factors are the same for both basic and acid catalysis. Subtracting Eq. (5.466) from Eq. (5.465) yields

$$\log(k_B'/k_{0,B}') = \log(k_A'/k_{0,A}') + \sigma_X(\rho_B - \rho_A) \qquad (5.467)$$

Assuming [346] that the nonpolar $\log(k_A'/k_{0,A}')$ measures nearly quantitatively the net kinetic energy steric effects, the equation

$$\log(k_A'/k_{0,A}') = E_s \qquad (5.468)$$

can be written. When there are unsaturated substituents conjugated with the carbonyl group, or when there are changes in attractive interactions between reactant and transition states (e.g., hydrogen bonding), exceptions occur. Taking $(\rho_B - \rho_A)$ as 2.48 for hydrolysis in aqueous–alcohol and aqueous–acetone solutions, Eqs. (5.467) and (5.468) yields

$$\log(k_B'/k_{0,B}') = E_s + 2.48\sigma^* \qquad (5.469)$$

where σ^* is a substituent constant dependent only upon the net polar effect of the substituent relative to that of the standard of comparison, corresponding to the rate constant k' compared to the rate constant k_0' selected for the standard comparison $(k_0', R = CH_3)$.

The constant σ^* is anologous to but different in nature and origin from, the Hammett [347] substituent constant σ, i.e., $\sigma \neq \log(k'/k_0')$ for the ionization of carboxylic acids. The polar effect dealt with by Taft is placed on a comparable scale to the Hammett σ values by introduction of the factor 2.48. According to Taft, σ^* is identical to σ_x, and according to Eq. (5.467) if $(\sigma_B - \sigma_A)$ equals 2.48 is given by

$$\sigma^* = \frac{1}{2.48}\left[\log\frac{k_B'}{k_{0,B}'} - \log\frac{k_A'}{k_{0,A}'} \right] \qquad (5.470)$$

For negligibly small stearic and resonance effects, Taft has correlated the rates of a number of reactions using the simple equation

$$\log(k'/k_0') = \sigma^*\rho^* \qquad (5.471)$$

which is formally like the Hammett equation. Here σ^* is the polar substituent constant for the R group relative to the standard CH_3 group and ρ^* is a constant representing the susceptibility of a given series of reactions to polar substituents; its value varies with the nature of the reacting center, the attacking reagent, etc. This equation applies to certain reactions of ortho-substituted benzene derivatives, as $o\text{-}X\text{—}C_6H_4\text{—}Y$, where X is the ortho substituent and Y is the reaction center. Values of k' and σ^* are generally related to the unsubstituted $(X=H)$ derivative.

Equation (5.471) is a free energy–polar energy relationship, since polar energies represented by σ^* have been separated from free energies. The equation is more limited than the Hammett equation since the former equation does not apply to reaction series in which there are appreciable steric and resonance effects upon reaction rates, conditions which often prevail for reaction series involving bulky and unsaturated constituents at the reaction center.

Except for ortho substitution σ^* values were determined using acetate esters as standards. For the ortho case benzoates were used. A liner plot of $\log(k'/k_0')$ versus σ^* values requires that only polar effects be operative. A nonlinear relation does not invalidate the proportional nature of polar effects but shows that these rates or equilibria are determined by influences other than polar effects according to Taft.

Many substituent constants both for the R—Y aliphatic series and for the ortho-substituted benzene derivatives have been tabulated [348]. In Table 5.20 are σ^* values for the group R relative to a CH_3 group calculated from

Table 5.20

Typical σ^* Values

R	σ^*	R	σ^*
CH_3N+	$+5.04$	CH_3CO	$+1.65$
CH_3SO_2	$+3.70$	HO	$+1.55$
$N\equiv C$	$+3.64$	$ClCH_2$	$+1.05$
F	$+3.08$	$C_6H_5OCH_2$	$+0.85$
HO_2C	$+2.94$	C_6H_5	$+0.60$
Cl	$+2.94$	$HOCH_2$	$+0.56$
Br	$+2.80$	H	$+0.49$
Cl_3I	$+2.65$	CH_3	$+0\,00$
I	$+2.38$	$(CH_3)_3C$	-0.30
Cl_2HC	$+1.94$	$(CH_3)_3Si$	-0.72

Eq. (5.470) using k_B' and k_A' as the rate constants for the normal basic and acid hydrolyses, respectively, of an ethyl ester of the type $RCO_2C_2H_5$ and $k_{o,B}'$ and $k_{o,A}'$ for the rate constants for the basic and acidic hydrolysis, respectively, of ethyl acetate. For a variety of reactions which are thought to be about free from steric and resonance contributions to the free energy term according to Eq. (5.471), these σ^* values correlate quantitatively the effects of structure on rates and equilibria. For some groups the σ^* values were calculated by multiplying by 2.8 the σ^* for the RCH_2 group. Derick [349] proposed a factor of 3. Some of the deviations of reactivity data not correlated by the Taft equation may be used as a measure of other variables [336] particularly resonance effects and nonclassical bonding interactions.

Substituent constitution has been qualitatively related to σ^* values. Qualitative correlation between the effects of structure on reactivity and the concepts of bonding and electronegatively are involved in these relationships.

If there are variable entropies of activation with changing substituents, and if the Hammett or Taft equation is not obeyed, there may be appreciable variation in steric hinderence of internal motions or steric hinderence of solvation within a reaction series. Solvent interaction with poles and dipoles may produce kinetic energy effects that are the principle cause of variable entropy terms in the reaction of meta- and para-substituted benzene-derivatives.

6. Hammett Equation

The art of relating chemical reactivity, including chemical reaction rates and ionization constants, to molecular structure has been highly developed especially in organic chemistry. Some of these efforts have correlated rather successfully large collections of data. The linear free-energy-relations discussed above and those to follow fall in this class. All of these correlations are based on the assumption that when a given molecule is involved in two similar rate or equilibrium processes the linear free energy changes involved in the two processes will depend in like manner on changes in structure.

It has been shown that the logarithms of the basic ionization constants of aromatics in strong acids such as H_2SO_4 and HF were linearly related to the logarithms of the rate of deuterium exchange of the aromatics with the solvent [347, 350, 351] and also with rate constants for the addition of free radicals to the aromatic [351]. The equilibrium for which the basicity constants for many aromatics in anhydrous HF was obtained was written

$$Ar + HF \rightarrow ArH^+ + F^- \tag{5.472}$$

The linearity between the logarithm of rate and ionization equilibrium constants indicated that the activation free energy changes ΔG^* for the rate processes and the standard free energy changes ΔG^0 for the ionization processes must be affected similarly by changes in structure. Thus the two kinds of free energy changes for the same two aromatics would be related by the equation

$$\Delta G_2{}^* - \Delta G_1{}^* = \rho(\Delta G_2{}^0 - \Delta G_1{}^0) \tag{5.473}$$

where ρ is a reaction constant depending on the reaction, the temperature, and the medium.

Now

$$\Delta G^* = -RT \ln k' \tag{5.474}$$

and

$$\Delta G^0 = -Rt \ln K \tag{5.475}$$

Therefore, Eq. (5.290) can be put in the form

$$\log(k'/k^{0\prime}) = \log(K'/K^{0\prime}) \tag{5.476}$$

Hammett [347, 352] selected the ratio $(K'/K^{0'})$ as a standard and assigned the effects on structural changes of meta and para substituents on benzene derivatives to the $\log(K'/K^{0'})$ term, and designated this as σ. Equation (5.476) became

$$\log(k'/k^{0'}) = \rho\sigma \qquad (5.477)$$

where σ now is a substituent constant depending on the substituent. It is clear why Eq. (5.477) and other similar correlative expressions are called linear free energy relationships.

Table 5.21

Values of σ from Data on the Ionization of Benzoic Acid in Water at 25°C[a]

R	Meta substituents σ	Para substituents σ	R	Meta substituents σ	Para substituents σ
$CH_2Si(CH_3)_2$	-0.19	-0.22	I	$+0.352$	$+0.18$
t-C_4H_9	-0.10	-0.197	Cl	$+0.373$	$+0.227$
CH_3	-0.069	-0.170	$COCH_3$	$+0.376$	$+0.502$
OCH_3	$+0.115$	-0.268	Br	$+0.391$	$+0.232$
OH	$+0.121$	-0.37	CF_3	$+0.43$	$+0.54$
OC_6H_5	$+0.250$	-0.320	CN	$+0.56$	$+0.660$
$COOCH_3$	$+0.39$	-0.31	SO_2NH_2	$+0.55$	$+0.62$
OCF_3	$+0.40$	-0.35	SO_2CH_3	$+0.56$	$+0.68$
SCF_3	$+0.40$	-0.50	NO_2	$+0.710$	$+0.778$
F	$+0.337$	-0.062	SO_2CF_3	$+0.79$	$+0.93$

[a] From Hammett [352].

Extensive compilations of σ values have been made [336, 353–356]. Table 5.21 contains a number of these values from data on ionization of benzoic acid in water at 25°C. See Fig. 5.14 for a plot of $\log k'$ versus σ for the ionization process listed. Thus far the rate constant for the reaction

$$p - BrC_6H_5CH_2Cl + KI \rightarrow p - BrC_6H_5CH_2BI + KCl \qquad (5.478)$$

Hammett's data gives $\log k^{0'}$ of Eq. (5.294) as 0.167, ρ as 0.785, and σ as 0.232. The reaction of the unsubstituted benzyl chloride with potassium iodide,

$$C_6H_5CH_2Cl + KI \rightarrow C_6H_5CH_2I + KCl \qquad (5.479)$$

yields the rate constant $k^{0\prime}$, ρ is the constant characteristic of the substituted phenyl chloride with potassium iodide

$$XC_6H_4CH_2Cl + KI \rightarrow XC_6H_4CH_2I + KCl \qquad (5.480)$$

and σ is a constant for para substitution of chlorine.

It was found that 38 reactions including both equilibrium and rate constants and involving derivatives of benzoic acid, phenol, aniline, benzene sulfonic acid, phenylboric acid, and phenylphosphine gave a mean value of the probable errors of the values of $\log k'$ calculated from Eq. (5.477), as compared to the corresponding experimental values to be only 0.067. These are the data represented in Fig. 5.14.

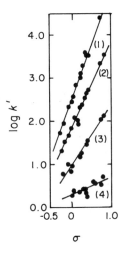

Fig. 5.14. Relationship between $\log k'$ and σ for various chemical processes. Curves: (1) acidity constants of substituted anilinium ions in water at 25°C, (2) ionization of substituted phenylboric acids in 25% ethyl alcohol at 25°C, (3) reaction of substituted benzoyl chlorides with methyl alcohol at 0°C, (4) base-catalyzed bromination of substituted acetophenones in acetic acid–water medium with sodium acetate as catalyst at 35°C.

The only available data give $\sigma\rho$ products. Hence some arbitrary value must be assigned σ or ρ. Hammett chose $\rho = 1$ for the ionization of substituted benzoic acids, and the difference in logarithms of substituted and unsubstituted benzoic acids gave σ for that substituent since in Eq. (5.477), $k^{0\prime}$ is the rate or equilibrium constant for the general reaction of an unsubstituted member of a class and k' is the rate or equilibrium constant for the same reaction for a substituted member of a class. There substituent values are used to obtain ρ values for other reactions, and these, in turn, can be used to obtain σ values for substituents the effects of which on the ionization of benzoic acid have never been determined values of ρ have been compiled [336, 355], and some examples are listed in Table 5.22.

Table 5.22

Values of ρ for Acid Dissociation in Water
at 25°C

Dissociating acid	ρ
$XC_5H_4NH^+$	6.11
XC_6H_5OH	2.26
$XC_6H_5NH_3^+$	2.94
$XC_6H_4CO_2H$	1.00
$XC_6H_4CH_2CO_2H$	0.56
$XC_6H_4CH_2CH_2CO_2H$	0.24

In Hammett's original proposal [347]

$$\log K = \log K^0 - \frac{A}{2.303} RTd^2 \left(\frac{B_1}{D} + B_2 \right) = \log K^0 + \sigma\rho \quad (5.481)$$

whence

$$\sigma = -A/2.303R$$

$$\rho = \frac{1}{d^2 T} \left(\frac{B_1}{D} + B_2 \right) \quad (5.482)$$

where d is the distance from the substituent to the reacting group, D is the dielectric constant of the medium in which the process occurs, and A, B_1, and B_2 are constants independent of the temperature and solvent. A and σ are constants depending on the substituent and its position in the ring relative to the reacting group while B_1 and B_2 depend only on the reaction, the constant ρ depends on the reaction, the medium, and the temperature. Hine [357] indicates that ρ generally increases with decrease in ion-solvating power and dielectric constant of the medium. According to Eq. (5.482), a decrease in D would cause an increase in ρ. This is illustrated by ρ values for the ionization of benzoic acid in Table 5.23.

Table 5.23

Values for the Ionization of Benzoic Acid at 25°C
in Different Solvents

	Solvent		
	Water	Methanol	Ethanol
D	78.55	31.5	24.33
ρ	1.000	1.537	1.957

The overall effect of the solvent on ρ is probably composed of several factors. Thus the electrostatic interactions between the substituents and the reaction centers increase as the dielectric constants of the solvent decreases. Also the nature of the rate or equilibrium process may change with changing solvent due to the difference in the nature of the active center solvated by different solvents. Thus the water-solvated and ethanol-solvated carboxylate anion group would probably differ considerably.

7. Steric Effects

Points for aliphatic compounds and for orthosubstituted benzene derivatives do not fall on the line with meta- and para-substituted benzene derivatives unless the reaction zone is farther removed from the reaction site. Thus in the triethylamine ester reaction the R group is farther removed from the reaction site than is the carbonyl carbon which is the reaction site, and the

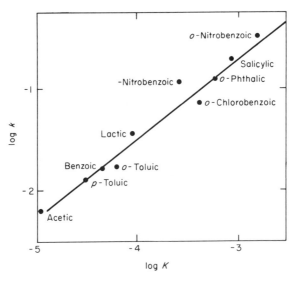

Fig. 5.15. Correlation of specific rate k' in the trimethylamine–ester reaction with the acidity constant K. Time in hours.

plot, Fig. 5.15, given by Hammett [353] correlates the rate constant k' with the acidity constant K both for ortho-, meta-, para-, nonsubstituted aromatic and aliphatic acids, and for aliphatic, aromatic and ortho-, meta-, and para-substituted benzoic ester. Nonconformity is cases when the R group is near the reaction site arises from short-range steric effects, at least to a large extent. These effects were first recognized by Meyer [358a] in the acid catalyzed esterification of carboxylic acids and called by him steric hinderance. Hammett [353] points out considerable evidence that an important factor in the steric effect is that the alteration in the structure in the vicinity of the reaction zone produce changes in the solute–solvent interactions. Thus plots of the relative acidity constants K/K_0 [353, 359] for a solvent composed of n-butanol containing 0.05 M lithium chloride versus the corresponding values for solvent water, K_0 being the acidity constant for solvent water, yielded a straight line of slope 1.5 for meta- and para-substituted benzoic acids. Except for the methoxy derivative, the points for the ortho-substituted benzoic acids lie to the right of this line, and the points for aliphatic acids not substituted in the α-position lie to the left. Other evidence is also cited such as the relative strengths of ortho-chlorobenzoic and ortho-nitrobenzoic acids being much less than expected in benzene when compared to the corresponding para-substituted acids, and the specialized effect of ortho-substitution on entropies and enthalpies of reaction or of activation.

8. Additivity Effect of Multiple Substitution

Meta- and para-substituted benzene derivatives show an additivity effect in multiple substitution. In 33 reactions the effect of 3,4 and 3,5 disubstitution and also 3,4,5 trisubstitution could be represented by the equation [355]

$$\log(k'/k_0') = \rho \sum \sigma \tag{5.483}$$

with a median deviation of 0.052 in the value of $\sum \sigma$.

In 24 reactions in which two rings occurred in the reactant, the equation

$$\log(k'/k_0') = \rho(\sigma_1 + \sigma_2) \tag{5.484}$$

was found to apply [353]. The equation had originally been applied to the reaction [357]

$$(C_6H_5)_2SeBr_2 \rightarrow (C_6H_5)_2Se + Br_2 \tag{5.485}$$

where as in the 24 cases cited above the substitutent constants σ_1 and σ_2 apply the one to the one ring and the other to the other ring.

For 46 cases of the reaction of meta- and para-substituted benzoic acids with *meta*- and *para*-diphenyldiazomethanes the equation

$$\log k' = -0.1089 - 1.620 \sum \sigma_D 2.37 \sum \sigma_B \tag{5.486}$$

was found to apply [360] with a standard deviation of 0.078.

9. Through Resonance

If the acidity constants of meta- and para-substituted phenols are plotted versus those of correspondingly substituted benzoic acid, the deviation of the para-substituted compounds from the straight line obtained arise from specialized long-distance interactions [353] termed through-resonance interactions [361]. Thus in p-nitrophenol

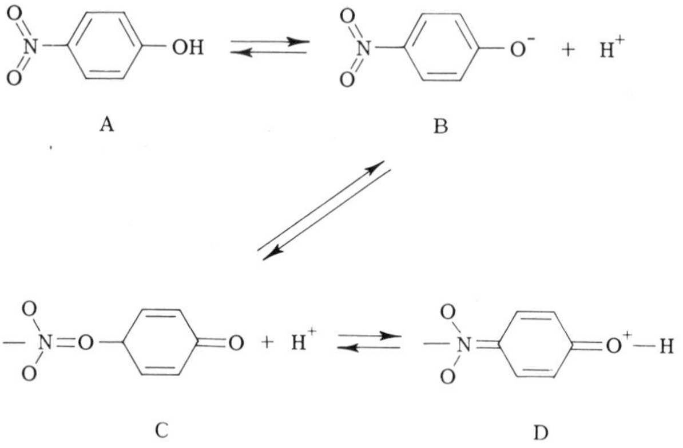

A ⇌ B + H⁺

$$(5.487)$$

C + H⁺ ⇌ D

structure C makes extensive contribution to the electron distribution in the anion, while the structure D makes a smaller contribution than that to p-nitrophenol itself. Thus the anion is stabilized and is strongly favored in equilibrium with the acid form. This is through-resonance; m-nitrophenol or its anion would not demonstrate this interaction.

Even though marked through-resonance is present with some substituents, linear free-energy relations apply to some, not necessarily closely similar groups of reactions. Hammett [353] uses compounds showing $+T$-type through resonance to illustrate this point. Substituents such as NH_2, OCH_3 which can supply electrons by resonance constitute $+T$ substituents and give normal substituent constants [360] σ^n applicable only to reactions, such as phenol ionization which are not anticipated to show through resonance with substituents of this kind [351]. In like manner $-T$ substituents in the para position give values for substitution limited only to reactions such as the ionization of benzoic acid which are anticipated to show through-resonance with substituents of this kind, for example, NO_2 and CN which can withdraw electrons.

In the determination [362] of σ^n values, the procedure was to start with σ values determined from acidity constants of m-CH$_3$, m-F, m-Cl, m-COCH$_3$, m-NO$_2$, p-COCH$_3$, and p-NO$_2$ substituted benzoic acid for which through-resonance is supposed to be minimal. From these σ values, ρ values were found for about 80 rate or equilibrium processes involving compounds having these substituents. Then σ^n values were calculated as average values using Eq. (5.477) for all substituents for which data of the effect on the ionization of C_6H_5–CH$_2$COOH were known and for which through resonance was unlikely. Substituents such as p-CN and p-COCH$_3$ were included, and σ^n values were obtained from the effect of these substituents on the ionization constant of phenyl boric acid, or on the specific rate of solvolysis of benzyl chloride. Values of σ^n were also found from the effect of such substituents as p-NH$_2$ and p-OCH$_3$ on the ionization constant of phenol.

The improved precision with which equilibrium and rate constants could be made was expressed in terms of the quantity $\bar{\sigma}$ defined as [363]

$$\bar{\sigma} = \log(k'/k_0')/\rho \qquad (5.488)$$

where ρ is determined using the relation between $\log k'$ and normal substituent constants. Illustrations of the difference in the highest and lowest value of $\bar{\sigma}$ calculated from different para substituents with and without the use of the limitation were given [353]. The standard deviation for 26 cases for which more than one $\bar{\sigma}$ value can be obtained for a given substituent ranges from 0.025 to 0.14 with only one value over 0.1 and seven over 0.05 to 0.14 when the limitation is applied. Even though seemingly random and in most instances well outside the experimental error, these deviations are felt to be significant [353]. Values of $\bar{\sigma}$ have been tabulated [353].

10. Other σ Values

Values for σ^0 have been calculated [363] and tabulated [353, 363]. Para values were calculated from only the ionization of substituted C$_6$H$_5$CH$_2$COOH and C$_6$H$_5$(CH$_2$)$_2$COOH and from the alkaline hydrolysis rates of substituted C$_6$H$_5$CH$_2$COOC$_2$H$_5$ and C$_6$H$_5$CH$_2$OCOCH$_3$.

While σ values are substituent constants, significant differences have been observed [363] for $\bar{\sigma}$ values when the chemical process occurred in highly polar media as compared to media of low polarity.

From F^{19} nuclear magnetic resonance, spectra of m- and p-substituted fluorobenzenes shows the dipolar resonance structure, e.g.,

$$^{+}F=\!\!\left\langle\!\!=\!\!\right\rangle\!\!=\!\!Y\overset{\displaystyle O^{-}}{\diagup}{}_{\diagdown} \qquad (5.489)$$

contribute to increasingly greater extents in more ionizing media [364].

It is concluded [363] that σ^0 values have general utility in identifying and studying specific polarization and Ar–Y resonance effects dependent on both solvent conditions and reaction type, and that deviations from the expression

$$\log(k/k_0) = \sigma^0\rho \qquad (5.490)$$

measure such effects [362, 365].

The values of σ^n discussed previously and σ^0 discussed here do not ordinarily differ significantly. There are significant differences however in the cases of

m- and p-N(CH$_3$)$_2$ and NH$_2$ substituents and p-F substituents. For these cases σ^0 has been preferred [363]. Specific interactions (especially hydrogen bonding) between the solvent and the substituent X can alter the Ar–Y inductive order [355, 366] and are taken into account by assigning a σ^0 value characteristic of a particular solvent class.

Two sets of σ values (designated as σ^-) have been selected [347] for substituents showing tendencies toward through-resonance, and based on available data on the p-nitro substituent. From the acidity constant for p-nitrobenzoic acid, the value of the p-nitro substituent was found to be 2.778 which applied to many reactions satisfactorily but failed markedly to predict the effect in reaction of phenols and aniline. These last two reactions only were accounted for acceptably by using a σ-value of 1.27 for the p-nitro group. For $-T$ substituents in the para position, the dual σ concept has been accepted and for seven such substituents dual values have been recorded [355]. The dual values for the reactions of phenols and anilines were designated by the symbol σ^*, but are now generally represented as σ^-. Thus from the acidity constants of substituted phenols in water the values for σ^- are 0.26, 0.57, 0.89, 1.25, and 1.36 for the respective substituents OCF$_3$, SCF$_3$, CN, NO$_2$, and SO$_2$CF$_3$, while the σ^- values for the same respective substituents are 0.27, 0.64, 1.00, 1.27, and 1.65 from the acidity constants of substituted anilinium ions.

For the first-order solvolysis of phenyldimethylcarbonyl chloride in 90% aqueous acetone a set of σ^+ constants have been calculated [367] for $+T$ substituents in the para position for which the transition state of this reaction should be stabilized [353] by the through resonance indicated below

$$\text{(5.491)}$$

a plot of $\log(k'/k_0')$ versus σ for this reaction is linear for meta substituents with a slope of -4.54. The value for the p-nitro group falls on this line, but in general $+T$-type para substituents vary widely. For these substituents, substitution of the $\log(k'/k_0')$ values in the reaction

$$\sigma^+ = -4.54 \log(k'/k_0') \qquad (5.492)$$

was used [353] to calculate values of σ^+. Log(k'/k_0') versus σ^+ values were linear. Such plots were obtained [367] for many reactions of the type in which strong through-resonance between the reaction zone and $a+T$ substituent would be expected. Figure 5.16 shows the correlation [367] of the specific rates of mercuration of substituted benzenes by mercuric acetate, in acetic acid with σ^+ constants.

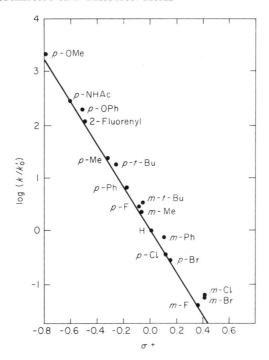

Fig. 5.16. Comparison of specific rates for mercuration of substituted benzenes with σ^+ constants. Mercuration $\rho = -4.0$.

A useful measure σ^r of the through-resonance effect can be defined [353] as

$$\sigma^r = \bar{\sigma} - \sigma^n \qquad (5.493)$$

11. Specific Cation Solvators and the Rates of Anion-Neutral-Molecule Reactions

Cation–anion interactions have a strong effect on anion-neutral molecule reactions when the solvating power of the solvent is small and especially when the dielectric constant is low. This interaction is generally more marked with the anion reactant than with the transition state, and the result is apt to be a relatively slow rate of reaction [353]. A tremendous increase in rate can be accomplished by transferring the reactants to a solvent which specifically solvates the cation strongly. Even relatively small proportions of the latter type solvent affect the reaction rates markedly. The relative rates at 25°C of the alkaylation of diethyl sodio-n-butylmalonate with n-butyl bromide in five pure solvents are as follows [368]: benzene = 1, tetrahydrofuran (THF) = 14, monoglyme = 80, dimethylformamide = 970, and dimethysulfoxide (DMSO) = 1420.

The relative rates at 37°C of the above reaction in benzene containing 1 M concentrations of strong cation solvators showed the following large effects [369]: ethanol = 6.3, DMF = 42, and N-methyl-2-pyridone = 153. It was found that the high cation solvation potential did not correlate with the dipole moment but rather with high electron density in a π orbital, e.g., that of oxygen in the amides [370]. For example, monoglyme could solvate sodium ion in a manner suggested [371] for the solvation of the lithium ion by that solvent:

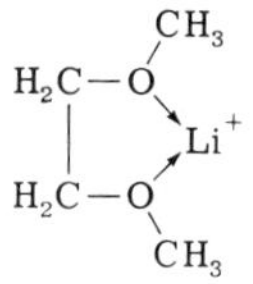

$$
\begin{array}{c}
\text{CH}_3 \\
\text{H}_2\text{C}-\text{O} \\
| \qquad \text{Li}^+ \\
\text{H}_2\text{C}-\text{O} \\
\text{CH}_3
\end{array}
\tag{5.494}
$$

For symmetrical ions, the smaller the ions the greater is the cation–anion interaction. The interaction is also affected by the nature of the ions in much the same way that ion-solvent interaction are affected. The order of nucleophilicity, or reactivity, of anions when unsolvated or unpaired increases in the order $I^- < Br^- < Cl^-$. There is the same order of increase in intensity of the interactions of the anions with either the solvent or with a cation, and this increase is often so rapid as to reverse the order of nucleophilicities in an anion-solvating solvent or when there is strong cation–anion interaction [353].

The specific rates of the reaction in acetone of isobutyl toluene sulfonate with tetrabutyl ammonium halides and with lithium halides are in the respective ratios [372], iodide:bromide:chloride = 1:4.9:18 and iodide: bromide:chloride = 1:0.92:0.16. This inversion of rates can be explained by assuming that there is no ion pairing with the lithium ion in the transition state, and by calculating using conductance data the extent of dissociation, which is smaller for smaller anions, of lithium halide ion pairs [353]. The smaller the anion, the stronger its solvation in anion solvating solvents. In water–dioxane mixture the reaction of ethyl toluenesulfonate with sodium or potassium halides give the ratio of reaction rates to be [373] iodide: bromide:chloride = 1:0.32:0.14. Dimethylformamide, which is a better cation solvator than acetone gives the ratio of specific reaction rates for the methyl toluene sulfonate–lithium halide reaction at zero salt concentration to be [374] iodide:bromide:chloride = 1:3.4:9.1. The specific rates of the chloride and iodide reactions are lowered respectively by factors of 24 and 2 by the addition of 5 moles of water per liter.

D. BRÖNSTED'S EQUATION FOR ACID- OR BASE-CATALYZED REACTIONS

After it became clear that acid catalysis could be separated into catalysis by hydrogen ion and one by undissociated acid, the correlation between rates of acid-catalyzed reactions and the strengths of the catalyzing acid was noted

[375]. The relative rates of catalysis by undissociated molecule and by hydrogen ions was expressed quantitatively by the equation [376]

$$\frac{k_m'}{k_H'} = \frac{C_H}{C_m^{1/2}} \tag{5.495}$$

where k_m' and k_H' are the catalytic specific rate constants for the undissociated acid molecules and for the hydrogen ions, respectively, and C_m and C_H are the respective concentrations of these two catalysts. A more adaptable equation [377] has largely superceded Eq. (5.495). In this equation the rate constant, k_{a0}', for an acid catalyzed reaction involving a one-ionizable hydrogen acid catalyst is given by

$$k_{a0}' = G_1 K_{a0}^\alpha \tag{5.496}$$

where G_1 is a constant for structurally similar acids, α is a positive fraction for a given reaction and solvent, and K_{a0} is the acid dissociation constant. K_{a0} can be defined by

$$K_{a0} = k_{dissoc}'/k_{assoc}' \tag{5.497}$$

for a typical acid-base reaction such as

$$AH + H_2O \rightleftarrows A^- + H_3O^+ \tag{5.498}$$

if the reaction from left to right be (arbitrarily) called a dissociation and from right to left an association, and if k_{dissoc}' and k_{assoc}' are the respective specific rate constants for the dissociation and association processes.

The equation for basic catalysis by a base having one point of attachment for a proton is

$$k_{b0}' = G_2 K_{b0}^\beta \tag{5.499}$$

where G_2 is a constant for structurally similar bases, β is a constant positive fraction for a given reaction and solvent and

$$K_{b0} = 1/K_{a0} \tag{5.500}$$

If there are p acid centers and q basic centers instead of 1 each, then Eqs. (5.496) and (5.499) become, respectively,

$$k_a' = G_1 K_a^\alpha p^{1-\alpha} q^\alpha \tag{5.501}$$

and

$$k_b' = G_2 K_b^\beta p^\beta q^{1-\beta} \tag{5.502}$$

Plots have been made [378] for the mutarotation of glucose in the presence of various acids and bases neglecting statistical effects and the influence of electrical types to be expected theoretically. The graph confirms that water acts as both an acidic and as a basic catalyst. Except for the doubly positively charged base, the points representing basic catalysts approximately fit a straight line. The curve for acid catalysis is not so well established as that for basic catalysis, but the increase with acid strength of the catalytic effect is plain. The plot covers a range of 10^{18} in K_a and 10^8 in K_b. The equations have been confirmed in numerous instances. Pfluger [379] has presented a plot of the catalytic constants of the base-catalyzed glucose metarotation against those for the nitramide reaction. A single straight line represents the data for a variety of catalysts. Several lines would be required to express the data using the Bronsted law. Two parallel lines have been used [79] to represent the base catalyzed decomposition of nitramide in aqueous solution at 18°C involving four different types of bases, of which three types B^0, B^{1-}, and B^{2-} fall on one line, and the type B^{2+} base falls on another line. The rate determining step is presumably the removal of the proton from the tautomeric azo acid, the mechanism being as follows:

$$\begin{array}{c} \text{H} \\ \quad \diagdown \text{N---N} \diagup \text{O} \\ \text{H} \diagup \qquad \diagdown \text{O} \end{array} \quad \underset{(2)}{\overset{(1)}{\rightleftarrows}} \quad \text{H---N} {=} \text{N} \begin{array}{c} \diagup \text{OH} \\ \diagdown \text{O} \end{array} \qquad (5.503)$$

$$\text{B} + \text{H---N} {=} \text{N} \begin{array}{c} \diagup \text{OH} \\ \diagdown \text{O} \end{array} \quad \underset{\text{slow}}{\overset{(3)}{\longrightarrow}} \quad \text{BH}^+ + \left[\text{N} {=} \text{N} \begin{array}{c} \diagup \text{OH} \\ \diagdown \text{O} \end{array} \right]^- \quad \overset{\text{fast}}{\longrightarrow} \quad \text{NNO} + \text{OH}^- \quad (5.504)$$

The logarithms of the catalytic rate constants for various bases in the decomposition of nitramide plotted against the basicity constants of these bases using Brönsted and Pedersen's [380] data fell [330] on a single straight line except for doubly positively charged base $[\text{M}(\text{CH}_3)_m(\text{H}_2\text{O})_n(\text{OH})]^{2+}$.

Reasons for the deviation of data from the Brönsted equation has been discussed and numerous applications of the theory made [381]. For aqueous solutions the anomalies may reside in the aqueous dissociation constants themselves, rather than in the kinetic data or the structure of the amines. Thus two straight lines, one for primary and one for tertiary amines, were obtained for the amine-catalyzed decomposition in anisole solution of nitramide, when plots were made of the logarithms of catalytic constants versus the logarithms of the dissociation constants in water. Plotting the logarithms of the catalytic constants in aqueous solutions against the logarithms of dissociation constants also in aqueous solutions resulted in a greater difference in this case, G_b for the tertiary amines being about four times as great as for primary amines, though the β-values were the same in the two cases. For kinetic decomposition data in anisole solvent, G_b for the tertiary amine was about ten times as great as for the primary amine, perhaps due to the much reduced solvation in anisole solution. The assumption that the kinetic anomalies observed with amines arise from the variable contribution of hydration to the dissociation constants in aqueous solutions was supported by the observation that catlytic constants compared with basic strengths measured in anisole or phenyl chloride solutions instead of water yielded data which obeyed a single equation for primary, secondary, and tertiary amines. Thus the above discussed deviations arose purely from solvent effects.

Deviation which arise from some stabilizing effect in the transition state which cannot occur either in the initial or final state has been discussed. These deviations are present in carboxylate anion catalyzed brominations of various ketones and esters [382]. Large groups such as halogens or hydrocarbon radicals near the seat of reaction in both substrate and catalyst gave positive deviations up to 300%. The cause of these deviations was assumed to be the proximity of these large groups in the transition state. The stabilization of the transition state may partly be due to the van der Waals forces between the group, but more responsible perhaps is the reduction in interaction between the water molecules produced by the necessity of making a smaller cavity in the solvent to accommodate the transition state with its closely spaced large groups than was necessary for the two reactants with large groups at greater distances from each other. The transition state separates fewer water molecules and is therefore stabilized. The problem might be further elucidated by a study of the energies and entropies of activation for the process. The properties of the solvent produces this effect which should be reduced in solvents with smaller interactions among their molecules.

A considerable difference in the charge distribution in the transition state and in the base of the conjugate acid (or vice versa) is another cause of both positive and negative deviations. However, if we are to compare the free energy changes for a series of similar reactions in terms of molecular structure all contributions to change in the standard free energy ΔG^0 which are accidental in nature such as symmetry change in reaction, and properties such as molecular weight, moment of inertia, and vibration must be eliminated [79]. The p and q corrections of Brönsted and the symmetry corrections prove to be very similar, with occasional differences arising from ambiguities in the use of p and q corrections. The use of p and q corrections has been discussed generally [383]. Hammett [353] had included in Eqs. (5.313) and (5.316) the Leffler–Grunwald [384] operator that describes the effect of a change in structure, in this case the structure of the catalyst on whatever quantities follows it. The catalytic equations for bases and acids become, respectively,

$$\delta_R \log k'_{b0} = \beta \delta_R K_{b0} \tag{5.505}$$

and

$$\delta_R \log k'_{a0} = \alpha \delta_R K_{a0} \tag{5.506}$$

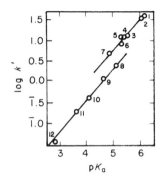

Fig. 5.17. A plot of the decadic logarithm of the catalytic rate constant versus pK_a for the amine-catalyzed decomposition of nitramide. (1) α-Picoline, γ-picoline, (3) dimethyl-o-toluidine, (4) isoquinoline, (5) pyridine, (6) dimethyl-m-toluidine, (7) quinoline, (8) p-toludine, (9) aniline, (10) p-chloroaniline, (11) m-chloroaniline, (12) o-chloroaniline.

In Fig. 5.17 are shown the data [385] for the kinetics of the amine-catalyzed decomposition of nitramide. Decadic logarithms of the catalytic rate constants versus the pK_a values for the amine cations are plotted. The separation of the two lines representing the primary and tertiary amines is not so great as when the reaction rates in anisol are used. For the same basic strengths the tertiary bases are more effective catalysts than the primary bases, because of the different effect on the dissociation constants of the tertiary and primary amines of hydrogen bonding between the amine cation and the water molecule. The same concept clarifies many anomalies in the effect of substituents on the basic strength of amines [386]. Amine ions are more highly solvated than free amines; and the amine ions, therefore, seem to be more stable when their dissociation constants are measured in water than when measured in an inert solvent. The order of stabilization of amine ions is primary > secondary > tertiary.

The statistical correction to the Brönsted equation is necessary since singly charged ions derived from dibacid bases or from dibasic acids would deviate markedly from the line determined by the ions from monoacid bases or monobasic acids if the statistical correction were neglected. Thus for the ion $HOOC(CH_2)_X COO^-$, K_{a0} is not the proper measure of its acidity since the ion has two equivalent ways of being formed from the completely ionized ion $^-OOC(CH)_2COO^-$, namely, by attachment of the proton to one or to the other carboxylate ion groups. Therefore the ion has twice the change of being formed as corresponds to the work of adding the proton, and the correct measure of the work is not K_{b0} but $2K_{b0}$. From similar considerations, the acidity of the ion is not measured by K_{a0} but by $\frac{1}{2}K_{a0}$.

Conclusions

For the linear free energy relations it is assumed that the free energy changes are linearly dependent upon the variables involved. The purposes in formulating the equations have been to correlate variations in substate, reagent, and reaction medium. The equations are basically empirical, though Wells [336] states that, when the variables can be identified, the parameters are useful theoretical quantities.

The Brönsted acid and base catalytic relationships have been more extensively and carefully studied than other free energy relationships [353]. The K in the Brönsted equations (5.505) and (5.506) is not identical with but is thermodynamically necessarily proportional to the equilibrium constant of the rate-determining proton transfer in the reaction being investigated.

The Hammett equation has been extended, modified, and increased in applicability by intensive studies, in particular those of Jaffe [355]. Except perhaps for the Brönsted, the Hammett equation remains the most satisfactory relationship under the most vigorous limitation of variations. The more limited Taft equation may be applicable with less stringent limitations.

Parameters correlating variations in solvent and reagents have been suggested for a few reactions, but doubtful significance must be assigned to the more adjustable parameters inserted in these equations in order to extend their applicability. Using sufficient parameters most data can be correlated; but, in general, the definiteness of the origin and significance of parameters decrease as their numbers increase.

If in a pair of reactions, the bond-making or bond-breaking processes vary widely, a linear free energy relationship will not apply. Nitromethane in which the proton donated to the reaction is originally attached to the carbon is a poorer acid catalyst by one or two orders of magnitude for the hydration of acetaldehyde than a carboxylic acids or phenols of the same acid strength [353, 387–389]. However, several ketoximes are better catalysts [353] by as much as two orders of magnitude than carboxylic acids or phenols of the same acid strength.

When nucleophilic atoms which become attached to carbon vary widely, the rates of the reactions involved present a confusing picture, showing little relation to the base strengths as measured by the pKs of the conjugate acids of the nucleophiles. There are strong solvent effects on the relative rates of such reactions which must be accounted for.

While the constants of some linear free energy relations show strong solvent dependence, there are in many cases a lack of consistency in the dependencies of a selected constant on a particular property of a series of solvents. Thus the b constant of Eq. (5.445) was shown [36] to demonstrate some consistency where plots were made of b versus the reciprocal of the dielectric constants for water–ethanol, water–acetone, and water–ethanol solvents in the case of solvolysis reactions. The data for the water ethanol and water–acetone solvents followed approximately the same shape of curve, but the curve for the water–methanol differed somewhat from the other two curves.

Wells suggests [336] that an analysis of the Hammett parameters in terms of solvent and reaction parameters could be one of the worthwhile new correlations which will undoubtedly be formulated.

A division into two distinctive classes of reactions based on whether σ^- had the value of 0.78 or 1.27 was negated by the observation [353, 362] that σ^- values for the *p*-nitro group varied in a regular way from 0.6 to 1.4 with no tendency to congregate in the vicinity of 0.78 or 1.27, while the σ^- values for the *p*-methoxy group varied almost as regularly from -0.8 to 0.

It was pointed out that different parameters are needed in the Swain and Scott and in the Edwards equations when going from protic solvents, where they are very successful for some reactions, to dipolar aprotic solvents.

The linear free energy relationships illustrate the oft repeated principle that exact numerical relationships can often be formulated when generally known observations become the object of careful investigation.

E. Reaction Rates and Chelate Solvation of Cations

In the alkylation of sodiomalonic ester derivatives and other cases, neutral molecule anion reactions are markedly faster in the ethylene glycol dimethyl ether than in tetrahydrofuran or in diethyl ether [390–392]. It was proposed [392] that the chelate coordination of the two oxygens in the glycol ether with the metal ion caused the effect. There is considerable entropy advantage to such a structure compared to the coordination of the ion to two separate oxygen possessing molecules [353]. This explanation is supported by conductance data on tetraphenyl borides in the two solvents [353, 393]. A combination of one sodium ion with two glycol ether molecules or with four tetrahydrofuran molecules, and a coordination of cesium ion with glycol ether but not with tetrahydrofuran are in accord with both the Stokes law radii and the equilibrium constants for association.

The rates of alkylation of butyrophenone anion in ether solutions increase by more than three powers of ten as the accompanying cation varied from lithium through sodium to potassium. The reactions were powerfully catalyzed by tetraalkylammonium salts. The ketonic product autocatalyzed the reaction. Thus when the product of the reaction, α-ethylbutyrophenone, was added in initial concentration of 0.40 M, the rate of the reaction was tripled. The other product of the reaction, sodium iodide, when added in large amounts initially did not alter the rate of the reaction. A reaction order greater than unity for alkyl halide suggested alkylation within a coordination complex, although no appreciable concentration of complex was indicated from vapor pressure measurements of the system methyl bromide–ethereal sodiobutyrophenone. However an association of three ion pairs in ether solution was indicated by ebulliometric studies of sodiobutyrophenone. It was suggested that alkylation could occur within the coordination complex as proposed [394] for the alkylation and acylation of β-dicarbonyl compounds:

$$RX \ + \ C_6H_2-\overset{\overset{\displaystyle ONa}{|}}{C}=CHR \ \rightleftharpoons \ \begin{array}{c} C_6H_5C\overset{O}{\underset{\parallel}{\diagdown}}Na\diagup OEt_2 \\ \underset{H}{R}\diagdown\underset{R}{\overset{\parallel}{C}}\diagup\overset{\uparrow}{X}^{OEt_2} \end{array} \tag{5.507}$$

$$\downarrow$$

$$C_6H_5\overset{\overset{\displaystyle O}{\parallel}}{C}-CHR_2 \ + \ NaX$$

The solvent ether is present in the coordination complex.

A six-centered reaction within a coordination complex would explain orders of reaction greater than unity in alkali halide. If more than one halide molecule were coordinated to sodium in the transition state, higher orders would be expected.

Solvated complexes of alkali-metal enolates have been isolated [395]. Chelation produced stability in these compounds. Formation constants in at least one such complex [396] decreased from lithium to cesium, which is opposite to the relative rates found by Zook and Gumby [392].

Free enolate ions are not involved in the rate determining step as is evidenced by the variation of the rate of alkylation with change of cation. This variation of rates with change of cation does not necessitate reaction within a coordination complex.

Definite compounds of fluorenyllithium and fluorenylsodium with polyglycol ethers $CH_3O(CH_2CH_2O)_xCH_3$, where x ranges from 1 to 4 has been reported [397] from spectroscopic and NMR evidence.

1. Further Cation Solvator Considerations

In the alkyl lithium and the sodio-malonic ester reactions involving ion aggregates, both reactions are first order in momomeric metal alkyl. Therefore, the metal ion, the anion, and the neutral reactant must be present in the transition states. The cation solvator must interact more strongly with the metal ion in the transition than in the reactant state, and in this way lower the barrier hindering reaction of the anion with the neutral molecule. The barrier would be heightened by an anion solvator [353].

Hammett [353] presents the following pertinent information about the lithium alkyl and the sodio-malonic ester reactions. In benzene the reaction

$$+ \ RLi \ \longrightarrow \ + \ RH \tag{5.508}$$

is first order with respect to the hydrocarbon but is of 0.18 order and 0.26 order, respectively, for n-butyllithium and t-butyllithium. These orders are about the same for the lithium alkyls reacting with 1,1-diphenylethelene [398]. Colligative properties show that in benzene the n-butyl compound is a hexemer [399], and the t-butyl compound is a tetramer. Thus from kinetic data the mechanism presumably involves the polymeric alkyl lithium in mobile equilibrium with a monomer, which in the transition state is associated with one molecule of hydrocarbon [353, 400].

In the alkylation of sodiomalonic ester derivatives, there is at least a forty-fold polymerization of sodium derivative in benzene [353]. Ignoring the complications in the time course of the reaction, we find the initial rate to be first order with respect to both the sodium compound and the alkyl halide. These observations were explained in either of two ways [353]. First, it was assumed there existed a transition state composed of one molecule each of monomeric sodium derivative and alkyl halide and also a mobile equilibrium

$$X\text{-polymer} \rightleftarrows (X-1) \text{ polymer} + \text{monomer} \qquad (5.509)$$

and that the concentration of the monomer determined by the equilibrium was approximately independent of X if X is large. Second, it was assumed that the average degree of polymerization was approximately independent of concentration, so that the number of sites available for the reaction, and the rate was nearly proportional to the total concentration of the sodium compound. With strong cation solvators present such as dimethylformamide, the alkylation is first order up to 80% completion. The additional catalyzed reaction produced by addition of the cation solvator is from the first to the three halves order with respect to the catalyst. From cryoscopic evidence, the addition of dimethylformamide to solutions of the sodium compound leaves the compound still highly polymerized.

If there is no formation of an impenetrable layer of reaction products on the surface of a solid, which would suppress any heterogeneous reaction at the surface, the reaction of the solid is essentially the same as that of a highly polymerized solute with a low surface to volume ratio. There is a large increase in available reaction sites and in the diffusion layer cross section through which a monomeric reactant must diffuse to reach the body of the solution, when a solid dissolves even as a polymer. Specific cation solvating solvents promote much faster reactions than inert solvents in such reactions as that of bromobenzene with potassium t-butoxide when the alkoxide is introduced as a solid [353]. This is reminescent of the effect of cation solvating solvents on the rate of reactions involving dissolved solutes.

An ambident ion, such as the β-naphthoxide ion, can react with an alkylating agent either at carbon to form a 1-alkyl-2-naphthol or at oxygen to form an alkyl naphthyl ether. Sodium β-naphthoxide reacting with benzyl bromide in the solvents dimethylformamide or dimethylsulfoxide shows no carbon alkylation, but in ethylene glycol dimethyl ether there is 22% carbon alkylation, in tetrahydrofuran there is 36%, in methanol 34%, in ethanol 28%, in 2,2,2-trifluorethanol 85%, and in water 84% [353, 401]. Similar solvent effects on product distribution are observed in the alkylation of sodium β-naphthoxide with methyl or normal propyl bromide or and in the reactions of alkali phenoxides with allyl chloride or bromide or with benzyl chloride [402]. The product distribution is independent of the cation when alkylations such as these are carried out in methanol and ethanol, but in diethyl ether, tetrahydrofuran, benzene, or toluene the relative amount of carbon alkylation decreases in going from lithium, to sodium, to potassium, to tetraalkyl cation. These data indicate that a phenoxide strongly paired with a small cation or strongly solvated reacts at carbon preferentially, thus indicating shielding of the oxygen atom of the phenoxide ion by cation pairing or solvation. Contrariwise, when paired with a bulky or strongly solvated cation of the phenoxide ion or when free, the phenoxide ion reacts at oxygen preferentially. The dielectric constants of ethyleneglycol dimethyl ether and tetrahydrofuran are nearly the same, 6.8 and 7.3, respectively, yet the former is markedly better cation solvator as is witnessed by the relative amounts of carbon alkylation in the two solvents [353].

2. Solvent Effect on the Isokinetic Relationship

For changes in composition and structure of the component and for changes in the solvent [353], the heat of evaporation $\bar{H} - \bar{H}^0$ of a solution component according to the Barclay–Butler rule [403] is linear with the entropy of evaporation $\bar{S} - \bar{S}^0$, the line having a slope of about 800. For no strong interactions, the rule applies, and for change in solvent the relation

$$\delta_M \bar{H}^0 = 800 \delta_M \bar{S}^0 \qquad (5.510)$$

holds. In this equation δ is the Leffler–Grunwald [384] solvent stabilization operator which describes the effect of solvent change on the quantity which it precedes. For a reaction in the solvent the enthalpy and entropy of activation should be related by the equation

$$\delta_M \Delta \bar{H}^* = 800 \delta_M \Delta \bar{S}^* \qquad (5.511)$$

For testing Eq. (5.511), no data are extant for reactions in systems where strong interactions are not present. The isokinetic relationship

$$\delta_M \Delta \bar{H}^0 = \beta \delta_m \Delta \bar{S}^0 \qquad (5.512)$$

does hold [404] for numerous reactions over considerable variation of solvent. In this equation β ranges from 300 to 400. In Fig. 5.18 the relationship is

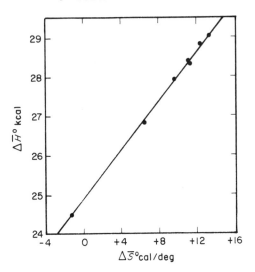

Fig. 5.18. Isokinetic relation for decompostion of $C_6H_5N = NC(C_6H_5)_3$. Solvents from left to right are cyclohexane, benzene, malonic ester, nitrobenzene, chlorobenzene, anisole, and benzonitrile. (Data of Alder and Leffler [4053].)

presented [405] for the case where $\beta = 320$. Twenty-five cases have been cited [384] where Eq. (5.512) applies with a correlation coefficient of 0.95 or better.

It has been suggested [353] that the isokinetic relationship could be applied in more instances by including those cases in which either $\delta_M \Delta \bar{H}^*$ or $\delta_m \Delta \bar{S}^*$ is so small that the ever present scatter obscures the enthalpy–entropy relationship but to which the equation

$$\delta_M \Delta \bar{S}^* = \left(1 - \frac{T}{\beta}\right)\delta_M \Delta \bar{H}^* \qquad (5.513)$$

or

$$\delta_M \Delta S^* = (T - \beta)\delta_M \Delta \bar{S}^* \qquad (5.514)$$

holds. Equations (5.513) and (5.514) are necessary carollaries of Eq. (5.512).

In general, according to Hammett,

$$\delta_M \Delta \bar{H}^* = \delta_M E_0^* + \delta_M \Delta \left(RT^2 \frac{\partial \ln Q}{\partial T}\right) \qquad (5.515)$$

and

$$\delta_M \Delta \bar{S}^* = \delta_M \Delta R \ln Q + \delta_M \Delta \left(RT \frac{\partial \ln Q}{\partial T}\right) \qquad (5.516)$$

Only when

$$\delta_M \left(\frac{\partial \ln Q}{\partial T} \right) = 0 \quad \text{and} \quad \left(\frac{\partial_M \ln Q}{\partial_M \Delta E_0} \right)$$

is independent both of solvent and of solute, which occurs when dispersion forces alone are involved, is the Barclay–Butler relation applicable. Provided that $(\partial_M \ln Q / \partial_M \Delta E_0)$ is the same for reaction and transition state in the reaction involved even though it is not generally independent of solute, and provided that $\delta_M \Delta(\partial \ln Q/\partial T) = 0$, the isokinetic relationship can occur with more specific solute–solvent interactions.

A critical degree of steric hinderance makes the molecular configuration that is ordinarily the more stable the less stable instead.

The isokinetic relationship holds with changing solvents if the solvents in a series perform closely similar roles in a reaction. If solvents perform different roles in a reaction the activation parameters fall on different lines. Thus the decomposition of triethylsulfonium bromide in hydroxylic solvents fall on one line and in nonhydroxylic solvents on another line. Differences in dielectric constant, fluidity parameters, hydrogen bonding, and quasi-crystalline vibrations of the solvent, or combinations of the variables, cause different solvents to play different roles in a reaction process and hence cause different isokinetic lines.

In the case of mixed solvents the isokinetic relationship often fails because of these stringent requirements. The isokinetic plots are not always straight lines, but are often hooked, N shaped, or otherwise shaped [384, 353].

3. Selective Solvation and Rates and Mechanisms

With respect to the bromoacetate ion–thiosulfate ion, tetraiodophenol-sulfonphthalein ion–hydroxide ion, and ammonium ion–cayanate ion reactions as the dielectric constants decrease to, say, 50 or less, the rates tend to vary in the direction of rates in the solvent of higher dielectric constant [33, 406]. Selective solvation of the reactant ions by the higher dielectric constant component of the solvent, in these cases water, has been suggested [47] as the cause of these deviations. Apparently only the rates and not the mechanisms of the reactions in all ranges of solvent composition were influenced by the solvent. Rather than random distribution at low dielectric constants of the solvents, there were congregation of the higher dielectric component of the solvent in the near vicinity of the ions, causing the rates in the low dielectric constant regions of the solvents to be more nearly equal to those in the higher dielectric solvent component. In the higher dielectric constant regions, the predominant influence of the solvent on the rates was that of modifying the electrostatic forces among reactant particles.

Throughout the solvent composition range the mechanisms of the reactions apparently remain unchanged. Thus for the tetraiodophenolsulfonphthalein ion–hydroxide ion reaction, the mechanism has been written [407]

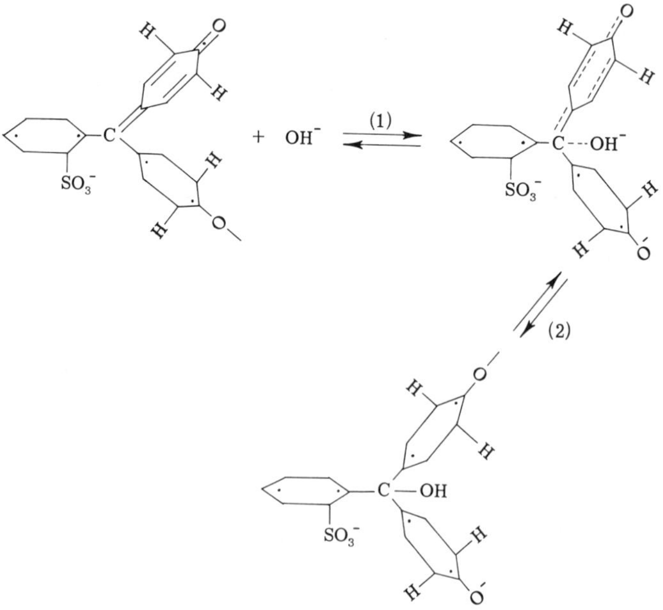

(5.517)

As to the ammonium cyanate reaction, while the mechanism apparently does not alter with change in composition of solvent, whether the mechanism is molecular or ionic cannot be determined from electrostatic effects. Because of the mobile equilibrium [408]

$$NH_4^+ + CNO^- \rightleftharpoons NH_3 + HCNO$$

$$NH_4CNO$$

(5.518)

the product of the concentration of the ions is proportional to the product of concentrations of the ammonia and the cyanic acid in solution; and thus the idea that the rate of formation of carbamide urea by ammonia and un-dissociated cyanic acid is apparently not inconsistent with observations that the rate of formation of carbamide is proportional to the concentrations of the two ions [409]. One mechanism proposed [410] for the reaction was the formation of the amide from the ammonia and isocyanic acid, which amide then undergoes a hydrogen atom shift to give carbamide as shown below:

$$NH_4 \cdot N{:}C{:}O \rightleftharpoons H \cdot N{:}C{:}O + NH_3 \rightleftharpoons H \cdot N{:}C\underset{NH_2}{\overset{OH}{\diagup}} \rightleftharpoons H_2N \cdot CO \cdot NH_2 \quad (5.519)$$

Lowry [411] supports this mechanism. The mechanism has also been written [412]

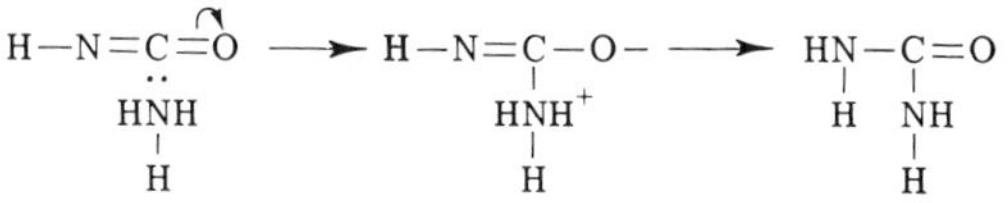

$$\underset{\underset{H}{|}}{\underset{HNH}{|}}H{-}N{=}C{=}\overset{\frown}{O} \longrightarrow \underset{\underset{H}{|}}{\underset{HNH^+}{|}}H{-}N{=}C{-}O{-} \longrightarrow \underset{\underset{H}{|}\;\underset{H}{|}}{\underset{H\;\;NH}{}}HN{-}C{=}O \quad (5.520)$$

After Chattaway [410] we have

$$K_b = \frac{C_{NH_4^+} \cdot C_{OH^-}}{C_{NH_3}} \quad (5.521)$$

$$K_a = \frac{C_{H^+} \cdot C_{NCO^-}}{C_{HNCO}} \quad (5.522)$$

and

$$K_a K_b = \frac{C_{NH_4^+} \cdot C_{NCO^-}}{C_{NH_3} \cdot C_{HNCO}} C_{H^+} \cdot C_{OH^-} \quad (5.523)$$

Hence

$$C_{NH_3} \cdot C_{HNCO} = \frac{K_w}{K_a K_b} \cdot C_{NH_4^+} \cdot C_{NCO^-} \quad (5.524)$$

Between neutral reactants the rate r of the reaction is given by

$$r = k_n' C_{NH_3} \cdot C_{HNCO} = \frac{k_n' K_w}{K_a K_b} \cdot C_{NH_4^+} \cdot C_{NCO^-} \quad (5.525)$$

and including activity coefficients f, Eq. (5.525) becomes

$$r = k_n' C_{NH_3} \cdot C_{HNCO} \frac{f_{NH_3} \cdot f_{HNCO}}{f_X} = \frac{k_n' K_w}{K_a K_b} C_{NH_4^+} \cdot C_{NCO^-} \frac{f_{NH_4^+} \cdot f_{NCO^-}}{f_X} \quad (5.526)$$

For ionic reactants the equation for the rate is

$$r = k_i' C_{NH_4^+} \cdot C_{NCO^-} \frac{f_{NH_4^+} f_{NCO^-}}{f_X} \quad (5.527)$$

In the above equations k_n' and k_i' are the respective specific velocity constants for the rates of reaction between the uncharged ammonia and isocyanic acid molecules and between the ammonium and cyanate ions. Hence

$$k_i' = \frac{k_n' K_w}{K_a K_b} \quad (5.528)$$

In terms of activity coefficients the effect of ionic strength upon reaction rates is given by

$$\frac{f_A f_B}{f_X} = \exp\left(\frac{2Z_A Z_B \varepsilon^2}{1 + \beta a_i \mu^{1/2}}\right), \tag{5.529}$$

and as shown previously the effect of dielectric constant upon the rates in terms of activity coefficients can be written

$$\frac{f_A f_B}{f_X} = \exp\left(\frac{Z_A Z_B \varepsilon^2}{kTr}\right)\left(\frac{1}{D_0} - \frac{1}{D}\right) \tag{5.530}$$

Hence from Eqs. (5.526)–(5.530) the molecular and ionic mechanisms cannot be distinguished in so far as the effects of dielectric constant and ionic strength on the activity coefficients are concerned.

The above discussion illustrates that in general mechanisms are not unique. Mechanisms are only hypothetical explanations of the rates and of stereochemical, collisional, isotopic, and other forms of data. For a given reaction new ideas or a different and/or more extensive experimental approach may lead to a different mechanism from the one commonly accepted. Only a few mechanisms are so simple and direct as to be generally accepted as correct.

Both the mechanism and the rate of some reactions are changed by the solvent. In this class are included certain nucleophilic reactions which in some solvents occur with retention of configuration, but in other solvents occur with no retention. Those nucleophilic reactions occurring with no retention of configuration have been explained using the $S_N i$ mechanism [413]. The reactions involving thionyl chloride were suggested as occurring by the internal decomposition of an intermediate chlorosulfite.

The thermal decomposition of secondary alkyl (2-butyl, 2-pentyl, and 2-octyl) chlorosulfites in dilute solutions in dioxane were found to be [414] first-order reactions yielding olefin and alkyl chloride as the principal products. The chloride was only slightly racemized and had the same configuration of the alcohol from which it was derived. In the absence of solvent and in "isooctane" solvents the reaction decomposition rate was slower and the chloride had the opposite configuration from the alcohol. The reaction was first order in both dioxane and isooctane solvents. The $S_N i$ mechanism would require a first-order reaction with retention of configuration. If reaction with retention of configuration occurs in isooctane, it is much slower in this solvent than in dioxane. That changing to a less polar solvent causes a reduction in rate, unless due to specific solvent effect, indicates that the transition state has a higher dipole moment than the normal chlorosulfite, and would involve the four structures for the transition state represented:

$$\underset{\underset{Cl}{|}}{R-O-\overset{\overset{O}{\|}}{S}} \longleftrightarrow \underset{\underset{Cl^-}{|}}{\overset{+}{R}O=\overset{\overset{O}{\|}}{S}} \longleftrightarrow \underset{\underset{Cl^-}{|}}{R-O-\overset{\overset{O}{\|}+}{S}} \longleftrightarrow \underset{\underset{Cl}{\diagdown}}{\overset{+}{R}O=\overset{\overset{O}{\|}}{S}} \quad (5.531)$$

The specific velocity constant for the decomposition of butyl chlorosulfite has been found to be [415] 56×10^{-4} sec^{-1} in dioxane at 99°C and 0.167×10^{-4} sec^{-1} in isooctane at 96°C. To give retention of configuration in dioxane a three-step process is proposed as the mechanism for the reaction: (1) Ionization of the chlorine–sulfur bond, and a resulting weakening of the carbon–oxygen bond. (2) The resulting carbonium ion is solvated by dioxane, and loses the now unnecessary sulfur dioxide of solvation. There is no racemization of these solvated ions since the solvation introduces an asymmetry which is present even if the carbonium ion itself is planar. (3) A neutral molecule is formed by the collapse of the carbonium ion. In this dioxane displacement from the oxonium ion or the reaction of chloride ion with solvated carbonium ion, the chlorine becomes bound to the carbon on the same side of the plane of H, CH_3, and H from which the SO_2 is disengated. Thus, configuration is unaltered. In the following representation of the mechanism, the cation in step (1) is represented as a resonance hybrid to show identity with a sulfur dioxide solvated carbonium ion.

$$(5.532)$$

Reversible reaction (1) can, but reaction (2) cannot take place in toluene which is not a good solvating agent for carbonium ion for any reason including the fact that it is not nucleophilic. As a rule the carbonium ion does not lose sulfur dioxide and attack on that side of the carbonium is not easy. In this solvent a slower reaction consisting of the relative motion of the ions within the ion pair can occur, followed by a rapid reaction analogous to (3) giving the inverted chloride. Thus

$$(5.533)$$

In other solvents studied such as acetonitrile and the ketones, the stereochemical results are nearer that in toluene than in dioxane. These solvents in general have higher dielectric constants than either dioxane or toluene, and the dissociation of ion pairs are no longer prohibitively difficult. The chain reaction

$$Cl^- + ROSOCl \rightarrow ClR + OSO + Cl^- \qquad (5.534)$$

can take place and results in inversion. The conditions under which $S_N i$ reactions occur have been generalized [416] and are (1) When a relatively stable carbonium ion is formed as when electron-releasing groups (e.g., phenyl) are attached to the carbon undergoing substitution, the $S_N i$ reaction has the best chance of predominating over competing reactions. (2) The added chloride ion seems to suppress the competing $S_N 1$ process and promotes the $S_N i$ reaction when phenyl alkyl carbinols are treated with the usual halogen substituting reagents in liquid sulfur dioxide. (3) In better ion-solvating media such as dioxane but not in poorer ion-solvating media such as isooctane, the $S_N i$ process predominates in the decomposition of chlorosulfites of ordinary dialkyl carbinols. (4) For the reactions of secondary alcohols with halogen-substituting reagents strongly basic media such as pyridine promote $S_N 1$ type processes and give inversion of configuration.

4. Organic Radical Exchange and Solvent Effects

The self exchange of $Cd(CH_3)_2$ in diethyl ether, the exchange of $Cd(CH_3)_2$ with $Ga(CH_3)_2$ in cyclopentane and in toluene, and the exchange of $Zn(CH_3)_2$ with $In(CH_3)_2$ in diethyl ether and in triethylamine have been investigated [417]. The activation energies and rate constants for each of these systems have been obtained. The results indicate that the rate of reaction depends on the coordinating ability of the solvent and the properties of the metal. Strongly coordinating solvents enhanced the exchange rate between the derivatives of group II metals but decreased the exchange rate when one of the species is from group III.

The rate of exchange was treated as follows [417]:

$$Cd(CH_3)_2 + M(CH_3)_n \underset{k_{-1}'}{\overset{k_1'}{\rightleftharpoons}} [AC] \xrightarrow{k_2'} \text{exchange products} \tag{5.535}$$

$$\text{Rate}_{exc} = k_2'[AC] \tag{5.536}$$

$$d[AC]/dt = k_1'[Cd(CH_3)_2][M(CH_3)_n] - k_{-1}'[AC] - k_2'[AC] \tag{5.537}$$

If $k_{-1}' = k_2'$, the steady state approximation yields

$$d[AC]/dt = 0 \tag{5.538}$$

and

$$2k_2'[AC] = k_1'[Cd(CH_3)_2][M(CH_3)_n] \tag{5.539}$$

and from Eqs. (5.353) and (5.356)

$$\text{Rate}_{exc} = \tfrac{1}{2}k_1'[Cd(CH_3)_2][M(CH_3)_n] = k_{obs}[Cd(CH_3)_2][M(CH_3)_n] \tag{5.540}$$

This rate expression assumes the formation of the activated complex in a simple bimolecular process and that once formed, it has a 50% chance of forming an exchange product, i.e., $k_{obs}' = \tfrac{1}{2}k_1'$.

Taking into consideration the number of methyl groups on each of the organometallic particles, the lifetime of each carbon–metal bond may be related to the rate of exchange. A carbon–metal bond will be broken, only $1/n$ times the number of exchanges in which the molecule is involved. Hence equations for the lifetime of all the species can be written as

$$\frac{1}{\tau_{Cd-CH_3}} = \frac{\tfrac{1}{2}k_1'[Cd(CH_3]_2[M(CH_3)_n]}{2[Cd(CH_3)_2]} = \frac{k_1'[M(CH_3)_n]}{4} \tag{5.541}$$

$$\frac{1}{\tau_{M-CH_3}} = \frac{\tfrac{1}{2}k_1'[Cd(CH_3)_2]}{n} = \frac{k_1'[Cd(CH_3)_2]}{2n} \tag{5.542}$$

in which n is the number of alkyl groups in MR_n.

Several papers have reported quantitative data on the self-exchange of $Cd(CH_3)_2$ [417–420]. Table 4.19 presents [417] data for the concentration dependence for the self-exchange of $Cd(CH_3)_2$ in ether, and shows unequivocally the second-order dependence of the exchange rate on $Cd(CH_3)_2$ concentration which has been implied by not previously reported. Figure 5.19 gives the data for the temperature dependence in both diethyl ether and triethylamine. Table 5.24 summarizes data dealing with self-exchange. The

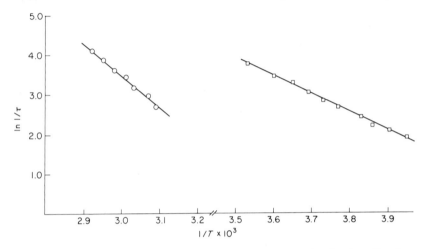

Fig. 5.19. The Arrhenius activation energy plots ($\ln 1/\tau$ versus $1/T$) for the self-exchange $Cd(CH_3)_2$ in diethylether □, $E_a^{\ddagger} = 5.6 \pm 0.2$, and triethylamine ○, $E_a^{\ddagger} = 6.9 \pm 0.2$ kcal/mole.

Table 5.24

Concentration Dependence of the Self-Exchange of Dimethylcadmium in Diethyl Ether at 33.3°C

$[Cd(CH_3)_2]$	$[Cd(CH_3)_2]_{eff}{}^a$	$1/\tau Cd\text{–}CH_3$ (sec^{-1})	k_1'(1./mole/sec)^b
0.681	0.172	2.20	51.2
0.860	0.217	2.80	51.6
1.07	0.270	3.46	51.3
1.20	0.302	4.24	56.2
1.36	0.333	4.40	52.9
1.90	0.488	5.34	43.8
			$k_{1\,av} = 51 \pm 2$

^{a} $[Cd(CH_3)_2]_{eff} = 0.252[Cd(CH_3)_2]$. This value is used in the determination of the rate constant because exchange with molecules of dimethylcadmium containing ^{112}Cd does not result in a change in the lineshape of the nmr resonance line.

^{b} $k_1' = 2k_{obs}$.

self-exchange of $Cd(CH_3)_2$ and the exchange of $Zn(CH_3)_2$ with $Cd(CH_3)_2$ are second order which is in harmony with a four-centered transition state. Thus [417]

$$H_3C-M \underset{\substack{\diagdown \\ \underset{H_3}{C}}}{\overset{\substack{\overset{H_3}{C} \\ \diagup}}{\diagup}} M'-CH_3 \qquad (5.543)$$

The energies of activation and rates of these reactions are comparable in nonpolar solvents [419, 421].

In more basic solvents such as THF, diethyl ether, and triethylamine, there is a dramatic drop of energy of activation to 5–7 kcal/mole. See Table 5.25 with pyridine which is still more basic, the activation energy for the self exchange apparently increases, but further data is needed to confirm this observation [417].

Table 5.25

Kinetic Parameters for the Self-Exchange of $Cd(CH_3)_2$ and for Its Exchange with $Zn(CH_3)_2$ in Various Solvents

Reaction	Solvent	$E_{a\,(kcal/mole)}$	$\Delta S(e.u.)^a$	k_1'(1./mole/sec) (25°C)
Self-exchange	$N(C_2H_5)_3{}^a$	6.9 ± 0.2	—	—
Self-exchange	$O(C_2H_5)_2{}^a$	5.6 ± 0.2	-38	38.5
Self-exchange	THF^b	6.8 ± 0.1	-28.3 ± 0.4	245
Self-exchange	py^c	13	—	—
Self-exchange	Toluene^c	16	—	—
Self-exchange	Neat^b	15.6 ± 0.1	-9.8	0.4
$Zn(CH_3)_2$–$Cd(CH_3)_2$	Methyl-cyclohexane^d	17.0 ± 1	-3	1.7

^{a} Soulati *et al.* [417].
^{b} Bremser *et al.* [419].
^{c} Jeffery and Mole [418] with added methanol giving $Cd(OCH_3)CH_3$ which catalyzes the exchange.
^{d} Henhold *et al.* [421].

Among the factors which affect the formation of the transition state are the reorganization energy of the metal orbitals, and the donor and acceptor character of the carbon and metal respectively. To evaluate the effect of these factors upon the rate and mechanism the structure and interaction of the solvent with alkyl derivatives must be considered. A group II metal alkyl can form either a mono- or dicoordinated derivative. Thus

$$
\begin{matrix} R \\ \diagdown \\ R \diagup \end{matrix} M - B \quad \text{or} \quad \begin{matrix} R \\ \diagdown \\ R \diagup \end{matrix} M \begin{matrix} \diagup B \\ \diagdown B \end{matrix} \tag{5.544}
$$

The metal retains an unoccupied orbital in the monocoordinated system, and the angles are about 120° instead of 180° as in the free alkyl; also, on the carbon bonded to the metal there is to some extent an enhancement of the electron density. These modifications are in harmony with or actually favor the formation of a four-centered transition state, especially since the C—M—C angle in stable complexes in between 100 and 120° [417, 422], and only a small reorganization energy is required to form the transition state. Also, the available electron density on the organic moiety entering into the bridged bond will lend assistance to the formation of the four-centered transition state. Opposing these favorable conditions will be steric crowding and decreased Lewis acidity of the central atom entering into the bridge.

Data indicate that these phenomena enhance the reactivity depending on the nature of the base involved. Stronger bases increase reactivity as indicated by the data for exchange on dimethylcadmium in solvents diethyl ether, THF, and triethylamine. Thus in weakly as compared to noninteracting solvating solvents, the energy of activation is smaller and the rate greater.

For a solvent not capable of forming a stable complex with any of the species present data for methyl derivatives give the following order of reaction rates

$$
\text{Ga} + \text{Zn} > \text{In} + \text{Zn} > \text{Ga} + \text{Cd} \sim \text{In} + \text{Cd} > \text{Zn} + \text{Cd} \gtrsim \text{Cd} + \text{Cd} \tag{5.545}
$$

As pointed out earlier, weakly interacting solvents enhance the exchanges between zinc and cadmium and the cadmium self-exchange, but slows down the exchange between group II and group III derivatives depending on the nature and strength of the solvent interaction with the vacant site on the group III compound, since the most probable transition state for a group II–group III exchange is

$$
\begin{matrix} R \cdot \diagdown \\ \\ R \diagup \end{matrix} M^{(III)} \begin{matrix} \diagup R \diagdown \\ \\ \diagdown R \diagup \end{matrix} M^{(II)} - R \tag{5.546}
$$

which completely utilized the vacant orbital on the group III derivatives. For a sufficiently strong donor solvent, a five-coordinate intermediate constitutes the transition state as was proposed [417, 418] to account for the exchange of alkyl groups between aluminum alkyls in pyridine. Thus

$$
\begin{array}{c}
\text{H}_3\text{C} \qquad\quad \text{C}\text{H}_3 \\
\text{B}\text{-}\text{-}\text{-}\text{M} \qquad \text{M}\text{---}\text{CH}_3 \\
\text{H}_3\text{C} \qquad\quad \text{C}\text{H}_3
\end{array}
\tag{5.547}
$$

This mechanism indicates a second-order mechanism, first order with respect to each of the two components, and also implies a large negative entropy of activation. Since experimental data show all the exchange reactions to be second order, a rate-controlling dissociation step can be ruled out in all cases studied to date.

Conductance methods support [423] the idea of selective solvation. The effects of change of solvent on free energies, enthalpies, and entropies of activation were found to be quite complex and often depended to the same extent on changes in solvation of reactants as on the solvation of the transition state [424].

XX. Solvation and Ionization Effects of the Solvent on Rates and Mechanisms

Since a solvent that promotes ionization will likewise show a strong tendency toward solvating solutes, ionization and solvation effects as presented here are largely synonymous. Only the solvation effect may be prominent in some cases, because the reactants exist as ions over a range of solvent compositions and the rate is affected only by change in solvation between the complex and reactant states. In contrast ionization-producing ionic reactants may take place in solvents of only certain ranges of polarity or of acid or base strength; and hence the solvent may influence the rate by both affecting the nature and extent of ionization and by solvation effects.

A qualitative theory of solvent effects has been proposed [425]. In summary it states that the creation and concentration of charges will be accelerated and the destruction and diffusion of charges inhibited by an increase in the solvating power of the medium. The expected effects of ionizing solvent on the rates of reaction according to the above theory are summarized in Table 5.21. Table 5.26 presents the data [426] for the increase of the velocity constant for the hydrolysis of t-butyl chloride as the percentage by volume of water in a water–ethanol solvent increases.

Prediction (a, 1) in Table 5.27 is in harmony with this increase of rate with increasing percentages of water in the water–ethanol solvent. Though the specific velocity constant changed with changing solvent, the values of

Table 5.26

Increase of Velocity Constant for Hydrolysis of t-Butyl Chloride with Increased Volume Percent of Water in $H_2O–E_2OH$.

Ethanol (% v/v)	[Halide] (M)	k_1' (hr^{-1})
90	0.0755	0.00616
80	0.0762	0.0329
70	0.0824	0.145
60	0.0735	0.453
50	0.0810	1.32
40	0.0306	4.66

Table 5.27

Expected Effect of Ionizing Media on Rates of Reaction

Type and mechanism	Charges concerned in rate-determining stage of reaction			Effect on charges of forming complex		Expected effect of ionizing media
	Factor	Complex	Products	Magnitude	Distribution	
(a)						
1 S_N^1	RX	$\overset{\delta+\ \ \delta-}{R\cdots X}$	$\overset{+}{R}+\overset{-}{X}$	Increase	—	Accelerate
2 S_N^2	$\overset{-}{Y}+RX$	$\overset{\delta+}{Y\cdots R}\overset{\delta-}{\cdots X}$	$YR+\overset{-}{X}$	No change	Dispersed	Retard
(b)						
3 S_N^1	RX	$\overset{\delta+\ \ \delta-}{R\cdots X}$	$\overset{+}{R}+\overset{-}{X}$	Increase	—	Accelerate
4 S_N^2	$Y+RX$	$\overset{\delta+}{Y}\cdots R\overset{\delta-}{\cdots X}$	$\overset{+}{YR}+\overset{-}{X}$	Increase	—	Accelerate
(c)						
5 S_N^1	$\overset{+}{RX}$	$\overset{\delta+\ \ \delta-}{R\cdots X}$	$\overset{+}{R}+X$	No change	Dispersed	Retard
6 S_N^2	$\overset{-}{Y}+\overset{+}{RX}$	$\overset{\delta+}{Y\cdots R}\overset{\delta-}{\cdots X}$	$YR+X$	Decrease	—	Retard

activation energies varied only slightly in the reactions presented, and hence the mechanism of the reaction it is assumed involves as its rate controlling step the ionization of *t*-butyl chloride:

$$t\text{-BuCl} \xrightarrow{\text{slow}} t\text{-Bu}^+ + Cl^- \tag{5.548}$$

$$t\text{-Bu}^+ + Cl^- + H_2O \xrightarrow{\text{fast}} t\text{-BuOH} + H^+, Cl^- \tag{5.549}$$

In agreement with prediction (a, 2), the rates of hydrolysis of methyl and ethyl iodides were observed to decrease on increasing the proportion of water in water–alcohol solvents [427]. Thus the mechanism is presumably

$$HO^- + RI \xrightarrow{\text{slow}} \overset{\delta-}{HO} \cdots R \cdots \overset{\delta-}{I} \xrightarrow{\text{fast}} HOR + I^- \qquad (5.550)$$

Also, the rates of racemization of secondary iodides in acetone decreased upon the addition of a small proportion of water [428] in agreement with prediction (a, 2). Menshutkin [429] as envisioned from prediction (b, 4) found that ethyl iodide combined with ethylamine more rapidly in alcohols than in hydrocarbons, the order being $MeOH > EtOH > Me_2CO > C_6H_6 > C_6H_{14}$. The mechanism presumably is

$$(C_2H_5)_3N + C_2H_5I \xrightarrow{\text{slow}} (C_2H_3)_3\overset{\delta+}{N} \cdots C_2H_5 \cdots \overset{\delta-}{I} \xrightarrow{\text{fast}} (C_2H_5)_4N^+ + I^-$$
$$(5.551)$$

The same general sequence for the reactions for other primary alkyl or methyl halides with other amines or sulfides has been found by later investigators [430–437].

The decomposition of triethylsulfonium bromide took place at a slower rate in alcohols than in acetone [438], and the rate of hydrolysis of the dimethyl-*t*-butyl sulfonium cation was decreased by increasing the proportion of water in water–alcohol solvent. Both of these results were in agreement with prediction (c, 5). The mechanism of the triethylsulfonium bromide reaction can be written

$$(C_2H_5)_3S^+ \xrightarrow{\text{slow}} (C_2\overset{\delta+}{H_5}) \cdots (C_2\overset{\delta+}{H_5})_2S \xrightarrow{\text{fast}} (C_2H_5)^+ + (C_2H_5)_2S$$
$$(5.552)$$

$$C_2H_5^+ + Br^- \xrightarrow{\text{fast}} C_2H_5Br \qquad (5.553)$$

As expected from prediction (c, 6) the rate of hydrolysis of the trimethylsulfonium cation was reduced by increasing the ratio of water to alcohol in water–alcohol solvent [439]. The mechanistic picture was given as

$$HO^- + (CH_3)S(CH_3)_2^+ \xrightarrow{\text{slow}} \overset{\delta-}{HO} \cdots (CH_3) \cdots \overset{\delta+}{S}(CH_3)_2$$

$$\xrightarrow{\text{fast}} CH_3OH + (CH_3)_2S \qquad (5.554)$$

It has been pointed out that not only the rate but also the order and the mechanism of a chemical reaction may be changed by changing the solvent [36, 425, 440]. Two mechanisms were perceived for the hydrolysis of secondary and tertiary alkyl halides [441]. These are a unimolecular S_N1 mechanism kinetically dependent on the ionization of the alkyl halide, and a bimolecular $S_N 2$ mechanism involving attack by the hydroxide ion. If R represents the alkyl radical and X represents the halide, these mechanisms can be represented as

$$RX \rightarrow R^+ + X^- \qquad (5.555)$$

followed by

$$R^+ + OH^- \rightarrow ROH \text{ (instantaneously)} (S_N 1) \qquad (5.556)$$

$$RX + OH^- \rightarrow ROH + X^- (S_N 2) \qquad (5.557)$$

The changeover from the unimolecular to the bimolecular mechanism was found to depend on the medium, the concentration, and the alkyl group involved. In dilute aqueous alcohol solutions the changeover was between the isopropyl and ethyl groups.

A case intermediate between $S_N 1$ and $S_N 2$ mechanisms has been pointed out [81] for the case where Y can facilitate the departure of X. The displacement reaction is ordinarily second order when Y represents an ion such as halide (X^-), hydroxide (OH^-), or alkoxide (RO^-); intermediate cases occur in the kinetics when Y is a solvent molecule behaving as a nucleophilic reagent and also simultaneously as an ionizing agent. In solvolysis, reactions the complexity of the molecular system as contrasted with the oversimplifications of the models used in explaining them, could cause much of the controversy in the interpretations of these reactions [81].

It is believed that solvation of the leaving group is necessary in all nucleophilic displacements of anions from saturated hydrocarbons [442]. The most effective solvation is one producing the quick formation of a coordinate covalent bond, for example, the solvation of bromide in methyl bromide by mercuric bromide, which in 0.05 M concentration increases by sevenfold the termolecular displacement of bromine in methyl bromide by pyridine in benzene solution. The next most effective type of solvation is ionic bonding such as hydrogen bonding by phenols. This principle is illustrated in the push–pull third-order mechanism involving the formation of the hydrogen bond in the case of the displacement of bromide in ethyl bromide by pyridine in the presence of phenol.
Thus

$$(5.558)$$

This reaction is superimposed on the ordinary, slower second-order mechanism that operates when there are no solvating agents present that are more effective than benzene as solvent. In the presence of pyridine the rate constant for the reaction of 2.6 M methyl bromide with 24.7 M methanol is five powers of ten greater than the rate constant in the absence of pyridine. When only polarization solvation by polarizable solvent molecules is possible may this type of solvation be important. The reaction in pure benzene of pyridine with methyl bromide is probably termolecular, but is second order due to large excess of solvent. This conclusion is supported by the fact that neither this nor any other nucleophilic reaction has been observed in the homogeneous gas phase. Also, except on the walls of the containing vessel or on the surfaces of precipitated crystals, the reaction does not occur in hexane or cyclohexane, which strongly resemble the gas phase in that polarization forces are very weak [81, 443].

The formation of an ionic dipole–dipole bond which facilitates the breaking of the old covalent bond constitutes the initial attack on the carbon, e.g., by pyridine. There is an immediate formation of the ionic solvation bond with carbon, and this ionic bond by a slow process may form a covalent bond. An intermediate "carbonium salt" in the case of methyl halides having a total of nearly four covalent bonds on only a minute fraction of a unit positive charge on the carbon may result from the quantitative difference in the rate of this changeover, while in the case of tertiary halides, the carbonium ion has mainly two ionic bonds and is relatively "free" and long-lived. In the tertiary case, the carbon-solvating reagent or solvent does not become extensively covalently bonded either in the rate-determining or the subsequent product-determining step.

The attack on carbon may be of different types in the rate determining step, involving covalent bonding in the case of primary halides, and solvation associated chiefly with ionic or ion-dipolar bonding in the case of tertiary halides. A carbon other than the one from which the leaving group is displaced may be attacked, as, for instance, in neopentyl chloride:

$$\begin{array}{ccccc} H_3C & H_3C & H & & \\ \diagdown & \diagdown & \diagup & & \\ O\cdots> & C\cdots>C-Cl\cdots>HOCH_3 & \xrightarrow{\text{slow}} \\ \diagup & \diagup\quad\diagdown & \diagdown & & \\ H & H_3C\quad CH_3 & H & & \end{array} \quad \begin{array}{ccccc} H_3C & H_3C & H & & \\ \diagdown & \diagdown & \diagup & & \\ O\cdots & C^{+}-C & +\ \overset{-}{Cl}\cdots HOCH_3. \\ \diagup & \diagup\quad\diagup & \diagdown & & \\ H & H_3C\quad H_3C & H & & \end{array}$$

$$(5.559)$$

The whole process including the Wagner–Meerwin rearrangement can involve only one step.

It is thought that the first-order racemizations in aqueous or alcoholic solutions may arise from the rapid formation of the disolvated carbonium ion, aided by the dissociation of the ion pair in these polar solvents and by the high concentration of hydroxylic reagent present. The hydroxide ion has a voluminous solvent shell in aqueous solution and is therefore stearically hindered from approaching closely to the central carbon atom in tertiary halides. Under these conditions the ion does not accelerate these reactions.

Ignoring finer details, the mechanism of acylation has been separated into two types [444]. Without distinguishing relative rates, which will depend on the system involved, the types are

$$R \cdot COX \underset{b}{\overset{a}{\rightleftharpoons}} R \cdot CO^+ + X^- (or\ R \cdot CO^+ X^-) \xrightarrow{S} Products$$

$$(5.560)$$

$$R \cdot COX + S \longrightarrow Products \qquad (5.561)$$

When reaction (a) alone in Eq. (5.560) is rate determining and especially when other molecules of the substrate A apart from those actually acelated can assist ionization, then only on the basis of particular definitions of molecularity and bonding change is a rigid distinction possible between mechanisms (5.560) and (5.561). Definitions are involved only in doubtful distinctions dealing with borderline cases [271]. The charge on the acelium ion RCO^+ is enhanced and cleavage promoted by an electron attracting group X which posses some stability as an ion X^-, but an opposite effect will be exerted by an electron-repelling group. The usual qualitative assignment of polar properties to a group are in harmony with less well-defined variations of the leaving group achieved within a given type of reagent, as well as the electron-attracting or electron-repelling properties [445]. The effect on reactivity of changes in X are more unmistakably clear than charges in R. The charge on the carbonyl carbon atom in both the acelium ion and in the polarized reagent will be reduced if R is an electron donor, but the charge will be increased and ionization hindered if R attracts electrons. The effect of changes in R is determined by whether the substrate reacts primarily with the ionized or unionized acetylating agent and on which step in the overall process is rate controlling. The factors important in acylation and other nucleophilic substitution have been discussed by many authors [271, 445–449].

In many cases of mechanism (5.561), subtrate S may add directly to the carbonyl group rather synchronously displacing X [444, 446]. For the substrate water has been written [443] the mechanism

$$R \cdot COX + H_2O \rightleftharpoons RC(OH)_2X \rightarrow RCO_2H + HX \qquad (5.562)$$

but varied situations may be made to comply with the individual steps of this scheme. The main reason for postulating such intermediates has been the concurrent occurrence of oxygen exchange and acelation, but such postulations have not been necessary in all cases, particularly those involving acid catalysis [450].

Some instances of water solvation are of particular interest. Rates of reaction have been correlated with a factor H_0, termed the acidity function, in up to 100% sulfuric acid [451]. The step-method of the relative basicity of indicators was used, and H_0 defined by the equation

$$H_0 = -\log a_{H^+} \frac{f_B}{f_{BH^+}} \tag{5.563}$$

where a_{H^+} is the activity of the hydrogen ion and f_B and f_{BH^+} the activity coefficients of the nonionized or neutral base indicator and of its conjugate acid, respectively. In dilute solutions $f_B = f_{BH^+} = 1$, and

$$H_0 = -\log a_{H^+} = pH \tag{5.564}$$

The function h_0 is sometimes used. It is defined as

$$h_0 = a_{H^+} \frac{f_B}{f_{BH^+}} \tag{5.565}$$

Hence

$$H_0 = -\log h_0 \tag{5.566}$$

It has been observed that, in acid-catalyzed hydrolysis of substituted ethylene oxides, logarithms of the first-order rate constants for the hydrolysis were linear with $-H_0$ [452].

In acid solution ranging from 0.2 to 3.6 M perchloric acid the acid-catalyzed iodination of acetophenone has been studied and the reaction rate found to be more nearly proportional to the hydrogen ion concentration than to the acidity function [453]. This observation combined with the presumably known mechanism of the reaction led to the proposal that acid-catalyzed reactions which include a water molecule in the transition state would depend on the concentration of hydrogen ion rather than the h_0 acidity function.

The keto–enol transformation of cyclohexane-1,2-diol was found to be sufficiently slow up to acidities of about 6 to 7 M mineral acid to be easily measurable by conventional means, and to reach a measurable equilibrium in aqueous solutions [454, 455] that was complicated by the fact that the ketone in these solutions was almost entirely in the form of its monohydrate.

The first-order rate constants for both the ketonization and enolization reactions were calculable from a combination of the kinetic and equilibrium data. The first-order specific rate constant for the acid-catalyzed approach to equilibrium was given by the equation

$$k'_{H^+} = k'_{obs} - k^w_{obs} \tag{5.567}$$

where k'_{obs} was the observed specific velocity constant for approach to equilibrium and k^w_{obs} was the water contribution to the rate. For catalysis by perchloric, sulfuric, and hydrochloric acids in the range 1 to 7 M, it was found [456] that $\log k'_{H^+}$ was linear with $-H_0$, thus

$$\log k'_{H^+} = -0.64 H_0 - 5.14 \tag{5.568}$$

If k_f' and k_r' are the first-order specific velocity constants for the acid catalyzed conversion of the enol to ketone and of the ketone to enol, respectively, then

$$k'_{H^+} = k_f' + k_r' \qquad (5.569)$$

Plots of $\log k_f'$ versus $-H_0$ and of $\log k_r'$ versus $-H_0$ were linear with slopes ranging from 0.5 to 0.7 and from 0.8 to 1.0, respectively, for the two types of plots depending on the acid catalyst. For the reaction of enol with hydronium ion the second-order specific velocity constant k_e' is related to k_f' by the equation

$$k_f' = k_e' C_{H_3O^+} \qquad (5.570)$$

and for the reaction of ketone with hydronium ion the second-order specific velocity constant k_k' is related to k_r' by the expression

$$k_r' = k_k' C_{H_3O^+} \qquad (5.571)$$

Straight lines were obtained from plots of both $\log k_e'$ and $\log k_k'$ versus $1/T$. From these plots the values of energies and entropies of activation were obtained. See Table 5.28.

Table 5.28a

Energies and Entropies of Activation for Bimolecular Reactions with Hydrogen Ion

Specific velocity constant[a]	Energy of activation, E^* (kcal/mole)	Entropy of activation, S^* (e.u.)
k_k' (enolization)	17.6 ± 1	-26 ± 2
k_e' (ketonization)	24.4 ± 1	-4 ± 2

[a] k' in liters/mole/sec.

Now only the ketone is significantly hydrated and a number of transition states were proposed as follows:

$$(5.572)$$

Table 5.28b

Amination of Mixtures of Chlorobenzene and Chlorobenzene-2-D

$d_0{}^a$	d^b	P^c	F_{calc}	$a_{calc}{}^d$	a_{obs}
0.407±0.003	0.350±0.008	0.452±0.006	0.844±0.011	—	Not enough product for analysis
0.816±0.019	0.414±0.014	0.718±0.006	0.870±0.005	0.416±0.004	0.441±0.000
0.594±0.014	0.290±0.005	0.819±0.012	0.868±0.001	0.283±0.004	0.326±0.005
0.396±0.009	0.227±0.009	0.783±0.007	0.874±0.005	0.185±0.005	0.204±0.003

^{a} d_0 = mole fraction of chlorobenzene-2-D at zero time.

^{b} d = mole fraction of chlorobenzene-2-D at time t.

^{c} P = mole fraction of starting material converted to chloride ion at time t.

^{d} a = mole fraction of aniline-D at time t. The maximum value was calculated from Eq. (5.661) using the minimum value of d and the maximum values of d_0, P and F. The maximum value of F was calculated from Eq .(5.661) using the minimum value of d and the maximum value of d_0 and P; the reverse combination gave the minimum value.

The main reaction is from unhydrated enol to hydrated ketone and hence the upper hydration equilibrium lies to the left and the lower one to the right. In the transition state two extremes of carbonyl hydration, one appropriate to reaction via the left-hand side of this equilibrium diagram and the other appropriate to reaction via the right-hand side were given as

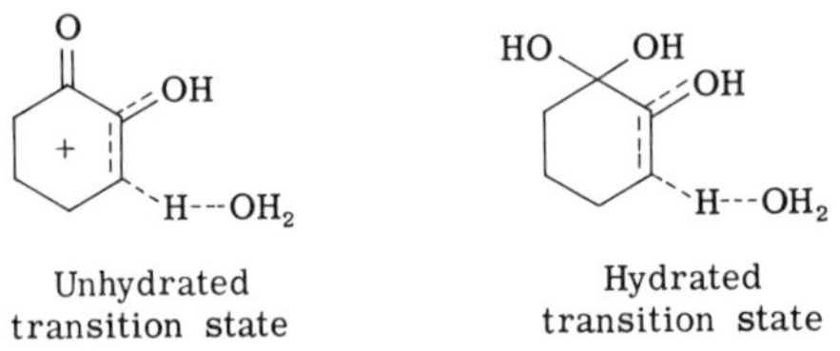

$$\text{(5.573)}$$

Unhydrated   Hydrated
transition state   transition state

The expected value of the entropy of activation for a second-order reaction when k' is in milliliters per mole per second is about -10 e.u. which is the value for a reaction of ordinary unhydrated ketones. The overall change in entropy is close to zero for a normal keto–enol reaction, and the S^* value for the reverse enol–ketone reaction should also be about -10 e.u. For the cyclohexe-1,2-dione reaction the S^* value is very abnormal, being 15 to 20 e.u. more negative than normal for the enol–ketone transformation, and 5 to 6 e.u. more positive than normal for the reverse keto–enol transformation. These phenomena are interpreted on the basis that the transition state holds comparatively firmly water of hydration [455, 456], comparing closely to the right-hand transition state shown above. Since these results are somewhat surprising, it is suggested that it might be profitable to examine further the entropies of activation for reactions of carbonyl compounds. In H_2O, the reaction of the enol is faster than in D_2O in harmony with the expected slow proton transfer mechanism for the enol.

It was thought [455, 456] that the slow proton transfers to the unhydrated enols would also very nearly correlate with h_0, perhaps with slightly higher slopes of $\log k'$ versus $-H_0$ than in the present case. This is particularly interesting since at lower acidities the enol reaction exhibited general acid catalysis.

A solvent water molecule of solvation of the cyclohexane-1,2-dione ketone causes the dependence of the rate of the establishment of the acid-catalyzed keto–enol equilibrium of this compound to be distinctly different from the normal keto–enol reaction.

The understanding of the acid-catalyzed keto–enol equilibrium of cyclohexane-1,2-dione relative to a consideration of the mechanistic implications of the hydration equilibrium and the Arrhenius parameters was not clarified by the introduction of water as a rather necessary variable and by characterizing a reaction by one or the other of two empirical factors w or w^* [457], where w is the slope of the $\log(k'_{H^+}/h_0)$ versus $\log a_{H_2O}$ plot and w^* is the slope of the $\log(k'_{H^+}/C_{H^+})$ versus $\log a_{H_2O}$ plot. It was assumed [457] that for any given reaction one or the other of these plots would be approximately linear. Omitting nonlinear plots, w approaches zero when the slope of $\log k'_{H^+}$ versus $-H_0$ is about unity, is negative when the slope is greater the unity, and is positive when the slope is less than unity. Similarly, w^* approaches zero when the plot of $\log k'_{H^+}$ versus $\log C_{H^+}$ is about unity. Hence a more complete classification of acid-catalyzed reactions is possible using the parmeters w and w^* than in using h_0 and C_{H^+} alone. For reactions of ketones to form enols, the w plots are ordinarily linear, with slopes ranging from 2 to 7. Sometimes the w^* plots are linear having slopes of the order of -1.5. If it is assumed that keto–enol equilibrium is approximately independent of electrolyte concentration, the same slope might be expected for ketonization of enols. The correlation of w and w^* with mechanism depended mainly on results from ketones. A chart was prepared [457] containing conclusions based on these correlations. The chart indicates that a w larger than 3.3 and a w^* larger than -2 indicates that water is acting as a proton transfer agent in the slow step.

For cyclohexane-1,2-dione, the w plots for the ketone reaction are straight, but w is low and is the range expected for water acting as a nucleophile. The slopes of the plots for the forward and reverse reactions are so different as to indicate that water was acting as a nucleophile for one direction, but a proton-transfer agent for the reverse. It has been suggested that this is an unreasonable conclusion [455, 456] and it has been indicated that since the keto–enol transformation of cyclohexane-1,2-dione involves a monohydrated ketone and perhaps also a monohydrated transition state, there is no real reason why this ketone should have identical behavior with that of other ketones.

The indicator acidity functions H_0 have been rationalized [458, 459] and it has been shown that strong hydration of the proton in concentrated acid solutions are the main reasons for the high acidities of these solutions.

By applying the above theory and calculating on the molar, mole fraction, and volume fraction scales, it was proven [460] that the hydration number of hydroxide ion was three, and that its structure could be represented by

$$\left[\begin{array}{c} H \\ | \\ H{\diagdown}O{-}H{\cdots}O{\cdots}H{-}O{\diagup}^{\,H} \\ | \\ H \\ | \\ O{\diagdown}_{H} \end{array}\right]^{-} \tag{5.574}$$

The effect of ionic hydration on the kinetics of base catalyzed reactions in concentrated hydroxide solutions was studied [461]. Derived equations depicted the relation between the observed rate constants of various base-catalyzed reactions and the indicator basicity of the solution expressed by the

H_- function. Using the trihydrated hydroxide ion the equations were also expressed in terms of the stoichiometric concentrations of the hydroxide ion C_{OH^-} and of "free water" C_{H_2O}. The conclusion reached was that the change in transition state with respect to the number of molecules of water involved in its formation was shown by the rate dependence in concentrated alkaline solution. For eight types of base-catalyzed reactions, the rate expressions were classified into three groups, of which two followed H_- or $C_{OH^-}/C_{H_2O}^4$, five followed $H_- + \log C_{H_2O}$ or $C_{OH^-}/C_{H_2O}^3$, and one followed $H_- + 2 \log C_{H_2O}$ or C_{OH^-}/C_{H_2O}. Complications may make the distinction between the three groups using the above criteria difficult in practice, as for example, when water is involved in the formation of the transition state in a manner not considered in these three groups. To illustrate, imagine the unimolecular decomposition of a conjugate base in which the transition state is stabilized by four molecules of hydration not present in the hydration shell of the substrate SH. As the following rate expression indicates, there would be no dependence on water concentration:

$$\frac{dC_{SH}}{dt} = k'' C_{S^-} C_{H_2O}^4 = k'' K_c C_{SH} \frac{C_{OH^-}}{C_{H_2O}^4} C_{H_2O}^4 = k'' K_c C_{OH^-} C_{SH} \quad (5.575)$$

where K_c is the equilibrium constant for the reaction of the substrate with the hydrated hydroxide ion,

$$SH + OH(H_2O)_3^- \rightleftharpoons S^- + 4H_2O \quad (5.576)$$

A rate law obeying $C_{OH^-}/C_{H_2O}^3$ was derived for a proton abstraction or for nucleophilic substitution by a nonhydrated hydroxide. A mechanism of proton abstraction was shown to be involved in the breakdown of serine phosphate in alkaline solution, and nucleophilic substitution by a nonhydrated hydroxide was illustrated by the hydrolysis of ethyl iodide, a S_N2 reaction.

In base-catalyzed reaction mechanisms H_- was defined as

$$H_- = -\log \frac{C_{BH}}{C_{B^-}} K_{BH} \quad (5.577)$$

where K_{BH} is the acid dissociation constant of indicator according to the equation

$$BH \rightleftharpoons B^- + H^+ \quad (5.578)$$

and $H_- = -\log h_-$, hence

$$h_- = \frac{C_{BH}}{C_B} K_{BH} = \frac{f_{B^-}}{f_{BH}} a_{H^+} = \frac{f_B \cdot a_{H_2O}}{f_{BH} \, a_{OH^-}} K_w \quad (5.579)$$

In alkaline solution, the dissociation of BH is given by

$$BH + OH^- \rightleftharpoons B^- + H_2O \qquad (5.580)$$

therefore, a quantity b_- was defined as

$$b_- = K_w/h_- = \frac{f_{BH}\, a_{OH^-}}{f_{B^-}\, a_{H_2O}} \qquad (5.581)$$

In dilute alkaline solutions in which the activity coefficients approach unity, b_- approaches C_{OH^-}. The quantity b_- is directly related to the OH^- activity and is the alkaline counterpart of h_0 in acid solution. From Eq. (5.581) the relation

$$\frac{f_{BH}}{f_{B^-}} = b_- \frac{a_{H_2O}}{a_{OH^-}} \qquad (5.582)$$

can be obtained and is basic to the formulation of the rate expression for the base catalyzed reactions.

In case II presented by Anbar *et al.* [461] the substrate SH is in rapid equilibrium with its conjugate base S^-, thus

$$SH + OH^- \rightleftharpoons S^- + H_2O, \qquad (5.583)$$

but S^- subsequently reacts in a bimolecular rate-determining step with another reactant Y to yield products

$$S^- + Y \xrightarrow{\text{slow}} \text{Products} \qquad (5.584)$$

Hence the rate expression is

$$\frac{dC_{SH}}{dt} = k' \frac{f_Y f_{S^-}}{f^*} C_Y C_{S^-} \qquad (5.585)$$

Using C_{S^-} from Eq. (5.583) and combining with a step in which S^- is converted to products in a unimolecular rate determining step thus:

$$S^- \xrightarrow{\text{slow}} \text{Products} \qquad (5.586)$$

one finds

$$C_{S^-} = C_{SH} \frac{K_{SH}}{K_w} \cdot \frac{f_{SH}}{f_{S^-}} \frac{a_{OH^-}}{a_{H_2O}} \qquad (5.587)$$

Substituting this value of C_{S^-} into Eq. (5.585) gives

$$-\frac{dC_{SH}}{dt} = k' \frac{K_{SH}}{K_w} \cdot \frac{f_Y f_{SH}}{f^*} \cdot \frac{a_{OH^-}}{a_{H_2O}} \cdot C_Y C_{SH} \qquad (5.588)$$

Now if there is similar interaction of the Y component in the transition and free states with the environment, then

$$f_{S^-}^* = f^*/f_Y \tag{5.589}$$

is the activity coefficient of the B^--like part of the transition state, provided it is assumed that

$$\frac{f_Y f_{SH}}{f^*} = \frac{f_{BH}}{f_{B^-}} \tag{5.590}$$

Combining this assumption and Eq. (5.582), then Eq. (5.588) can be used to derive the expression

$$k_{obs}' = -\frac{1}{C_Y C_{SH}} \frac{dC_{SH}}{dt} = \text{const}\, b_- \tag{5.591}$$

Applying the actual prequilibrium Eq. (5.576), and using no further assumptions, the equation

$$-\frac{dC_{SH}}{dt} = k'' C_{S^-} C_Y = k'' K_c \frac{C_{OH^-}}{C_{H_2O}^4} C_Y C_{SH} \tag{5.592}$$

can be derived and gives

$$k_{obs}' = \text{const} \frac{C_{OH^-}}{C_{H_2O}^4} = \text{const}\, b_- \tag{5.593}$$

which is identical with Eq. (5.591). Case I in which the same preequilibrium equation (5.583) was postulated but in which S^- is converted to reaction products in a unimolecular rate determining step as shown in Eq. (5.586) resulted in exactly the same equation.

An example of case II is the formation of hydrazine from chloramine

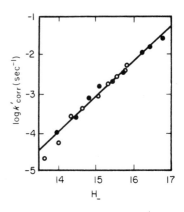

Fig. 5.20. Base-catalyzed chloramine– ammonia reaction. $[NH_2Cl] = 2 \times 10^{-3}$, $[NH_3] = 1.0$, $t = 27.3°$;   ○: sodium hydroxide;   ●: potassium hydroxide.

(NH_2Cl) and ammonia in strongly alkaline aqueous solutions. The first-order reaction in b_- had a slope of 0.90 [462]. Figure 5.20 represents the data for this reaction, where k'_{corr} was given by the relation

$$k'_{corr} = k'_{obs} - k_n'$$ (5.594)

In this equation k'_{obs} was the observed pseudounimolecular rate constant at high alkalinities resulting from the large excess of ammonia, and k_n' was the alkaline independent pseudounimolecular rate constant obtained at pH 13.7 or less. From Eqs. (5.577), (5.579), (5.581), (5.593), and (5.594) $\log k'_{obs}$ or $\log k'_{corr}$ would be linear with $\log b_-$, $\log h_-$, and with H_-.

Pulse radiolysis studies of organic and aquoorganic systems have been made [463–470]. A peak at 7000 Å in deaerated water and in various aqueous solutions was observed [463] using a linear accelerator giving 2-μm sec pulses of approximately 1.8-MeV electrons, and with the peak current in the pulse being variable up to 0.5 A. The band was attributed to the hydrated electron. The rate of decay of the absorption and the effect of various anions, cations, and dissolved gases were studied. Similar absorption spectra-produced irradiation have been observed in concentrated solutions of ammonia and of methylamine. These absorption spectra are similar to known absorption spectra of solvated electrons in liquid ammonia or in liquid methylamine.

The kinetics of the reactions of two transient species, the hydrated electron and the α-ethanol radical, were studied in irradiated deaerated aqueous ethanol solutions using the pulse radiolysis technique [467]. The solution was subjected to a 0.4-μm sec, 15-MeV electron pulse from a linear accelerator after which the kinetics of the hydrated electron were measured.

In both neutral and acid solutions, the hydrated electron e_{aq} in the bulk of the solution in these high intensity experiments reacted according to the equation

$$e_{aq} + H^+_{aq} \rightarrow H$$ (5.595)

The absolute rate for the reaction was obtained by observing the decay of the hydrated electron at 577–579 mμm. In the 0.5 M ethanol solution used, the hydroxyl radical which together with the hydrogen atom are formed as transient species in water, is removed by the reaction

$$OH + C_2H_5OH \rightarrow CH_3-\dot{C}HOH + H_2O$$ (5.596)

and thus is prevented from competing for the hydrated electron. Also the hydrogen atom is removed by the reaction

$$H + C_2H_5OH \rightarrow CH_3-\dot{C}HOH + H_2$$ (5.597)

The reaction equation (5.595) was made to be pseudo-first order by regulating the hydrogen ion concentration and the pulse current so that $[H^+_{aq}]/[e^-_{aq}] \gg 10$.

The absolute second-order rate constant was $(2.26 \pm 0.21) \times 10^{10}/\text{mole/sec}$ at 23°C as found from the slope of the plot of the logarithm of the optical density versus time and the known hydrogen ion concentration, corrected for the small additional amount formed by the pulse itself. In comparison the rate constant for the hydrated proton reacting with the hydrated hydroxide ion

$$H_{aq}^{+} + OH_{aq}^{-} \rightarrow H_2O \qquad (5.598)$$

was found [471] to be $1.3 \times 10^{11}/M/\text{sec}$. This high rate constant has been explained on the basis of a proton jump mechanism resulting in a phenomenalogical interaction distance of 8 Å. No explanation was given [467] for the difference in the rate constants for the reactions depicted in Eqs. (5.595) and (5.598). A bimolecular reaction was discussed [466] accounting for the disappearance of the α-ethanol radicals shown in Eqs. (5.596) and (5.597).

For steady-state radiolysis depending on the intensity, the reaction given in Eq. (5.595) will make a much smaller contribution at much lower intensities, and the first-order decay of the electron represented by the equation

$$e_{aq}^{-} + H_2O \rightarrow H + OH^{-} \qquad (5.599)$$

will increase in importance. The plot of the logarithm of the optical density versus time yielded an excellent straight line from the slope of which the first-order rate constant k' was found to be $\leqslant 4.4 \times 10^4/\text{sec}$. Since at low initial concentrations of the hydrated electron of less than 1 micromole, trace impurities such as oxygen at concentrations less than 0.1 micromole might make a significant contribution to the rate only an upper limit of k' was given.

The reaction equation (5.595) is very small or even negligible in amount in basic solution, depending on pH and on intensity, and the reaction equation (5.599), and

$$e_{aq}^{-} + e_{aq}^{-} \rightarrow H_2 + 2OH^{-} \qquad (5.600)$$

will predominate. The latter reaction is bimolecular and a plot of the reciprocal of the optical density versus time is a straight line up to the time that at least 65% of the initially formed hydrated electrons have disappeared. The straight line portion of the curve yields only the upper limit of the specific velocity constant since there may be small contributions to the rate of electron disappearance arising from the reaction of the electron with small amounts of hydrogen peroxide formed as a residual molecular yield and from the unimolecular decay under the conditions used of the hydrated electron as represented by Eq. (5.599). This upper limit of k' is given by

$$2k' \leqslant 1.32 \times 10^6 \, \varepsilon_{577} \qquad (5.601)$$

where ε_{577} is the molar extinction of the hydrated electron at the specified wavelength. From initial optical densities read from the rate curves in basic solution together with the appropriate yield of the electron, the value of ε_{577} may be found. The value is given [467] as 2.7 ± 0.2 molecules/100 eV for $G_{e_{aq}}$, the hydrated electron portion of the total yield of reducing species in the bulk of the solution after correcting the initial optical densities for the electron disappearance during the pulse. This correction was small, since in basic solution the halflife of the hydrated electron is an order of magnitude greater than the pulse width. The value $(9.7 \pm 1.5) \times 10^3/M$/sec was found for ε_{577}. Hence at 23°C the specific velocity constant for the reaction equation 5.600 has the upper limit given by

$$2k' < (1.3 \pm 0.2) \times 10^{10}/\text{mole/sec} \qquad (5.602)$$

where k' is defined by the differential equation for the second-order rate expression

$$-\frac{d[e_{aq}^-]}{dt} = 2k'[e_{aq}^-]^2 \qquad (5.603)$$

The absolute rate constants for the reactions of some metal ions with hydrated electrons have been determined [472] by following the decay of the absorption of the hydrated electron at 7000 Å. In the presence of the metal ions, the absorption was produced by a 2-μsec pulse of 4-MeV electrons, and a sensitive photomultiplier/oscilloscope technique was used to follow its decay, the rate of which was considerably increased by the presence of metal ions. The decay of the hydrated electron in pure water under conditions obtained is negligible compared to the decay in the presence of 50–200 micro-mole metal ions at which concentration the decay is a first-order process. For the electron-ion reaction, the bimolecular rate constants were obtained and had the values of 0.17×10^{10}, 1.35×10^{10}, 2.3×10^{10}, 3.0×10^{10}, 5.8×10^{10}, $>9 \times 10^{10}$, respectively, for the ions Zn^{2+}, Co^{2+}, Ni^{2+}, Cu^{2+}, Cd^{2+}, and $Co(NH_3)_6^{3+}$. The Mn^{2+} reaction was observed to be an order of magnitude less than the slowest measured other than that of Mn^{2+}.

In these solutions transient absorption was observed at 3000 Å and was attributed to unstable valency states produced in these reactions, e.g., Zn^+. This was confirmed using the photomultiplier technique which indicated a build-up of these absorptions in the time period observed for the enhanced electron decay. The build-up of the hydrated electron during the pulse confirmed the activity of the ions as listed above. It was suggested that the transient ultraviolet absorptions are the charge-transfer spectra of the reduced ions, the molar extinction coefficient of the Zn^+ species being at least 1.3×10^3.

The reactivities of more than 700 inorganic and organic compounds with the hydrated electron, e_{aq}^-, have been described [473]. A range of 10^9 is found in their reaction rates, from those immeasurably slow to those that are diffusion controlled. Halide ions unable to accommodate an additional electron, and the perchlorate, sulfate, phosphate and other highly oxidized ions are among the unreactive inorganic substances. The highly reactive inorganic reactants include the single electron acceptors such as MnO_4^-, CrO_4^{2-}, and $IrCl_6^{2-}$, and also Cd^{2+} and Pb^{2+} which are not ordinarily thought of as efficient oxidizing agents. Oxidized unreactive organic species, such as alcohols and carboxylate ions, as well as the very reactive ethylene derivatives rarely regarded as typical oxidants are included.

Two principal mechanistic descriptions [474] are used for the transfer of an electron from its site in the solvent into the orbital of the substrate. The "classical" mechanism involves the orbital overlap of the substrate with those of e_{aq}^- and a subsequent rapid electron transfer. This orbital overlap state involves changes in the atomic configuration of the substrate molecules and also in the hydration sphere of e_{aq}^- and may be thought of as a transition state in equilibrium with both reactants. The "nonclassical" or nonadiabatic electron transfer [475] mechanism is nonclassical electron tunneling through a potential barrier which separates the hydrated electron from an electron incorporated in the acceptor molecule, There is no requirement for an intra-molecular atomic movement preceding this process, but the mechanism may involve electron polarization of the acceptor molecule concurrently with the electron transfer.

There are exceptions, but in general there is an increased reactivity within a subgroup in the periodic table. The reactivity, however, is not in all cases correlated with the oxidation state of the substrate. Thus the activity of $IO_4^- > IO_3^-$, but the activity of $ClO_3^- > ClO_4^-$. There are several other examples. Since in a homologous series of organic compounds, the changes in electronic structure and atomic configuration are small and gradual, the reactivity of e_{aq}^- is easily estimated.

The reaction mechanism of e_{aq}^- with a compound probably involves an electron transfer process with an invariant energy of activation, but with a free energy of activation which varies with the reaction.

The Marcus [476] theory is in qualitative agreement with the dependence of reactivity on electron affinity. This theory attributes the increase in free energy of activation in slower reactions to the reorganization energy necessary to adapt the coordination sphere of the electron acceptor for electron transfer, in harmony with the Franck-Condon limitation [475–477]. The overall gain in free energy in the reaction compensates for the reorganization energy. A slow ($> 10^{13}$ sec) approach to a transition state characterizes this mechanism which predicts a correlation between the free energy of reaction, ΔF^0, and that of activation ΔF^*. The latter occurs in the equation for the rate constant k' as given below

$$k' = 10^{11} e^{-\Delta F^*/RT,}$$

(5.604)

and, if w is the work to bring the reactants together, and λ is the reorganization factor, then ΔF^0 and ΔF^* are related by the equation

$$\Delta F^0 = w + \frac{\lambda}{4}\left(1 + \frac{\Delta F^0 - w}{\lambda}\right)^2 \qquad (5.605)$$

The Marcus theory, however, does not agree quantitatively with the experimental observations [474].

Even when the redox potential is not appreciably altered, the ligand strongly affects the activity of the metal ion, particularly when the last water molecule is substituted [474]. Other chemical phenomena has been found to be profoundly affected when the last layer of water is breached, perhaps not when the last water molecule is replaced, by organic ligand [478]. Since the redox potential of the electron acceptor plays an important role in these nonaquo complexes, these complexes follow a trend paralleling the aquocomplexes.

The primary products of the reaction of e_{aq}^- is the species into a vacancy of which an electron transfers from its site in the solvent. This vibrationally excited primary product species is formed without violation of the Franck–Condon principle which necessitates that the atoms involved do not change their positions during the instant ($< 10^{14}$ sec) of the actual electron transfer. These states become deexcited within the time of a few vibrations ($< 10^{12}$ sec) which lifetimes are practically always neglected in stoichiometric chemistry, but which must be considered when the detailed mechanism of electron transfer is being formulated. These activated species may be deactivated in steps without undergoing further chemical change or dissociation, their lifetimes being limited by their own concentrations, but their survival rarely being longer than milliseconds. Eventually, stable secondary or tertiary products are formed.

Out of four possible mechanisms, the one favored [473] was one in which a nonadiabatic pathway was followed in the electron transfer from e_{aq}^-, the difference in the aquo and other complexes lying in the width of the potential barrier. The excited products arise from the exothermal nature of all reported electron transfer reactions, in which case there is no necessity for postulating an adiabatic electron transfer and an activated complex. This mechanism favors the kinetic behavior of organic compounds.

One approach to the problem of slow hydrated electron reactions is from the consideration of the potential energy profile of the system; the rate of electron tranfer being little affected by the potential energy barrier between reactants and products [473]. About 40 kcal of hydration energy holds the electron in its trap. Also, there are the repulsive force of the electrons belonging to the acceptor molecule, the outer sphere water molecules, and gegen ions creating potential barriers. The resulting barrier is over an order of magnitude greater than the observed practically invariant 3.5 kcal/mole, and it is concluded that the electron tunnels through this energy barrier without forming a transition state in overall exothermal electron-transfer processes.

The diffusion coeffiient $(4.9 \times 10^{-5}$ cm^2/sec) of the hydrated electron is equal to that of OH$^-$ in water, but much higher than anticipated for a negative monovalent ion of 3-Å radius. Hence the diffusion of e_{aq}^- must be by some extraordinary mechanism, as the OH$^-$ and H$_3$O$^+$ ions diffuse by the Grotthuss process [473]. It was concluded [473] that if no cavity exists, the electron creates one in the water matrix, and that its solvent interaction consists mainly of polarization. In water, diffusing from trap to trap over considerable distances without passing through a transition state, the hydrated electron perhaps also reacts with an electron acceptor in a nonadiabatic process.

For those instances where there is insufficient orbital overlap (geometrical hindrance or thickness of barrier) between e_{aq}^- and the acceptor orbital to permit reaction, the transmission coefficient for the tunneling process may be less than unity and strongly dependent on the gain of free energy upon electron transfer [473]. The electron tunnels with a transmission coefficient greater than unity over distances longer than those required for electron overlap in the ground state, when the gain in ΔF^0 is substantial.

Conclusions from a large number of diffusion-controlled reactions support the tunneling mechanism [479]. Eighty percent of 60 diffusion controlled reactions studied [473] agreed in rates with those predicted by the Smoluchowsky–Debye formula [473, 480].

$$k_{\text{diff}} = \frac{4\pi r D N}{1000} \tag{5.606}$$

where r is the sum of the radii of the hydrated electron and of the solute particle X, D is the sum of the diffusion constants of the hydrated electron and the solute X, and N is Avogadro's number. The cross sections for the reactions r_x were "normal" and agreed with those estimated from crystallographic data. Quite a number of electron acceptors, however, demonstrated a higher reactivity than predicted by Eq. (5.606). These highly reactive substances include a number of organic and inorganic compounds which are excellent oxidizing agents.

Large cross-sections, two to three times those necessary for diffusion controlled reactions, are assumed in explaining these abnormally high rate constants. This assumption is in agreement with the tunneling mechanism, which permits greater electron transfer distances than those determined by the electron orbital overlap of reactants in their ground states for reactions with large gains in free energy and insufficient time for rearrangement of the

acceptor molecule prior to electron transfer. The transfer occurs before the
acceptor molecule "perceives" the presence of e_{aq}^-.

In the slower reactions it is also proposed [473] that the electron is transferred to the substrate without forming a transition state but by tunneling
through the potential barrier, and thus permitting no significant increase of
the activation energy. The height of the potential barrier depends on the free
energy change and its width on e_{aq}^--substrate distance of closest approach,
and these two properties of the barrier determine the probability of electron
transfer.

The tunneling hypothesis is supported [473] by the very comparable rates
of aromatic compounds in water and methanol [481] and of other compounds
in water and ice [482]. Different potential barriers in each of these media
would have been necessitated by a transition state because of the differences
in solvation of reactants and products.

The "primary" product of a hydrated electron reaction after deexcitation
is a particle containing an additional electron. This particle may be a stable
species as $CO(NH_3)_6^{2+}$ or it may be an unstable species which may undergo:
(1) a monomolecular dissociation, (2) a proton transfer from the solvent,
(3) protonation accompanied by hydration or dehydration, (4) association,
(5) association with nonreduced substrate, (6) charge transfer to other
solutes, and (7) disproportionation [473]. An example of each of these types
of reaction is given below [473].

(1) $Fe(CN)_6^{3-} \rightarrow Fe(CN)_5^{2-} + CN^-$ (5.607)

(2) $(CH_3)_2CO^- + H_2O \rightarrow (CH_3)_2COH + OH^-$ (5.608)

(3) $NO_3^{2-} + H_2O \rightarrow (NO_2)_{aq} + 2OH^-$ (5.609)

(4) $Cu^+ + Cu^+ \rightarrow Cu_2^{2+}$ (5.610)

(5) $Ag^0 + Ag^+ \rightarrow Ag_2^+$ (5.611)

(6) $Zn^+ + CH_3Br \rightarrow Zn^{2+} + CH_3Br^-$ (5.612)

(7) $(CH_3)_2COH + (CH_3)_2COH \rightarrow (CH_3)_2CHOH + (CH_3)_2CO$ (5.613)

Hydrogen atoms may formally be considered as the conjugate acid of e_{aq}^-
because of the acid-base equilibrium reactions:

$$e_{aq}^- + H_2O \rightleftharpoons OH^- + H \qquad (5.614)$$

$$e_{aq}^- + H_3O^+ \rightleftharpoons H_2O + H \qquad (5.615)$$

However, H and e_{aq}^- are not similar chemically, the reaction of hydrogen
atoms with aromatic compounds proving the hydrogen atoms are electrophilic
rather than nucleophilic reactants [473, 483].

The positronium perhaps has a substantial hydration shell and perhaps exists as $(e^+ - e^-)_{aq}$ since its rates of diffusion estimated from its rates of reaction are similar to those of the e^-_{aq}. The positronium is more reactive than the H-atom, and its rates of reaction with many substrates in solution are diffusion controlled $(10^9 - 10^{11}/M/sec^{-1})$ [473]. If the positronium were unhydrated or only slightly hydrated, its rate of diffusion would be greater by at least an order of magnitude than that of the hydrogen atom [473].

This section indicates that solvation and ionization phenomena are closely intertwined. Strong solvent interactions are involved in ionization mechanism, though solvation does not necessarily cause ionization. It is advantageous, however, to discuss the two phenomena in the same section.

It is evident that solvation and ionization affect rates and mechanisms in numerous ways, and that both of these solvent effects are highly important influences of solvents on rates and mechanisms of reactions in solution. Many of the ideas concerning the effects of solvents through solvation and ionization on rates and mechanisms of reactions have been verified experimentally and theoretically, but many, while reasonable, are conjectural. Some of these phenomena remain items of polemical discussions. We have tried to present the material without bias, and to leave the reader free to make interpretations of the data in a manner satisfactory to himself. The reader is to be congratulated who can formulate sane, sound, and logical conclusions on material pertinent to his own fields of scientific interests.

XXI. Solvolysis and the Effect of Solvent on Rates

In this section we extend the discussion of solvolysis reactions. A solvolysis is a reaction in which the attacking group Y is a solvent molecule in the first-order nucleophilic substitution or $S_N 1$-type reaction. The reaction is designated as a hydrolysis if water is the solvent.

A mechanism has been discussed [484, 485] for the hydrolysis of alkyl and aryl halides in water solvent in which the ionization of the organic compound is the rate determining step followed by a fast reaction with the nucleophilic reactant. Let RX represent the alkyl or aryl halide, then the hydrolysis may be represented as

$$RX \underset{k_2'}{\overset{k_1'}{\rightleftharpoons}} R^+ + X^- \tag{5.616}$$

$$R^+ + H_2O \underset{k_4'}{\overset{k_3'}{\rightleftharpoons}} ROH_2^+ \underset{k_6'}{\overset{k_5'}{\rightleftharpoons}} ROH + H^+ \tag{5.617}$$

The ionization rate is so slow, and the rate of reaction of R^+ is so much faster with water than with halide ion, that the simple equation

$$-\frac{d(\text{RX})}{dt} = k_1{}'(\text{RX}) \qquad (5.618)$$

can be used to represent the rates of these hydrolyses reactions, which are strongly dependent on the organic group R and upon the solvent. That the mechanism does not involve polar or ionic intermediates is supported by the observation that these reactions do not take place in the gas phase.

The acetolysis of the *p*-toluene sulfonates of the stereoisomers of 3-phenyl-2-butanol (3-phenyl-2-butyltosylate) in glacial acetic acid were studied [486]. In the acetolysis of the erythro compound there was almost complete retention of optical activity, however, optically active thero compounds gave the racemic thero acetate. An intermediate cyclic carbonium ion in a Wagner–Meerwein rearrangement was used in interpreting the results. In the acetolysis of the erythryo compound the carbonium ion was symmetric, but was asymmetric in the acetolysis of the thero compound. A simultaneous loss of HOTs⁻ group and the backside bridging of the phenyl group to form a carbonium ion was proposed. The identical optically active ethro product will result from the attachment of an ester group to either of the two bridged carbon atoms in the carbonium ion. Thus

$$(5.619)$$

With the therotosylate, however, the attachment of an ester group in a statistical distribution will result in a racemic thero product.

$$
\begin{array}{c}
\underset{\substack{H_3C \\ H\text{---}C^1\text{---}^2C\text{---}CH_3 \\ C_6H_5 \qquad H}}{\overset{OTs}{}} \xrightarrow{H^+}
\underset{\substack{H_3C \\ H\text{---}C^1\rightleftharpoons^2C\text{---}CH_3 \\ C_6H_5 \qquad H}}{\overset{\overset{\oplus}{O}=TsH}{}}
\xrightarrow{-HOTs}
\left[\begin{array}{c}
\underset{\substack{H_3C \\ H \qquad\quad H}}{C\text{===}C{-}CH_3} \\
\\
H
\end{array} \right]^+
\end{array}
$$

$$\text{AcOH} \qquad \qquad (5.620)$$

$$
\underset{\substack{H_3C\text{---}C^1\text{---}^2C\text{---}H \\ H \qquad C_6H_5}}{\overset{\overset{H}{\underset{\oplus}{Ac\text{---}O}} \quad CH_3}{}}
\quad + \quad
\underset{\substack{H\text{---}C^1\text{---}^2C\text{---}CH_3 \\ C_6H_5 \qquad H}}{\overset{H_3C \quad \overset{H}{\underset{\oplus}{O\text{---}Ac}}}{}}
$$

$$-H^+$$

$$
\underset{\substack{H_3C\text{---}C^1\text{---}^2C\text{---}H \\ H \qquad C_6H_5}}{\overset{AcO \quad CH_3}{}}
\quad \text{and} \quad
\underset{\substack{H\text{---}C^1\text{---}^2C\text{---}CH_3 \\ C_6H_5 \qquad H}}{\overset{H_3C \quad OAc}{}}
$$

Extensive studies have been made [487–492] of the hydrolysis of ethylene halohydrins, CH_2OHCH_2X, and of ethylene oxide, CH_2OCH_2. The rates of reaction with hydroxide ion are exceedingly rapid compared to the rates with water for the reactions of ethylene bromohydrin, ethylene chlorohydrin, and ethylene fluorohydrin. Ethylene chlorohydrin is greater than the fluorohydrin in reactivity, but is less in reactivity than the ethylene bromohydrin or ethylene iodohydrin [489]. Ethylene oxide in aqueous hydrochloric has been found to react by four different paths. Thus

$$
\underset{\substack{H_2C}}{\overset{H_2C}{}}\!\!\!\diagdown\!\!O + H_2O \longrightarrow \underset{CH_2OH}{\overset{CH_2OH}{|}} \qquad (5.621)
$$

$$
\underset{\substack{H_2C}}{\overset{H_2C}{}}\!\!\!\diagdown\!\!O + H_2O \xrightarrow{H_3O^+} \underset{CH_2OH}{\overset{CH_2OH}{|}} \qquad (5.622)
$$

$$
\underset{\substack{H_2C}}{\overset{H_2C}{}}\!\!\!\diagdown\!\!O + Cl^- + H_2O \longrightarrow \underset{CH_2OH}{\overset{CH_2Cl}{|}} + OH^- \qquad (5.623)
$$

$$
\underset{\substack{H_2C}}{\overset{H_2C}{}}\!\!\!\diagdown\!\!O + Cl^- + H_3O^+ \longrightarrow \underset{CH_2OH}{\overset{CH_2Cl}{|}} + H_2O \qquad (5.624)
$$

Direct and hydrogen ion catalyzed hydrolysis constitute the first and second paths. The four steps in the entire process can be represented by the rate equation

$$-\frac{1}{[C_2H_4O]}\frac{d[C_2H_4O]}{dt} = k_0' + k_0'[H_3O^+] + k_2'[Cl^-] + k_2'[Cl^-][H_3O^+]$$

(5.625)

Other nucleophilic reagents than water or chloride ion will react in a manner similar to the reactions given for water and chloride ion.

If as in the case of acetolysis of 3-phenyl-2-butyltosylate, hydrogen ion is first added, presumably an oxonium ion complex

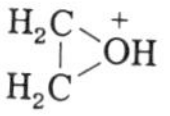

$$\begin{array}{c} H_2C \\ | \;\; \overset{+}{>}OH \\ H_2C \end{array}$$

(5.626)

is formed with which the nucleophilic reagent will react from 500 to 10,000 times as fast as with the neutral oxide, since the weaker base ROH rather than the strong base RO$^-$ is being displaced from its carbon bonding. The rate becomes faster as the attacking reagent becomes more nucleophilic since the rate is determined more by the nucleophilicity of the attacking reagent than by its basic strength.

Using a diltometric procedure the acid-catalyzed hydrolysis at 0°C of substituted ethylene oxides were studied [493] in aqueous perchloric acid ranging up to 3.5 M. The rates were markedly changed by substitution. For the same acid concentrations isobutylene oxide reacted about 10^4 times faster than epibromohydrin. On the basis of ^{18}O studies in basic hydrolysis, and because the logarithm of the first-order rate constant for the hydrolysis proved to be linear with $-H_0$, it was decided that the reaction occurred by a A-1 mechanism via a carbonium ion intermediate as

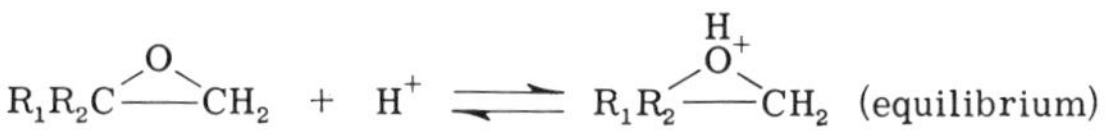

$$R_1R_2C\overset{O}{-\!\!\!\triangle\!\!\!-}CH_2 + H^+ \rightleftharpoons R_1R_2\overset{\overset{+}{O}-H}{-\!\!\!\triangle\!\!\!-}CH_2 \quad \text{(equilibrium)}$$

(5.627)

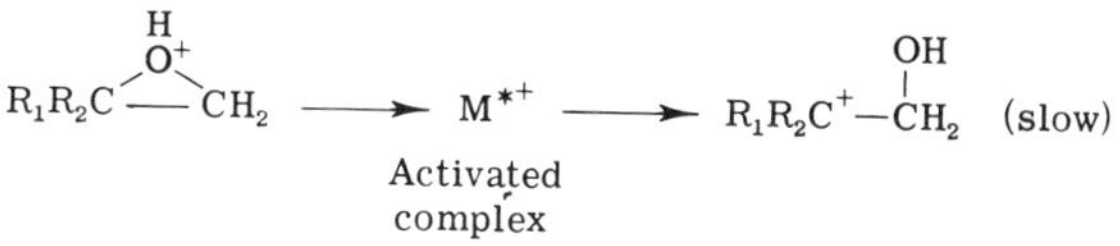

$$R_1R_2C\overset{\overset{H}{\overset{+}{O}}}{-\!\!\!-\!\!\!-}CH_2 \longrightarrow M^{*+} \longrightarrow R_1R_2C^+\overset{OH}{-}CH_2 \quad \text{(slow)}$$
$$\text{Activated}$$
$$\text{complex}$$

(5.628)

$$R_1R_2C^+\overset{OH}{-}CH_2 + H_2O \longrightarrow R_1R_2C\overset{OH}{\underset{OH}{-}}CH_2 + H^+ \quad \text{(fast)}$$

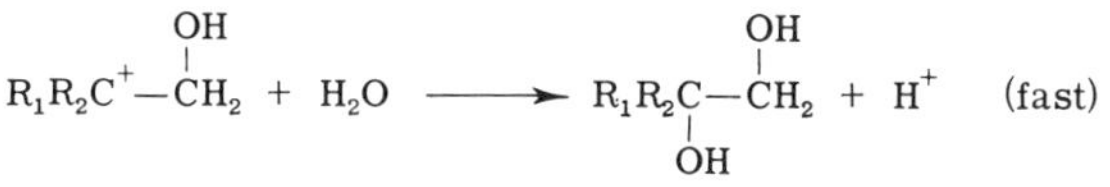

(5.629)

For an A-2 mechanism, an attack on the conjugate acid by a water molecule with its resulting inclusion in the activated complex would have been the rate-controlling step; and the rates would have closely paralleled the hydrogen ion concentration and not the acidity function. In the A-1 mechanism, however, the rates parallel the acidity function, and the solvolytic action of the solvent is not rate controlling but is critical in the product-forming step.

For the hydrolysis of substituted ethylene oxides, the rate of reaction was given

$$\text{Rate} = k_1' C_{\text{oxide}} = k_{\text{bi}}' C_{\text{oxide}} C_{\text{H}^+}$$

$$= k_{\text{bi}}^{\circ\prime} C_{\text{oxide}} \frac{C_{\text{H}^+} f_{\text{H}^+} f_{\text{oxide}}}{f_{\text{M}^{*+}}} \tag{5.630}$$

where $k_{\text{bi}}^{\circ\prime}$ is a true constant and the f's are the activity coefficients of the species indicated. Hence

$$\log k_1' = \log \frac{C_{\text{H}^+} f_{\text{H}^+} f_{\text{oxide}}}{f_{\text{M}^{*+}}} + \text{const} \tag{5.631}$$

and addition of Eqs. (5.563) and (5.631) gives

$$\log k_1' = -H_0 + \log \frac{f_{\text{oxide}}}{f_{\text{M}^{*+}}} - \log \frac{f_{\text{B}}}{f_{\text{BH}^+}} + \text{const} \tag{5.632}$$

In dilute solutions, all activity coefficients become unity, and the observed linearity with unit slope of $\log k_1'$ with $-H_0$ is evidence that

$$\frac{f_{\text{oxide}}}{f_{\text{M}^{*+}}} = \frac{f_{\text{B}}}{f_{\text{BH}^+}} \tag{5.633}$$

which is reasonable for an A-1 mechanism, since the activated complex and the conjugate acid of the oxide reactant are quite similar structurally.

From the acidic, basic, and neutral hydrolysis of propylene and iso-butylene oxides in ^{18}O-labeled water, it was observed [494] that the base-catalyzed reaction in the case of propylene oxide occurs predominantly at the primary carbon atom, and exclusively so for the isobutylene oxide. In contrast, the acid-catalyzed reaction occurs only at the branched chain for isobutylene oxide and preponderantly so for propylene oxide.

Principally, for steric reasons nucleophilic attacks by the hydroxide ion at the primary carbon atom of the epoxide ring is involved in the base-catalyzed hydrolysis for which the mechanism would be

$$\text{Me}_2\text{C}\underset{\text{O}}{-}\text{CH}_2 + \text{OH}^- \longrightarrow \text{Me}_2\text{C}(\text{O}^-)-\text{CH}_2\text{OH} \quad (\text{slow}) \tag{5.634}$$

$$\text{Me}_2\text{C}(\text{O}^-)-\text{CH}_2\text{OH} + \text{H}_2\text{O} \longrightarrow \text{Me}_2\text{C}(\text{OH})-\text{CH}_2\text{OH} + \text{OH}^- \quad (\text{fast}) \tag{5.635}$$

Ammonolysis which resembles hydrolysis may be illustrated by the reaction [495]

$$\underset{\substack{|\\ H}}{C}\!\!\!\overset{OR}{=}\!\!O \;+\; NH_3 \;\rightleftharpoons\; HC\!\!\overset{OR}{\underset{NH_2}{-}}\!\!OH \;\rightleftharpoons\; ROH \;+\; HC\!\!\overset{=O}{\underset{NH_2}{}} \qquad (5.636)$$

For higher esters ammonolysis may become very slow, and it was reported [496] that at room temperature isobutrate reacts only after about eight weeks.

As in the aqua series, salts of weak acids and of weak bases are hydrolyzed, so in the ammonia series ammonolysis occur [497]. If the metal in the compound is not exercising its maximum coordination number, coordination with one or more molecules of water usually precedes hydrolysis. In liquid ammonia, it is thought [498, 499] ammonolysis occurs in a manner involving similar steps. The hydrolysis of silicon tetrachloride is given [498]

$$SiCl_4 \;+\; 2\,H_2O \;=\; \underset{H_2O}{\overset{H_2O}{\diagdown}}\!\!SiCl_4 \;\longrightarrow\; 2\,HCl \;+\; Si(OH)_2Cl_2 \qquad (5.637)$$

$$Si(OH)_2Cl_2 \;+\; 2\,H_2O \;=\; \underset{H_2O}{\overset{H_2O}{\diagdown}}\!\!Si(OH)_2Cl_2 \longrightarrow 2\,HCl \;+\; Si(OH)_4 \qquad (5.638)$$

For the Group IVA halides, the degree of ammonolysis decreases regularly down the series $SiCl_4$, $TiCl_4$, $ZrCl_4$, and $ThCl_4$ [498]. Apparently $SiCl_4$ was completely ammonolyzed to $Si(NH_2)_4$ [499], while $TiCl_4$ and $ZrCl_4$ form the amidochlorides $Ti(NH_2)_3Cl$ and $Zr(NH_2)Cl_3$ with liquid ammonia while $ThCl_4$ gives the hexaammoniate $ThCl_4 \cdot 6NH_3$ [498]. A similar order of reactivity is shown in hydrolysis and ammonolysis [498, 499], and the mechanisms of alcoholysis and ammonolysis are perhaps similar. Thus, for alcoholysis [498, 500]

$$ROH \;+\; MCl_4 \;\longrightarrow\; \underset{\substack{H\\ \delta+}}{\overset{R}{\underset{}{\diagdown}}}\!\!O\overset{\delta-}{\underset{\delta+}{-}}MCl_3 \;\longrightarrow\; \overset{R}{\underset{}{\diagdown}}O\!-\!M\!-\!Cl_3 \;+\; H^+ \;+\; Cl^- \qquad (5.639)$$

and for ammonolysis

$$NH_3 \;+\; MCl_4 \;\longrightarrow\; \underset{\substack{H\\ \delta+}}{\overset{H}{\diagdown}}H\!-\!N\overset{\delta-}{\underset{\delta+}{-}}MCl_3 \;\longrightarrow\; NH_2\!-\!MCl_3 \;+\; H \;+\; Cl^- \qquad (5.640)$$

Supposedly the replacement of chlorine atoms by amino groups should occur most readily when the M—Cl bond is markedly ionic in character, but the fractional ionic character of the Group IV halides increases regularly from silicon to thorium tetracloride. However, by taking into consideration the radius of the different atoms, the calculated specific charge density is in the reverse order. The coordination energy available for the ionization of M—Cl and N—H bonds, and hence the observed orders of reactivity and the degree of displacement are as should be anticipated.

An alternate mechanism involving the reaction of the amide ion with the halide molecule has been suggested [498]. The chloride ion is subsequently discharged and the amide ion retained. Thus

$$NH_2^- + MCl_4 \rightarrow [NH_2 \cdots MCl_3 \cdots Cl]^- \rightarrow NH_2 \cdot MCl_3 + Cl^-$$
$$(5.641)$$

Increased ammonolysis upon addition to the solution of amide ions in the form of alkali metal amides support this mechanism, while somewhat opposed to it is the fact that the variation of temperature, which should influence the autoionization, shows no effect. Ionization of the $M = Cl$ is predicted by both schemes. The mechanism of Bradley *et al.* [500] assumes the initial co-ordination of the ammonia molecule followed by the ionization of the M—Cl and N—H bonds, while the mechanism of Fowles and Pollard [498] involves the initial coordination of an amide ion, followed by the transmission of the amide ion charge to the chlorine atom which splits off as an ion.

An amide group replaces another atom or group, for example, a halogen atom in a molecule, in both ammonolysis and amination which are thus synonymous. In 1936 aromatic halides were first animated using alkali metal halides in liquid ammonia [501]. Chloro-, bromo-, and iodobenzene were converted into aniline and several byproducts, including diphenylamine, triphenylamine, and *p*-phenylphenylamine. No reaction was obtained with fluorobenzene. Using the above procedure meta-amino ethers were prepared from ortho-halogenated ethers [502]. The entering amide displaces a hydrogen atom ortho to the halogen atom; also ether solutions of aryl halides were attacked by lithium dialkylamides [503]. The reaction was reviewed [504] in 1951, and a mechanism proposed for the rearrangement in which the migration of the hydride ion from carbon 2 to carbon 1 and its consequent displacement of the halide ion is forced by the attack by the amide ion on carbon 2 of a halo-benzene as

$$(5.642)$$

Since about equal portions of aniline-1-^{14}C and aniline-2-^{14}C were produced when chlorobenzene-1-^{14}C was aminated with potassium amide in liquid ammonia, it was suggested that benzyne was a probable intermediate in the amination of unsubstituted halobenzenes [504]. Thus

$$\text{(5.643)}$$

$$\text{(5.644)}$$

The same results were observed with chlorobenzene-1-^{14}C, bromobenzene-1-^{14}C, and iodobenzene-1-^{14}C, and in deuterated bromobenzene it was found that the percentage of deuterium increased with percentage of reaction [505]. The ratio of the velocity constants for the amination of the undeuterated and deuterated material was $k_H'/k_D' = 5.5$. In dilute solution for partial reaction chlorobenzene-2-D showed a small increase in deuterium content, but showed a decrease in deuterium content in concentrated solutions. This was explained on the basis of exchange of deuterium by chlorobenzene-2-D with the solvent at a rate comparable with the amination. For chlorobenzene-2-D, k_H'/k_D' was 2.70. Fluorobenzene-2-D exchanged deuterium with the solvent very rapidly [506]. The mechanism given for the amination and exchange of deuterium in the case of chlorobenzene was given:

$$\text{(5.645)}$$

The procedure used in calculating k_H'/k_D' was as follows. The equations for the rates of disappearance of chlorobenzene-2-D (D) and ordinary benzene (H) were given as

$$-\frac{d[D]}{dt} = [k_D'+k_H'][D][NH_2^-]^n \qquad (5.646)$$

$$-\frac{d[H]}{dt} = 2k_H'[H][NH_2^-]^n \qquad (5.647)$$

Taking n as unity, dividing Eq. (5.646) by Eq. (5.647), rearranging the quotient, and integrating k_H'/k_D' becomes

$$k_H'/k_D' = \tfrac{1}{2}\{\ln([D]_t/[D]_0)/\ln([H]_t/[H]_0)\} - 1 \qquad (5.648)$$

It was assumed in this derivation that all the reactions were constant volume processes, the reactions were first order with respect to halobenzenes, in the overall amination reactions only hydrogen or deuterium ortho to the halogens are involved, and the reaction of benzyne to give products is neither rate determining nor reversible.

A kinetic rate expression for mechanism 5.645 was derived [505] to account for the change in the deuterium content of chlorobenzene during amination. Chlorobenzene-2-D disappeared at the rate

$$-\frac{d[D]}{dt} = (k_H'+k_D')[D][NH_2^-] - k_{-1}'[\text{anion IV}] \qquad (5.649)$$

and applying the steady state assumption

$$[\text{anion IV}] = \frac{k_H'[D][NH_2^-]}{k_{-1}'+k_2'} \qquad (5.650)$$

Substituting Eq. (5.650) into Eq. (5.649) yields

$$-\frac{d[D]}{dt} = (k_H'+k_D')[NH_2^-] - \frac{k_{-1}'k_H'}{k_{-1}'+k_H'}[D][NH_2^-] \qquad (5.651)$$

Similarly the rate of disappearance of ordinary chlorobenzene is found to be

$$-\frac{d[H]}{dt} = 2k_H'[H][NH_2^-] - \frac{k_{-1}'}{k_{-1}'+k_2'}(2k_H'[H][NH_2^-]+k'[D][NH_2^-]) \qquad (5.652)$$

Setting the fraction of the intermediate anion returning to initial reaction $= F = k_{-1}'/(k_{-1}'+k_2')$, and the isotope effect $= i = k_H'/k_D'$ gives

$$\frac{d[D]}{d[H]} = \frac{[i(1-F)+1][D]}{2i[(1-F)[H]-F[D]} \qquad (5.653)$$

Dividing Eq. (5.653) by [D] and assuming that $[i(1-F)+1]$, $2i(1-F)$, and $-F$ const, yields after integration

$$\ln[D]\Big|_{D_o}^{D_r} + \frac{1+(1-F)i}{1-(1-F)i}\ln\left[(-1+(1-F)i\frac{[H]}{[D]}-F\right]\Bigg|_{[H]_o/[D]_o}^{[H]_t/[D]_t} = 0$$

(5.654)

using the symbol d for the fraction of the chlorobenzene deuterated at time t and the symbol P for the fraction of starting material that has reacted in time t, Eq. (5.654) becomes

$$\ln\frac{d}{d_0}(1-P) = \frac{i(1-F)+1}{i(1-F)-1}\ln\frac{(1/d)-1-[F/\{i(1-F)-1\}]}{(1/d_0)-1-[F/\{i(1-F)-1\}]}$$

(5.655)

The change of concentration of deuterium in unreacted chlorobenzene is given by the above equation. To calculate the aniline composition indicated by mechanism (5.645), deuterated and undeuterated aniline were respectively represented [506] by A_D and A_H. Then

$$-\frac{d[A_D]}{dt} = k_2'[\text{anion IV}]$$

(5.656)

Dividing Eq. (5.656) by Eq. (5.649) yields

$$-\frac{d[A_D]}{d[D]} = \frac{k_2'[\text{anion IV}]}{(k_H'+k_D')[D][NH_2^-]-k_{-1}'[\text{anion IV}]}$$

(5.657)

Substituting the steady-state expression for [anion IV] and introducing i and F where feasible yields

$$-\frac{d[A_D]}{d[D]} = \frac{i(1-F)}{i(1-F)+1}$$

(5.658)

If $[A_D] = 0$, when $[D] = [D]_0$, Eq. (5.658) upon integration becomes

$$[A_D] = \frac{i(1-F)}{i(1-F)+1}([D]_0-[D])$$

(5.659)

Letting

$$P = \frac{[A_D]+[A_H]}{[D]_0+[H]_0}$$

and

$$(1-P) = \frac{[D]+[H]}{[D]_0-[H]_0}$$

Eq. (5.659) can be written

$$\frac{[A_D]}{[A_D]+[A_H]} = \frac{1}{P}\frac{i(1-F)}{i(1-F)+1}\frac{[D]_0}{[D]_0+[H]_0} - (1-P)\frac{[D]}{[D]+[H]}$$

(5.660)

If a equal the fraction of analine and d the fraction of chlorobenzene, respectively, which are deuterated, Eq. (5.660) simplifies to

$$a = \frac{i(1-F)}{i(1-F)+1} \frac{d_0 - d(1-P)}{P} \qquad (5.661)$$

Both the unreacted starting material and the product aniline were analyzed for deuterium when mixtures of ordinary chlorobenzene and chlorobenzene-2-D were subjected to partial amination by sodamide in a liquid ammonia [506]. The deuterium in the aniline was found to be distributed almost equally between the ortho and meta positions. This data supports mechanism (5.645) in which the slow step is the formation of an intermediate, for example, benzyne symmetric with respect to carbon atoms 1 and 2. For data see Table 5.28 where a was calculated applying Eq. (5.661).

It was found [507] in the ammonolysis of ethylene chlorohydrin (I) that the yield varied directly with the NH_3/I ratio, but indicated an optimum at approximately 8. The optimum temperature appeared to be 90°C. The reaction was independent of the ammonia concentration for any NH_3/I ratio. It was found that liquid ammonia alone was a far less effective reagent in ammonolysis than aqueous ammonia.

Employing a mercury lamp and passing a stream of ammonia through the reaction flask at a pressure, the photochemical ammonolysis, and p-chloro-phenetol, o-chloroanisole, o-bromoanisole, and p-bromoanisole in 25% methanol was studied [508] at 30 to 40°. The reaction yielded primary aromatic amines and was zero order. The primary process was photolysis and the secondary process ammonolysis since the reaction was the same in nitrogen as in ammonia stream.

$Na_4P_3O_9NH_2$ was prepared [509] by treating solutions of $Na_3P_3O_9$ with concentrated aqueous ammonia. Using 0.27 M $Na_3P_3O_9$ and 5.8 M NH_3 the formation at 26°C of $Na_4P_3O_9NH_2$ was followed by nuclear magnetic resonance measurements. The cyclic P_3O_9 was regenerated and NH_4^+ was produced when solutions of $NA_4P_3O_9NH_2$ were acidified.

The ammonolysis rates of methyl-β-D-glucuronoside-3,6-lactone (I), methyl-α-D-glucuronoside-3,6-lactone (II), and 1,2-O-isopropylidene-α-D-glucuronoside-3,6-lactone (III) were found [510] to decrease in the order II > III > I. Ammonolysis has been used in the production of alkylamines [511], and in the conversion of rosin acids to nitrides [512].

Included among reactions in liquid ammonia are those in which nitrogen from ammonia is exchanged with the nitrogen in carboxylic acid amides for example [513]. In this reference the aminolysis of p-nitrobenzamide, p-$O_2NC_6H_4CONH_2$ in liquid ammonia was studied kinetically, and at 20°C the first-order rate constant was found to be $k' = 1.27 \times 10^{-8}/sec$. Without the presence of NH_4Cl as a catalyst, higher temperatures are necessary. For this experiment the p-nitrobenzamide was 0.13 M, the ammonium chloride was 3.33 M, and nitrogen 15 constituted 12.23% of the nitrogen in the p-nitrobenzamide. The reaction is a form of ammonolysis in which H^+ ion and NH_3 from ammonium ion are added to the acidamide as follows [513]:

$$R-\overset{O}{\overset{\|}{C}}\underset{^{15}NH_2}{} + H^+ \rightleftharpoons \left[R-\overset{\overset{+}{O}H}{\underset{^{15}NH_2}{C}} \longleftrightarrow R-\overset{OH}{\underset{\overset{+}{^{15}NH_2}}{C}} \; ; \; R\overset{O}{\overset{\|}{}}\underset{\overset{+}{^{15}NH_3}}{} \right] + NH_3$$

(5.662)

$$\left[R-\overset{\overset{+}{O}H_2}{\underset{^{15}NH_2}{\overset{|}{\underset{|}{C}}}}-NH_2 \; ; \; R-\overset{OH}{\underset{NH_2}{\overset{|}{\underset{|}{C}}}}-{}^{15}\overset{+}{NH_3} \right] \xrightarrow{\begin{array}{c} -NH_4^+ \\ \\ \\ -{}^{15}\overset{+}{NH_4} \end{array}} \begin{array}{c} R-\overset{O}{\overset{\|}{C}}\underset{^{15}NH_2}{} \\ \\ R-\overset{O}{\overset{\|}{C}}\underset{NH_2}{} \end{array}$$

The kinetics of the reaction of benzylchloride, p-nitrobenzylchloride, and methyl and ethyl iodides with fourteen nitrogen bases in nitrobenzene–ethyl alcohol mixtures were studied [514]. The rate constants of these S_N2 reactions could be fitted by the Swain–Scott and Edward's relation.

In a study of the effects of temperature pressure, and starting composition on the rate of hydrogenative amination of acetone and isopropyl alcohol over nickel catalyst, it was found [515], that for the acetone reaction hydrogenation of the amine was the rate determining step; while for the isopropyl alcohol its dehydrogenation, which has a high activation energy, limited the rate.

In amination and ammonolysis, as in other reactions, the solvent or media in general is indeed influential, for example, the presence of water or of a salt as NH_4Cl influences the effectiveness and the speed of the reaction.

Two other examples of solvolysis will be cited to emphasize how solvolysis may alter the nature of the reacting species and thus determine the mechanism and rate of a reaction.

The kinetics of the reaction

$$U(IV) + Tl(III) \rightarrow U(VI) + Tl(I) \tag{5.663}$$

was studied [516] in aqueous perchloric acid solution and found to comply with the equation

$$-\frac{d[U(IV)]}{dt} = [U^{4+}][Tl^{3+}](k_1'[H^+]^{-1} + k_2'[H^+]^{-2}) \tag{5.664}$$

The two reaction paths

$$U^{4+} + Tl^{3+} + H_2O \xrightleftharpoons{k_1'} (U \cdot OH \cdot Tl)^{6+} + H^+ \qquad (5.665)$$

$$U^{4+} + Tl^3 + H_2O \xrightleftharpoons{k_2'} (U \cdot O \cdot Tl)^{5+} + 2H^+ \qquad (5.666)$$

involving activated complexes of the compositions $(U \cdot OH \cdot Tl)^{6+}$ and $(U \cdot O \cdot Tl)^{5+}$ were identified with the rate constants in Eq. (5.664).

The rate data for the same reaction in 25% by weight methanol in water could be reproduced using the equation [517]

$$-\frac{dU(IV)}{dt} = k_t'[U^{4+}][Tl^{3+}][H^+]^{-1} + k_2'[U^{4+}][Tl^{3+}][H^+]^{-2}$$

$$(5.667)$$

which can be converted to

$$-\frac{d[U(IV)]}{dt}\left[\frac{([H^+]+K_U)([H^+]+K_{Tl})^2}{[U(IV)][Tl(III)]^2[H^+]}\right] = k_1'\left[\frac{[H^+]+K_{Tl}}{[Tl(III)]}\right] + k_2'$$

$$(5.668)$$

Plotting the left-hand side of Eq. (5.668) versus $([H^+]+K_{Tl})/[Tl(III)]$ yielded a straight line which, at a given temperature, gives k_1' as the slope and k_2' as the intercept. The values of K_U and K_{Tl}, the hydrolysis constants of U^{4+} and Tl^{3+}, were available in the literature. For 25% by weight methanol,

$$U^{4+} + Tl^{3+} + H_2O \xrightleftharpoons{k_1} (U \cdot OH \cdot Tl)^{6+} + H^+ \qquad (5.669)$$

$$U^{4+} + Tl^{3+} + 2H_2O \xrightleftharpoons{k_2'} (Tl \cdot HO \cdot U \cdot OH \cdot Tl)^{8+} + 2H^+ \quad (5.670)$$

Table 5.29 contains the data for the 25% MeOH–75% H_2O and for pure water. From the rate expressions, and the data in Table 5.29a, b it is evident that changing the solvent from pure water to 25% methanol–75% water by weight changed the rates, the mechanisms and the thermodynamic quantities for the chemical reaction. The basic differences in the rates, the mechanisms

Table 5.29a

Kinetic Data for the Oxidation of U(IV) by Tl(III), Ionic Strength Is 2.9 in 25% Methyl Alcohol–75% Water Media

Path[a]	15°C	20°C	25°C	Z	ΔH^* (cal/mole)	ΔS^* (e.u.)	ΔF^* (cal/mole)
1	0.0037	0.0040	0.0041	0.79	3112	−59.0	20,694
2	0.75	0.85	0.98	1871	4474	−43.6	17,467

[a] Path 1 refers to k_1' and path 2 to k_2'. Units of k_1' and k_2' are sec^{-1}.

Table 5.29b

Kinetic Data for the Oxidation of U(IV) by Tl(III), Ionic Strength 2.9
in Water Media

Path[a]	k'			ΔH^* (cal/mole)	ΔS^* (e.u.)	ΔF^* (cal/mole)
	16°C	20°C	25°C			
1	0.57	1.02	2.11	24,600	16	19,700
2	0.67	1.17	2.13	21,700	7	19,700

[a] Path 1 refers to k_1' and path 2 to k_2'. Units of k_1' and k_2' are sec^{-1}.

and the thermodynamic quantities is the difference in the solvolytic processes for the two cases, Eqs. (5.665) and (5.666) for the one case and Eqs. (5.669) and (5.670) for the other case. Increased weight percent of methanol in the solvent induced still greater changes in the kinetic factors of this reaction. As the dielectric constant of the media is lowered by increasing amounts of methanol, forces between oppositely charged ions, e.g., OH$^-$ and U^{4+} and OH$^-$ and Tl^{3+} will enhance the formation of more complex intermediates, as is implied in the two activation processes in the two media.

The kinetics for the reduction of neptunium(VI) to neptunium(V) with, EDTA, ethylenediaminetetraacetic acid [518], and with oxalic acid [519] in aqueous perchloric acid have been studied. A solvolytic step involving solvent water and the reactants was rate-controlling in each case, also in each case the reaction was found to be first order with respect to each of the reactants, but inversely 1.5 order and first order, respectively, with respect to hydrogen ion. As to stoichiometry, 1 mole of ethylene diaminetetraacetic acid was found to react with six moles of neptunium(VI), and 1 mole of oxalic acid was found to react with 4 moles of neptunium(VI).

That initial complexes were formed between the reactants in the two reactions were evidenced by the color change in each case upon mixing the reactants. Neptunium(VI) has a pink color in perchloric acid, but when mixed with ethylenediaminetetraacetic acid the color became yellow, as was also the case when it was mixed with oxalic acid.

The equations used for calculating the apparent second-order rate constants at constant acidity for the ethylenediametetraacetic acid and for oxalic were respectively

$$k' = \frac{1}{t} \frac{2.303 \times 6}{a - 6b} \log \frac{6b(a-x)}{a(6b-x)} \qquad (5.671)$$

and

$$k' = \frac{1}{t} \frac{2.303 \times 4}{(a-4b)} \log \frac{4b(a-x)}{a(4b-x)} \qquad (5.672)$$

where in both equations a is the initial concentration of neptunium(VI), and x is the concentration of neptunium(V) produced in time t. In Eqs. (5.671) and (5.672), b is the respective initial concentrations of ethylenediammetetra-acetic acid and of oxalic acid.

The respective mechanisms for the two reactions were written

$$NpO_2^{2+} + EDTA + H_2O \xrightarrow{\text{slow}} NpO_2(OH)EDTA^+ + H^+ \quad (5.673)$$

$$NpO_2^{2+} + EDTA + 2H_2O \xrightarrow{\text{slow}} NpO_2(OH)_2EDTA + 2H^+ \quad (5.674)$$

$$10NpO_2^{2+} + NpO_2(OH)EDTA^+ + NpO_2(OH)_2EDTA \xrightarrow{\text{fast}} \text{products} \quad (5.675)$$

and

$$NpO_2^{2+} + H_2C_2O_4 + H_2O \xrightarrow{\text{slow}} NpO_2(OH)H_2C_2O_4^+ + H^+ \quad (5.676)$$

$$3NpO_2^{2+} + NpO_2(OH)H_2C_2O_4 \xrightarrow{\text{fast}} 4NpO_2^+ + 2CO_2 + 3H^+ + 0.5O_2 \quad (5.677)$$

$$4NpO_2^{2+} + H_2C_2O_4 + H_2O \longrightarrow 4NpO_2^+ + 2CO_2 + 4H^+ + 0.5O_2 \quad (5.678)$$

In the last mechanism, carbon dioxide was detected as a product of the

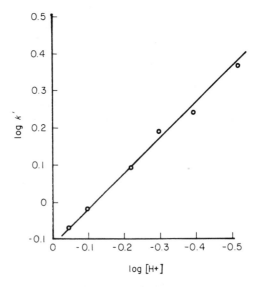

Fig. 5.21. Effect of hydrogen ion concentration on the apparent rate constant k' at 25°C and $\mu = 1$ in the reduction of neptunium(VI) with oxalic acid.

reaction, but oxygen, while not detected, was included as a product in order to account for the stoichiometry of the reaction. Even at complete reaction oxygen would have been of the order of 10^{-3} M, and it would have been difficult to distinguish it from atmospheric oxygen although no runs free of atmospheric oxygen were attempted. The product Np(V) was used in the spectrophotometric analysis of the reaction as a function of time by observing the growth of the characteristic Np(V) peak at 983 mμm using constant slit width. Figure 5.21 shows the effect of the hydrogen ion concentration on the rate constant at 25°C and ionic strength $\mu = 1$. The range of hydrogen ion concentration under these conditions within which this reaction can be studied is limited to between 0.3 and 0.9 M. If the hydrogen ion is decreased below 0.3 M, complex formation between neptunium(V) and potassium oxalate affects the spectrophotometric analysis; at hydrogen ion concentration greater than 0.9 M oxalic acid precipitates. Table 5.30 gives the rate constant

Table 5.30

Specific Velocity Constants at 25°C and $\mu = 1$ in the Reduction of Neptunium(VI) with Oxalic Acid in Perchloric Acid

10^3 [Np-(VI)](M)	10^3 [H$_2$-C$_2$O$_4$](M)	10^3 [H$^+$] (M)	k' (mole/sec)	k (sec^{-1})
2.72	20	402	0.029	0.012
1.36	20	402	0.028	0.011
10.96	20	402	0.032	0 013
5.44	20	402	0.028	0.011
2.72	20	402	0.028	0.011
2.72	15	402	0.028	0.011
2.72	10	402	0.030	0.012
2.72	5	402	0.027	0.011
2.72	0.2	402	0.024	0.011
2.72	20	900	0.014	0.013
2.72	20	800	0.016	0.013
2.72	20	600	0.020	0.013
2.72	20	500	0.026	0 013
2.72	20	300	0.038	0.012
				Av 0.012±0.001

k'/mole/sec and k'/sec at 25°C and $\mu = 1$ as a function of the concentration of reactants and of hydrogen ion. The pseudounimolecular rate constant k'/sec values were calculated since oxalic acid was usually in excess especially when stoichiometry is considered. The energy of activation for the reaction was 15.5 kcal/mole.

XXII. Nitration of Aromatic Compounds and the Solvent

It is important to discuss the nitration of aromatic compounds with respect to the effect of the solvent upon rates and mechanisms since (1) the reaction has been more extensively studied than any other reaction of aromatic compounds, (2) in the rate and mechanism of the reaction the solvent plays such an important role, and (3) there has been a rather complete clarification of the mechanism.

The nitronium ion, NO_2^+, was suggested [520] as the reactive species in nitration as early as 1903. That the substitution in the ring of electron withdrawing groups such as the nitro group reduced the ease with which nitration occurred, suggested that a positive ion or a strong dipole was an active intermediate which reacted with the benzene ring in nitration. Based on an incorrect value of about 3 for the Van't Hoff factor, i, it was concluded [521] that $H_2NO_3^+$ and $H_3NO_2^{2+}$ were the ions present in nitric acid dissolved in sulfuric acid. In addition to freezing-point depression data, these conclusions were based on ultraviolet absorption and on conductance. Orientation by substitutients was used to support the positive-ion mechanism [522], by calculating fairly accurately the ortho–para to meta ratio for the nitration of several substituted benzenes on the basis of electrostatic interactions, assuming a positive ion reactant and using bond distance and dipole moment data. Not present in the Raman spectra of the pure substances but found [523] in mixtures of nitric and sulfuric acids were two Raman spectrum lines at 1050/cm and 1400/cm. Perchloric and selenic acid solutions of nitric acid showed [524] the 1400/cm line, which must be common to a substance present in all strong acid solutions of nitric acid. Thus the two lines do not originate from a common molecular source. A careful study showed that the source of the 1400/cm frequency has no other line in its spectrum. The nitronium ion is the unambiguous source of the 1400/cm frequency since no other structured derivable from the elements in nitric acid would yield the observed spectral characters as produced by the use of sulfuric acid; there are several other lines in the Raman spectrum of the 1050/cm frequency, and these lines were found to identify the hydrogen sulfate ion HSO_4^-. From freezing point data it was found [525] that nitric acid reacts with sulfuric acid quantitatively as shown by the equation

$$HNO_3 + H_2SO_4 \rightarrow NO_2^+ + H_3O^+ + 2HSO_4^- \qquad (5.679)$$

with the equilibrium

$$H_3O^+ + HSO_4^- \rightleftharpoons H_2O + H_2SO_4 \qquad (5.680)$$

being established, because water is only a semistrong base and a fraction of the H_3O^+ and HSO_4^- ions recombine, and hence the Van't Hoff factor is not 4 as would be the case if Eq. (5.679) were completely quantitative. The factor was found to have a mean value of 3.77 over the range of concentrations used.

The existence of its stable salts [524, 526–528] further substantiates the actuality of the nitronium ion. From Raman spectra of the solid material, N_2O_5 has been identified as NO_2NO_3.

Either strong acid solutions of nitric acid or nitronium salts can be used as a source of the nitration reagenet, i.e., of nitronium ion. The rates of nitration by nitric acid of aromatic compounds in sulfuric acid–water mixtures increase with increasing sulfuric acid up to about 90% sulfuric acid [529] and then decrease with further increase of sulfuric acid.

The ionization of nitric acid in sulfuric acid to form nitronium ion is analogous to the ionization of a triarylcarbinol indicator [530]. The increase of nitration of nitrobenzene by a factor of about 3000 when the sulfuric acid is increased from 80 to 90% parallels the ionization of trinitrotriphenylcarbinol in the same media, but does not parallel the ionization of anthraquinone which does not ionize in sulfuric acid in a manner similar to that of nitric acid when the latter forms the nitronium ion [530].

In the region of 90% sulfuric acid, nitric acid, trinitrophenylcarbinol, and anthraquinone ionize, respectively, as shown [525]:

$$HNO_3 + 2H_2SO_4 \rightarrow NO_2^+ + H_3O^+ + 2HSO_4^- \qquad (5.681)$$

$$(O_2NC_6H_4)_3COH + 2H_2SO_4 \rightarrow (O_2NC_6H_4)_3C^+ + H_3O^+ + 2HSO_4^-$$
$$(5.682)$$

$$(5.683)$$

The similarity in the ionization of nitric acid and of trinitrotriphenylcarbinol is evident. The nitronium ion in the case of nitric acid is replaced by the triphenylcarbonium ion in the case of trinitrotriphenylcarbinol. There is no similarity, however, in the ionization of nitric acid and anthraquinone. With increasing concentration of sulfuric acid the rate of nitration parallels the ionization of trinitrotriphenylcarbinol, but not that of the ionization of anthraquinone. Thus that the rate of nitration paralleled the increase in nitronium ion concentration is a logical assumption. The rate of nitration reaches a maximum at that concentration of sulfuric acid where the ionization to nitronium ion is complete, and further increase in sulfuric acid concentration causes no further increase in rate of nitration. A decrease in rate after the maximum was explained [531] on the assumption that the removal or the proton was part of the rate controlling step:

$$NO_2^+ + ArH + SO_4H^- \rightarrow ArNO_2 + H_2SO_4 \qquad (5.684)$$

Thus the composition of the media in which the rate of nitration is a maximum is presumably at that point where there is a balance between the acid function of the solvent and its basic function as a proton acceptor. There is no unanimity of opinion concerning the removal of the proton being part of the rate controlling step [532, 533]. If the rate controlling step involves the removal of the proton as given in Eq. (5.684), then for solutions richer than 90% in sulfuric acid, the concentration of HSO_4^- should decrease rapidly with increasing sulfuric acid concentration since the bisulfate ion is produced chiefly by the loss of a proton to water to produce hydronium ion and bisulfate ion. Based on this theory addition of bisulfate ion should accelerate the rate of nitration in concentrated sulfuric acid solutions, but should decelerate the rate of nitration in solutions more dilute in sulfuric acid than the optimum amount [530, 531]. The acidity of the medium as measured by the H_0 function is the determining factor for nitration in sulfuric acid [530]. Based on the hypothesis that the maximum rate occurs at a definite value of H_0, the acidity, rather than a definite composition of acid, the influence of the bisulfate ion on the rate of nitration is understandable. Thus, when the acidity is less than the optimum, addition of bisulfate ion, which lowers the acidity, will cause a decrease in the rate; but when the acidity is greater than the optimum, bisulfate ion will increase the rate. Dinitrobenzene, phosphorus pentoxide, and other substances which do not appreciably affect the acidity of the medium, have little effect upon the rate.

Some objections have been raised [412] to the assumption that the removal of the proton is involved in the rate determining step. First, such a hypothesis would predict that the maximum in the rate would be a function only of the solvent and independent of the aromatic compound involved. Experimentally, this is not quite true. Second, such an assumption would imply the constancy, independent of solvent composition, of the ratio of rate constants for two compounds which is far from true. Also, it was shown [532] that in nitration, tritium was displaced from the ring as readily as was hydrogen, which is a more direct proof that the removal of the proton is not part of the rate determining step. If such were the case, the rate of nitration would be slower in the tritium substituted compound since the tritium ion would be less readily removed than the proton. The existence of a dissociation equilibrium cannot explain the observations since there is no exchange of hydrogen with the acids and the water. It is concluded [532] that the rate determining step is the addition of the nitronium ion which is followed by a rapid splitting off of the proton.

The formation of unreactive positively charged or hydrogen-bonded species is apparently the most acceptable explanation of the decrease in rate at higher percentages of sulfuric acid [412]. The species $C_6H_5NO_2H^+$, $C_6H_5NO_2H_2SO_4$, $CH_3C_6H_4NO_2H^+$, and $CH_3NO_2H^+$ have been discussed [534–536]. It has been suggested [412] that a relation exists between the basicity of the organic molecule and the rate of fall in the rate of nitration after the maximum in the rate is attained. The more basic a compound, the steeper is the maximum. The work of Baker and Hey [537] wherein the increase of meta nitration for benzaldehyde, acetophenone, and ethyl benzoate with increase of sulfuric acid in the nitrating mixture is explained as due to the conversion of these substances to positively charged ions, thus enhancing the tendency toward meta orientation.

Further it is argued [412] that since sulfuric acid is more polar than water, the charge would be spread out in the activated complex for a rate-determining step for reaction between a neutral molecule and a positive ion in going to the more polar solvent, and that hence there would be a decrease in rate with increasing sulfuric acid. The destruction of charge would cause a greater decrease in rate upon going to a more polar solvent.

A mechanism for the nitration of aromatic compounds when nitronium ion is formed was written [533]

$$2HNO_3 \underset{k_2'}{\overset{k_1'}{\rightleftharpoons}} H_2NO_3^+ + NO_3^- \text{ (fast)} \qquad (5.685)$$

$$O_2N \cdot H_2O \underset{k_4'}{\overset{k_3'}{\rightleftharpoons}} O_2N^+ + OH_2 \text{ (slow)} \qquad (5.686)$$

$$NO_2^+ + ArH \xrightarrow{k_5'} Ar^+\!\!\begin{smallmatrix}H\\ \\NO_2\end{smallmatrix} \qquad \text{(slow)} \qquad (5.687)$$

$$Ar^+\!\!\begin{smallmatrix}H\\ \\NO_2\end{smallmatrix} \xrightarrow{k_6'} ArNO_2 + H^+ \text{ (fast)} \qquad (5.688)$$

To date this mechanism seems to be in agreement with facts. Unlike the mechanism depicted in Eq. (5.684), the above mechanism does not show a high solvent sensitivity. While the process represented by rate constant k_2' Eq. (5.685) would be solvent sensitive as would be the reaction rate depicted by Eq. (5.684), in the above mechanism, the reverse step in Eq. (5.685) is not rate controlling. The reaction shown in Eq. (5.687), in agreement with observation, shows a slight solvent effect due to the spreading of the charge in the transition state. One of the nitric acid molecules in Eq. (5.685) might be replaced by a molecule of any strong acid, the anion of which would replace the nitrate ion in the reaction. The observed hyperbolic repression of the rate of nitration by nitrate ions (rate $\propto C^{-1}$) as would be expected from the reverse reaction in Eq. (5.685), without disturbing the reaction order even when the order is zero, further supports the above mechanism.

While exact solutions for Eqs. (5.685)–(5.688) cannot be realized, approximate solutions for the limiting cases in which k_5' is much less than k_3' or k_4' and in which k_5' is much greater than k_3' or k_4' have been given [412].

Only in the presence of relatively large amounts of water does the reverse step in Eq. (5.686) occur, and then the zeroth order reaction becomes a first-order reaction in which the kinetic effect of water on first-order nitration is similar to nitrate ion [538]. The specific retardation by nitrate ions from ionizing nitrates and the acceleration by sulfuric acid of aromatic nitration in nitric acid, nitromethane, or acetic acid, with a constant excess of nitric acid, obey limiting linear laws, and occur alike with zeroth and first-order reactions, always without disturbance of the reaction order. Thus in the conversion of nitric acid to nitronium ion, one nitrate ion is eliminated and one proton used. These steps precede the step in which nitronium ion is produced since a zeroth order nitration only maintains its kinetic order where the latter step is not reversed.

Kinetically aromatic nitration is second order in sulfuric acid, first order in nitric acid, and is zeroth order for sufficiently reactive aromatic compounds in either nitromethane or acetic acid with nitric acid in constant excess. The formation and effectiveness of the nitronium ion, NO_2^+, is conclusively proven by the zeroth order kinetics in organic solvents, and its transition to first-order kinetics with decreasing aromatic activity. The mechanism represented by Eqs. (5.685)–(5.688) are supported by all the observations.

The nitration of 2,4,6-trinitroaniline by nitric acid in 90.57 to 97.47% sulfuric acid at $25.0 \pm 1°C$ was studied [539] and the product found to be N-nitro-2,4,6-trinitroaniline. The reaction was reversible, and the nitrating agent was protonized molecules of nitric acid. The equilibrium constant and rate constants for the reaction were determined. In this range of sulfuric acid concentration, the rate constant decreased rapidly with sulfuric acid concentration.

The rate of aromatic nitration in acetic anhydride was enormously accelerated if the nitrating solution is prepared from fuming nitric acid (when HNO_2 is developed), rather than pure nitric acid (HNO_2 concentration about $10^{-4} M$) [540]. It was suggested that nitration via nitrosation is accompanied by another process of nitration.

Aromatic nitration shows a varied dependence of rates and mechanisms on the solvent. Hydrogen bonding, complex formation, polarization, ionization, and the presence of nitrous acid are among the effects which the medium exerts on this type of reaction. The solvent can influence mechanism to the extent of causing reversibility in at least one step of the mechanism and of changing the order of the reaction.

XXIII. Aqueous Proton Transfers and the Solvent

In recent years several extensive discussions on various aspects of proton transfer reactions in aqueous solutions have appeared [541–546], including an important Faraday Society discussion of the whole subject [544].

Alberty [546] discusses principally the role of the solvent in proton transfer reactions as disclosed by (a) relaxation and electrochemical methods, (b) nuclear magnetic resonance, (c) Raman broadening, and (d) isotope effects in isotopically mixed solvents. He also discusses stepwise and concerted mechanisms in aqueous proton transfers. In a proton transfer of the type

$$H^+ + S \rightarrow SH^+ \tag{5.689}$$

alternative transition states A and B

$$\tag{5.690}$$

are presented for evaluation, and potential energy surfaces are given for a proton transfer of the type

$$AH + B \rightarrow A + BH \tag{5.691}$$

The Grothuss mechanism for proton mobility and the transfer of a proton from $H_9O_4^+$ to O or N on a base where O or N forms strong hydrogen bonds with water are considered under the type transfer given in Eq. (5.691). For the Grotthuss mechanism, the stepwise mechanism was selected over the concerted mechanism as the simpler hypothesis on the basis of the shapes of the energy profiles for the two cases.

The following changes were considered [546] for all the Grotthuss mechanisms for hydrogen transport:

or

$$(D) \quad H-\overset{\delta+}{\underset{\underset{H}{|}}{O_1}} \cdots H \cdots \overset{\delta+}{\underset{\underset{H}{|}}{O_2}} \cdots H \cdots \overset{\delta+}{\underset{\underset{H}{|}}{O_3}}-H$$

$$H-\underset{\underset{H}{|}}{O_1} \cdots H-\underset{\underset{H}{|}}{O_2} \cdots H-\underset{\underset{H}{|}}{O_3^+}-H$$

$$(5.692)$$

$$\underset{\underset{H}{|}}{O_1}-H \qquad H-\underset{\underset{H}{|}}{O_2} \cdots H-\underset{\underset{H}{|}}{O_3^+}-H$$

Since on the schematic energy surface the ridge between A and B has been shown with a higher energy of activation than the ridge between B and C, apparently the rotation of the water molecule is the rate determining step in proton conductance. That the rate of the reaction

$$H_3O^+ + OH^- \rightarrow 2H_2O \qquad (5.693)$$

in ice is between 10^{13} and $10^{14}/M/\text{sec}$ supports this model.

In the proton transfer from $H_9O_4^+$ to O or N on a base, where O or N forms strong hydrogen bonds with water, if the base has a $pK_A > 2$, the reaction is downhill. Base B replaces O_3—H in the mechanism Eq. (5.692), and the rate

$$|$$
$$H$$

determining step is the formation of the hydrogen bonded complex between the approaching $H_9O_4^+$ unit and the base. In this case the concerted mechanism for the final proton transfer is more likely, because of the symmetry involved [546].

There are cases, for example the "pseudoacids" which involve the ionization of a C—H bond, and for which there is the transfer of a proton to a receiving atom which does not form a hydrogen bond with the solvent. The ionization of the C—H bond can occur if the bond is adjacent to a group such as =C=O, —C≡N, or —NO$_2$, which can remove the negative charge from the carbon atom and locate it on the more electronegative oxygen or nitrogen atoms. A reaction of this type is [546, 547]

$$CH_3-\underset{\underset{O}{\|}}{C}-CH=C-CF_3 + H^+ \xrightarrow{k'} CH_3-\underset{\underset{O}{\|}}{C}-CH_2-\underset{\underset{O}{\|}}{C}-CF_3 \qquad (5.694)$$

for which the equilibrium and reaction rate constants are respectively $K = 2 \times 10^{-5}$ M, and $k' = 7.6 \times 10^2/M/\text{sec}$. The solvent is hydrogen bonded to the oxygen atom in the anion rather than to the carbon atom to which the proton is to become attached. Since there are no stable hydrogen bonds between the carbon and the incoming proton, there are no minima represented by depressions in the potential energy surface as is the case for other types of proton transfers. The distance on the potential surface between the carbon and the approaching proton cannot be shortened until the proton is actually being transferred from its position in the complex to its position on the carbon accepting the proton. It is thought [546] that H_3O^+ is the proton donor in proton transfers to nonhydrogen-bonded substrates, while $H_9O_4^+$ is the proton donor to hydrogen-bonded substrates.

In the protonation by $H_9O_4^+$, the removal of water from the secondary hydration sphere of $H_9O_4^+$, will require energy. Also more energy will be required in the final step to take the charge from adjacent electronegative group, destroy the water dipole lone pair interactions around this group, and move the electrons to the incoming group, which may all amount to a considerable activation energy, so that the reactions, while spontaneous, will not approach in rate diffusion controlled reactions. In fact, a reaction of the type represented by Eq. (5.694) may in extreme cases have a rate of the order of 10^8 less than that for reactions in which proton transfers do take place to bases which hydrogen bond with the solvent.

Further discussions of hydrated proton transfers have been [see Eq. (5.598)] presented in other sections of this book, however, for more details concerning the solvent effect on this type of reaction the reader is referred to the article by Alberty [546].

References

1. E. S. Amis, "Solvent Effects on Reaction Rates and Mechanisms," pp. 43–46. Academic Press, New York, Second Printing, 1967.
2. G. Åkerlöf, *J. Amer. Chem. Soc.* **48**, 3046 (1926); **49**, 2950 (1927); **50**, 1272 (1928).
3. A. J. Parker, *Advan. Phys. Org. Chem.* **5**, 8, 192 (1967).
4. G. Scatchard, *Chem. Rev.* **10**, 229 (1932).
5. J. N. Brönsted, *Z. Phys. Chem. Stoechim. Verwandschaftslehre* **102**, 169 (1922); **115**, 337 (1925).
6. K. J. Laidler and H. Eyring, *Ann. N. Y. Acad. Sci.* **39**, 303 (1940).
7. E. S. Amis and V. K. LaMer, *J. Amer. Chem· Soc.* **61**, 905 (1939).
8. E. S. Amis, "Kinetics of Chemical Change in Solution," Chapter IV. Macmillan, New York, 1949.
9. J. A. Christiansen, *Z. Phys. Chem. Stoechim. Verwandschaftslehre* **113**, 15 (1924).
10. E. S. Amis and J. E. Price, *J. Phys. Chem.* **47**, 338 (1943).
11. W. J. Svirbely and A. Shramm, *J. Amer. Chem. Soc.* **60**, 330 (1938).
12. J. Lander and W. J. Svirbely, *J. Amer. Chem. Soc.* **60**, 1613 (1938).
13. K. J. Laidler and H. Eyring, *Ann. N. Y. Acad. Sci.* **39**, 303 (1940).
14. K. J. Laidler, "Chemical Kinetics," pp. 213–214. McGraw-Hill, New York, 1965.

15. E. A. Moelwyn-Hughes, "Physical Chemistry," p. 1268. Pergamon, Oxford, Reprinted Second Edition, 1964.
16. W. J. Svirbely and J. C. Warner, *J. Amer. Chem. Soc.* **57**, 1883 (1935).
17. V. K. LaMer, *J. Franklin Inst.* **225**, 709 (1938).
18. E. S. Amis and F. C. Holmes, *J. Amer. Chem. Soc.* **63**, 2231 (1941).
19. E. S. Amis and J. E. Potts, *J. Amer. Chem. Soc.* **63**, 2833 (1941).
20. E. S. Amis, *J. Amer. Chem. Soc.* **63**, 1606 (1941).
21. V. K. LaMer and M. E. Kamner, *J. Amer. Chem. Soc.* **47**, 2662 (1935).
22. E. A. Moelwyn-Hughes, *Proc. Roy. Soc. Ser. A* **155**, 308 (1936); **157**, 667 (1936).
23. E. S. Amis and G. Jaffé, *J. Chem. Phys.* **10**, 646 (1942).
24. J. A. Christiansen, *Z. Phys. Chem. Stoechim. Verwandschaftslehre* **113**, 35 (1924).
24a. P. T. Stroup and V. W. Meloche, *J. Amer. Chem. Soc.* **53**, 3331 (1931).
24b. W. F. Timofeew, G. E. Muchin, and W. G. Gurewitsch, *Z. Phys. Chem. Stoechim. Verwandschaftslehre* **115**, 161 (1925).
24c. F. Daniels and E. H. Johnston, *J. Amer. Chem. Soc.* **43**, 43 (1921).
24d. F. H. MacDougall, "Physical Chemistry," p. 404. Macmillan Company, New York, 1936.
24e. C. N. Hinshelwood and R. E. Burk, *Proc. Roy. Soc. Ser. A* **106**, 284 (1924).
24f. A. R. Choppin, H. A. Frediani, and G. F. Kirby, Jr., *J. Amer. Chem. Soc.* **61**, 3176 (1939).
24g. M. Bodenstein, *Z. Phys. Chem. Stoechim. Verwandschaftslehre* **29**, 295 (1899).
24h. H. S. Taylor and H. A. Taylor, "Elementary Physical Chemistry," p. 242. Van Nostrand-Reinhold, Princeton, New Jersey, 1937.
24i. K. H. Geib and P. Harteck, *Z. Phys. Chem. Bodenstein Festband* 849–862 (1931).
25. E. S. Amis and S. Cook, *J. Amer. Chem. Soc.* **63**, 2621 (1941).
26. G. Scatchard, *Ann. N. Y. Acad. Sci.* **39**, 341 (1940).
27. G. Scatchard, *J. Chem. Phys.* **7**, 657 (1939).
28. V. K. LaMer, *Chem. Rev.* **10**, 179 (1932).
29. J. N. Brönsted, *Z. Phys. Chem. Stoechim. Verwandschaftslehre* **102**, 169 (1922); **115**, 337 (1925).
30. N. Bjerrum, *Z. Phys. Chem. Stoechim. Verwandschaftslehre* **108**, 82 (1924); **118** 251 (1925).
31. V. K. LaMer, *Chem. Rev.* **10**, 179 (1932).
32. A. Weller, *Progr. React. Kinet.* **1**, 189 (1961).
33. E. S. Amis, "Kinetics of Chemical Change in Solution," Macmillan, New York, 1949.
34. A. R. Olson and J. R. Simonson, *J. Chem. Phys.* **17**, 1167 (1949).
35. C. W. Davies, *Progr. React. Kinet.* **1**, 167 (1961).
36. E. S. Amis, "Solvent Effects on Reaction Rates and Mechanism," pp. 258–264. Academic Press, New York, Second Printing, 1967.
37. P. A. H. Wyatt and C. W. Davies, *Trans. Faraday Soc.* **46**, 140 (1950).
38. V. K. LaMer and R. W. Fessenden, *J. Amer. Chem. Soc.* **54**, 2351 (1932).
39. A. von Kiss and P. Vass, *Z. Anorg. Allg. Chem.* **217**, 305 (1934).
40. A. Indelli and E. S. Amis, *J. Amer. Chem. Soc.* **82**, 332 (1960).
41. V. Carasitti and C. Dejak, *Atti Acad. Sci. Ist. Bologna Cl. Sci. Fis. Rend.* **11**, 2 (1955).
42. W. J. Howells, *J. Chem. Soc. London* p. 203 (1946).
43. P. G. M. Brown and J. E. Prue, *Proc Roy. Soc. Ser. A* **232**, 320 (1955).
44. E. A. Guggenheim, *Discuss. Faraday Soc.* **24**, 53 (1957).
45. K. H. Stern and E. S. Amis, *Chem. Rev.* **59**, 1 (1956).
46. J. O. Wear and E. S. Amis, *J. Inorg. Nucl. Chem.* **24**, 903 (1962).
47. J. K. Laidler and H. Eyring, *Ann. N. Y. Acad. Sci.* **39**, 303 (1940).

48. E. Hückel, *Phys. Z.* **26**, 93 (1935).
49. J. G. Kirkwood, *J. Chem. Phys.* **2**, 351 (1934).
50. P. Debye and J. McAulay, *Phys. Z.* **26**, 22 (1925).
51. C. C. J. Fontein, *Rec. Trav. Chim. Pays-Bas* **47**, 635 (1928).
52. K. J. Laidler and P. A. Landskroener, *Trans. Faraday Soc.* **52**, 200 (1956).
53. K. J. Laidler, "Chemical Kinetics," 2nd ed., pp. 225–230. McGraw-Hill, New York, 1965.
54. J. G. Kirkwood, *J. Chem. Phys.* **2**, 35 (1934).
55. J. G. Kirkwood and F. H. Westheimer, *J. Chem. Phys.* **6**, 506 (1938).
56. J. N. Brönsted and W. K. Wynne-Jones, *Trans. Faraday Soc.* **25**, 59 (1929).
57. R. H. Stokes and R. A. Robinson, *J. Amer. Chem. Soc.* **70**, 1870 (1948); R. A. Robinson and R. H. Stokes, "Electrolyte Solutions," p. 246. Butterworth, London, 1959.
58. F. A. Long and W. F. McDewit, *Chem. Rev.* **51**, 119 (1952).
59. E. S. Amis and G. Jaffe, *J. Chem. Phys.* **10**, 598 (1942).
60. L. C. Bateman, M. G. Church, E. D. Hughes, C. K. Ingold, and N. A. Thaer, *J. Chem. Soc. London* p. 979 (1940).
61. L. Onsager, *J. Amer. Chem. Soc.* **58**, 1486 (1936).
62. E. S. Amis, *J. Chem. Educ.* **29**, 337 (1952); **30**, 351 (1953).
63. J. E. Quinlan and E. S. Amis, *J. Amer. Chem. Soc.* **71**, 4187 (1955).
64. C. P. Smythe, "Dielectric Constant and Moleculat Structures," pp. 193–199. Chem. Catalog Co. (Tudor), New York, 1931.
65. G. J. Nolan and E. S. Amis, *J. Phys. Chem.* **65**, 1556 (1961).
66. N. W. Sidgwick, *Trans. Faraday Soc.* **30**, Appendix (1934).
67. R. J. W. LaFerte, and H. Vine, *J. Chem. Soc. London* p. 1795 (1938).
68. H. L. Bender, *J. Amer. Chem. Soc.* **73**, 1626 (1951).
69. J. L. Hockersmith and E. S. Amis, *Anal. Chim. Acta* **9**, 101 (1953).
70. E. S. Amis, *Anal. Chem.* **27**, 1672 (1955).
71. E. S. Amis, G. Jaffee, and R. I. Overman, *J. Amer. Chem. Soc.* **66**, 1823 (1944).
72. W. J. Broach and E. S. Amis, *J. Chem. Phys.* **22**, 39 (1954).
73. W. S. Nathan and H. B. Watson, pp. 217, 890 (1933).
74. E. A. Moelwyn-Hughes, "Physical Chemistry," 2nd ed., pp. 1281–1286. Pergamon, Oxford, reprinted with corrections, 1964.
75. C. K. Ingold and W. S. Nathan, *J. Chem. Soc. London* p. 222 (1936).
76. D. P. Evans, J. J. Gordon, and H. B. Watson, *J. Chem. Soc. London* p. 1430 (1937).
77. C. K. Ingold and W. S. Nathan, *J. Chem. Soc. London* p. 222 (1936).
78. R. A. Ogg and M. Polanyi, *Trans. Faraday Soc.* **31**, 482 (1935).
79. D. N. Glew and E. A. Moelwyn-Hughes, *Proc. Roy. Soc. Ser. A* **211**, 254 (1952)
80. B. Jones, *J. Chem. Soc. London* pp. 1006, 1073 (1928).
81. S. W. Benson, "The Fundamentals of Chemical Kinetics," p. 536. McGraw-Hill, New York, 1960.
82. R. P. Bell, *Trans. Faraday Soc.* **31**, 1557 (1935).
83. A. R. Martin and B. Collie, *J. Chem. Soc. London* p. 2658 (1932).
84. A. R. Martin, *Trans. Faraday Soc.* **30**, 759 (1934).
85. A. R. Martin and C. M. George, *J. Chem. Soc. London* p. 1413 (1933).
86. A. R. Martin, *Trans. Faraday Soc.* **33**, 191 (1937).
87. A. R. Martin and A. C. Brown, *Trans. Faraday Soc.* **34**, 742 (1938).
88. H. G. Grimm, H. Ruf, and H. Wolff, *Z. Phys. Chem.* **813**, 301 (1931).
89. N. Menschutkin, *Z. Phys. Chem. Stoechim, Verwandschaftslehre* **6**, 41 (1890).
90. N. Bjerrum and E. Jozefawicz, *Z. Phys. Chem. Abt. B* **13**, 301 (1931).

91. P. Walden, *Izv. Imp. Akad. Nauk* **7**, 427 (1913).
92. A. F. Stearn and H. Eyring, *J. Chem. Phys.* **5**, 113 (1937).
93. E. S. Amis, *J. Chem. Educ.* **30**, 351 (1953); Anal. Chem. **27**, 1672 (1955).
94. E. D. Hughes, *J. Chem. Soc. London* p. 1925 (1935).
95. R. H. Wiswal, Jr., and C. P. Smythe, *J. Chem. Phys.* **9**, 356 (1941).
96. S. G. Entelis, R. P. Tiger, E. Ya. Nevel' skü, and J. V. Epelbaum, *Izv. Akad. Nauk SSSR, Otd. Khim. Nauk* p. 245 (1963).
97. S. W. Benson, "The Foundation of Chemical Kinetics," pp. 534–535. McGraw-Hill, New York 1960.
98. H. S. Frank and M. W. Evans *J. Chem. Phys.* **13**, 507 (1945); H. S. Frank and W. Y. Wen, *Discuss. Faraday Soc.* **24**, 133 (1957).
99. K. J. Laidler, *Can. J. Chem.* **37**, 138 (1959); K. J. Laidler and C. Pegis, *Proc. Roy. Soc. Ser A* **241**, 80 (1957).
100. E. Sacher and K. J. Laidler, *Trans. Faraday Soc.* **59**, 396 (1963).
101. A. J. Parker, Rates of bimolecular substitution reactions in protic and dipolar aprotic solvents. *In* "Recent Advance in Physical Organic Chemistry" (V. Gold, ed.), pp. 173–235, Academic Press, New York 1967.
102. J. Miller and A J. Parker. *J. Amer. Chem. Soc.* **83**, 117 (1961).
103. C. Reichardt, *Angew. Chem. Int. Ed. Engl.* **4**, 29 (1965).
104. K. J. Laidler, "Chemical Kinetics," 2nd ed., pp. 209–210. McGraw-Hill, New York 1965.
105. J. E. Leffler and E Gruenwald, "Rates and Equilibrium of Organic Reactions," pp. 263–314. Wiley, New York, 1963.
106. K. E. Wieburg, "Physical Organic Chemistry," p. 374–396. Wiley, New York, 1964.
107. E. M. Arnett, W. G. Bentrude J. J. Burke, and P. McC. Duggleby, *J. Amer. Chem. Soc.* **87**, 1541 (1965).
107. E. M. Arnett, W. G. Bentrude, J. J. Burke, and P. McC. Duggleby, *J. Amer. Chem. Soc.* **87**, 1541 (1965).
108 A. J. Parker, *J. Chem. Soc. A* p. 220 (1966).
109. E. M. Arnett and D. R. McKelvey, *J. Amer. Chem. Soc.* **87**, 1515 (1966).
110. A. J. Parker, *Quart. Rev. Chem. Soc.* **16**, 163 (1962).
111. A. J. Parker, *Advan. Org. Chem.* **5** (1965).
112. J. P. Evans and A. J. Parker, *Tetrahedron Lett.* p. 163 (1966).
113. E. Grunwald and S. Winstein, *J Amer. Chem. Soc.* **70**, 846 (1948).
114. S. Winstein, B. Apple, R. Baker, and A. Diaz, *Chem. Soc. Spec. Publ.* **19**, 109–130 (1965).
115. A J Parker, *Aust. J. Chem.* **16**, 585 (1963).
116. A. J. Parker and D. Brody, *J. Chem. Soc.* p. 220 (1966).
117. J. O. Edwards and R. G. Pearson, *J. Amer. Chem. Soc.* **84**, 16 (1962).
118. R. G. Pearson and J. Söngstad, *J. Amer. Chem. Soc.* **89**, 1827 (1967).
119. I. M. Kolthoff and S. Bruckenstein, "Treatise in Analitical Chemistry," Vol. I, Pt. I, p. 492. Wiley (Interscience), New York 1959.
120. I. M. Kolthoff, *J. Polarog. Soc.* **10**, 22 (1964).
121. H. Stehlow, *Z. Elektrochem.* **56**, 825 (1952).
122. E. Grunwald, G. Baughman, and G. Kohnstam, *J. Amer. Chem. Soc.* **82**, 5801 (1960).
123. H. M. Koepp, H. Wendt, and H. Strehlow, *Z. Elektrochem.* **64**, 483 (1960).
124. J. M. Kolthoff and F. G. Thomas, *J. Phys. Chem.* **69**, 3049 (1965).
125. V. A. Pleskov, *Usp. Khim.* **16**, 254 (1947).
126. V. A. Pleskov, *Zh. Fiz. Khim.* **22**, 351 (1948).
127. I. V. Nelson and R. J. Iwamoto, *Anal. Chem.* **33** 1795 (1961).

128. J. F. Coetzee, D. K. McGuire, and J. L. Hedrick *J. Phys. Chem.* **67** 1814 (1963).
129. R. Alexander and A. J. Parker, *J. Amer. Chem. Soc.* **89**, 5549 (1967).
130. Y. C. Ure and H. L. Friedman *J. Phys Chem.* **70** 501 (1966).
131. B. O. Coniglio, D. E. Giles, W. R. McDonald, and A. J. Parker, *J. Chem. Soc. B* p. 152 (1966).
132. U. Belluco, M. Martelli, and A. Orio, *Inorg. Chem.* **5**, 582 (1966).
133. U. Belluco, A. Orio, and M. Martelli, *Inorg. Chem.* **5**, 1370 (1966).
134. Y. C. Mac, W. A. Millen, A. J. Parker, and D. W. Watts, *J. Chem. Soc. B* p. 525 (1967).
135. J. F. Bunnett, *Annu. Rev. Phys. Chem.* **14**, 271 (1963).
136. A. J. Parker, J. Chem. Soc. *London* p. 1238 (1961).
137. C. G. Swain and E. R. Thornton, *J. Amer. Chem. Soc.* **84**, 822 (1962).
138. S. Popovici and M. Pop, *C. R. Acad. Sci.* **245**, 846 (1957).
139. S. R. Palit, *J. Org. Chem.* **12**, 752 (1947).
140. E. F. Caldin and J. Peacock, *Trans. Faraday Soc.* **51**, 1217 (1955).
141. C. K. Ingold, "Structure and Mechanism in Organic Chemistry," Bell, London, 1953.
142. C. A. Kingsbury, *J. Org. Chem.* **29**, 3262 (1964).
143. S. Yoneda, J. Marishima, K. Fukui, and Z. Yoshidi, *Kogyo Kagaku Zasshi* **68**, 1077 (1965).
144. Y. Pocker and A. J. Parker, *J. Org. Chem.* **31**, 1526 (1966).
145. E. A. Cavell and J. A. Speed, *J. Chem. Soc. London* p. 226 (1961).
146. A. J. Parker and D. Brody, *J. Amer. Chem. Soc.* p. 4061 (1963).
147. J. Hine, "Physical Organic Chemistry," pp. 159–162, 294. McGraw-Hill, New York, 1962.
148. E. G. Gould, "Mechanism and Structure in Organic Chemistry," Holt, New York, 1962.
149. D. L. Hill, K. C. Ho, and J. Miller, *J. Chem. Soc. B* p. 299 (1966).
150. R. E. Davis, *J. Amer. Chem. Soc.* **87**, 3010 (1965).
151. J. G. Swain and C. B. Scott, *J. Amer. Chem. Soc.* **75**, 141 (1953).
152. J. O. Edwards and R. G. Pearson, *J. Amer. Chem. Soc.* **84**, 16 (1962).
153. B. W. Clare, D. Cook, E. C. F. Ko, Y. C. Mac, and A. J. Parker, *J. Amer. Chem. Soc.* **88**, 1911 (1966).
154. D. Cook, I. P. Evans, E. C. F. Ko, and A. J. Parker, *J. Chem. Soc. G* p. 404 (1966).
155. D. C. Luehrs R. T. Inomoto, and J. Kleinburg, *Inorg. Chem.* **5**, 201 (1966).
156. R. Alexander, E. C. F. Ko, A. J. Parker, and T. J. Broxton, *J. Amer. Chem. Soc.* **90**, 5049 (1968).
157. A. J. Parker and R. Alexander, *J. Amer. Chem. Soc.* **90**, 3313 (1968).
158. E. D. Hughes, C. K. Ingold, and J. D. H. Mackie, *J. Chem. Soc. London* p. 3173 (1955).
159. E. D. Hughes, C. K. Ingold, and J. D. H. Mackie, *J. Chem. Soc. London* p. 3177 (1955).
160. P. B. D. de la Mare, *J. Chem. Soc. London* p. 3180 (1955).
161. L. Fowden, E. D. Hughes, and C. K. Ingold, *J. Chem. Soc. London* p. 3187 (1955).
162. L. Fowden, E. D. Hughes, and C. K. Ingold, *J. Chem. Soc. London* p. 3193 (1955).
163. P. B. D. de la Mare, *J. Chem. Soc. London* p. 3196 (1955).
164. P. B. D. de la Mare, L. Fowden, E. D. Hughes, C. K. Ingold, and J. D. H. Mackie, *J. Chem. Soc. London* p. 1208 (1955).
165. D. Cook and A. J. Parker, *J. Chem. Soc. B* p. 142 (1968).
166. S. Winstein, S. Smith, and D. Darwish, *Tetrahedron Lett.* **16**, 24 (1959).
167. Farhat-Aziz and E. A. Moelwyn-Hughes, *J. Chem. Soc. London* p. 2635 (1959).
168. E. D. Hughes and D. G. Whittingham, *J. Chem. Soc. London* p. 806 (1960).
169. C. K. Ingold, *Quart. Rev. Chem. Soc.* **11**, 1 (1957).
170. J. F. Hinton and E. S. Amis, to be published.

171. J. F. Hinton, E. S. Amis, and W. Mettetal, *Spectrochim. Acta Part A* **25**, 119 (1969).
172. E. S. Amis, *Inorg. Chim. Acta* **3**, 7 (1969).
173. J. F. Hinton and E. S. Amis, *Chem. Rev.* **67**, 367 (1967).
174. K. Dimroth, C. Richardt, T. Siepmann, and F. Bohlmann, *Ann. Chem.* **661**, 1 (1963).
175. E. M. Kosower, *J. Amer. Chem. Soc.* **80**, 3253 (1958).
176. E. Grunwald and S. Weinstein, *J, Amer. Chem. Soc.* **70**, 846 (1948).
177. J. Berson, J. Hamlet, and W. A. Muller, *J. Amer. Chem Soc*, **84**, 297 (1962).
178. S. Brownstein, *Can. J. Chem.* **38**, 1590 (1960).
179. S. G. Smith, A. H. Fainberg, and S. Winstein, *J. Amer. Chem. Soc.* **83**, 618 (1961).
180. E D Hughes and C, K, Ingold, *J, Chem, Soc, London* p, 244 (1935).
181. Y. Pocker, *Progr. React. Kinet.* **1**, 215–260 (1961).
182. E. R. Swart and L. J. LeRoux, *J. Chem. Soc. London* p. 406 (1957).
183. S. F. Acree, *Amer. Chem. J.* **48**, 352 (1912).
184. C. C. Evans and S. Sugden, *J. Chim. Phys. Physicochim. Biol.* **45**, 147 (1948); *J. Chem. Soc. London* p. 270 (1949).
185a. A. R. Olson, L. D. Frashier, and F. J. Spieth, *J. Phys. Colloid Chem.* **55**, 860 (1951); A R, Olson and J, Koency, *J. Amer. Chem. Soc.* **75**, 5801 (1953).
185b. R. D. Heyding and C. A. Winkler, *Can. J. Chem.* **29**, 790 (1951).
185c. A. Brandström, *Ark. Kemi* **11**, 567 (1957).
185d. E. A. Moelwyn-Hughes, *Trans. Faraday Soc.* **45**, 167 (1949).
185e. N. N. Lichten and K. N. Rao, Abstracts. *Amer. Chem. Soc. Meeting, Chicago, Illinois, September 7–12, 1958*, p. 128.
186. S. W. Winstein, L. G. Savedoff, S. Smith, I. D. R. Stevens, and J. S. Gall, *Tetrahedron Lett.* **9**, 24 (1960).
187. R. F. Rodewald, K. Mahendran, J. L. Bear, and R. Fuchs, *J. Amer. Chem. Soc.* **90**, 6698 (1968).
188. A. G. Grimm, H. Ruf, and H. Wolff, *Z. Phys. Chem. Abt. B* **13**, 301 (1937).
189. N. Menshutkin, *Z. Phys. Chem. Stoechim. Verwandschaftslehre* **6**, 41 (1890).
190. J. G. Kirkwood, *J. Chem. Phys.* **2**, 351 (1934).
191. R. G. Pearson and D. G. Voglsong, *J. Amer. Chem. Soc.* **80**, 1038 (1958).
192. C. G. Swain and R. W. Eddy, *J. Amer. Chem. Soc.* **79**, 2989 (1948).
193. Y. Pocker, *J. Chem. Soc. London* p. 1279 (1957).
194. E. D. Hughes and D. J. Whittingham, *J. Chem. Soc. London* p. 806 (1960).
195. Y. Pocker and A. J. Parker, *J. Org. Chem.* **31**, 1252 (1966).
196. Y. C. Mac, W. A. Millen, A. J. Parker, and D. W. Watts, *J. Chem. Soc. London* p. 525 (1967).
197. S. Glasstone, K. J. Laidler, and H. Eyring, "The Theory of Rate Processes," McGraw-Hill, New York, 1941.
198. M. Richardson and R. G. Soper, *J. Chem. Soc. London* p. 1873 (1929).
199. A. P. Stefani, *J. Amer. Chem. Soc.* **90**, 1694 (1968).
200. W. D. Harkins D. S. Davis, and G. Clark, *J. Amer. Chem. Soc.* **39**, 555 (1917).
201. J. Stefan, *Ann. Phys. Leipzig* [3] **29**, 655 (1886).
202. S. Glasstone, *J. Chem. Soc. London* p. 723 (1936).
203. R. A. Ogg, Jr., and O. K. Rice, *J. Chem. Phys.* **5**, 140 (1937).
204. J. H. Hildebrand and R. L. Scott, "Regular Solutions," Prentice-Hall, Englewood Cliffs, New Jersey, 1962.
205. P. S. Disson, A. P. Stefani, and M. Scwarc, *J. Amer. Chem. Soc.* **85**, 2551, 3344 (1963).
206. R. A. Marcus, *J. Chem. Phys.* **20**, 364 (1952).
207. O. K. Rice, *J. Phys. Chem.* **65**, 1588 (1961).

208. J. N. Bradley, *J. Chem. Phys.* **35**, 748 (1961).
209. J. A. Kerr and A. J. Trotman-Dickenson, *Prog. Chem. Kinet.* **1**, 110, 135 (1961).
210. J. M. Bradley, *J. Chem. Phys.* **35**, 748 (1961).
211. A. P. Stefani, G. F. Thrower, and C. F. Jordan, *J. Phys. Chem.* **73**, 1057 (1969).
212. M. G. Evans and M. Polanyi, *Trans. Faraday Soc.* **31**, 875 (1935); **32**, 1333 (1936).
213. J. H. van't Hoff, "Vorlesungen über Theoretische und Physikalische Chemie," Vol. 1. Braunsweig, 1901.
214. E. A. Moelwyn-Hughes, "Physical Chemistry," Chapter XXIV. Pergamon, Oxford, 1961.
215. J. J. van Laar, *Z. Phys. Chem. Stoechim. Verwandschaftslehre* **15**, 457 (1894); and succeeding papers up to 1929.
216. A. L. T. Moesveld, *Z. Phys. Chem. Stoechim. Verwandschaftslehre* **105**, 442, 450, 455 (1923).
217. M. W. Perrin, *Trans. Faraday Soc.* **34**, 144 (1938).
218. A. P. Harris and K. E. Weale, *J. Chem. Soc. London* p. 146 (1961).
219. A. E. Nicholson and R. G. W. Norrish, *Discuss. Faraday Soc.* **22**, 97 (1956).
220. M. G. Evans and M. Polyanyi, *Trans. Faraday Soc.* **31**, 875 (1935); **32**, 1333 (1936).
221. M. G. Evans, *Trans. Faraday Soc.* **34**, 49 (1938).
222. E. Whalley, *Trans. Faraday Soc.* **55**, 798 (1959).
223. J. Koskikallio and E. Whalley, *Trans. Faraday Soc.* **55**, 809 (1959).
224. J. Koskikallio and E. Whalley, *Trans. Faraday Soc.* **55**, 815 (1959).
225. C. K. Ingold, "Structure and Mechanism in Organic Chemistry," Cornell Univ. Press, Ithaca, New York, 1953.
226. M. J. Mackinnon, L. A. Lateef, and J. B. Hyne, *Can. J. Chem.* **48**, 2025 (1970); J. B. Hyne, H. S. Golinkin, and W. G. Laidlaw, *J. Amer Chem Soc* **88**, 2104 (1966); D. B. McDonald and J. B. Hyne, *Can. J. Chem.* **48**, 2494 (1970).
227. H. S. Golmkin, I. Lee, and J. B. Hyne, *J. Amer. Chem. Soc.* **89**, 1307 (1967).
228. D. L. Gay and E. Whalley, *Can. J. Chem.* **48**, 2021 (1970).
229. R. G. W. Norrish, *Trans. Faraday Soc.* **34**, 154 (1938).
230. W. R. Angus, *Trans. Faraday Soc.* **34**, 153 (1939).
231. G. E. Edwards, *Trans. Faraday Soc.* **33**, 1294 (1937).
232. D. M. Newitt and A. Wasserman, *Trans. Chem. Soc.* p. 735 (1940).
234. I. W. Sizer, *Advan. Enzymol.* **3**, 35 (1943).
235. M. Dixon and E. C. Webb, "Enzymes," Chapter 4. Longmans, Green, New York, 1964.
236. K. J. Laidler, "The Chemistry of Enzyme Action," Oxford Univ. Press, London and New York, 1958.
237. L. Geller and R. A. Alberty, *in* "Physical Chemical Aspects of Enzyme Kinetics" (G. Ported, ed.), Chapter IX. Pergamon, Oxford, 1961.
238. O. D. Trapp and G. S. Hammond, *J. Amer. Chem. Soc.* **81**, 4867 (1959).
239. R. M. Noyes, *J. Amer. Chem. Soc.* **86**, 4529 (1964).
240. S. F. Nelsen and P. D. Bartlett, *J. Amer. Chem. Soc.* **88**, 143 (1966).
241. H. Kipfer and T. G. Taylor, *J. Amer. Chem. Soc.* **89**, 6667 (1967).
242. W. Braun, L. Rajbenbach, and F. R. Eurich, *J. Phys. Chem.* **66**, 1591 (1962).
243. S. Kodama, *Bull. Chem. Soc. Jap.* **35**, 685, 827 (1962).
244. R. M. Noyes, *Progr. React. Kinet.* **1**, 129 (1961).
245. A. F. Stefani, G. F. Thrower, and C. P. Jordan, *J. Phys. Chem.* **73**, 1257 (1969).
246. D. Booth and R. M. Noyes, *J. Amer. Chem Soc* **82**, 1868 (1960).
247. W. A. Pryor and K. Smith, *J Amer. Chem. Soc.* **89**, 1741 (1967).
248. J. Koenig and M. Deinzer, *J. Amer. Chem. Soc.* **88**, 4518 (1966).

249. W. Hecht and M. Conrad, *Ann. Phys. Chem.* **3**, 450 (1889).
250. E. J. Bowen and F. Wokes, "Fluorescence of Solutions," Longmans, Green, New York, 1953.
251. G. Oster and A. H. Adelman, *J Amer. Chem. Soc.* **78**, 913 (1956).
252. L. Umberger and U. K. LaMer, *J. Amer. Chem. Soc.* **67**, 1099 (1945).
253. L. E. Medows and R. M. Noyes, *J. Amer. Chem. Soc.* **82**, 1872 (1960).
254. D. Booth and R. M. Noyes, *J. Amer. Chem. Soc.* **82**, 1868 (1960).
255. L. E. Medows and R. M. Noyes, *J. Amer. Chem. Soc.* **82**, 1872 (1960).
256. E. Rabinowich and W. C. Wood, *Trans. Faraday Soc.* **32**, 1381 (1936).
257. J. O. Hirschfelder, C. F. Curtiss, and R. B. Bird, "Molecular Theory of Gases and Liquids," Wiley, New York, 1964.
258. M. Ross and J. H. Hildebrand, *J. Chem. Phys.* **40**, 2397 (1964).
259. E. W. Haycock, B. J. Alder, and J. H. Hildebrand, *J. Chem. Phys.* **21**, 1601 (1953).
260. H. Watts, B. J. Alder, and J. H. Hildebrand, *J. Chem. Phys.* **23**, 659 (1955).
261. G. C. Pimental and A. L. McClellan, "The Hydrogen Bond," Freeman, San Francisco, California, 1960.
262. L. Hunter, *J. Chem. Soc. London* p. 806 (1945).
263. A. Sherman, *J. Phys. Chem.* **41**, 117 (1937).
264. G. V. Tsitsishvili, *Zh. Fiz. Khim.* **15**, 1082 (1941).
265. A. R. Ubbelohde, *J. Chim. Phys. Physicochim. Biol.* **46**, 429 (1949).
266. K. Wirtz, *Z. Naturforsch. A* **2**, 264 (1947).
267. W. Gordy and S. C. Standford, *J. Chem. Phys.* **9**, 204 (1941).
268. W. Gordy and S. C. Standford, *J. Chem. Phys.* **8**, 170 (1940).
269. Y. Sato and S. Nagakura, *Sci. Light (Tokyo)* **4**, 120 (1955).
270. H. Tsubomura, *Bull. Chem. Soc. Jap.* **27**, 445 (1954).
271. J. Hine, "Physical Organic Chemistry," pp. 33–34. McGraw-Hill, New York, 1962.
272. H. Hart, F. A. Cassis, and J. J. Bordeaus, *J. Amer. Chem. Soc.* **76**, 1639 (1954).
273. W. J. Svirbely and J. F. Roth, *J. Amer. Chem. Soc.* **75**, 3106 (1953).
274. M. Szwarc and J. Smid, *J. Chem. Phys.* **27**, 421 (1951).
275. S. Palit, *J. Org. Chem.* **12**, 752 (1957).
276. P. S. Rees, D. P. Long, and G. T. Young, *J. Chem. Soc. London* p. 662, (1954).
277. C. G. Swain, *J. Amer. Chem. Soc.* **79**, 1119 (1948).
278. C. G. Swain and R. W. Eddy, *J. Amer. Chem. Soc.* **70**, 2989 (1948).
279. G. Aksnes, *Acta Chem. Scand.* **14**, 1526 (1960).
280. H. Euler and A. Olander, *Z. Phys. Chem. Stoechim, Verwandschafts lehre* **131**, 107 (1928).
281. S. Ono, K. Hiromi, and Y. Sano, *Bull. Chem. Soc. Jap.* **36**, 431 (1963).
282. K. Hiromi, *Bull. Chem. Soc. Jap.* **33**, 1257 (1960).
283. K. Hiromi, *Bull. Chem. Soc. Jap.* **33**, 1264 (1960).
284. W. H. R. Shaw and D. G. Walker, *J. Amer. Chem. Soc.* **79**, 4329 (1957).
285. J. L. Rabinowitz and E. C. Wagner, *J. Amer. Chem. Soc.* **73**, 3030 (1951).
286. N. L. Miller and E. C. Wagner, *J. Amer. Chem. Soc.* **76**, 1847 (1954).
287. M. E. Runner, M. L. Kilpatrick, and E. C. Wagner, *J. Amer. Chem. Soc.* **69**, 1406 (1969).
288. M. E. Runner and E. C. Wagner, *J. Amer. Chem. Soc.* **74**, 2529 (1952).
289. G. Vavon and P. Montheard, *Bull. Soc. Chim. Fr.* **7**, 551, 560 (1940).
290. M. J. Astle and M. J. McConnell, *J. Amer. Chem. Soc.* **65**, 35 (1943).
291. M. J. Astle and M. P. Cooper, *J. Amer. Chem. Soc.* **65**, 2395 (1943).
292. M. J. Astle and S. T. Stevenson, *J. Amer. Chem. Soc.* **65**, 2399 (1943).
293. M. J. Astle, *Trans. Electrochem. Soc* **.92**, 473 (1943).

294. L. Pekkerinen, *Ann. Acad. Sci. Fenn. Ser. A2* No. **62,** 77 (1954).
295 D. G. Knorre and N. M. Emanuel, *Vop. Khim. Knet. Katal. Reakts. Sposobnosti Dokl. Vses. Soveshch.* p. 106 (1955).
296. D. G. Knorre and N. M. Enamuel, *Zh. Fiz Khim.* **26,** 425 (1952).
297. D. G. Knorre and N. M. Emanuel, *Dokl. Akad. Nauk SSSR* **91,** 1163 (1953).
298. D. G. Knorre and N M. Emanuel, *Usp. Khim.* **24,** 275 (1955).
299. L. Grossman, D. Stallar, and K. Herrington, *J. Chim. Phys. Physicochim. Biol.* **58,** 1078 (1961).
300. L. G. Augustine, C. A. Gibson, K. L. Grist, and R. Mason, *Proc. Nat. Acad Sci. U. S.* **47,** 1733 (1961).
301. L. DeMaeyer, M. Eigen, and J. Saurez, *J. Amer. Chem. Soc.* **90,** 3157 (1968).
302. M. Eigen, G. Moss, W. Müller, and D. Porschke, to be published.
303. R. W. Hay and K. R. Tate, *Aust. J. Chem.* **23,** 1407 (1970).
304. C. G. Swaim, R. F. W. Baser, R. M. Esteve, and R. N. Griffin, *J. Amer. Chem. Soc.* **83,** 1951 (1961).
305. E. O. Wiig, *J. Phys. Chem.* **32,** 961 (1928).
306. G. A. Hall and E. S. Hanrahan, *J. Phys. Chem.* **69,** 2402 (1965).
307. B. R. Brown, *Quart Rev. Chem. Soc.* **5,** 141 (1951).
308. A. Lapworth, *Nature (London)* **115,** 625 (1925).
309. C. K. Ingold, *Rec. Trav. Chim. Pays-Bas* **48,** 797 (1929); *J. Chem. Soc. London* p. 1120 (1933).
310. J. F. Bunnett, *Annu. Rev. Phys. Chem.* **14,** 271 (1963).
311. C. K. Ingold, *Chem. Rev.* **15,** 265 (1934).
312. C. K. Ingold, "Structure and Mechanism in Organic Chemistry," p. 200. Cornell Univ. Press, Ithaca, New York, 1953.
313. J. F. Bunnett and R. H. Zahler, *Chem. Rev* **49,** 273 (1951).
314. C. G. Swain and C. B. Scott, *J. Amer. Chem. Soc.* **75,** 141 (1953).
315. J. N. Brönsted, *Rec. Trav. Chim. Pays-Bas* **42,** 718 (1923).
316. G. N. Lewis, *J. Franklin Inst.* **226,** 293 (1938).
317. A. J. Parker, *Proc. Chem. Soc. London* p. 371 (1961).
318. N. Kornblum, R. A. Smiley, R. K. Blockwood, and D. C. Iffland, *J. Amer. Chem. Soc.* **77,** 6269 (1955).
319. G. E. K. Branch and M. Calvin, "The Theory of Organic Chemistry," pp. 408–423. Prentice-Hall, Englewood Cliffs, New Jersey, 1941.
320. J. O. Edwards and R. G. Pearson, *J. Amer. Chem. Soc.* **84,** 16 (1962).
321. R. F. Hudson, *Chimia* **16** 173 (1962).
322. J. Fosteni D. H. Rosenblatt, and M. M. Damek, *J. Amer. Chem. Soc.* **78,** 341 (1956).
323. W. P. Jencks and J. Carrivolo, *J. Amer. Chem. Soc.* **82,** 1778 (1960).
324. R. F. Hudson, *Chimia* **16,** 173 (1962).
325. J. F. Bunnett, *J. Amer. Chem. Soc.* **79,** 5969 (1957).
326. J. D. Reinheimer and J. F. Bunnett, *J. Amer. Chem. Soc.* **81,** 315 (1959).
327. K. S. Pitzer, "Quantum Chemistry," Prentice-Hall, Englewood Cliffs, New Jersey, 1953.
328. J. F. Bunnett and J. F. Reinheimer, *J. Amer. Chem. Soc.* **84,** 3284 (1962).
329. C. G. Swain and C. G. Scott, *J. Amer. Chem. Soc.* **75,** 141 (1953).
330. C. G. Swain, R. B. Mosley, and D. E. Brown, *J. Amer. Chem. Soc.* **77,** 3731 (1955).
331. J. Hine, "Physical Organic Chemistry," p. 137. McGraw-Hill, New York 1956.
332. C. G. Swain, D. C. Dittmar, and L. E. Kaiser, *J. Amer. Chem. Soc.* **77,** 3737 (1955).
333. J. O. Edwards, *J. Amer. Chem. Soc.* **76,** 1540 (1954).
334. J. O. Edwards, *J. Amer. Chem. Soc.* **78,** 1819 (1956).

335. M. I. Hawthorne and D. I. Cram, *J. Amer. Chem. Soc.* **77**, 486 (1955).
336. P. R. Wells, *Chem. Rev.* **63**, 171 (1963).
337. J. Hansson, *Sv. Kem. Tidskr.* **66**, 351 (1954).
338. R. E. Parker and N. S. Isaacs, *Chem. Rev.* **59**, 766 (1959).
339. J. Hansson, *Sv. Kem. Tidskr.* **67**, 246 (1955).
340. E. Grunwald and S. Winstein, *J. Amer. Chem. Soc.* **70**, 846 (1948).
341. S. Winstein, E. Grunwald, and H. W. Jones, *J. Amer. Chem Soc* **73**, 2700 (1951).
342. C. G. Swain and R. B. Mosely, *J. Amer. Chem. Soc.* **77**, 3727 (1955).
343. A. H. Fainberg and S. Winstein, *J. Amer. Chem. Soc.* **79**, 1597, 1602, 1608 (1957).
344. S. Winstein, A. H. Fainberg, and E. Grunwald, *J. Amer. Chem. Soc.* **79**, 4146 (1957).
345. R. W. Taft, *J. Amer. Chem. Soc.* **74**, 3120 (1952).
346. C. K. Ingold, *J. Chem. Soc. London* p. 1032 (1930).
347. L. P. Hammett, *J. Amer. Chem. Soc.* **59**, 96 (1937).
348. R. W. Taft, Jr., *in* "Steric Effects in Organic Chemistry" (M. S. Newman, ed.) Chapter 13. Wiley, New York, 1956.
349. C. G. Derick, *J. Amer. Chem. Soc.* **33**, 1181 (1911).
350. E. L. Mackor, P. J. Smit, and J. H. van der Waals, *Trans. Faraday Soc.* **53**, 1309 (1957).
351. E. L. Mackor, A. Hofstra, and J. H. van der Waals, *Trans. Faraday Soc.* **54**, 66 (1958).
352. L. P. Hammett, *Chem Rev.* **17**, 125 (1935); *Trans. Faraday Soc.* **34**, 156 (1938).
353. L. P. Hammett, "Physical Organic Chemistry," 2nd ed. McGraw-Hill, New York, 1970.
354. G. Berlin and D. D. Perrin, *Quart. Rev. Chem. Soc.* **20**, 75 (1966).
355. H. H. Jaffe, *Chem. Rev.* **53**, 191 (1953).
356. C. D. Richie and W. F. Sager, *Progr. Phys. Org. Chem.* **2**, 323 (1964).
357. J. Hine, "Physical Organic Chemistry," Chapter 4, p. 98. McGraw-Hill, New York, 1962.
358. C. K. Hancock and J. S. Westmorland, *J. Amer. Chem. Soc.* **80**, 545 (1958).
358a.V. Meyer, *Ber.* **27**, 510 (1894).
359. J. P. McCullough and B. A. Eckerson, *J. Amer. Chem. Soc.* **67**, 707 (1945); J. P. McCullough and M. K. Barsh, *Ibid.* **71**, 3029 (1949).
360. C. K. Hancock and J. S. Westmoreland, *J. Amer. Chem. Soc.* **80**, 545 (1958).
361. J. Clark and D. D. Perrin, *Quart. Rev. Chem. Soc.* **18**, 295 (1964).
362. H. van Bekkum, E. P. Verkade, and B. M. Wepster, *Rec. Trav. Chim. Pays-Bas* **79**, 815 (1959).
363. R. W. Taft, Jr., *J. Phys. Chem.* **64**, 1805 (1960).
364. R. W Taft, Jr., R. E. Glick, I. C. Lewis, I. Fox, and S. Ehrenson, *J. Amer.* Chem. Soc. **82**, 756 (1960).
365. R. W. Taft and I. C. Lewis, *J. Amer. Chem. Soc.* **81**, 5343 (1959).
366. H. K. Hall, Jr., *J. Amer. Chem. Soc.* **79**, 5441 (1957).
367. H. C. Brown and G. Goldman, *J. Amer. Chem. Soc.* **84**, 1650 (1962).
368. H. E. Zaugg, *J. Amer. Chem. Soc.* **83**, 837 (1961).
369. H. E. Zaugg, B. W Horrom, and S. Borgwardt, *J. Amer. Chem. Soc.* **82**, 2895 (1960).
370. H. E. Zaugg, *J. Amer. Chem. Soc.* **82**, 2903 (1960).
371. G. Wittig and E Stahnecker, *Ann* **605**, 69 (1957).
372. S. Winstein, L G Savedoff, S Smith, I. D. R. Stevents, and J. S. Gall, *Tetrahedron Lett.* **9**, 24 (1960).
373. H. R. McCleary and L. P. Hammett, *J. Amer. Chem. Soc.* **63**, 2254 (1941).
374. W. M. Weaver and J. D. Hutchinson, *J. Amer. Chem. Soc* **86**, 261 (1964).
375. H. C. S. Snethlage, *Z. Elektrochem.* **18**, 539 (1913); *Z. Phys. Chem. Stoechim Verwand. schaftslehre* **85**, 211 (1913).

376. H. S. Taylor, *Medd. Vetenskapsakad. Nobelinst.* **2** 1 (1913); *Z. Elektrochem.* **20** 201 (1914).
377. J. N. Brönsted, *Trans Faraday Soc.* **24**, 630 (1928); *Chem. Rev.* **5**, 231 (1928).
378. J. N. Brönsted and E. A. Guggenheim, *J. Amer. Chem. Soc.* **49**, 2554 (1928).
379. H. L. Pfluger, *J. Amer. Chem. Soc.* **60**, 1513 (1938).
380. J. N. Brönsted and K. Pedersen, *Z. Phys. Chem. Stoechim. Verwandschaftslehre* **108**, 2554 (1924).
381. R. P. Bell and E. C. E. Higgins, Proc. Roy Soc. *Ser A* **197**, 141 (1949; R. P. Bell, *J. Phys. Colloid Chem.* **55**, 885 (1951); R. P. Bell and J. C. Clunie, *Proc. Roy. Soc. Ser A* **212**, 33 (1952).
382. R. P. Bell, E. Gelles, and E. Moller, *Proc. Roy. Soc. Ser A* **198**, 308 (1949).
383. S. W. Benson, *J. Amer. Chem. Soc.* **80**, 5151 (1951).
384. J. E. Leffler and E. Grunwald, "Rates and Equilibria of Organic Reactions," p. 26. Wiley, New York, 1963.
385. R. P. Bell and G. L. Wilson, *Trans. Faraday Soc.* **46**, 407 (1950).
386. A. F. Trotman-Dickenson, *J. Chem. Soc. London* p. 1923 (1949).
387. R. P. Bell and W. C. E. Higginson, Proc. Roy. Soc. **A197**, 141 (1948).
388. R. P. Bell and R. G. Pearson, J. Chem. Soc. *London* p. 1953 (1943).
389. R. P. Bell, "The Proton in Chemistry." Cornell Univ. Press, Ithaca, New York, 1959.
390. G. Wittig and E. Stahnecker, Ann. **605**, 69 (1957).
391. H. D. Zook and T. Rosso, *J. Amer. Chem. Soc.* **82**, 1258 (1960).
392. H. D. Zook and W. L. Gumby, *J. Amer. Chem. Soc.* **82**, 1386 (1960).
393. C. Carvajal, K. J. Tolle, J. Smid, and M. Swarc, *J. Amer. Chem. Soc.* **87**, 5548 (1965).
394. A. Brändström, *Ark. Kemi* **6**, 955; **7**, 81 (1954).
395. N. V. Sidgwick and F. M. Brewer, *J. Chem. Soc. London* p. 2379 (1925).
396. W. C. Fernelius and L. G. Van Uitert, *Acta Chem. Scand.* **8**, 1726 (1954).
397. L. L. Chan and J. Smid, *J. Amer. Chem. Soc.* **89**, 4547 (1967).
398. A. G. Evans and D. B. George, *J. Chem. Soc. London* p. 4653 (1961); A. G. Evans and N. H. Rees, *Ibid.* p. 6039 (1963); R. A. H. Casling, A. G. Evans, and N. H. Rees, *J. Chem. Soc. B* p. 519 (1966).
399. D. Margerison and J. P. Newport, *Trans. Faraday Soc.* **59**, 2058 (1963).
400. M. Weiner, G. Vogel, and R. West, *Inorg. Chem.* **1**, 654 (1962).
401. N. Kornblum, R. Seltzer, and P. Haberfield, *J. Amer. Chem. Soc.* **85**, 1148 (1963).
402. N. Kornblum, P. J. Berrigan, and W. J. de Noble, *J. Amer. Chem. Soc.* **85**, 1142 (1963).
403. I. M. Barclay and J. A. V. Butler, *Trans. Faraday Soc.* **34**, 1445 (1938).
404. J. E. Leffler, *J. Org. Chem.* **20**, 1201 (1955).
405. M. G. Alder and J. E. Leffler, *J. Amer. Chem. Soc.* **76**, 1425 (1954).
406. V. K. LaMer, *J. Franklin Inst.* **225**, 709 (1938).
407. E. S. Amis, *Anal. Chem.* **27**, 1672 (1955).
408. F. D. Chattaway, *J. Chem. Soc. London* **101**, 170 (1912).
409. J. Walker and J. Hambly, *J. Chem. Soc., Trans.* **67**, 746 (1895); **71**, 489 (1897).
410. F. D. Chattaway, *J. Chem. Soc. London* **101**, 170 (1912).
411. L. M. Lowry, *Trans. Faraday Soc.* **30**, 375 (1934).
412. A. A. Frost and R. G. Pearson, "Kinetics and Mechanism," pp. 307–315. Wiley, New York, 1961.
413. W. A. Cowdrey, E. D. Hughes C. K. Ingold, S. Masterman, and A. D. Scott, *J. Chem. Soc. London* p. 1267 (1937).
414. E. S. Lewis and C. E. Boozer, *J. Amer. Chem. Soc.* **74**, 308 (1952).
415. C. E. Boozer and E. S. Lewis, *J. Amer. Chem. Soc.* **75**, 3182 (1953).
416. D. J. Cram, *J. Amer. Chem. Soc.* **75**, 332 (1953).

417. J. Soulati, K. L. Henold, and J. P. Oliver, private communication, 1971.
418. E. A. Jeffery and T. Mole, *Aust. J. Chem.* **21**, 1187 (1968).
419. W. Bremser, M. Winokur, and J. D. Roberts, *J. Amer. Chem. Soc.* **92**, 1080 (1970).
420. C. R. McCoy and A. L. Allred, *J. Amer. Chem. Soc.* **84**, 912 (1962).
421. K. L. Henold, J. Soulati, and J. P. Oliver, *J. Amer. Chem. Soc.* **91**, 3171 (1969).
422. G. E. Coates, M. L. H. Green, and K. Wade, "Organometallic Compounds," 3rd ed., Vol. I, Chapters 2 and 3. Methuen, London, 1967.
423. E. S. Amis, *J. Phys. Chem.* **60**, 428 (1956).
424. S. Winstein and A. H. Fainberg, *J. Amer. Chem. Soc.* **79**, 5937 (1957).
425. E. D. Hughes and C. K. Ingold, *J. Chem. Soc. London* p. 255 (1935).
426. E. D. Hughes, J. Chem Soc. *London* p. 255 (1935).
427. C. A. L. De Bruyn and A. Steger, *Rec. Trav. Chim. Pays-Bas* **18**, 41, 311 (1899).
428. E. Bergman, M. Polanyi, and A. Szabo, *Z. Phys. Chem. Stoechim. Verwandschaftslehre* **5**, 589; **6**, 41 (1890).
429. N. Menshutkin, *Z. Phys. Chem. Stoechim. Verwandschaftslehre* **5**, 589; **6**, 41 (1890).
430. G. Carrera, *Gazetta* **241**, 180 (1894).
431. A. Hamptinne and A. Bakaert, *Z. Phys. Chem. Stoechim. Verwandschaftslehre* **28**, 225 (1899).
432. H. von Halban, *Z. Phys. Chem. Stoechim, Verwandschaftslehre* **84**, 428 (1913).
433. H. E. Cox, *J. Chem. Soc.* **119**, 142 (1921).
434. J. A. Hawkins, *J. Chem. Soc.* **121**, 1170 (1922).
435. G. E. Muchin, R. B. Ginsberg, and Kh. M. Moissezera, *Ukr. Khem, Zh.* **2**, 136 (1926).
436. H. McCombie, H. A. Scarborough, and F. F. R. Smith, *J. Chem. Soc. London* p. 802 (1927).
437. H. Essex and O. Gelormini, *J. Amer. Chem. Soc.* **48**, 882 (1926).
438. H. Von Halban, *Z. Phys. Chem.* Heft 2, **67**, 129 (1929).
439. J. L. Gleave, E. D. Hughes, and C. K. Ingold, *J. Chem. Soc. London* p. 236 (1935).
440. E. D. Hughes, *J. Amer. Chem Soc.* **57**, 708 (1935).
441. E. D. Hughes, C. K. Ingold, and C S Patel, *J. Chem. Soc. London* p. 526 (1933).
442. C. W. Swain and R. W. Eddy, *J. Amer. Chem. Soc.* **70**, 2989 (1948).
443. N. J. T. Pickles and C. N. Hinshelwood, *J. Chem. Soc. London* p. 231 (1932).
444. D. P. N. Satchell, *Quart. Rev. Chem. Rev.* **17**, 160 (1963).
445. C. K. Ingold, "Structure and Mechanism in Organic Chemistry," Bell, London, 1953.
446. M. L. Bender, *Chem. Rev.* **60**, 53 (1960).
447. C. W. Gould, "Mechanism and Structure in Organic Chemistry," Holt, New York, 1959.
448. A. Streitweiser, Jr., *Chem. Rev.* **56**, 571 (1956).
449. J. P. M. Satchell, *J. Chem. Soc. London* p. 5404 (1951).
450. D. P. N. Satchell, *J. Chem. Soc. London* p. 555 (1963).
451. L. P. Hammett and A. J. Dyrup, *J. Amer. Chem. Soc.* **54**, 2721 (1932).
452. J. G. Pritchard and F. A. Long, *J. Amer. Chem. Soc.* **78**, 2667 (1956).
453. L. Zucker and L. P. Hammett, *J. Amer. Chem. Soc.* **61**, 2791 (1939).
454. G. Swartzenback and C. Wittwer, *Helv. Chim. Acta.* **30**, 663 (1947).
455. F. A. Long and R. Bakule, *J. Amer. Chem. Soc.* **85**, 2313 (1963).
456. R. Bakule and F. A. Long, *J. Amer. Chem. Soc.* **85**, 2309 (1963).
458. K. N. Bascombe and R. P. Bell, *Discuss. Faraday Soc.* **24**, 158 (1957).
459. R. P. Bell, "The Proton in Chemistry," p. 81 ff. Cornell Univ. Press, Ithaca, New York, 1959.
460. G. Yagil and N. Anbar, *J. Amer. Chem. Soc.* **85**, 2376 (1963).

461. M. Anbar, M. Bobtelsky, D. Samuel, B. Silver, and G. Yagil, *J. Amer. Chem. Soc.* **85**, 2380 (1963).
462. G. Yagil and M. Anbar, *J. Amer. Chem. Soc.* **85**, 2376 (1963)
463 E. J. Hart and J. W. Boag, *J. Amer. Chem. Soc.* **84**, 4090 (1962).
464. J. W. Boag and E. W. Hart, *Nature London* **197**, 45 (1963).
465. L. M. Dorfman, I. A. Taub, and R. E. Buehler, *J. Chem. Phys.* **36**, 3051 (1962).
466. I. A. Taub and L. M. Dorfman, *J. Amer. Chem. Soc.* **84**, 4053 (1962)
467. L. M. Dorfman and I. A. Taub., *J. Amer. Chem. Soc.* **85**, 2370 (1963).
468. M. J. Bronskill, R. K. Wolf, and J. W. Hunt, *J. Chem. Phys.* **53**, 4201 (1970).
469. R. K. Wolf, M. J. Bronskill, and J. W. Hunt, *J. Chem. Phys.* **53**, 4211 (1970).
470. J. H. Fendler and L. K. Patterson, *J. Phys. Chem.* **74**, 4608 (1970).
471. M. Eigen and L. DeMaeyer, *Z. Elektrochem.* **59**, 986 (1955).
472. J. H. Baxendale, E. M. Fielding, and J. P. Keene, *Proc. Chem. Soc. London* p. 242–243 (1963).
473. E. J. Hart and M. Anbar, "The Hydrated Electron," Chapters V, VI, and VII. Wiley (Interscience), New York, 1970.
474. E. J. Hart and M. Anbar, "The Hydrated Electron," Chapter VIII. Wiley (Interscience), New York, 1970.
475. N. Sutin, "Exchange Reactions," p. 7. IAEA, Vienna, 1965.
476. R. A. Marcus, *Advan Chem.* **50**, 138 (1965).
477. R. A. Marcus, *J. Chem. Phys.* **43**, 3477 (1965).
478. E. S. Amis, *Inorg. Chim. Acta Rev.* **3**, 7 (1969).
479. M. Anbar and E. J. Hart, *Advan. Chem.* **81**, 79 (1968).
480. M. V. Smoluchowski, *Z. Phys. Chem. Stoechim Verwandschaftslehre* **92**, 129 (1917).
481. M. Anbar and E. J. Hart, Unpublished results, 1966.
482. L. Kevan, *J. Amer. Chem. Soc.* **89**, 4238 (1967).
483. M. Anbar, D. Meyenstein, and P. Neta, *Nature (London)* **203**, 1348 (1966).
484. E. D. Hughes, C. K. Ingold, and C. S. Patel, *J. Chem. Soc. London* p. 526 (1933).
485. C. K. Ingold, "Structure and Mechanism in Organic Chemistry," Cornell Univ. Press, Ithaca, New York, 1953.
486. D. J. Cram. *J. Amer. Chem. Soc.* **71**, 3863 (1949); **74**, 2149 (1952).
487. S. Winstein and R. B. Henderson, Ethylene and triethylene oxides. *In* "Heterocyclic Compounds" (R. C. Elderfield, ed.), Vol I. Wiley, New York, 1950.
488. J. N. Bronsted, M. L. Kilpatrick, and M. Kilpatrick, *J. Amer. Chem. Soc.* **51**, 428 (1929).
489. C. L. McCabe and J. C. Warner, *J. Amer. Chem. Soc.* **70**, 4031 (1948).
490. S. Winstein and H. J. Lucas, *J. Amer. Chem. Soc.* **61**, 1576 (1939).
491. J. A. Long and J. G. Prichard, *J. Amer. Chem. Soc.* **78**, 2263 (1958).
492. R. E. Parker and N. S. Isaacs, *Chem. Rev.* **59**, 737 (1959).
493. J. G. Prichard and F. A. Long, *J. Amer. Chem. Soc.* **78**, 2667 (1956).
494. F. A. Long and J. G. Prichard, *J. Amer. Chem. Soc.* **78**, 1668 (1958).
495. F. C. Whitmore, "Organic Chemistry," p. 339. Van Nostrand-Reinhold, Princeton, New Jersey, 1937.
496. C. D. Hurd, "Pyrolysis of Organic Compounds," A.C.S. Monograph. Van Nostrand-Reinhold, Princeton, New Jersey, 1929.
497. E. de B. Barnett and C. L. Wilson, "Inorganic Chemistry," Longmans, Green, New York, 1958.
498. G. W. A. Fowles and F. H. Pollard, *J. Chem. Soc. London* p. 4128 (1953).
499. H. A. Emeleus, *Chem. Ind. (London)* p. 813 (1937).

500. D. C. Bradley, F. M. Abd-el Halim, and W. Wardlaw, *J. Chem. Soc. London* p. 3450 (1950).
501. F. W. Bergstrom, R. E. Wright, C. Chandler, and W. L. Gilkey, *J. Org. Chem.* **1**, 170 (1936); R. E. Wright and J. W. Berstrom, *Ibid.* **1**, 179 (1936).
502. S. Gilman and S. Avakian, *J. Amer. Chem. Soc.* **67**, 349 (1945).
503. R. A. Benkeser and W. E. Bunting, *J. Amer. Chem. Soc.* **74**, 3011 (1952); R. A. Benkeser and R. G. Severson **71**, 3838 (1949); R. A. Benkeser and G. Schroll **78**, 3196 (1953).
504. J. D. Roberts, H. E. Simmons, Jr., L. A. Carlsmith, and C. W. Vaughan, *J. Amer. Chem. Soc.* **75**, 3290 (1953).
505. J. D. Roberts, D. A. Semenow, H. E. Simmons, Jr., and L. A. Carlsmith, *J. Amer. Chem. Soc.* **78**, 601 (1956).
506. G. E. Dunn, P. J. Krueger, and W. Rodewald, *Can. J. Chem.* **39**, 180 (1961).
507. V. A. Krishnamurthy and N. R. A. Rao, *J. Indian Inst. Sci.* **40**, 145 (1958).
508. J. Szychlinski, *Rocz. Chem.* **33**, 443 (1959).
509. O. L. Quimby and I. J. Flautt, *Z. Anorg. Allg. Chem.* **296**, 294 (1958).
510. H. Wedmann, K. Dax, and D. Wewerka, *Monatsh. Chem.* **101**, 1831 (1970).
511. A. S. Basov, Sh. A Zelenoya, J. M. Bolotin, A. A. Pavlov, N. V. Gushchin, and N. K. Petryakova, *Neftepererab. Neftekhim (Moscow)* p. 15 (1970).
512. I. I. Bradiphev, V. Ya. Paderin, and G. D. Atasnanchukov, *Gidroliz. Lesokhim. Prom.* **23**, 4 (1970).
513. K. H. Dyns, R. Brockman, and A. Roggenbuck, *Ann. Chem.* **614**, 97 (1958).
514. P. S. Radhakrishnamurti and G. P. Pawgraki, *J. Indian Chem. Sci.* **46**, 41, 318 (1969).
515. J. Pasek and P. Richter, *Chem. Prum.* **19**, 63 (1969).
516. A. C. Harkness and J. Halpern, *J. Amer. Chem. Soc.* **81**, 3526 (1959).
517. F. A. Jones and E. S. Amis, *J. Inorg. Nucl. Chem.* **26**, 1045 (1964).
518. N. K. Shastri and E. S. Amis, *Inorg. Chem.* **8**, 2484 (1969).
519. N. K. Shastri and E. S. Amis, *Inorg. Chem.* **8**, 2487 (1969).
520. H. Euler, *Ann. Chem.* **330**, 286 (1903).
521. A. Hantzsch, *Z. Phys. Chem. Stoechim, Verwandschaftslehre* **61**, 257 (1907); **65**, 41 (1908); *Ber. Deut. Chem. Ges. B* **58**, 941 (1925).
522. T. Ri and H. Eyring, *J. Chem. Phys.* **8**, 433 (1940).
523. J. Chedin, *C. R. Acad. Sci.* **200**, 1937 (1935).
524. C. K. Ingold, D. J. Millen, and H. G. Poole, *J. Chem. Soc. London,* p. 2576 (1950).
525. R. J. Gillespi, J. Graham, E. D. Hughes, C. K. Ingold, and E. A. R. Peeling, *J. Chem. Soc.* London p. 2504 (1950).
526. D. R. Goddard, E. D. Hughes, and C. K. Ingold, *J. Chem. Soc. London* p. 2559 (1950).
527. G. Olah, S. Kuhn, and A. Mlinko, *J. Chem. Soc. London* p. 2504 (1950).
528. D. J. Millen, *J. Chem. Soc. London* p. 2606 (1950).
529. R. J. Gillespie and D. J. Millen, *Quart. Rev. Chem. Rev.* **2**, 227 (1948).
530. F. H. Westheimer and M. S. Kharasch, *J. Amer. Chem. Soc.* **68**, 1871 (1946).
531. G. M. Bennett, J. C. D. Brand, D. M. James, J. G. Saunders, and G. Williams, *J. Chem. Soc. London* p. 474 (1947).
532. L. Melander, *Nature London* **163**, 599 (1949).
533. R. J. Gillespie, E. D. Hughes, C. K. Ingold, D. J. Millen, and R. I. Reed, *Nature London* **163**, 599 (1949).
534. R. J. Gillespie, *J. Chem. Soc. London* p. 2542 (1950).
535. E. Cherbuliez, *Helv. Chim. Acta* **6**, 281 (1923).
536. J. Masson, *J. Chem. Soc. London* p. 3200 (1931).
537. J. W. Baker and L. Hey, *J. Chem. Soc. London* pp. 1227, 2917 (1932).

538. E. D. Hughes, C. K. Ingold, and R. J. Reed, *J. Chem. Soc. London* p. 2400 (1950).
539. Zh. E Grabovskaya and M. I. Vinnik, *Zh. Fiz. Khim.* **43**, 901 (1969)
540 J G. Hoggett and R. B. Moodie, *J. Chem. Soc. D* p. 605 (1969).
541. R. P. Bell, "The Proton in Chemistry," Methuen, London, 1959.
542. M. Eigen, W. Kruse, G. Maas, and L. DeMaeyer, *Prog. React. Kinet.* **2**, 287 (1964).
543. *Discuss Faraday Soc.* **39**, 223 et seq (1965).
544. *Discuss Faraday Soc.*, **39**, 194ff (1965).
545. W. J. Alberty, *Ann. Rep. Prog. Chem.* (*Chem. Soc. London*) **60**, 40 (1963).
546. W. J. Alberty, *Prog. React. Kinet.* **4**, 353 (1967).
547. J. C. Reed and J. C. Calvin, *J. Amer. Chem. Soc.* **72**, 2948 (1950).

AUTHOR INDEX

Numbers in parentheses are reference numbers and indicate that an author's work is referred to although his name is not cited in the text. Numbers in italics show the page on which the complete reference is listed.

451

SUBJECT INDEX